Methods in Enzymology

Volume 409
DNA REPAIR
PART B

METHODS IN ENZYMOLOGY

EDITORS-IN-CHIEF

John N. Abelson   Melvin I. Simon

DIVISION OF BIOLOGY
CALIFORNIA INSTITUTE OF TECHNOLOGY
PASADENA, CALIFORNIA

FOUNDING EDITORS

Sidney P. Colowick and Nathan O. Kaplan

Methods in Enzymology

Volume 409

DNA Repair
Part B

EDITED BY

Judith L. Campbell

BRAUN LABORATORIES
CALIFORNIA INSTITUTE OF TECHNOLOGY
PASADENA, CALIFORNIA

Paul Modrich

HOWARD HUGHES MEDICAL INSTITUTE
DEPARTMENT OF BIOCHEMISTRY
DURHAM, NORTH CAROLINA

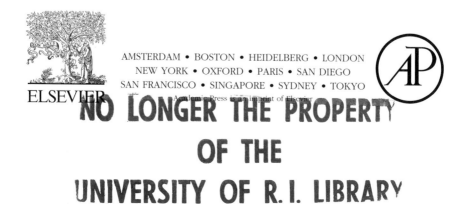

AMSTERDAM • BOSTON • HEIDELBERG • LONDON
NEW YORK • OXFORD • PARIS • SAN DIEGO
SAN FRANCISCO • SINGAPORE • SYDNEY • TOKYO

Academic Press is an imprint of Elsevier

ELSEVIER

Academic Press is an imprint of Elsevier
525 B Street, Suite 1900, San Diego, California 92101-4495, USA
84 Theobald's Road, London WC1X 8RR, UK

This book is printed on acid-free paper. ⊗

For information on all Academic Press publications
visit our Web site at www.books.elsevier.com

ISBN-13: 978-0-12-182814-1
ISBN-10: 0-12-182814-X

PRINTED IN THE UNITED STATES OF AMERICA
06 07 08 09 9 8 7 6 5 4 3 2 1

Working together to grow
libraries in developing countries

www.elsevier.com | www.bookaid.org | www.sabre.org

ELSEVIER   BOOK AID
            International   Sabre Foundation

Table of Contents

Contributors to Volume 409

HIROYUKI ABURATANI (23), *Genome Science Division, Research Center for Advanced Science and Technology, University of Tokyo, Meguro-Ku, Tokyo, Japan*

GENEVIEVE ALMOUZNI (21), *Laboratory of Nuclear Dynamics and Genome Plasticity, Institut Curie UMR 218 CNRS, Paris, France*

JEAN-CHRISTOPHE AMÉ (29), *Département Intégrité du Génome, CNRS Laboratoire Conventionne avec le Commissariat à l'Energie Atomique, Ecole Supérieure de Biotechnologie de Strasbourg, Illkirch-Cedex, France*

CSANAD BACHRATI (4), *Cancer Research UK Laboratories, Weatherall Institute of Molecular Medicine, University of Oxford, John Radcliffe Hospital, Oxford, United Kingdom*

VLADIMIR I. BASHKIROV* (10), *Section of Microbiology, University of California, Davis, California*

AARON BENSIMON (26), *Genome Stability Unit, Department of Structure and Dynamics of Genomes, Institut Pasteur, Paris, France*

SARA K. BINZ (2), *Department of Biochemistry, Roy J. and Lucille A. Carver College of Medicine, University of Iowa, Iowa City, Iowa*

VILHELM. A. BOHR (4), *Laboratory of Molecular Gerontology, National Institute of Aging-IRP, National Institutes of Health, Baltimore, Maryland*

WILLAM M. BONNER (14), *Laboratory of Molecular Pharmacology, Center for Cancer Research, National Cancer Institute, National Institutes of Health, Bethesda, Maryland*

CHARLES BOONE (13), *Department of Medical Genetics and Microbiology, University of Toronto, Toronto, Ontario, Canada*

DAMIEN BREGEON (20), *Laboratoire "Instabilite genetique et cancer," Institut Gustav Roussy–PR2, Villejuif cedex, France*

CLAIRE BRESLIN (24), *Genome Damage and Stability Center, University of Sussex, Brighton, United Kingdom*

ROBERT M. BROSH, JR. (4), *Laboratory of Molecular Gerontology, National Institute of Aging-IRP, National Institutes of Health, Baltimore, Maryland*

GRANT BROWN (13), *Department of Biochemistry, University of Toronto, Toronto, Ontario, Canada*

PETER M. J. BURGERS (1), *Department of Biochemistry and Molecular Biophysics, Washington University School of Medicine, Saint Louis, Missouri*

ALEX B. BURGIN, JR. (30), *deCODE Biostructures, Bainbridge Island, Washington*

GÖRAN O. BYLUND (1), *Department of Biochemistry and Molecular Biophysics, Washington University School of Medicine, Saint Louis, Missouri*

*Present address: Applied Biosystems, Foster City, California.

ix

KEITH W. CALDECOTT (24), *Genome Damage and Stability Centre, University of Sussex, Brighton, United Kingdom*

MICHAEL CHANG (13), *Biochemistry, University of Toronto, Toronto, Ontario, Canada*

XI CHEN (3), *Radiation Oncology, Research Laboratory and the Marlene and Stewart Greenebaum Cancer Center, University of Maryland School of Medicine, Baltimore, Maryland*

PAULA M. CLEMENTS (24), *Genome Damage and Stability Centre, University of Sussex, Brighton, United Kingdom*

CHIARA CONTI (26), *Genome Stability Unit, Department of Structure and Dynamics of Genomes, Institut Pasteur, Paris, France*

CECILIA COTTA-RAMUSINO (26), *F.I.R.C. Institute of Molecular Oncology Foundation and DSBB-University of Milan, Milan, Italy*

CHARMAIN T. COURCELLE (25), *Department of Biology, Portland State University, Portland, Oregon*

JUSTIN COURCELLE (25), *Department of Biology, Portland State University, Portland, Oregon*

FRANÇOISE DANTZER (29), *Département Intégrité du Génome, CNRS Laboratoire Conventionne avec le Commissariat à l'Energie Atomique, Ecole Supérieure de Biotechnologie de Strasbourg, Illkirch-Cedex, France*

ANNE M. DICKSON (2), *Department of Biochemistry, Roy J. and Lucille A. Carver College of Medicine, University of Iowa, Iowa City, Iowa*

PAUL W. DOETSCH (20), *Department of Biochemistry, Emory University School of Medicine, Rollins Research Center, Atlanta, Georgia*

HARALD DÜRR (22), *Gene Center and Department of Chemistry and Biochemistry, Ludwig Maximilians University of Munich, Munich, Germany*

SHERIF F. EL-KHAMISY (24), *Genome Damage and Stability Centre, University of Sussex, Brighton, United Kingdom; Department of Biochemistry, School of Pharmacy, Ain Shams University, Cairo, Egypt*

TOM ELLENBERGER (3), *Department of Biological Chemistry and Molecular Pharmacology, Harvard Medical School, Boston, Massachusetts*

MARCO FOIANI (26), *F.I.R.C. Institute of Molecular Oncology Foundation and DSBB-University of Milan, Milan, Italy*

PATRICIA L. FOSTER (12), *Department of Biology, Indiana University, Bloomington, Indiana*

STEVEN S. FOSTER (16), *Institute for Ageing and Health, Henry Wellcome Laboratory for Biogerontology Research, University of Newcastle, Newcastle upon Tyne, United Kingdom*

ANNABELLE GÉRARD (21), *Laboratory of Nuclear Dynamics and Genome Plasticity, Institut Curie, UMR 218 CNRS, Paris, France*

CHRIS GILBERT (8), *Department of Biochemistry, National University of Ireland, Galway, Ireland*

CATHERINE GREEN (8), *Department of Biochemistry, National University of Ireland, Galway, Ireland*

MURIEL GRENON (8), *Department of Biochemistry, National University of Ireland, Galway, Ireland*

EDWIN HAGHNAZARI[†] (10), *Section of Microbiology, University of California, Davis, California*

STUART J. HARING (2), *Department of Biochemistry, Roy J. and Lucille A. Carver*

†*Present address: Scios Inc., Freemont, California.*

College of Medicine, University of Iowa, Iowa City, Iowa

HILDUR R. HELGADOTTIR[‡] (31), Molecular Biology Program, Memorial Sloan Kettering Cancer Center, New York, New York

KRISTINA HERZBERG (10), Section of Microbiology, University of California, Davis, California

WOLF-DIETRICH HEYER (10), Sections of Microbiology and of Molecular and Cellular Biology, Center for Genetics and Develoment, University of California, Davis, California

IAN D. HICKSON (5), Cancer Research UK Laboratories, Weatherall Institute of Molecular Medicine, University of Oxford, John Radcliffe Hospital, Oxford, United Kingdom

KARL-PETER HOPFNER (22), Gene Center and Department of Chemistry and Biochemistry, Ludwig-Maximilians University of Munich, Munich, Germany

CHIH-LIN HSIEH (17), Departments of Pathology, Biochemistry and Molecular Biology, Microbiology, and Biology, University of Southern California Keck School of Medicine, Los Angeles, California

FENG-TING HUANG (18), Norris Comprehensive Cancer Center, Departments of Pathology, Biochemistry and Molecular Biology, Microbiology, and Biology, University of Southern California Keck School of Medicine, Los Angeles, California

NATASHA ILES (24), Genome Damage and Stability Centre, University of Sussex, Brighton, United Kingdom

MARIA JASIN (31), Molecular Biology Program, Memorial Sloan Kettering Cancer Center, New York, New York

MIHOKO KAI (11), Department of Pathology, Stanford University School of Medicine, Stanford, California

KIYOFUMI KANESHIRO (23), Genome Science Division, Research Center for Advanced Science and Technology, University of Tokyo, Meguro-ku, Tokyo, Japan

YUKI KATOU (23), Laboratory of Genome Structure and Function, Division for Gene Research, Center for Biological Resources and Informatics, Tokyo Institute of Technology, Yokohama City, Kanagawa, Japan

RICHARD D. KOLODNER (27), Ludwig Institute for Cancer Research, University of California San Diego School of Medicine, La Jolla, California

KEN KREUZER (28), Department of Biochemistry, Duke University Medical Center, Durham, North Carolina

GIORDANO LIBERI (26), F.I.R.C. Institute of Molecular Oncology Foundation and DSBB-University of Milan, Milan, Italy

MICHAEL LICHTEN (14), Laboratory of Biochemistry, Center for Cancer Research, National Cancer Institute, National Institutes of Health, Bethesda, Maryland

MICHAEL R. LIEBER (17, 18), Norris Comprehensive Cancer Center, Departments of Pathology, Biochemistry and Molecular Biology, Microbiology, and Biology, University of Southern California Keck School of Medicine, Los Angeles, California

MASSIMO LOPES (26), Institute of Molecular Cancer Research, University of Zürich, Zürich, Switzerland

NOEL LOWNDES (8), Department of Biochemistry, National University of Ireland, Galway, Ireland

[‡]Present address: The Salk Institute, La Jolla, California.

DAVID LYDALL (16), *Institute for Ageing and Health, Institute for Cell and Molecular Biosciences, Henry Wellcome Laboratory for Biogerontology Research, University of Newcastle, Newcastle upon Tyne, United Kingdom*

JERZY MAJKA (1), *Department of Biochemistry and Molecular Biophysics, Washington University School of Medicine, Saint Louis, Missouri*

LAURA MARINGELE (16), *Institute for Ageing and Health, Henry Wellcome Laboratory for Biogerontology Research, University of Newcastle, Newcastle upon Tyne, United Kingdom*

JOSIANE MENISSIER-DE MURCIA (29), *Département Intégrité du Génome, CNRS Laboratoire Conventionne avec le Commissariat à l'Energie Atomique, Ecole Supérieure de Biotechnologie de Strasbourg, Illkirch-cedex, France*

GILBERT DE MURCIA (29), *Département Intégrité du Génome, CNRS Laboratoire Conventionne avec le Commissariat à l'Energie Atomique, Ecole Supérieure de Biotechnologie de Strasbourg, Illkirch-cedex, France*

ASAKO NAKAMURA (14), *Laboratory of Molecular Pharmacology, Center for Cancer Research, National Cancer Institute, National Institutes of Health, Bethesda, Maryland*

JUN NAKAMURA (29), *Department of Environmental Sciences and Engineering, University of North Carolina, Chapel Hill, North Carolina*

KOJI NAKANISHI (31), *Molecular Biology Program, Memorial Sloan Kettering Cancer Center, New York, New York*

AISLING O'SHAUGHNESSY (8), *Department of Biochemistry, National University of Ireland, Galway, Ireland*

PATRICIA L. OPRESKO (4), *University of Pittsburgh Graduate School of Public Health, Department of Environmental and Occupational Health, Pittsburgh, Pennsylvania*

RITU PABLA (6), *Department of Cell Biology and Genetics, University of North Texas Health Science Center, Fort Worth, Texas*

AINSLIE B. PARSONS (13), *Department of Medical Genetics and Microbiology, University of Toronto, Toronto, Ontario, Canada*

JOHN PASCAL (3), *Department of Biological Chemistry and Molecular Pharmacology, Harvard Medical School, Boston, Massachusetts*

VAIBHAV PAWAR (6), *Department of Cell Biology and Genetics, University of North Texas Health Science Center, Fort Worth, Texas*

VINCENT PENNANEACH (27), *Ludwig Institute for Cancer Research, University of California San Diego School of Medicine, La Jolla, California*

EVA PETERMANN (24), *Genome Damage and Stability Centre, University of Sussex, Brighton, United Kingdom*

PETER PETRINI (15), *Memorial Sloan Kettering Cancer Center, New York, New York*

DUANE R. PILCH (14), *Laboratory of Molecular Pharmacology, Center for Cancer Research, National Cancer Institute, National Institutes of Health, Bethesda, Maryland*

JENNIFER REINEKE POHLHAUS (28), *Department of Biochemistry, Duke University Medical Center, Durham, North Carolina*

SOPHIE POLO (21), *Laboratory of Nuclear Dynamics and Genome Plasticity, Institut Curie, UMR 218 CNRS, Paris, France*

CHRISTOPHER D. PUTNAM (27), *Ludwig Institute for Cancer Research, University of*

California San Diego School of Medicine, La Jolla, California

SATHEES C. RAGHAVAN (17), Departments of Pathology, Biochemistry and Molecular Biology, Microbiology, and Biology, University of Southern California Keck School of Medicine, Los Angeles, California

AMY C. RAYMOND (30), Molecular Biology Program, Sloan-Kettering Institute, New York, New York

CHRISTOPHE REDON (14), Laboratory of Molecular Pharmacology, Center for Cancer Research, National Cancer Institute, National Institutes of Health, Bethesda, Maryland

MICHAEL A. RESNICK (19), Chromosome Stability Section, Laboratory of Molecular Genetics, National Institute of Environmental Health Sciences, National Institutes of Health, Research Triangle Park, North Carolina

DANIELE ROCHE (21), Laboratory of Nuclear Dynamics and Genome Plasticity, Institut Curie, UMR 218 CNRS, Paris, France

DEEPANKAR ROY (18), Norris Comprehensive Cancer Center, Departments of Pathology, Biochemistry and Molecular Biology, Microbiology, and Biology, University of Southern California Keck School of Medicine, Los Angeles, California

VALÉRIE SCHREIBER (29), Département Intégrité du Génome, CNRS Laboratoire Conventionne avec le Commissariat à l'Energie Atomique, Ecole Supérieure de Biotechnologie de Strasbourg, Illkirchcedex, France

ROBERT SCHROFF (14), Laboratory of Biochemistry, Center for Cancer Research, National Cancer Institute, National Institutes of Health, Bethesda, Maryland

OLGA A. SEDELNIKOVA (14), Laboratory of Molecular Pharmacology, Center for Cancer Research, National Cancer Institute, National Institutes of Health, Bethesda, Maryland

BILAL H. SHEIKH (13), Department of Medical Genetics and Microbiology, University of Toronto, Toronto, Ontario, Canada

KATSUHIKO SHIRAHIGE (23), Laboratory of Genome Structure and Function, Division for Gene Research, Center for Biological Resources and Informatics, Tokyo Institute of Technology, Yokohama City, Kanagawa, Japan

KRISTINA H. SHMIDT (27), Ludwig Institute for Cancer Research, University of California San Diego School of Medicine, La Jolla, California

WOLFRAM SIEDE (6), Department of Cell Biology and Genetics, University of North Texas, Health Science Center, Fort Worth, Texas

NATASHA I. SINOGEEVA (14), Laboratory of Molecular Pharmacology, Center for Cancer Research, National Cancer Institute, National Institutes of Health, Bethesda, Maryland

JOSE' SOGO (26), Institute of Cell Biology, Zürich, Switzerland

FRANCESCA STORICI (19), Chromosome Stability Section, Laboratory of Molecular Genetics, National Institute of Environmental Health Sciences, National Institutes of Health, Research Triangle Park, North Carolina

LORENA TARICANI (11), Department of Pathology, Stanford University School of Medicine, Stanford, California

JAN-WILLEM F. THEUNISSEN (15), Memorial Sloan Kettering Cancer Center, New York, New York

DAVID P. TOCZYSKI (9), Cancer Research Institute, Department of Biochemistry and Biophysics, University of California, San Francisco, California

GERALDINE W.-L. TOH (8), *Department of Biochemistry, National University of Ireland, Galway, Ireland*

ALAN E. TOMKINSON (3), *Radiation Oncology Research Laboratory and The Marlene and Stewart Greenbaum Cancer Center, University of Maryland School of Medicine, Baltimore, Maryland*

ALBERT TSAI (17), *Departments of Pathology, Biochemistry and Molecular Biology, Microbiology, and Biology, University of Southern California Keck School of Medicine, Los Angeles, California*

SANGEETHA VIJAYAKUMAR (3), *Radiation Oncology Research Laboratory and The Marlene and Stewart Greenbaum Cancer Center, University of Maryland School of Medicine, Baltimore, Maryland*

ALEXEY S. VLASENKO (10), *Section of Microbiology, University of California, Davis, California*

TERESA S.-F. WANG (11), *Department of Pathology, Stanford University School of Medicine, Stanford, California*

DAVID M. WEINSTOCK (31), *Department of Medicine, Memorial Sloan Kettering Cancer Center, New York, New York*

GERALD M. WILSON (3), *Department of Biochemistry and Molecular Biology and the Marlene and Stewart Greenbaum Cancer Center, University of Maryland School of Medicine, Baltimore, Maryland*

MARC S. WOLD (2), *Department of Biochemistry, Roy J. and Lucille A. Carver College of Medicine, University of Iowa, Iowa City, Iowa*

XIAOHONG HELENA YANG (7), *Massachusetts General Hospital Cancer Center, Harvard Medical School, Charlestown, Massachusetts*

KEFEI YU (18), *Norris Comprehensive Cancer Center, Departments of Pathology, Biochemistry and Molecular Biology, Microbiology, and Biology, University of Southern California Keck School of Medicine, Los Angeles, California*

HONG ZHANG (6), *Department of Cell Biology, Capital University of Medical Sciences, Beijing, China*

LEE ZOU (7), *Massachusetts General Hospital Cancer Center, Harvard Medical School, Charlestown, Massachusetts*

MIKHAJLO K. ZUBKO (16), *Institute for Ageing and Health, Henry Wellcome Laboratory for Biogerontology Research, University of Newcastle, Newcastle upon Tyne, United Kingdom*

METHODS IN ENZYMOLOGY

VOLUME 262. DNA Replication
Edited by JUDITH L. CAMPBELL

VOLUME 263. Plasma Lipoproteins (Part C: Quantitation)
Edited by WILLIAM A. BRADLEY, SANDRA H. GIANTURCO, AND JERE P. SEGREST

VOLUME 264. Mitochondrial Biogenesis and Genetics (Part B)
Edited by GIUSEPPE M. ATTARDI AND ANNE CHOMYN

VOLUME 265. Cumulative Subject Index Volumes 228, 230–262

VOLUME 266. Computer Methods for Macromolecular Sequence Analysis
Edited by RUSSELL F. DOOLITTLE

VOLUME 267. Combinatorial Chemistry
Edited by JOHN N. ABELSON

VOLUME 268. Nitric Oxide (Part A: Sources and Detection of NO; NO
Synthase)
Edited by LESTER PACKER

VOLUME 269. Nitric Oxide (Part B: Physiological and
Pathological Processes)
Edited by LESTER PACKER

VOLUME 270. High Resolution Separation and Analysis of Biological
Macromolecules (Part A: Fundamentals)
Edited by BARRY L. KARGER AND WILLIAM S. HANCOCK

VOLUME 271. High Resolution Separation and Analysis of Biological
Macromolecules (Part B: Applications)
Edited by BARRY L. KARGER AND WILLIAM S. HANCOCK

VOLUME 272. Cytochrome P450 (Part B)
Edited by ERIC F. JOHNSON AND MICHAEL R. WATERMAN

VOLUME 273. RNA Polymerase and Associated Factors (Part A)
Edited by SANKAR ADHYA

VOLUME 274. RNA Polymerase and Associated Factors (Part B)
Edited by SANKAR ADHYA

VOLUME 275. Viral Polymerases and Related Proteins
Edited by LAWRENCE C. KUO, DAVID B. OLSEN, AND STEVEN S. CARROLL

VOLUME 276. Macromolecular Crystallography (Part A)
Edited by CHARLES W. CARTER, JR., AND ROBERT M. SWEET

VOLUME 277. Macromolecular Crystallography (Part B)
Edited by CHARLES W. CARTER, JR., AND ROBERT M. SWEET

VOLUME 278. Fluorescence Spectroscopy
Edited by LUDWIG BRAND AND MICHAEL L. JOHNSON

VOLUME 279. Vitamins and Coenzymes (Part I)
Edited by DONALD B. MCCORMICK, JOHN W. SUTTIE, AND CONRAD WAGNER

[1]   Overproduction and Purification of RFC-Related Clamp Loaders and PCNA-Related Clamps from *Saccharomyces cerevisiae*

By Göran O. Bylund, Jerzy Majka, and Peter M. J. Burgers

Abstract

The replication clamp PCNA and its loader RFC (Replication Factor C) are central factors required for processive replication and coordinated DNA repair. Recently, several additional related clamp loaders have been identified. These alternative clamp loaders contain the small Rfc2–5 subunits of RFC, but replace the large Rfc1 subunit by a pathway-specific alternative large subunit, Rad24 for the DNA damage checkpoint, Ctf18 for the establishment of sister chromatid cohesion, and Elg1 for a general function in chromosome stability. In order to define biochemical functions for these loaders, the loaders were overproduced in yeast and purified at a milligram scale. To aid in purification, the large subunit of each clamp loader was fused to a GST-tag that, after purification could be easily removed by a rhinoviral protease. This methodology yielded all clamp loaders in high yield and with high enzymatic activity. The yeast 9-1-1 checkpoint clamp, consisting of Rad17, Mec3, and Ddc1, was overproduced and purified in a similar manner.

Introduction

The proliferating cell nuclear antigen PCNA is the processivity clamp that organizes and stabilizes DNA replication complexes and most DNA repair complexes on the DNA (reviewed in Majka and Burgers, 2004). Its function as a sliding clamp that encircles the DNA requires that PCNA is loaded around DNA by a clamp loader, RFC. The heteropentameric RFC uses the energy of ATP to open PCNA, bind the effector DNA, and reclose PCNA around this DNA. RFC consists of four small subunits of 36–40 kDa in all eukaryotes, Rfc2, Rfc3, Rfc4, and Rfc5, and one large subunit ∼95–145 kDa in size, Rfc1. The 4-subunit Rfc2–5 complex is designated as the core complex because it functions in at least three other clamp loader complexes.

These alternative eukaryotic clamp loaders are designated Rad24-RFC, Ctf18-RFC and Elg1-RFC. Rad24-RFC, a complex of Rad24 and the Rfc2–5 core, functions in the DNA damage checkpoint, and this complex

METHODS IN ENZYMOLOGY, VOL. 409
0076-6879/06 $35.00
DOI: 10.1016/S0076-6879(05)09001-4

loads an alternative clamp, the 9-1-1 complex (Rad17/3/1 in *S. cerevisiae*), at sites of DNA damage in order to mediate checkpoint activation (Bermudez *et al.*, 2003a; Ellison and Stillman, 2003; Majka and Burgers, 2003; Zou *et al.*, 2003). Ctf18-RFC, a 7-subunit complex of Rfc2–5, Ctf18, Dcc1, and Ctf8, is required for the establishment of sister chromatid cohesion (Hanna *et al.*, 2001; Mayer *et al.*, 2001). Ctf18 functions as the Rfc1 homolog in this complex. Biochemically, Ctf18-RFC can both load and unload PCNA (Bermudez *et al.*, 2003b; Bylund and Burgers, 2005; Shiomi *et al.*, 2004). A distinct biochemical function for the 5-subunit Elg1-RFC complex, with Elg1 replacing Rfc1, has not yet been defined (Bylund and Burgers, 2005).

The discovery of these clamp loaders has created a demand for large quantities of highly pure and active complexes in order to facilitate biochemical studies. An important consideration during the development of our strategy was that overproduction of clamp loaders in *E. coli* met with varying success. While the canonical RFC could be overexpressed and purified from *E. coli* in high yield and with high activity, a similar strategy failed for the Rad24- and Ctf18-related loaders (Gomes *et al.*, 2000). Whereas it was relatively easy to overproduce these alternative clamp loaders in *E. coli*, and obtain them in pure form, the purified complexes lacked any detectable clamp loading activity (unpublished results from our laboratory). Therefore, all alternative clamp loader complexes were purified from yeast overproduction systems.

In this report, we describe a robust procedure to overproduce and purify the alternative clamp loaders, Rad24-RFC, Ctf18-RFC, and Elg1-RFC and the damage response clamp, Rad17/3/1, yielding milligram amounts of pure complexes. One protein in each of the complexes is expressed as an N-terminal fusion to a protease cleavable form of glutathione S-transferase (GST) and the complexes are affinity purified by the use of glutathione sepharose chromatography (Smith and Johnson, 1988). The advantage of using GST-tagged fusion proteins and glutathione sepharose chromatography is that protein with a purity of >80% can readily be obtained in this single purification step that does not require harsh elution conditions. The GST-tag is cleaved off by treatment with a human rhinovirus protease, which has a highly unique substrate recognition site, and the complexes are further purified by FPLC either on a heparin, monoS or monoQ column (Cordingley *et al.*, 1990). Neither the protease nor the cleaved GST moiety bind to these columns and are therefore readily washed away. The Ssa1 chaperone protein appears as a consistent contaminant during these purification procedures; however, it can be eluted from the columns by washing with a Mg-ATP buffer prior to elution of the complex.

Overproduction of Complexes in Yeast

Expression Plasmids and Strains

In our laboratory, we have successfully used the pRS424-GAL (*TRP1*), pRS425-GAL (*LEU2*), and pRS426-GAL (*URA3*) plasmids for the over-production of a variety of replication proteins in yeast, including clamps and clamp loaders (Fig. 1A) (Burgers, 1999). These plasmids contain a 2μ origin for high copy maintenance in yeast and a Bluescript SKII$^+$ backbone for amplification in *E. coli*. Expression of genes cloned into these plasmid vectors is driven from the bidirectional *GAL1–10* promoter. The expression of genes under *GAL1–10* control is high when galactose is present in the growth medium and low in its absence. Glucose represses expression from the *GAL1–10* promoter and must therefore be excluded from the growth medium, or kept at a low concentration ($\leq 0.1\%$) in order to achieve rapid and efficient induction of expression upon addition of galactose (Johnston *et al.*, 1994).

We have extended the utility of these vectors by making GST-fusion derivatives. Vectors pRS424-GALGST, pRS425-GALGST, and pRS426-GALGST all contain a GST-tag (encoding the 26.6 kDa **g**lutathione **S**-transferase from *Schistosoma japanicum*) cloned into the respective GAL vector. The GST tag is separated from the target gene by a recognition sequence for the human rhinovirus protease (Fig. 1B). Vectors and

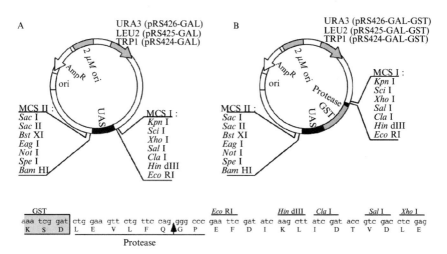

FIG. 1. Vectors used to overexpress alternative clamp loaders and clamps in *S. cerevisieae*. A. GAL vectors. B. GAL-GST vectors. The sequence around the protease cleavage site is shown. The vectors are available upon request from the author (burgers@biochem.wustl.edu).

sequences are available from the corresponding author upon request. In order to overproduce a complex in yeast, we routinely clone one of the genes of the particular complex into the GAL–GST vector via an N-terminal fusion to GST, while the other genes are cloned without modification into the regular GAL vectors. If desired, multiple genes can be cloned into a single GAL vector by cloning multiple *GAL1–10* sites into the plasmid (e.g., see Gerik *et al.*, 1997). Alternatively, or in addition, genes can be cloned into multiple 2μ plasmids, each with a different selectable marker, as these are readily maintained in yeast, (e.g., see Bylund and Burgers, 2005).

Overexpression is routinely carried out in the protease-deficient strain BJ2168 (*MAT*a, *ura3–52*, *trp1–289*, *leu2–3*, *112*, *prb1–1122*, *prc1–407*, *pep4-3*). In this strain, the *GAL1–10* promoter is strongly induced by the addition of galactose to the growth medium. However, using galactose as the only carbon source gives very poor growth. Instead, a combination of glycerol and lactate, allowing slow but satisfactory cell growth, is used as the carbon source prior to induction.

Media

Indicated quantities are per liter of media. Solid media contain in addition 20 g of agar. SCD: 1.7 g of yeast nitrogen base without amino acids and ammonium sulfate (Difco), 5 g ammonium sulfate, 1 g amino acid drop-out mix (from a blended mixture of equal quantities of alanine, arginine, asparagine, aspartic acid, cysteine, glutamine, glutamic acid, glycine, histidine, inositol, isoleucine, lysine, methionine, phenylalanine, proline, serine, threonine, tyrosine, and valine), 20 g glucose. When needed, 20 mg each of uracil and tryptophan, 50 mg of adenine, and 100 mg of leucine are added. SCGL: 1.7 g of yeast nitrogen base without amino acids and ammonium sulfate, 5 g ammonium sulfate, 1 g glucose, 30 ml glycerol, 20 ml lactic acid, and amino acids as for SCD. Prior to autoclaving the pH of the media is adjusted to 5–6 with concentrated sodium hydroxide. YPGLA: 10 g yeast extract, 20 g peptone, 2 g glucose, 30 ml glycerol, 20 ml lactic acid, 20 mg adenine. Prior to autoclaving, the pH of the media is adjusted to 5–6 with concentrated sodium hydroxide. If bacterial contamination of the media is a concern, kanamycin can be added to a final concentration of 50 μg/ml. Kanamycin will not affect the growth of yeast.

Procedure

Strain BJ2168 is transformed with the appropriate plasmids using the LiAc procedure. Transformants are selected on selective SCD plates and purified by restreaking on the same medium.

A single colony is streaked out onto an entire selective SCD plate and grown for 1–2 days at 30°.

A 150 ml starter culture of selective SCGL media is inoculated with a large amount of cells from the SCD plate and grown overnight with shaking at ~200 rpm.

The next morning, the starter culture generally has reached an OD_{660} of about 5. The starter culture is split over 6×1 liter selective SCGL, in 3 liter flasks, and growth is continued overnight with shaking.

Next morning the cultures generally have reached an $OD_{660} = 2$–3. To each flask, one liter of YPGLA is added, and growth is continued for another 2–3 h.

Expression is induced by adding 40 g of solid galactose to each 2-liter culture, and shaking is continued for 3–5 h.

The cells are harvested by centrifugation at 4000g for 10 min and washed with 500 ml of ice-cold water.

The cell paste is resuspended in 1–2 ml of ice-cold water per 5 g of wet cells, and frozen by slowly pouring with stirring into liquid nitrogen. This popcorn yeast is stored at $-70°$. Yields are typically 50–100 g of cells, wet weight, per 12 liters of culture.

Extract Preparation and Purification

Buffers and Inhibitors

Hepes-based buffers containing sodium or potassium chloride are used. For the purification of the checkpoint clamp/clamp loader complexes (Rad17/3/1 and Rad24-RFC, respectively) usage of the KCl buffer results in a better yield, whereas for the Ctf18-RFC and Elg1-RFC complexes, the NaCl buffer proved to be more suitable.

Composition of buffer $1 \times HEP_{100}$: 50 mM Hepes-NaOH/KOH (pH 7.5), 100 mM NaCl/KCl, 10% glycerol, 0.1% Tween-20, 3 mM DTT. The suffix refers to the mM NaCl/KCl concentration. For the preparation of the extract the buffer is made as $2\times HEP_{100}$ (contains all components at twice the concentration of buffer $1\times HEP_{100}$) supplemented with a mixture of protease and phosphatase inhibitors (final concentration in $1\times HEP_{100}$: 1 mM EDTA, 1 mM EGTA, 2.5 mM Na-pyrophosphate, 1 mM β-glycerophosphate, 10 mM Na-bisulfite, 2.5 mM benzamidine, 5 μg/ml chymostatin, 5 μM pepstatin A, 10 μM leupeptin, and 1 mM PMSF).

Breakage of Cells with Dry Ice

In order to reduce the activity of proteases the frozen yeast cells are cracked open by blending the popcorn yeast with dry ice in a commercial blender (Waring Commercial Laboratory blender, Torrington, CT).

The stainless steel chamber is filled to 1/4-1/3 volume with crushed dry ice. Blending is started and continued until the dry ice is a fine powder.

Thereafter 2× HEP$_{100}$ buffer, containing inhibitors, is poured on top of the dry ice in 20–30 ml portions and blending is continued between additions of buffer. The total volume of 2× buffer added should be equal to the amount of popcorn yeast in grams that will be cracked open. 70–100 grams of popcorn yeast plus buffer and dry ice is the maximum load for the model 51BL30 blender.

After addition of the popcorn yeast, blending is continued for 10 min. When the blender motor gives off a high whining noise, this is an indication that the blade is no longer properly blending the mixture, and manual mixing of the chamber contents becomes necessary. In general, every 20–30 sec, the blender is turned off and the sides scraped with a steel spatula. It also helps to knock the walls of the chamber with the spatula while blending. More dry ice is added if the chamber appears to warm up (melting on the sides).

Ammonium Sulfate Fractionation

The dry ice mixture is thawed at room temperature in a glass beaker, which is submerged in water. From now on all steps are performed at +4° or on ice.

Ammonium sulfate (4 M) is added to a final concentration of 200–300 mM to the extract. This step prevents the coprecipitation of DNA binding proteins with the nucleic acids during the next step.

Nucleic acids are precipitated from the cell extract by the addition of 45 μl of 10% Polymin P per ml cell extract, and the mixture gently stirred on ice for ∼10 min.

Precipitated nucleic acids are pelleted at 40,000g for 45 min, in several tubes in a Sorvall SS34 rotor, and the pellets discarded.

The volume of the supernatant is measured and proteins are precipitated by slowly adding 0.35 g fine powdered ammonium sulfate per ml of extract while stirring on ice. When all of the ammonium sulfate has gone into solution, stirring on ice is continued for another 10 min.

The precipitated protein is pelleted at 40,000g for 45 min. The supernatants are discarded and the tubes are spun for an extra 5 min for residual supernatant removal. At this point the protein pellets can be frozen at −70°.

Glutathione Sepharose Chromatography

Glutathione Sepharose 4B (Amersham Biosciences), ∼1 ml per 50 g of cells, is transferred to a column that should be able to hold at least 10 ml, and washed with 10 vol of MQ water.

The glutathione sepharose is equilibrated with at least 10 vol of HEP_{150}.

The protein pellets are carefully dissolved in a total of 25–50 ml of HEP_0 using a wide-bore pipette to stir and suck the suspension up and down. Avoid foaming of the protein solution.

Thereafter, the protein solution is diluted with HEP_0 until it reaches a conductivity of HEP_{150}. The final volume is usually 150–250 ml.

Equilibrated glutathione sepharose is transferred to the protein solution and GST-fusion proteins are batch bound to the glutathione-sepharose for 1–3 h by gentle rotation at 4°.

The sepharose is pelleted briefly at $300g$ for 1 min in a clinical centrifuge (equipped with a swing-out rotor) and the supernatant discarded (Fig. 2, lane 1).

The sepharose is transferred back into the column and washed with at least 10 vol of HEP_{150} (lane 2).

GST-fusion protein-containing complex is eluted with HEP_{150} supplemented with 0.05% ampholytes pH 3–10, 20 mM glutathione (reduced form), and protease inhibitors, omitting benzamidine and PMSF. Since glutathione is acidic, 5 M NaOH is added to adjust the pH to 8.0. The

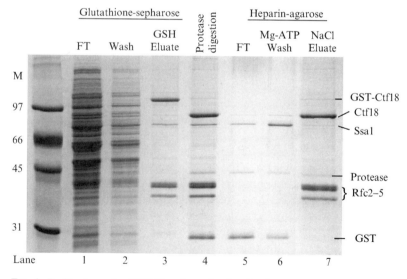

FIG. 2. Purification of a RFC-like clamp loader. During the course of purifying the 5-subunit Ctf18-RFC, samples were collected from the different purification steps and separated on a 13% polyacrylamide-SDS gel. FT, Wash, and GSH Eluate denote the flow through, wash, and elution, respectively, from the glutathione sepharose column (lanes 1–3). Lane 4, after PreScission protease treatment. Lanes 5–7 are the flow through, Mg-ATP wash, and peak eluate fraction, respectively, of the heparin column. Molecular weight marker proteins are indicated in the left lane of the gel.

HEP buffers are supplemented with 0.05% ampholytes to stabilize the clamp loaders (Gomes *et al.*, 2000). Benzamidine is omitted to allow detection of protein peaks at A_{280} during the subsequent FPLC purification step. PMSF is omitted since it has a tendency to precipitate on FPLC columns resulting in clogging of the columns.

To improve the efficiency of elution, 1 ml of elution buffer is mixed by pipetting with 1 ml of resin. After sitting for 15 min the column is drained, and the next ml of elution buffer mixed in. This procedure is repeated 4 times. Alternatively, the column can be hooked up to a peristaltic pump and ~5 ml of elution buffer very slowly pumped through at 100 μl/min (lane 3).

Proteolytic Removal of the GST-Tag

In order to cleave off the GST-tag from the GST-fusion proteins the glutathione column eluates are treated with human rhinovirus protease (commercially available as PreScission[TM] protease, Amersham Biosciences). The cleavage efficiency differs between different GST-fusion proteins and the optimal amount must be determined in a small-scale experiment. Meanwhile, the bulk of the preparation can be frozen on dry ice and stored at $-70°$.

Typically, a 5 ml fraction is treated with 5–25 units of PreScission protease overnight on ice.

Purification of Ctf18-RFC and Elg1-RFC

After overnight proteolytic cleavage, the treated fractions are pooled and diluted with HEP_0 to HEP_{100} (lane 4).

The protein is loaded on a 1 ml monoS column (Amersham Biosciences), equilibrated in HEP_{100}, for purification of the 7-subunit Ctf18-RFC complex or Elg1-RFC. The 5-subunit Ctf18-RFC is loaded on a 5 ml HiTrap Heparin column (Amersham Biosciences), equilibrated in HEP_{100}. Most of the GST and PreScission protease flow through (lane 5). The column is washed with 5–10 column vol of HEP_{100}.

The yeast chaperone Ssa1 purifies together with the overexpressed complexes both during glutathione-sepharose chromatography and MonoS or heparin chromatography. It is removed during the latter step by washing the column with 5–10 column vol of HEP_{100} containing 5 mM Mg-acetate and 100 μM ATP (lane 6). Subsequently the column is washed with 2 vol of HEP_{100}.

Protein is eluted from the MonoS column with a 15–20 ml linear gradient of 100–500 mM NaCl, Ctf18-RFC eluting at ~0.35 M NaCl and Elg1-RFC at ~0.3 M NaCl. The 5-subunit Ctf18-RFC complex is eluted from the heparin column with a 20–30 ml linear gradient from 100–1000 mM NaCl, the complex eluting at ~0.5 M NaCl. 0.3–0.5 ml fractions are collected, analyzed by SDS-PAGE (lane 7), and stored frozen at $-70°$.

Complex-containing fractions are generally divided in smaller aliquots since activity is gradually lost upon repeated freezing and thawing. A tube is usually discarded after five uses.

Purification of Rad24-RFC and Rad17/Mec3/Ddc1

After overnight proteolytic cleavage, the treated fractions are pooled and diluted with HEP$_0$ to HEP$_{125}$ (KCl).

The protein is loaded on a 1 ml monoQ column (Amersham Biosciences), equilibrated in HEP$_{125}$. Most of the GST and PreScission protease flow through. The column is washed with 10 column vol of HEP$_{125}$, followed by 5–10 column vol of HEP$_{125}$ containing 5 mM Mg-acetate and 100 μM ATP, and 2 column vol of HEP$_{125}$.

Protein is eluted with a 15 ml linear gradient of 125–500 mM KCl. Rad24-RFC elutes at ~0.3 M KCl and Rad17/3/1 at ~0.25 M KCl.

Discussion

Saccharomyces cerevisiae has proved to be a very useful organism for the overproduction of not only yeast proteins but also other eukaryotic proteins, which in other organisms are hard or impossible to overexpress or purify. Available protease deficient yeast strains and the wide variety of inducible promoters, together with the development of several affinity chromatography techniques in recent years, has simplified the purification of eukaryotic multi-protein complexes. Of the several methodologies available to prepare extracts, we prefer the dry ice method because of the ease with which this procedure can be carried out on a relatively large scale while inhibiting proteolysis. The only caveat of this method is that it is associated with a drop in pH, generally to 6.5–7, however, buffers can be used to limit this pH drop (Burgers, 1999). Using this methodology together with a GST-affinity purification step, all known yeast clamp loader complexes were purified to close to homogeneity

The use of N-terminal GST-protein fusions for the overproduction and purification of various proteins/multi-protein complexes is a very valuable tool in the biochemical research field. The conditions used for column chromatography and protein elution are very mild, and are compatible with reducing agents such as DTT. This is an important consideration because the RFC-like complexes are rapidly inactivated in the absence of reducing agent. Generally, we use 3 mM DTT in our buffers. However, there are some possible limitations to this approach that have to be taken into consideration.

First, the presence of the GST tag can alter the properties of the protein of interest that it is fused to. These include the size of GST itself (26.6 kDa) and the ability of GST to form dimers (Kaplan *et al.*, 1997). To overcome these potential problems GST can be proteolytically removed by the human rhinovirus protease, whose recognition site is often embedded in a linker in the cloning vectors, including the pRS420-GALGST series (Fig. 1B).

Second, the human rhinovirus protease has a unique recognition sequence. It cleaves between the glutamine and glycine in the amino acid sequence: Leucine-glutamic acid-X-leucine-phenyl alanine-glutamine-glycine-proline, where X is valine, alanine, or threonine. None of the many complexes we have purified thus far contained an internal cleavage site. Upon digestion of the GST-fusion protein, the released protein moiety of interest will, at the minimum, contain an N-terminal extension of GPEF, if the EcoRI site is used for cloning (Fig. 1B).

Third, when GST-fusion proteins are overexpressed in yeast, the Ssa1 chaperone often copurifies with these proteins. In order to remove this contamination, an ATP-Mg wash step can be included in the purification procedure, as indicated.

Acknowledgments

This work was supported in part by Grant GM32431 from the National Institutes of Health. G.O.B. was supported in part by a stipend from the Washington University-Umeå University Exchange program.

References

Bermudez, V. P., Lindsey-Boltz, L. A., Cesare, A. J., Maniwa, Y., Griffith, J. D., Hurwitz, J., and Sancar, A. (2003a). Loading of the human 9-1-1 checkpoint complex onto DNA by the checkpoint clamp loader hRad17-replication factor C complex *in vitro*. *Proc. Natl. Acad. Sci. USA* **100,** 1633–1638.

Bermudez, V. P., Maniwa, Y., Tappin, I., Ozato, K., Yokomori, K., and Hurwitz, J. (2003b). The alternative Ctf18-Dcc1-Ctf8-replication factor C complex required for sister chromatid cohesion loads proliferating cell nuclear antigen onto DNA. *Proc. Natl. Acad. Sci. USA* **100,** 10237–10242.

Burgers, P. M. (1999). Overexpression of multisubunit replication factors in yeast. *Methods* **18,** 349–355.

Bylund, G. O., and Burgers, P. M. (2005). Replication protein A-directed unloading of PCNA by the Ctf18 cohesion establishment complex. *Mol. Cell. Biol.* **25,** 5445–5455.

Cordingley, M. G., Callahan, P. L., Sardana, V. V., Garsky, V. M., and Colonno, R. J. (1990). Substrate requirements of human rhinovirus 3C protease for peptide cleavage *in vitro*. *J. Biol. Chem.* **265,** 9062–9065.

Ellison, V., and Stillman, B. (2003). Biochemical characterization of DNA damage checkpoint complexes: Clamp loader and clamp complexes with specificity for 5′ recessed DNA. *PLoS. Biol.* **1,** 231–243.

Gerik, K. J., Gary, S. L., and Burgers, P. M. (1997). Overproduction and affinity purification of *Saccharomyces cerevisiae* replication factor C. *J. Biol. Chem.* **272,** 1256–1262.

Gomes, X. V., Gary, S. L., and Burgers, P. M. (2000). Overproduction in Escherichia coli and characterization of yeast replication factor C lacking the ligase homology domain. *J. Biol. Chem.* **275,** 14541–14549.

Hanna, J. S., Kroll, E. S., Lundblad, V., and Spencer, F. A. (2001). *Saccharomyces cerevisiae* CTF18 and CTF4 are required for sister chromatid cohesion. *Mol. Cell. Biol.* **21,** 3144–3158.

Johnston, M., Flick, J. S., and Pexton, T. (1994). Multiple mechanisms provide rapid and stringent glucose repression of GAL gene expression in *Saccharomyces cerevisiae. Mol. Cell. Biol.* **14,** 3834–3841.

Kaplan, W., Husler, P., Klump, H., Erhardt, J., Sluis, C. N., and Dirr, H. (1997). Conformational stability of pGEX-expressed Schistosoma japonicum glutathione S-transferase: A detoxification enzyme and fusion-protein affinity tag. *Protein Sci.* **6,** 399–406.

Majka, J., and Burgers, P. M. (2003). Yeast Rad17/Mec3/Ddc1: A sliding clamp for the DNA damage checkpoint. *Proc. Natl. Acad. Sci. USA* **100,** 2249–2254.

Majka, J., and Burgers, P. M. (2004). The PCNA-RFC families of DNA clamps and clamp loaders. *Progr. Nucl. Acids Res. Mol. Biol.* **78,** 227–260.

Mayer, M. L., Gygi, S. P., Aebersold, R., and Hieter, P. (2001). Identification of RFC(Ctf18p, Ctf8p, Dcc1p): An alternative RFC complex required for sister chromatid cohesion in S. cerevisiae. *Mol. Cell.* **7,** 959–970.

Shiomi, Y., Shinozaki, A., Sugimoto, K., Usukura, J., Obuse, C., and Tsurimoto, T. (2004). The reconstituted human Chl12-RFC complex functions as a second PCNA loader. *Genes Cells* **9,** 279–290.

Smith, D. B., and Johnson, K. S. (1988). Single-step purification of polypeptides expressed in Escherichia coli as fusions with glutathione S-transferase. *Gene* **67,** 31–40.

Zou, L., Liu, D., and Elledge, S. J. (2003). Replication protein A-mediated recruitment and activation of Rad17 complexes. *Proc. Natl. Acad. Sci. USA* **100,** 13827–13832.

[2] Functional Assays for Replication Protein A (RPA)

By Sara K. Binz, Anne M. Dickson,
Stuart J. Haring, and Marc S. Wold

Abstract

Replication protein A (RPA) is a heterotrimeric, single-stranded DNA-binding protein. RPA is conserved in all eukaryotes and is essential for DNA replication, DNA repair, and recombination. RPA also plays a role in coordinating DNA metabolism and the cellular response to DNA damage. Assays have been established for many of these reactions. This

METHODS IN ENZYMOLOGY, VOL. 409   0076-6879/06 $35.00
DOI: 10.1016/S0076-6879(05)09002-6

chapter provides an overview of the methods used for analyzing RPA-DNA interactions, RPA-protein interactions, and functional activities of RPA. Methods are also discussed for visualizing RPA in the cell and analyzing the effects of RPA function on cell cycle progression in mammalian cells.

Introduction

RPA is the prototypic eukaryotic single-stranded DNA-binding protein. RPA homologues have been found in all eukaryotes examined and are essential for multiple processes in cellular DNA metabolism. RPA is absolutely required for DNA replication, multiple DNA repair pathways, and recombination. RPA has two primary activities: it binds single-stranded DNA (ssDNA) with high affinity and low specificity and interacts specifically with multiple proteins (Iftode et al., 1999; Wold, 1997). RPA is phosphorylated during DNA replication and in response to DNA damage. Phosphorylation modulates RPA interactions with both DNA and proteins and appears to be involved in coordinating the cellular DNA damage response (Binz et al., 2003; Oakley et al., 2003; Vassin et al., 2004). There have been several comprehensive reviews of RPA structure and function (Iftode et al., 1999; Wold, 1997) and regulation of RPA by phosphorylation (Binz et al., 2004).

RPA is composed of three subunits of approximately 70, 32, and 14 kDa. The subunit composition and domain structure of the subunits are generally conserved through evolution, suggesting that there is conservation of function (Ishibashi et al., 2001). The 70-kDa subunit of RPA is composed of four structurally related DNA binding domains (DBDs). Each of these domains has a similar core structure, an oligosaccharide/oligonucleotide-binding fold (OB-fold), commonly found in single-stranded DNA binding proteins. Although all of the DBDs in RPA70 interact with DNA, the central two DBDs form the high affinity ssDNA-binding core of the RPA complex. The N-terminal DBD interacts with ssDNA and seems to be involved in interactions with partially duplex DNA, while the C-terminal DBD has a preference for binding to ssDNA containing certain forms of damaged DNA (Lao et al., 1999, 2000). The 32-kDa subunit is made up of three domains: an N-terminal flexible phosphorylation domain, which is the predominant site of phosphorylation in RPA, a central DNA binding domain, and a C-terminal winged helix domain. The N-terminal phosphorylation domain is involved in regulating the RPA complex through interdomain interactions (Binz et al., 2004). The central domain binds both DNA and proteins and is the essential component of this complex. The C-terminal winged helix domain participates in multiple protein-protein interactions and has been recently shown to be important for initiation of viral DNA replication (Arunkumar *et al.*, 2005). The 14-kDa subunit of RPA is composed entirely of a single OB-fold domain. There is no evidence that this domain interacts with DNA. The primary function of the 14-kDa subunit is thought to be structural; it is required to form a stable RPA complex.

The large number of reactions involving RPA precludes complete descriptions of all functional assays for RPA. Therefore, this chapter will focus on the key assays for RPA activity. The chapter begins with the discussion of the purification of recombinant RPA from bacteria. Methods for analyzing RPA binding to DNA and to proteins are then described. An overview is provided for assays used to determine RPA function in DNA replication, repair, and recombination. The final section of the chapter focuses on visualizing and assaying RPA function in cells.

Purification of Recombinant RPA

RPA is an abundant protein that can be purified directly from most cell types. A detailed procedure for purifying RPA from human extracts has been described previously by Brush *et al.* (1995). Fully functional recombinant RPA can be isolated from both insect cells (Stigger *et al.*, 1994) and *E. coli* (Henricksen *et al.*, 1994). Because the recombinant proteins have been found to be fully active, this section focuses on purification of recombinant RPA protein expressed in *E. coli*. There are a number of plasmids that can be used to direct the expression of RPA in *E. coli*. Most are derived from the pET series of vectors (Studier *et al.*, 1990) and have all three subunits of RPA under the control of an inducible T7 promoter. Human RPA (p11d-tRPA; Henricksen *et al.*, 1994) and yeast RPA (Ishiai *et al.*, 1996; Philipova *et al.*, 1996; Sibenaller *et al.*, 1998) have been made in this way. In addition, plasmids exist that express individual subunits and/or mutant forms of human (for example (Henricksen *et al.*, 1994; Lao *et al.*, 2000)), yeast (Bastin-Shanower and Brill, 2001), rice (Ishibashi *et al.*, 2001), and zebrafish (Lee and Lee, 2002) RPA. The procedure for purifying human RPA expressed from the plasmid p11d-tRPA is described later and is based on the method originally described by Henricksen *et al.* (1994). This method has been used to purify both wild-type and mutant forms of human and yeast RPA.

Buffers

The buffers used in the purification are based on a core buffer containing HEPES and inositol (HI buffer): 30 mM HEPES (from a 1 M stock at pH 7.8), 0.25 M EDTA, 0.25% (w/v) myo-inositol, 1 mM dithiothreitol (DTT), and 0.01% (v/v) Nonidet-P40. The inositol and detergent act as protein stabilizers and can be replaced by 10% glycerol and other non-ionic detergents (e.g., Tween-20), respectively. The protein stabilizers are important to prevent protein aggregation during freeze/thaw cycles. HI buffer is supplemented with various salts as indicated later. In the case of HI-phosphate buffers, two complete buffers containing the desired final concentration of either monobasic phosphate salt or dibasic phosphate salt are made and then mixed to obtain the final buffer at a pH of 7.5.

Purification of RPA

Growth and Expression. Expression of RPA in *E. coli* is toxic, so this procedure is optimized to minimize the number of generations (and time in stationary phase) between transformation and induction. The extended growth time needed for large volume fermentors has been observed to reduce RPA expression so multiple one-liter cultures are grown. Starter cultures are also not used because they significantly reduce expression. LB (1 liter – 5 g Bacto tryptone, 2.5 g yeast extract, 2.5 g NaCl, 0.2 ml 10 N NaOH) or TB (1 liter – 5 g Bacto tryptone, 2.5 g NaCl, 0.2 ml 10 N NaOH) media with the desired antibiotic (100 μg/ml ampicillin for p11d-tRPA) is used. p11d-tRPA (10–50 ng) is used to transform BL21(DE3) cells. One-liter cultures are then inoculated with a single colony of freshly transformed cells (1–2 days old is optimal) late in the day and incubated at 37° overnight without aeration or shaking. The next morning, an OD at 600 nm is taken, and then the cells are shaken (200–250 rpm) at 37° until they reach an OD_{600} of ∼0.5–0.8. Protein expression is induced by the addition of isopropyl-β-D-thiogalactopyranoside (IPTG) to a final concentration of 0.3 mM, and incubated with shaking for 2–3 h. The cells are pelleted by centrifugation at 4000 rpm at 4°, then resuspended in HI buffer containing final concentrations of 1 mM phenylmethylsulfonyl fluoride (PMSF), 1 mM DTT, and 2 μl/ml of bacterial protease inhibitor cocktail (Sigma, catalogue number P8465, St. Louis, MO). Cells can either be frozen at −80° or lysed. All subsequent steps are performed at 4°.

Extract Preparation. Cells can be lysed by several methods, including French press, EmulsiFlex (Avestin, Inc., Ottawa, Canada), and sonication. For French press, the cell solution is processed 3 times with a pressure limit of 1100 psi. For EmulsiFlex, the sample is processed 3 times at 10,000–15,000 psi. French press and EmulsiFlex methods have been found to give more reproducible lysis but all methods work well. The lysate is centrifuged at 14,000 rpm in a Sorval SS34 rotor for 35 min at 4°, and the supernatant is then subjected to purification or frozen at −80°. If the lysate is frozen, an additional 1 mM PMSF, 1 mM DTT, and bacterial protease inhibitor cocktail are added to the lysate upon thawing.

Affi-Gel Blue Chromatography. Affi-Gel Blue matrix (Bio-Rad, Hercules, CA) is a crosslinked agarose gel with a Cibacron® Blue dye covalently attached. This affinity column has been used to purify many nucleotide-binding proteins and binds RPA with very high affinity. Lysate is loaded at 10 mg of total protein per ml of Affi-Gel Blue resin (100–200 mesh). The column is washed in water and with 3 column vol of HI buffer. Lysate is then applied to the column at a maximum flow rate of 2.5 ml/min. The column is washed sequentially with 4 column vol of HI-80 mM KCl, HI-800 mM KCl, HI-500 mM NaSCN, and HI-1.5 M NaSCN. The column is then regenerated by washing in HI-2 M NaSCN buffer. RPA complex elutes in HI-1.5 M NaSCN. This fraction contains approximately 5–10% of the total protein eluted and should contain RPA that is ~50–70% pure. NaSCN is a chaotropic salt that causes RPA to dissociate from the matrix. NaSCN is thought to cause partial, reversible denaturation of RPA; therefore, the 1.5 M NaSCN fraction should be loaded onto the next column as rapidly as possible to minimize loss of protein activity. Freezing at this stage of the purification is usually avoided because it reduces protein yield.

Hydroxylapatite Chromatography. The hydroxylapatite column is first equilibrated in HI-30 mM KCl. The Affi-Gel Blue 1.5 M NaSCN fraction is loaded directly onto a hydroxylapatite column at 5 mg of protein/ml of resin. Hydroxylapatite is usually run at the maximum practical flow rate during loading and elution (up to 50 ml/hr/cm^2). The column is washed sequentially with 3 column vol of HI-buffer, HI-80 mM KPO$_4$, and HI-500 mM KPO$_4$. RPA usually elutes in the 80 mM phosphate wash, but can elute in the HI-buffer wash in some preparations. The 500 mM phosphate wash also contains RPA but this fraction is discarded because it contains low molecular weight contaminants that are difficult to remove in other steps. The HI-buffer or 80 mM phosphate fractions generally contain ~40% of the eluted protein and are composed of ~80–90% RPA. Protein from either fraction is active; although, 0 mM fractions contain residual NaSCN that can disrupt functional assays if not removed. These RPA-containing fractions can be safely frozen at −80°.

Mono-Q Chromatography. Hydroxylapatite purified fractions are usually concentrated and trace contaminants removed by Mono-Q (Pharmacia) chromatography. Up to 8 mg of total protein can be loaded on a Mono-Q 5/5 (0.8 ml) column. For larger amounts of protein either multiple separations or a Mono-Q 10/10 (8 ml) column can be used. HAP fractions containing RPA are diluted 1:3 (80 mM fractions) or 1:5 (0 mM phosphate fractions) into HI buffer and loaded onto a Mono-Q column equilibrated with HI-50 mM KCl at the maximum practical flow rate. The column is washed sequentially with 4 column vol of HI-50 mM KCl and HI-100 mM KCl, and then eluted with a 10 column vol linear salt gradient in HI-buffer with from 200–400 mM KCl. RPA elutes as a single peak at ~300 mM KCl. The peak fractions should be >90% pure. The usual yield of RPA is 0.5–1 mg of RPA per liter of culture. Mono-Q fractions are stored at −80° and are generally stable through multiple freeze/thaw cycles.

DNA-Binding Assays

RPA has been shown to interact with single-stranded, partially duplex, damaged and, under some conditions, duplex DNA (Iftode *et al.*, 1999; Wold, 1997). In addition, RPA can destabilize double-stranded DNA promoting denaturation to single-stranded DNA (Iftode *et al.*, 1999; Wold,1997). RPA has high affinity for single-stranded DNA (ssDNA) (~10^{10} M^{-1}) and substantially lower affinity for double-stranded DNA (dsDNA) and RNA (10^6–10^7 M^{-1}). DNA of a variety of lengths and compositions has been used to monitor RPA binding. RPA has an occluded binding site on DNA of 30 nucleotides (nt). Therefore, short oligonucleotides of approximately 30 nt are usually used as substrates in binding assays. RPA complexes with DNA fragments 10 nt or shorter are less stable than complexes with longer probes.

There have been multiple detailed methodologies discussing how to analyze protein interactions with single-stranded DNA. These include filter-binding (Wong and Lohman, 1993), fluorescence quenching (Lohman and Mascotti, 1992), and calorimetry (Kozlov and Lohman, 1998). RPA-DNA interactions can be analyzed with these methodologies; however, since they have been described in detail previously, this chapter will focus on assay methods that have been documented less extensively in the literature. This section will focus on gel based assays, surface plasmon resonance, and fluorescence polarization. The general properties of these methods are summarized in Table I.

Gel-Binding Assays (Gel Mobility Shift and Helix-Destabilization)

RPA-DNA interactions can be analyzed with several gel-based assays: gel mobility shift assay (GMSA; also known as electrophoretic mobility shift assay) and helix destabilization assay. The GMSA relies on the protein-DNA complex having reduced mobility relative to free DNA during gel electrophoresis. The principles and general applications of this method have been reviewed previously (Lane *et al.*, 1992; Molloy, 2000), and a general protocol has been described (Ausubel *et al.*, 2001). The free and bound DNA can be separated by a variety of electrophoretic methods

TABLE I
COMPARISON OF METHODS USED TO ANALYZE RPA BINDING TO DNA

	GMSA	Surface plasmon resonance (SPR)	Florescence polarization (FP)
Useful range	10^5–10^{10}	10^5–10^{12}	10^6–10^9
RPA needed (concentration)	5 pMoles – 0.5 μg (pM-nM)	20 pMoles – 0.1–2 μg (pM-nM)	50 pMoles – 5 μg (nM-μM)
Substrates	ssDNA, dsDNA, plasmids	ssDNA, dsDNA (must be linked to chip–usually through a biotin moiety)	Fluorescently labeled ssDNA, dsDNA
Advantages	Technically simple assay; no special equipment needed; purified protein not required	Rapid, direct assay, can be done under wide range of buffer conditions	Rapid, direct assay, can be done under wide range of buffer conditions
Disadvantages	Indirect assay, depends on complex stability in gel	Requires high salt concentrations for equilibrium binding; subject to experimental artifacts; interpretation can be demanding	Requires high salt concentrations for equilibrium binding; fluorescence of fluorescein is pH dependent; nonspecific light scattering and reagents with high background fluorescence

(agarose or native polyacrylamide gels) and visualized by radioisotope detection or fluorescent imaging. This method can also be used with a variety of different DNA probes: including short or long, single-stranded or double-stranded, or even a plasmid DNA. For proteins that bind specific sequences, GMSA also has the advantage that it can be used to analyze binding activity in extracts and partially purified fractions (Table I). In a helix destabilization assay, a duplex DNA is incubated with RPA and conversion to single-stranded DNA monitored. Although the ability of RPA to destabilize DNA is dependent on its ssDNA binding ability, there are forms of RPA that have decreased helix destabilization activity but wild-type ssDNA binding affinity (Binz et al., 2003; Lao et al., 1999). Helix destabilization can also be assayed with short dsDNA substrates or utilizing an oligonucleotide annealed to the M13 plasmid.

Buffers

- 10× T4 polynucleotide kinase buffer: 700 m*M* Tris-HCl, 100 m*M* MgCl$_2$, 50 m*M* DTT, pH 7.6
- 10× Klenow buffer: 100 m*M* Tris-HCl, 50 m*M* MgCl$_2$, 75 m*M* DTT, pH 7.5
- 1× Tris borate/EDTA (1× TBE): 89 m*M* Tris, 89 m*M* boric acid, 2 m*M* EDTA
- PAGE elution buffer: 0.5 *M* ammonium acetate, 25 m*M* Tris-HCl pH 7.5, 1 m*M* EDTA
- 10× FBB: 300 m*M* HEPES (diluted from 1 *M* stock at pH 7.8), 1 *M* NaCl, 50 m*M* MgCl$_2$, 5% inositol, 10 m*M* DTT
- 0.1× TAE: 4 m*M* Tris acetate and 0.2 m*M* EDTA
- HI-30: 30 m*M* HEPES (diluted from 1 *M* stock at pH 7.8), 1 m*M* dithiolthreitol, 0.25 m*M* EDTA, 0.5% (w/v) inositol and 0.01% (v/v) Nonidet-P40, and 30 m*M* KCl
- 10× loading buffer: 40% glycerol, 0.2% bromophenol blue and 2% SDS.

GMSA Assay

Oligonuclotides are radiolabeled by incorporating ^{32}P from γ-ATP using 25 units of T4 polynucleotide kinase, which catalyzes the transfer of the radioactive gamma-phosphate from ATP to the 5′-OH group of single- and double-stranded DNAs. The labeling reaction includes 25 pmol DNA, 5 μCi [γ-^{32}P]ATP (6000 μCi/mmol), and 25 units of T4 polynucleotide kinase in 50 μL T4 polynucloetide kinase buffer) is incubated 1–3 h at 37°. The labeled DNA is separated from free ATP by size exclusion chromatography (e.g., P-30 Tris column (Bio-Rad) following the manufacturer's specifications).

Double-stranded DNA obtained from a PCR reaction or cut from a plasmid can be radiolabeled by incorporating ^{32}P from [α-^{32}P]dNTP using a 3′ fill-in reaction with Klenow fragment. The fill-in reaction contains 1 μg DNA substrate, five units of Klenow fragment, and 33 μM of each dNTP in 1× Klenow buffer. The reaction is incubated for 30 min at 25° and terminated by the addition of EDTA to a final concentration of 10 m*M* or heat inactivation at 75° for 10 min. DNA fragments are separated by a 6% 1× TBE polyacrylamide gel, visualized by autoradiography, and excised from the gel. The resulting gel fragment is cut into little pieces in 400 μl PAGE elution buffer and incubated over night at 42°. The polyacrylamide fragments are then removed from the DNA using a spin column (e.g., Spin-X, Costar, Cambridge, MA) and ethanol precipitated. Alternatively, the DNA substrate may be separated using agarose gel electrophoresis and purified using standard gel extraction methods

(Ausubel *et al.*, 1989). Annealing reactions for the creation of dsDNA substrates are discussed later under helix destabilization.

Radiolabel incorporation is determined by thin layer chromatography on Polyethyleneimine strips (PEI Cellulose F, Merck, distributed by EMD Chemicals Inc., Gibbstown, NJ). 1 μl of labeling reaction is spotted near the bottom of a 1 × ~6 cm strip of PEI cellulose, dried, and developed in 1 M HCl and 0.1 M sodium pyrophosphate. Chromatography is stopped when the liquid phase is within 1 cm of the top of the strip. The strips are then dried, cut into equal thirds, and each counted by a scintillation counter. DNA remains at the bottom while nucleotides and free phosphate migrate with the solvent. The percent incorporation is the amount of radioactive phosphate on the bottom of the PEI strip divided by the total amount of radioactive phosphate of the strip. This quantitation also allows the determination of the specific activity of the labeled DNA (radioactivity incorporated per mole DNA).

GMSA analysis of RPA binding to different DNA substrates has been described previously (Kim *et al.*, 1992; Lao *et al.*, 1999). In each GMSA experiment an increasing amount of protein is added to a constant amount of DNA in a total reaction volume of 15 μl (Fig. 1). The normal range of concentrations used for a wild-type RPA titration is $10^{-11} - 10^{-7}$ M. For even spacing, concentrations should be increased logarithmically (e.g., 1 and 3.16 units per order of magnitude). With oligonucleotide DNA substrates, each reaction should contain 0.2 fmoles of ssDNA (13 nM DNA). The affinity of RPA is in the nM range, so higher concentrations of DNA result in stoichiometric binding conditions that prevent accurate determination of binding constants. (Under stoichiometric binding conditions, the apparent binding constant represents a lower limit estimate of the true equilibrium binding constant.) Higher DNA concentrations (2 fmoles or more) can be used with oligonucleotides shorter than 20 base pairs in length or with double-stranded DNA because RPA has a reduced affinity for these substrates. Binding to double-stranded DNA is weak enough that reactions are usually performed in low ionic strength buffer (HI-30). Reducing KCl and MgCl$_2$ concentrations cause the apparent affinity of RPA for dsDNA to increase 300-fold (Lao *et al.*, 1999).

A DNA cocktail containing 150 μg/ml BSA, 3 × FBB, and DNA substrate is assembled at 25°. 5 μl of the cocktail is added to reaction tubes containing RPA in a total of 10 μl of HI buffer with 30 mM KCl. The 15 μL reactions are then incubated for 20 min at 25° before being brought to a final concentration of 4% glycerol and 0.01% bromophenol blue. The mixture is then separated at 100 V on a 1% agarose gel in 0.1× TAE buffer. Separation is complete when the bromphenol blue is approximately half way down the gels. To prevent loss of DNA during drying, the gels are dried on Whatman grade DE81 paper (Fisher Scientific, Pittsburgh, PA)

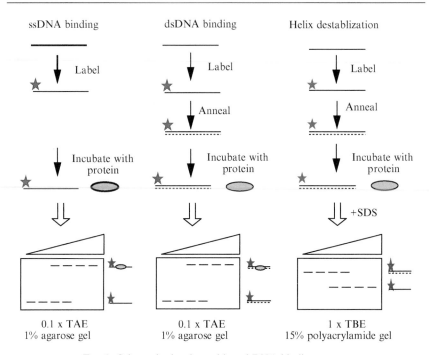

FIG. 1. Schematic showing gel-based DNA-binding assays.

supported by a sheet of Whatman 3-mm paper and radioactive bands visualized by autoradiography. The radioactivity in each band is quantitated using an instant imager or phosphorimager. The binding isotherms are generated by plotting the fraction DNA in a complex verses the concentration of RPA (M; on a log scale). The intrinsic binding constants can be calculated by fitting the data to the Langmuir binding equation. The Langmuir binding equation is modified using mass balance and equilibrium binding equations to obtain an equation dependent on the total concentration of RPA and DNA: concentration of RPA·DNA complex = $[(kD + kR + 1) - [(kD + kR + 1)^2 - (4k^2RD)]^{0.5}]/(2k)$, where k = association constant, R = total molar RPA, and D = total molar DNA (Kim et al., 1994). Fractional saturation of the DNA equals the concentration of RPA·DNA complex divided by the total DNA.

Helix Destabilization Assay

In this reaction, double-stranded DNA is incubated with increasing concentrations of RPA ($10^{-12} - 10^{-8}$ M) in HI-30 buffer and the amount of ssDNA present monitored by gel electrophoresis (Fig. 1). Substrates are short dsDNA fragments 30–50 bp in length that are either completely complementary (fully-duplex DNA) or contain partially noncomplementary sequences (2–8 nt) in the middle of the sequence (bubble DNA). The bubble substrates act as a pseudo-origin substrate (Iftode and Borowiec, 1997) and provide information about binding steps that occur after formation of the initial RPA-DNA complex (Binz *et al.*, 2003; Iftode and Borowiec, 2000; Lao *et al.*, 1999).

Oligonucleotides are labeled as described previously. Then 100 fmol/μl labeled oligonucleotide and 100 fmol/μl cold complementary oligonucleotide are annealed in 50 μl 20 mM MgCl$_2$, 50 mM NaCl, and 10 mM Tris-HCl (pH 7.5) in a thermocycler (95° for 1 min followed by a ramped cooling to 22° at a rate of 0.01° per s with a hold temperature of 4°). Annealing is monitored by 15% polyacrylamide gel electrophoresis (1× TBE). Greater than 95% of labeled DNA should be in double-stranded form after annealing.

Helix destabilization by RPA is very sensitive to ionic strength so it is essential to minimize the contributions of salt from the RPA proteins fractions. RPA is usually dialyzed against HI-30 prior to being used in this assay. The reaction for helix destabilization is identical to that described previously except that the final reaction is in HI-30 buffer and the reactions are terminated by adding 2 μl of a 10× loading buffer containing SDS to disrupt RPA-DNA complexes. Reaction products are separated on a 15% polyacrylamide gel with 1× TBE at 200 v until the dye is near the bottom of the gel. The polyacrylamide gel can be dried on Whatman 3-mm chromatography paper or wrapped in plastic wrap and quantitated directly and/or visualized by autoradiography. The percentage of dsDNA is graphed against the molar amount of RPA and then fit to a Langmuir binding equation (Lao *et al.*, 1999). Although the bimolecular Langmuir binding equation does not represent the multistep reactions that occur during destabilization of DNA by RPA, the equation precisely determines the midpoint of the transition between dsDNA and ssDNA. The midpoint, with units of M^{-1}, is used as a value for comparing the activities of RPA forms in this assay.

Surface Plasmon Resonance

In surface plasmon resonance (SPR), polarized light is reflected off a gold layer positioned between a liquid solution and a glass surface in a sensor chip. The chip allows liquids to be passed across the gold surface. Usually a smaller component (ligand), attached to the sensor surface, is bound by the larger component (analyte), flowing over the surface. A small electric field is generated when the light interacts with the gold. Changes in the local environment caused by analyte binding change the electric field (through plasmon resonance) which causes the changes in the reflected light. These changes, measured in resonance units (RU), are directly proportional to the amount of analyte bound and allow quantitation of both the kinetics and affinity of the binding reaction ([Lin et al., 2005]; see also Biacore 3000 Instrument Handbook 1999 Biacore AB, [1999]). The collection of RU data is plotted in a sensorgram, a real time graph of RU versus time. SPR can be used to analyze interactions between almost any two macromolecules. In the case of RPA:DNA interactions, biotinylated oligonucleotides are attached to the streptavidin biosensor chip (SA-chip). SPR has several advantages over other DNA binding assays. With SPR, equilibrium binding constants can be measured directly and obtained quickly. In addition, SPR can simultaneously determine rate constants. SPR can also be performed in a variety of buffer conditions to ensure that equilibrium binding conditions can be obtained. However, SPR binding reactions are subject to a number of experimental artifacts so care must be taken to ensure that appropriate experimental conditions, controls, and data analyses are used (discussed later).

Buffers

HBS-EP buffer (BIAcore): 10 mM HEPES, pH 7.4, 150 mM NaCl, 3 mM EDTA and 0.005% polysorbate–20.

Surface Plasmon Resonance Procedure

• Initial Setup: The HBS-EP buffer should be degassed, the sensor chip acclimated to room temperature, inserted into the Biacore instrument, and washed with the HBS-EP buffer. These tasks prevent gas bubbles from accumulating on the sensor surface and interfering with the protein:DNA signal. A slope of >10 RU/min on the sensorgram indicates that the chip is not equilibrated with temperature or buffer. A continual drift may also indicate that a new flow cell needs to be used (Myszka, 1999).

• Attaching DNA to Biosensor Chip: For RPA-DNA interactions, biotinylated oligonucleotides are attached to the streptavidin biosensor chip (SA-chip from BIAcore). The biosensor chip has four flow cells and the streptavidin biosensor surface of one flow cell is prepared by manually injecting biotinylated DNA diluted to 0.5 nM in 10 mM sodium acetate (pH 4.8) with 1.0 M NaCl. The change in RU upon addition of DNA to the chip should be kept below 40. It is important to limit the amount of DNA attached to the chip to prevent steric hindrance between attached DNA and to prevent the binding reaction being driven by mass transfer, the effect observed when the rate of diffusion to and from the sensor surface is limiting. For a dsDNA substrate, a complementary nonbiotinylated strand, diluted in the sodium acetate buffer, is passed over the flow cell until the RU has doubled. The oligonucleotides are biotinylated at the 5′ end to allow maximum DNA access for the 5–3′ polarity of the RPA high affinity binding sites. There does not appear to be a minimum or maximum DNA length for SPR.

• Regeneration of Biosensor Chip: After RPA binding the sensor surface is regenerated by injecting 20 μl of 100 mM NaOH at 10 μl/min or 30 s of 0.25% SDS (Schubert et al., 2003; Wang et al., 2000). Both are capable of disrupting protein-DNA interactions and are unlikely to interfere with dsDNA structure. Buffers with low pH will lead to a slight loss of biotinylated ligand from the surface while strong bases will disassociate dsDNA (BIA applications Handbook [1999]). To confirm that the protein has been stripped from the DNA and, if applicable, that the dsDNA is still intact, the RU of the sensor surface should be reduced to that of the DNA alone. A signal higher than the DNA indicates protein still bound while a signal lower than the DNA would indicate a loss of the complementary strand from the dsDNA.

• Preliminary Experiments: To obtain accurate equilibrium and kinetic constants with SPR technology there are certain experimental conditions that need to be met (Table II) (Karlsson and Falt, 1997; Van Regenmortel, 2003). (i) First, unless experiments are performed to examine mass transfer effects, kinetic rates determined at low flow rates, especially for large proteins (slowly diffusing) with high affinity, almost certainly have mass transfer limitations and do not represent the actual rates. As a corollary, unless the flow rate at which the experiments are performed is reported, the kinetic rates determined cannot be validated. Two methods that maximize the efficient delivery of analyte to the surface are used to reduce mass transfer limitations. First, the flow rate of analyte can be increased. This increases delivery of analyte to the surface. Alternatively (or in addition) the ligand concentration on the sensor surface can be decreased to reduce the amount of analyte required. Mass transfer is not limiting if the sensorgram tracings have similar curvatures and the measured on-rate is constant as the flow rate is increased (analyte [protein] contact time is held constant). Mass transfer effects can be ignored if the initial binding rate (on-rate) does not increase more than 5–10% when the flow rate is increased from 15 to 75 μl/min. If mass transfer effects are still observed using low ligand concentrations and high flow rates, the data can be fit to a modified Langmuir binding equation that accounts for mass transfer (part of Biacore analysis software). Experiments from this lab have indicated that RPA binding is limited by mass transfer effects at low flow rates (Binz, unpublished). (ii) The binding reaction must be bimolecular. This can be

TABLE II

ESSENTIAL PARAMETERS FOR SURFACE PLASMON RESONANCE ASSAYS

Preliminary experiments	Vary flow rate to prove mass transfer is not limiting
	Vary protein concentrations to prove bimolecular reaction
Parameters necessary to evaluate data (should be reported)	
Experimental	Temperature
	Reference surface
	RU of ligand
	Flow rate
	Type of injection
	Concentration of protein
	Time of association and disassociation
	Sensorgram showing binding response at equilibrium
Analysis	Equation used
	Ratio of Chi2 to Rmax value
	Residuals
	ln dR/dt plot

determined by varying the analyte concentration over a 10-fold range of concentrations around the anticipated binding constant. After fitting to a Langmuir binding equation, the resulting equilibrium response (R_{eq}) should be graphed as R_{eq}/concentration vs R_{eq}. A linear response indicates that the binding reaction is indeed bimolecular and, thus, the equilibrium and kinetic constants are valid. A linear response of the association and disassociation phase in the ln dR/dt plot also indicates a bimolecular reaction.

• Equilibrium vs. Kinetic Analysis: For large proteins like RPA that have mass transfer limitations at low flow rates, there are two sets of possible experimental conditions: equilibrium and kinetic. Equilibrium experiments are done at low flow rates (10 μl/min) while kinetic experiments, which simultaneously determine equilibrium binding constants and kinetic rates, are performed at high flow rates (75 μl/min). The equilibrium binding constant may be accurately determined at low flow rates, even with possible mass transport effects, only if the binding response reaches equilibrium. At equilibrium the effect of diffusion is negligible as mass transport causes both the kinetic constants to be underestimated. However, accurate kinetic rates can only be obtained when mass transfer effects are limited by high flow rates. In addition, the association rate is only accurately determined if it is not faster than 4 s (Biacore Handbook). Decreasing the protein concentration of the analyte will decrease the measured association rate. Furthermore, the determination of affinity constants will be affected by inaccurate protein concentrations or the presence of impurities. Other experimental conditions, such as temperature, also affect binding so it is important that they are kept constant and reported (Table II) (Van Regenmortel, 2003).

• RPA Binding Studies: RPA is diluted in HBS-EP buffer containing 1 mM DTT and the appropriate amount of NaCl. RPA is loaded by the kinject option, which reduces noise in the dissociation phase (Biacore Handbook). For wild-type RPA, equilibrium is reached with an analyte concentration of 16 nM at low flow rates and with 128 nM RPA at a flow rate of 75 μl/min. A reference surface is required as a control. Subtracting the reference from the experimental channel will correct for bulk shift (changes in refractive index due to buffer changes), nonspecific binding, injection noise, and baseline drift. The reference surface is typically an adjoined flow cell on the sensor chip. Each experiment should be repeated at least twice in randomized order. To verify that the regeneration steps have not affected the binding surface, the RPA form initially bound to the surface should be repeated at the end of the series of experiments. If the second trial is not consistent with the first trial, it indicates that the regeneration has affected the binding surface. Data is analyzed with the BIA Evaluation program and fit to a bimolecular Langmuir binding curve. Unless the data indicates that the binding reaction is not bimolecular, the data should always be fit to the simplest model. The fit is assessed by observing several parameters (Biacore Handbook): (i) Is the experimental binding constant consistent with other studies? (ii) The residual plot of the association and disassociation phase should be $\pm$ 2 RU with no systematic variation pattern. (iii) The chi^2 value should be $\leq$10% of the maximum RU signal, R_{max}. (iv) The plot ln dR/dt of the association and disassociation phase should be assessed: a sloped line indicates bimolecular binding while a concave and convex curve indicates a higher order reaction and mass transfer effects respectively. (v) Finally, the T value, which is used to describe the fit sensitivity, should be >10 (>100 is preferred). The T value is calculated by taking a ratio of the fitted parameter (i.e., R_{max}, k_a, k_d) over the standard error (SE) of the parameter.

• Equilibrium Binding: RPA binds with high affinity ($K_a \sim 10^{10}$ M^{-1}). Thus under physiological solution conditions, RPA binding is usually stoichiometric. Therefore, in order to obtain valid binding parameters for RPA, it is necessary to demonstrate that binding is occurring under equilibrium not stoichiometric binding conditions. This is done by carrying out binding reactions under a variety of salt concentrations (e.g., from 100 mM to 2 M NaCl). Under stoichiometric binding conditions the apparent association constant will not vary (or only vary minimally) with the log of the NaCl concentration. In contrast, under equilibrium binding conditions, the log of the association constant will decrease linearly with the log of the NaCl concentration. We have found that equilibrium binding for RPA can be obtained in HBS buffer containing 1–1.5 M NaCl. Several of the published reports analyzing RPA binding with SPR were carried out under physiological solution conditions and thus probably do not reflect equilibrium binding.

Fluorescence Polarization

Fluorescence polarization (FP) measures the change in anisotropy (rotation rate) of a fluorescently labeled DNA probe when it is bound to a protein (Lundblad *et al.*, 1996). In these assays fluorescently labeled ssDNA is incubated with increasing concentrations of RPA. After excitation with polarized light, the rotation rate of the DNA is measured. Free ssDNA rotates faster than ssDNA·RPA complexes. A binding isotherm can be generated by plotting protein concentration versus changes in the anisotropy of the DNA. FP has the advantage of being a rapid, direct assay that does not require radioactivity. Solution conditions can be varied to ensure that binding is occurring under equilibrium conditions. The disadvantages of this method are that FP is less sensitive than GMSA; it requires several hundred fold higher DNA concentrations (Table I). Because of the high DNA concentration, this assay must be performed at high salt and protein concentrations to obtain equilibrium binding.

This method can be carried out with any spectrophotometer that can monitor fluorescence anisotropy. The methods described later are for the Wallac Victor^2 1420 system (Perkin Elmer, Welesley, MA) which monitors anisotropy in black microtiter plates and can be automated for multipoint titrations. These characteristics make it well suited for RPA binding analysis.

Eighty microliter reactions containing buffer, RPA, and fluorophore-conjugated ssDNA are assembled at room temperature in the wells of black 96-well FluoroNunc plates. As with SPR, it is essential to carry out binding reactions at a variety of salt concentrations to demonstrate that equilibrium binding is occurring. Generally we have found that equilibrium binding occurs at concentration of between 1.25–1.5 M KCl. RPA is titrated from 0–1000 nM, and additional salt-containing buffer is added to correct for salt concentration differences between the samples. DNA is added last to a final concentration of 5 nM. Fluorescein-conjugated deoxythymidine 30 nucleotides in length is a standard probe, although other length and base compositions can be used. When not in use, the tube of DNA is protected from light and stored at −80°. The plate is covered with aluminum foil and gently rocked for 10 min. The plate is then uncovered and monitored for polarization. For fluorescein, the excitation wavelength is 480 nm with a 30 nm bandwidth and the emission wavelength is 535 nm with a 40 nm bandwidth. Data is collected with a measurement time of 0.2 s. Anisotropy, r or $R = (I_{vv} - I_{vh})/(I_{vv} + 2GI_{vh})$, where I_{vv} is the vertically excited and vertically emitted light, I_{vh} is the vertically excited and horizontally emitted light, and G is the noise of the instrument. Polarization, $P = 3r/2 + r$. Data is analyzed by fitting to the Langmuir binding equation as described previously for GMSA. At 1.25 M KCl, RPA typically has an association constant of 10^8 M^{-1} using FP.

Other DNA-Binding Assays

NMR has been used to directly analyze the interactions of individual domains of RPA with DNA and with protein fragments. The advantage of this method is that it can be used to directly monitor RPA-DNA (or protein) interactions and map the interacting residues. The disadvantages are that large amounts of ^{15}N or ^{13}C labeled protein are needed and that the experiments are technically challenging and require specialized expertise. In addition, NMR spectra become more complex as the size of the protein becomes larger. This means that for a protein the size of RPA (110 kDa), studies are limited to subcomplexes or domains. For example, current analyses of RPA have been limited to analysis of DNA binding to the N-terminal domains of RPA70 (Daughdrill et al., 2003; Jacobs et al., 1999) and to the central high affinity DNA binding core of RPA70 (Arunkumar et al., 2003). The protein winged helix domain of RPA32 has also been analyzed by NMR (Arunkumar et al., 2005).

Recently Shell and coworkers demonstrated that biotin-lysine protein footprinting can be used to map residues in RPA that interact with DNA (Shell et al., 2005). In this method, RPA is biotinylated at lysine residues and individual residues identified by mass spectrometry. DNA binding sites can then be mapped by identifying residues that are protected from modification after DNA-binding.

RPA-Protein Interactions (Enzyme-Linked Immunosorbent Assay)

Enzyme-linked immunosorbent assay (ELISA) is used to examine interactions between purified proteins. For a general discussion of this method see Chapter 11 of Current Protocols in Molecular Biology (Ausubel et al., 2001). Protein-protein interactions are detected by utilizing antibodies conjugated to an enzyme that coordinates an observable color change. ELISAs are advantageous because they are quickly performed, reproducible, and lack the requirement for sophisticated equipment.

In addition, there are a variety of antibody-conjugated enzymes that may be used, such as horseradish peroxidase, akaline phosphatase, β-galactosidase, glucoamylase, and urease. Many of the enzymes also have a multitude of substrates that produce detectable products. Some products require excitation for detection (fluorogenic) while others are seen visibly (chromogenic).

Buffers

- 1× phosphate buffered saline (PBS: 137 mM NaCl, 2.7 mM KCl, 4.3 mM Na$_2$HPO$_4$·7H$_2$O, 1.4 mM KH$_2$PO$_4$).

Experimental Procedure

Wells in microtiter plates are coated with 1 μg of RPA or the desired protein in 50 μl of H$_2$O and incubated for 1 h at room temperature. Plates are washed with 1× PBS with 0.2% Tween-20 three times to remove unbound protein. Plates are then blocked with 300 μl of 5% nonfat-dried milk in 1× PBS for 10 min and washed. Various amounts of the secondary protein or BSA (the negative control) in 50 μl of PBS with 5% milk are added to each well, incubated for 1 h, and washed. The amount of secondary protein necessary must be determined empirically. We have found that most RPA interacting proteins are detected when 10–70 nM are added. Primary antibodies for the secondary protein (diluted appropriately in 1× PBS with 5% nonfat-dried milk) are added to the plates, incubated for 30 min, and washed. A secondary antibody with a peroxidase conjugate diluted in 50 μl of 1× PBS with 5% milk are added to the plates, incubated for 30 min, and washed. Plates are developed using 200 μl of 0.8 mg/ml o-phenylenediamine (200 μmole per tablet) in 0.05 M phosphate-citrate buffer with 0.03% sodium perborate at pH 5.0 in 100 ml water (Sigma-Aldrich, St. Louis, MO). With this system an interaction is indicated when the buffer changes from yellow to orange. After 20–60 min, the interactions are quantitated at OD$_{450}$ (absorption maximum at 492 nm) using a microtiter plate reader. The o-phenylenediamine is a light sensitive substrate that will eventually turn orange so the reaction must be monitored. The reaction may be stopped with 100 μl 2.5 N H$_2$SO$_4$ or 20 μl of 6 N HCl. For each experiment, backgrounds are determined for wild-type and all mutant forms of RPA using BSA as the secondary protein. This is a control for nonspecific interactions with the antibodies.

To perform a successful ELISA several parameters need to be considered. The ELISA is only as sensitive and specific as the antibody used. The antibody must detect the native structure of the secondary protein. To confirm the primary antibody is effective, a control experiment is carried out in which the secondary protein is adhered to the microtiter wells and then directly probed with primary antibody. If RPA is used as the second protein, the RPA antibody must interact with the mutant forms of RPA to the same extent as wild-type RPA.

Functional Assays for RPA

RPA is involved in multiple reactions in the cells. These include DNA replication, DNA repair, and recombination. In this section, we will give an overview of the types of *in vitro* assays used to examine RPA function in these processes. Because of the number of assays and the complexity of each, detailed protocols will not be provided. Rather a brief discussion of the assay and appropriate references will be given.

DNA Replication

RPA was originally identified as a component necessary for SV40 replication *in vitro*. The SV40 viral system remains one of the best assays for looking at RPA function in DNA replication. A detailed description of the SV40 system has been presented by Brush *et al.* (1995).

The SV40 replication reaction is normally carried out in 25 μl and contains the following components: 30 mM HEPES·HCl (from 1 M stock, pH 7.8), 7 mM MgCl$_2$, 40 ng DNA (supercoiled plasmid), 40 mM creatine phosphate, 2.5 μg creatine kinase, 4 mM ATP, 0.2 mM each of CTP, GTP, UTP, 0.1 mM each of dATP, dGTP, dTTP, 0.05 mM dCTP with α ^{32}P-dCTP $\sim$1000 cpm/pmol, 0.5–1 μg SV40 T-antigen, replication proteins (either 75–100 μg cytoplasmic extract or a combination of required protein fractions). RPA function is assayed by supplying all other required proteins in excess as purified or partially purified fractions (described in Brush *et al.*, 1995; Henricksen *et al.*, 1994). For example, replication can be reconstituted with RPA, SV40 large T-antigen, and cytoplasmic protein fractions CFIBC and CFII (Brush *et al.*, 1995; Henricksen *et al.*, 1994). An alternate method of making SV40 replication dependent on RPA is to subject cellular cytoplasmic extracts to sequential ammonium sulfate precipitations (Kenny *et al.*, 1989; Wobbe *et al.*, 1987). Extract is precipitated with ammonium sulfate at 35% saturation and the resulting supernatant precipitated with ammonium sulfate at 65% saturation. The resulting 35–65% pellet is collected and resuspended in one-fifth initial vol 50 mM Tris-HCl, pH 7.8, 1 mM dithiothreitol, 0.1 mM EDTA, 10% glycerol, and dialyzed to remove residual ammonium sulfate. The resulting fraction, called AS65, contains all of the proteins necessary to support SV40 DNA replication except for RPA and SV40 T antigen. SV40 replication is monitored by quantifying the incorporation of radioactive nucleotides into DNA by either gel electrophoresis or trichloroacetic acid precipitation (Brush et al., 1995; Henricksen et al., 1994).

A second model system for examining RPA function in DNA replication is the cell-free Xenopus extract system (Walter and Newport, 2000; Walter et al., 1998). This system utilizes Xenopus egg extracts to create synthetic nuclei, which are then crushed to generate an activated nucleoplasmic extract (NPE). When the NPE is mixed with cytoplasmic extract and DNA, origin independent DNA synthesis occurs. This system has DNA replication properties similar to metazoan cells and is absolutely dependent on RPA (Walter and Newport, 2000).

DNA Repair Assays

RPA is involved in nucleotide excision repair, base excision repair, mismatch repair, and recombination repair processes. *In vitro* assays have been established for all of these reactions.

Nucleotide excision repair involves a multistep process in which damaged DNA is recognized and then excised on a short single strand oligonucleotide. This results in a short ssDNA gap that is then filled in by DNA polymerases. RPA is required throughout this process (Riedl et al., 2003). RPA has been shown to be involved in damage recognition (He et al., 1995) and assembly of the excision complex (De Laat et al., 1998; Wakasugi and Sancar, 1999). RPA is also required for reconstituted nucleotide excision repair reactions (Aboussekhra et al., 1995; Mu et al., 1995). General methods for nucleotide excision repair have been described previously (Biggerstaff and Wood, 1999; Shivji et al., 1999; and elsewhere in this volume).

Base excision repair removes damaged base residues to create an abasic site in the DNA. The abasic site is then repaired by the action of specific endonucleases and polymerase beta (short patch repair) or a modified replication complex (long patch repair) (Krokan et al., 2000; Kunkel and Erie, 2005). RPA interacts with uracil DNA glycolsylase (Nagelhus et al., 1997) and plays a role in DNA synthesis in long patch repair (DeMott et al., 1998; Krokan et al., 2000). Base excision repair has also been reconstituted *in vitro* (Kubota et al., 1996; Nicholl et al., 1997).

Mismatch repair occurs to repair nucleotides that are incorrectly incorporated during DNA replication. RPA appears to be essential for multiple processes in human mismatch repair (Ramilo et al., 2002). A cell-free assay system has been described for mismatch repair (Ramilo et al., 2002).

Recombination

RPA is absolutely required for recombination. In recombination RPA interacts with RAD51 and RAD52 and is essential for the formation of heteroduplex DNA (synapsis) and recombination of homologous sequences. Assays have been established to look at each of these processes in cell-free systems. Most involve analyzing both formation of hybrid or recombined molecules in gels (for example see Kantake *et al.*, 2003) or by monitoring changes in intermolecular interactions (for example by fluorescence energy transfer, see Gupta *et al.*, 1998). See also other chapters in this volume.

In Vivo Localization

Over the last several years RNA interference (RNAi) has been used widely for the knock down of proteins in eukaryotic cells (Campbell and Choy, 2005; Hannon, 2002; Hannon and Rossi, 2004). RNAi has been used successfully to dramatically reduce the cellular concentration of both the 70- and 32-kDa subunits of RPA (Dodson *et al.*, 2004; Ishibashi *et al.*, 2005; Miyamoto *et al.*, 2000; Vassin *et al.*, 2004). RNAi methodology allows the examination of the function of RPA and, when coupled in appropriate expression systems, specific RPA mutants in human cells. This section will discuss the methods used to visualize RPA in mammalian cells and prepare cells for cell cycle analysis.

In Vivo Localization by Indirect Immunofluorescence

There are numerous reports using immunofluorescence to examine RPA localization in cells. These include examining total protein in fixed cells, chromatin-associated RPA in cells extracted with detergent, and RPA localization and abundance after RNAi knock down (for examples see Brénot-Bosc *et al.*, 1995; Dimitrova and Gilbert, 2000; Vassin *et al.*, 2004). The procedure described later provides a general method that has been found to be effective for looking at total RPA or chromatin associated RPA in mammalian cells.

Buffers

- PBS: 137 mM NaCl, 2.7 mM KCl, 4.3 mM Na$_2$HPO$_4$·7H$_2$O, 1.4 mM KH$_2$PO$_4$
- CSK buffer: 10 mM HEPES pH 7.4, 300 mM sucrose, 100 mM NaCl, 3 mM MgCl$_2$·6H$_2$O.

- 0.5% Triton solution: 0.5% Triton X-100, 1 mM phenylmethyl sulfonyl fluoride (PMSF),1 μg/ml aprotinin, 1 mM Na$_3$VO$_4$, CSK buffer
- 4% formaldehyde solution: 4% formaldehyde, PBS
- 0.5% NP-40 solution: 0.5% NP-40, PBS
- Blocking solution I: 2% bovine serum albumin (BSA), 1% normal goat serum (NGS; for goat secondary antibody), PBS
- Blocking solution II: 2% BSA, PBS
- DNA staining solution: 1 μg/ml 4′,6-diamidino-2-phenylindole (DAPI), PBS
- Mounting medium: 1,4 phenylene diamine, 58.5% glycerol, 100 mM NaHCO$_3$ pH 9.0.

Methods

HeLa cells are seeded at 2×10^5 cells/well (1×10^5 cells/ml) and are grown as a monolayer on 18-mm circular coverslips in a 6-well tissue culture dish at 37° (5% CO$_2$) for 3–5 days. The coverslips are then removed and placed cell-side up in a 12-well tissue culture dish. The coverslips are washed twice for 2 min with 1 ml cold CSK buffer. RPA is found as both chromatin- and nonchromatin-associated fractions within the cell (Fig. 2) (Brénot-Bosc *et al.*, 1995; Dimitrova and Gilbert, 2000; Vassin *et al.*, 2004). To localize the chromatin-associated fraction, the cells are treated with 0.5% Triton solution for 5 min at room temperature (RT). The cells on the coverslips are then fixed with 1 ml 4% formaldehyde solution for 20 min at RT. Once fixed, the coverslips are washed once with 1 ml PBS for 5 min at RT. The cells on the coverslips are treated with 1 ml 0.5% NP-40 solution for 5 min at RT. The coverslips are then washed three times with 1 ml PBS at RT.

Before immunostaining, coverslips are incubated in 1 ml blocking solution I for 30 min at RT. After removal of the blocking solution, the RPA antibody is diluted appropriately in blocking solution II, 500 μl of diluted primary antibody is added to each well, and the coverslips are incubated at RT for 2–4 h. The coverslips are washed in 1 ml PBS three times for 5 min at RT. The secondary antibody is diluted appropriately in blocking solution II, 500 μl of diluted secondary antibody is added to each well, and the coverslips are incubated at RT for 1 h in the dark. The coverslips are washed in 1 ml DNA staining solution for 5 min at RT. After removal of the DNA staining solution, the coverslips are washed in 1 ml PBS twice for 5 min, then with 1 ml sterile H$_2$O twice for 2 min. After completing the washes, 5 μl of mounting medium is added to each of the coverslips, and the coverslips are mounted onto slides. Clear nail polish is used to seal each coverslip.

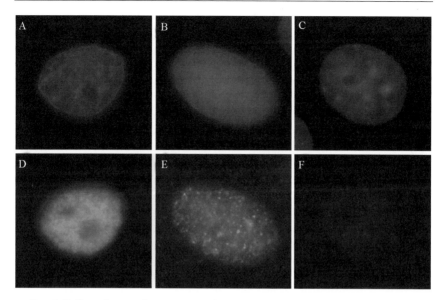

Fig. 2. Indirect immunofluorescence staining of RPA70. DAPI staining of an untransfected cell (A), an untransfected cell that has been treated with 0.5% Triton X solution (chromatin associated fraction) (B), and a RPA70 siRNA transfected cell (C). (D–F) Indirect immunostaining of the same cells in (A–C), respectively, with α-RPA70 antibody. (See color insert.)

Flow Cytometry to Analyze Effects of RPA on the Cell Cycle

Depletion or mutation of RPA can effect cell cycle progression. An effective way of examining this is to analyze the cell cycle distribution of cells by flow cytometry. Described later is one method that has been found to be effective for isolating cells for analysis of cell cycle distribution. This method can be carried out after RNAi depletion or introduction of a specific expression vector.

Buffers

- 70% methanol solution: 70% methanol, PBS
- RNase A solution: 1 mg/ml RNase A, PBS
- Propidium iodide solution: 100 μg/ml propidium iodide, PBS.

Methods

HeLa cells are seeded at 2×10^5 cells/well and grown as a monolayer in the well of a 6-well tissue culture dish at 37° (5% CO_2) for 3–5 days.

The supernatant from each well is collected, and the attached cells are treated with 0.7 ml 1× trypsin for 3 min to release them from the well and transferred to a microfuge tube. The trypsinized cells are pelleted for 5 min at 4000 RPM and the supernatant is added to the tube. The cells are pelleted again and are washed twice with PBS. To fix the cells, they are resuspended in 1 ml of 70% methanol, and incubated at 4° for at least 1 h.

To examine the effects of RNAi knockdown of RPA on the cell cycle, the cells are immunostained for RPA. After fixation, the cells are pelleted and washed twice with 1 ml PBS for 15 min. The cells are then incubated in 150 μl blocking solution I at room temperature for 30 min. After dilution of the primary antibody in blocking solution I, 150 μl antibody solution is added directly to the cells, without removing the previous blocking solution. The cells are incubated at RT for 1–2 h, with a 3 s vortex every 30 min. The cells are then pelleted and washed twice with 1 ml PBS for 5 min at RT. After the washes the cells are pelleted, the secondary antibody is appropriately diluted in blocking solution I, and 150 μl diluted secondary antibody is added to the cells. The cells are then incubated for 1 h at RT in the dark. After the secondary antibody incubation, the cells are pelleted and washed twice with 1 ml PBS for 5 min. The cells are again pelleted and incubated in 150 μl RNase A solution for at least 30 min at 4°. Propidium iodide solution (150 μl) is then added to the cells in the RNase A solution, and the cells are incubated at 4° in the dark for at least 1 h prior to flow cytometry. Immunostaining for RPA allows for selective analysis of RPA knockdown cells.

Summary and Conclusion

RPA is involved in all aspects of DNA metabolism in the cell. As such, it is required for many different reactions involving DNA. In this chapter, we have focused on the methods used to analyze RPA's two main activities, DNA binding and protein-protein interactions. In addition, we have given an overview of several functional assays that are used to investigate RPAs function in DNA replication, repair, and recombination.

References

Aboussekhra, A., Biggerstaff, M., Shivji, M. K. K., Vilpo, J. A., Moncollin, V., Podust, V. N., Protic, M., Hübscher, U., Egly, J.-M., and Wood, R. D. (1995). Mammalian DNA nucleotide excision repair reconstituted with purified protein components. *Cell* **80,** 859–868.

Arunkumar, A. I., Klimovich, V., Jiang, X., Ott, R. D., Mizoue, L., Fanning, E., and Chazin, W. J. (2005). Insights into hRPA32 C-terminal domain-mediated assembly of the simian virus 40 replisome. *Nat. Struct. Mol. Biol.* **12,** 332–339.

Arunkumar, A. I., Stauffer, M. E., Bochkareva, E., Bochkarev, A., and Chazin, W. J. (2003). Independent and coordinated functions of replication protein A tandem high affinity single-stranded DNA binding domains. *J. Biol. Chem.* **278**, 41077–41082.

Ausubel, F. M., Brent, R., Kingston, R. E., Moore, D. D., Seidman, J. G., Smith, J. A., and Struhl, K. (1989). "Current Protocols in Molecular Biology." John Wiley & Sons, New York.

Ausubel, F. M., Brent, R., Kingston, R. E., Moore, D. D., Seidman, J. G., Smith, J. A., Struhl, K., Albright, L. M., Coen, D. M., and Varki, A. (2001). "Current Protocols in Molecular Biology." John Wiley & Sons, New York.

Biacore 3000 Instrument Handbook. Biacore International AB Uppsala, Sweden (1999).

Bastin-Shanower, S. A., and Brill, S. J. (2001). Functional analysis of the four DNA binding domains of replication protein A: The role of RPA2 in ssDNA binding. *J. Biol. Chem.* **276**, 36446–36453.

Biggerstaff, M., and Wood, R. D. (1999). Assay for nucleotide excision repair protein activity using fractionated cell extracts and UV-damaged plasmid DNA. *Methods Mol. Biol.* **113**, 357–372.

Binz, S. K., Lao, Y., Lowry, D. F., and Wold, M. S. (2003). The phosphorylation domain of the 32-kDa subunit of replication protein A (RPA) modulates RPA-DNA interactions: Evidence for an intersubunit interaction. *J. Biol. Chem.* **278**, 35584–35591.

Binz, S. K., Sheehan, A. M., and Wold, M. S. (2004). Replication protein A phosphorylation and the cellular response to DNA damage. *DNA Repair (Amst)* **3**, 1015–1024.

Brénot-Bosc, F., Gupta, S., Margolis, R. L., and Fotedar, R. (1995). Changes in the subcellular localization of replication initiation proteins and cell cycle proteins during G1- to S-phase transition in mammalian cells. *Chromosoma* **103**, 517–527.

Brush, G. S., Kelly, T. J., and Stillman, B. (1995). Identification of eukaryotic DNA replication proteins using simian virus 40 *in vitro* replication system. *Methods Enzymol.* **262**, 522–548.

Campbell, T. N., and Choy, F. Y. (2005). RNA interference: Past, present and future. *Curr. Issues Mol. Biol.* **7**, 1–6.

Daughdrill, G. W., Buchko, G. W., Botuyan, M. V., Arrowsmith, C., Wold, M. S., Kennedy, M. A., and Lowry, D. F. (2003). Chemical shift changes provide evidence for overlapping single-stranded DNA- and XPA-binding sites on the 70 kDa subunit of human replication protein A. *Nucleic Acids Res.* **31**, 4176–4183.

De Laat, W. L., Appeldoorn, E., Sugasawa, K., Weterings, E., Jaspers, N. G. J., and Hoeijmakers, J. H. J. (1998). DNA-binding polarity of human replication protein A positions nucleases in nucleotide excision repair. *Genes Dev.* **12**, 2598–2609.

DeMott, M. S., Zigman, S., and Bambara, R. A. (1998). Replication protein A stimulates long patch DNA base excision repair. *J. Biol. Chem.* **273**, 27492–27498.

Dimitrova, D. S., and Gilbert, D. M. (2000). Stability and nuclear distribution of mammalian replication protein A heterotrimeric complex. *Experimental Cell Research* **254**, 321–327.

Dodson, G. E., Shi, Y., and Tibbetts, R. S. (2004). DNA replication defects, spontaneous DNA damage, and ATM-dependent checkpoint activation in replication protein A-deficient cells. *J. Biol. Chem.* **279**, 34010–34014.

Gupta, R. C., Golub, E. I., Wold, M. S., and Radding, C. M. (1998). Polarity of DNA strand exchange promoted by recombination proteins of the RecA family. *Proc. Natl. Acad. Sci. USA* **95**, 9843–9848.

Hannon, G. J. (2002). RNA interference. *Nature* **418**, 244–251.

Hannon, G. J., and Rossi, J. J. (2004). Unlocking the potential of the human genome with RNA interference. *Nature* **431**, 371–378.

He, Z., Henricksen, L. A., Wold, M. S., and Ingles, C. J. (1995). RPA involvement in the damage-recognition and incision steps of nucleotide excision repair. *Nature* **374**, 566–569.

Henricksen, L. A., Umbricht, C. B., and Wold, M. S. (1994). Recombinant replication protein A: Expression, complex formation, and functional characterization. *J. Biol. Chem.* **269**, 11121–11132.

Iftode, C., and Borowiec, J. A. (1997). Denaturation of the simian virus 40 origin of replication mediated by human replication protein A. *Mol. Cell. Biol.* **17**, 3876–3883.

Iftode, C., and Borowiec, J. A. (2000). $5'$-> $3'$ molecular polarity of human replication protein A (hRPA) binding to pseudo-origin DNA substrates. *Biochemistry* **39**, 11970–11981.

Iftode, C., Daniely, Y., and Borowiec, J. A. (1999). Replication protein A (RPA): The eukaryotic SSB. *Crit. Rev. Biochem.* **34**, 141–180.

Ishiai, M., Sanchez, J. P., Amin, A. A., Murakami, Y., and Hurwitz, J. (1996). Purification, gene cloning and reconstitution of the heterotrimeric single-stranded DNA binding protein from Schizosaccharomyces pombe. *J. Biol. Chem.* **271**, 20868–20878.

Ishibashi, T., Kimura, S., Furukawa, T., Hatanaka, M., Hashimoto, J., and Sakaguchia, K. (2001). Two types of replication protein A 70 kDa subunit in rice, Oryza sativa: Molecular cloning, characterization, and cellular and tissue distribution. *Gene* **272**, 335–343.

Ishibashi, T., Koga, A., Yamamoto, T., Uchiyama, Y., Mori, Y., Hashimoto, J., Kimura, S., and Sakaguchi, K. (2005). Two types of replication protein A in seed plants. *FEBS J.* **272**, 3270–3281.

Jacobs, D. M., Lipton, A. S., Isern, N. G., Daughdrill, G. W., Lowry, D. F., Gomes, X., and Wold, M. S. (1999). Human replication protein A: Global fold of the N-terminal RPA-70 domain reveals a basic cleft and flexible C-terminal linker. *J. Biomol. NMR* **14**, 321–331.

Kantake, N., Sugiyama, T., Kolodner, R. D., and Kowalczykowski, S. C. (2003). The recombination-deficient mutant RPA (rfa1-t11) is displaced slowly from single-stranded DNA by Rad51 protein. *J. Biol. Chem.* **278**, 23410–23417.

Karlsson, R., and Falt, A. (1997). Experimental design for kinetic analysis of protein-protein interactions with surface plasmon resonance biosensors. *J. Immunol. Methods* **200**, 121–133.

Kenny, M. K., Lee, S.-H., and Hurwitz, J. (1989). Multiple functions of human single-stranded-DNA binding protein in simian virus 40 DNA replication: Single-strand stabilization and stimulation of DNA polymerases a and d. *Proc. Natl. Acad. Sci. USA* **86**, 9757–9761.

Kim, C., Paulus, B. F., and Wold, M. S. (1994). Interactions of human replication protein A with oligonucleotides. *Biochemistry* **33**, 14197–14206.

Kim, C., Snyder, R. O., and Wold, M. S. (1992). Binding properties of replication protein A from human and yeast cells. *Mol. Cell. Biol.* **12**, 3050–3059.

Kozlov, A. G., and Lohman, T. M. (1998). Calorimetric studies of E-coli SSB protein single-stranded DNA interactions. Effects of monovalent salts on binding enthalpy. *J. Mol. Biol.* **278**, 999–1014.

Krokan, H. E., Nilsen, H., Skorpen, F., Otterlei, M., and Slupphaug, G. (2000). Base excision repair of DNA in mammalian cells. *FEBS Lett.* **476**, 73–77.

Kubota, Y., Nash, R. A., Klungland, A., Schär, P., Barnes, D. E., and Lindahl, T. (1996). Reconstitution of DNA base excision-repair with purified human proteins: Interaction between DNA polymerase b and the XRCC1 protein. *EMBO J.* **15**, 6662–6670.

Kunkel, T. A., and Erie, D. A. (2005). DNA mismatch repair. *Annu. Rev. Biochem.* **74** (epub ahead of print).

Lane, D., Prentki, P., and Chandler, M. (1992). Use of gel retardation to analyze protein-nucleic acid interactions. *Microbiol. Rev.* **56**, 509–528.

Lao, Y., Gomes, X. V., Ren, Y. J., Taylor, J. S., and Wold, M. S. (2000). Replication protein A interactions with DNA. III. Molecular basis of recognition of damaged DNA. *Biochemistry* **39**, 850–859.

Lao, Y., Lee, C. G., and Wold, M. S. (1999). Replication protein A interactions with DNA. 2. Characterization of double-stranded DNA-binding/helix-destabilization activities and the role of the zinc-finger domain in DNA interactions. *Biochemistry* **38**, 3974–3984.

Lee, J. S., and Lee, Y. S. (2002). Cloning and characterization of replication protein A p32 complementary DNA in zebrafish (Danio rerio). *Mar. Biotechnol. (NY)* **4**, 1–5.

Lin, L. P., Huang, L. S., Lin, C. W., Lee, C. K., Chen, J. L., Hsu, S. M., and Lin, S. (2005). Determination of binding constant of DNA-binding drug to target DNA by surface plasmon resonance biosensor technology. *Curr. Drug Targets Immune Endocr. Metabol. Disord.* **5**, 61–72.

Lohman, T. M., and Mascotti, D. P. (1992). Nonspecific ligand-DNA equilibrium binding parameters determined by fluorescence methods. *Methods Enzymol.* **212**, 424–458.

Lundblad, J. R., Laurance, M., and Goodman, R. H. (1996). Fluorescence polarization analysis of protein-DNA and protein-protein interactions. *Mol. Endocrinol.* **10**, 607–612.

Miyamoto, Y., Saito, Y., Nakayama, M., Shimasaki, Y., Yoshimura, T., Yoshimura, M., Harada, M., Kajiyama, N., Kishimoto, I., Kuwahara, K., Hino, J., Ogawa, E., Hamanaka, I., Kamitani, S., Takahashi, N., Kawakami, R., Kangawa, K., Yasue, H., and Nakao, K. (2000). Replication protein A1 reduces transcription of the endothelial nitric oxide synthase gene containing a-786T ->C mutation associated with coronary spastic angina. *Hum. Mol. Genetics* **9**, 2629–2637.

Molloy, P. L. (2000). Electrophoretic mobility shift assays. *Methods Mol. Biol.* **130**, 235–246.

Mu, D., Park, C.-H., Matsunaga, T., Hsu, D. S., Reardon, J. T., and Sancar, A. (1995). Reconstitution of human DNA repair excision nuclease in a highly defined system. *J. Biol. Chem.* **270**, 2415–2418.

Myszka, D. G. (1999). Improving biosensor analysis. *J. Mol. Recognit.* **12**, 279–284.

Nagelhus, T. A., Haug, T., Singh, K. K., Keshav, K. F., Skorpen, F., Otterlei, M., Bharati, S., Lindmo, T., Benichou, S., Benarous, R., and Krokan, H. E. (1997). A sequence in the N-terminal region of human uracil-DNA glycosylase with homology to XPA interacts with the c-terminal part of the 34-kDa subunit of replication protein A. *J. Biol. Chem.* **272**, 6561–6566.

Nicholl, I. D., Nealon, K., and Kenny, M. K. (1997). Reconstitution of human base excision repair with purified proteins. *Biochemistry* **36**, 7557–7566.

Oakley, G. G., Patrick, S. M., Yao, J., Carty, M. P., Turchi, J. J., and Dixon, K. (2003). RPA phosphorylation in mitosis alters DNA binding and protein-protein interactions. *Biochemistry* **42**, 3255–3264.

Philipova, D., Mullen, J. R., Maniar, H. S., Gu, C., and Brill, S. J. (1996). A hierarchy of SSB promoters in replication protein A. *Genes Dev.* **10**, 2222–2233.

Ramilo, C., Gu, L., Guo, S., Zhang, X., Patrick, S. M., Turchi, J. J., and Li, G. M. (2002). Partial reconstitution of human DNA mismatch repair in vitro: Characterization of the role of human replication protein A. *Mol. Cell. Biol.* **22**, 2037–2046.

Riedl, T., Hanaoka, F., and Egly, J. M. (2003). The comings and goings of nucleotide excision repair factors on damaged DNA. *EMBO J.* **22**, 5293–5303.

Schubert, F., Zettl, H., Hafner, W., Krauss, G., and Krausch, G. (2003). Comparative thermodynamic analysis of DNA–protein interactions using surface plasmon resonance and fluorescence correlation spectroscopy. *Biochemistry* **42**, 10288–10294.

Shell, S. M., Hess, S., Kvaratskhelia, M., and Zou, Y. (2005). Mass spectrometric identification of lysines involved in the interaction of human replication protein a with single-stranded DNA. *Biochemistry* **44**, 971–978.

Shivji, M. K., Moggs, J. G., Kuraoka, I., and Wood, R. D. (1999). Dual-incision assays for nucleotide excision repair using DNA with a lesion at a specific site. *Methods Mol. Biol.* **113**, 373–392.

Sibenaller, Z. A., Sorensen, B. R., and Wold, M. S. (1998). The 32- and 14-kDa subunits of replication protein A are responsible for species-specific interactions with ssDNA. *Biochemistry* **37**, 12496–12506.

Stigger, E., Dean, F. B., Hurwitz, J., and Lee, S.-H. (1994). Reconstitution of functional human single-stranded DNA-binding protein from individual subunits expressed by recombinant baculoviruses. *Proc. Natl. Acad. Sci. USA* **91**, 579–583.

Studier, F. W., Rosenberg, A. H., Dunn, J. J., and Dubendorff, J. W. (1990). Use of T7 RNA polymerase to direct expression of cloned genes. *Methods Enzymol.* **185**, 60–89.

Van Regenmortel, M. H. (2003). Improving the quality of BIACORE-based affinity measurements. *Dev. Biol. (Basel)* **112**, 141–151.

Vassin, V. M., Wold, M. S., and Borowiec, J. A. (2004). Replication protein A (RPA) phosphorylation prevents RPA association with replication centers. *Mol. Cell. Biol.* **24**, 1930–1943.

Wakasugi, M., and Sancar, A. (1999). Order of assembly of human DNA repair excision nuclease. *J. Biol. Chem.* **274**, 18759–18768.

Walter, J., and Newport, J. (2000). Initiation of eukaryotic DNA replication: Origin unwinding and sequential chromatin association of Cdc45, RPA, and DNA polymerase a. *Mol. Cell* **5**, 617–627.

Walter, J., Sun, L., and Newport, J. (1998). Regulated chromosomal DNA replication in the absence of a nucleus. *Mol. Cell* **1**, 519–529.

Wang, M., Mahrenholz, A., and Lee, S. H. (2000). RPA stabilizes the XPA-damaged DNA complex through protein-protein interaction. *Biochemistry* **39**, 6433–6439.

Wobbe, C. R., Weissbach, L., Borowiec, J. A., Dean, F. B., Murakami, Y., Bullock, P., and Hurwitz, J. (1987). Replication of simian virus 40 origin-containing DNA *in vitro* with purified proteins. *Proc. Natl. Acad. Sci.* **84**, 1834–1838.

Wold, M. S. (1997). Replication protein A: A heterotrimeric, single-stranded DNA-binding protein required for eukaryotic DNA metabolism. *Annu. Rev. Biochem.* **66**, 61–92.

Wong, I., and Lohman, T. M. (1993). A double-filter method for nitrocellulose-filter binding: Application to protein-nucleic acid interactions. *Proc. Natl. Acad. Sci. USA* **90**, 5428–5432.

[3] Human DNA Ligases I, III, and IV—Purification and New Specific Assays for These Enzymes

By Xi Chen, John Pascal, Sangeetha Vijayakumar, Gerald M. Wilson, Tom Ellenberger, and Alan E. Tomkinson

Abstract

The joining of DNA strand breaks by DNA ligases is required to seal Okazaki fragments during DNA replication and to complete almost all DNA repair pathways. In human cells, there are multiple species of DNA ligase encoded by the *LIG1*, *LIG3*, and *LIG4* genes. Here we describe protocols to overexpress and purify recombinant DNA ligase I, DNA ligase IIIβ, and DNA ligase IV/XRCC4 and the assays used to purify and distinguish between these enzymes. In addition, we describe a fluorescence-based ligation assay that can be used for high throughput screening of chemical libraries.

Introduction

The fractionation of extracts from bovine thymus provided the first evidence for the existence of multiple DNA ligase activities (Soderhall and Lindahl, 1973; Tomkinson *et al.*, 1991). Subsequently, three human *LIG* genes have been cloned (Barnes *et al.*, 1990; Chen *et al.*, 1995; Wei *et al.*, 1995). The ATP-dependent human DNA ligases share a conserved catalytic domain that is composed of three subdomains: DNA binding, adenylation and OB-fold (Pascal *et al.*, 2004), and utilize the same three-step ligation reaction as the NAD-dependent *E. coli* DNA ligase and the ATP-dependent T4 DNA ligase (Lehman, 1974). The first two steps of the ligation reaction are shared with other nucleotidyl transferases, RNA ligases, and mRNA capping enzymes (Shuman and Schwer, 1995).

Unlike the *LIG1* and *LIG4* genes, the *LIG3* gene encodes more than one species of DNA ligase. Specifically, nuclear and mitochondrial forms of DNA ligase IIIα are generated by alternative translation initiations (Lakshmipathy and Campbell, 1999). In addition, a germ cell-specific form, DNA ligase IIIβ, is generated by alternative splicing (Mackey *et al.*, 1997). The different phenotypes of cell lines with reduced activity of one of the DNA ligases provide compelling evidence that these enzymes have distinct cellular functions (Barnes *et al.*, 1992; Caldecott *et al.*, 1995; Riballo *et al.*, 1999). The biological specificity of the human DNA ligases appears to be

METHODS IN ENZYMOLOGY, VOL. 409
0076-6879/06 $35.00
DOI: 10.1016/S0076-6879(05)09003-8

conferred by interactions with different protein partners mediated by the unique sequences flanking the conserved catalytic domain (Caldecott *et al.*, 1994; Chen *et al.*, 2000; Dimitriadis *et al.*, 1998; Grawunder *et al.*, 1997; Levin *et al.*, 1997; Mackey *et al.*, 1999).

In the later section, expression systems and purification schemes for recombinant versions of DNA ligase I, III, and IV are described. Next, assays that either discriminate between the DNA ligases or detect functional interactions between the DNA ligase and their interacting proteins will be detailed. Finally, we will describe ligation assays with fluorescence-labeled DNA substrates that have been designed to screen for DNA ligase inhibitors.

Overexpression and Purification of Human DNA Ligases

Plasmid and baculovirus expression vectors have been constructed for the overexpression of human DNA ligases in *E. coli* and insect cells (Chen *et al.*, 2000; Mackey *et al.*, 1999; Prasad *et al.*, 1996; Teraoka *et al.*, 1993; Wang *et al.*, 1994). The purification of the overexpressed recombinant DNA ligases is monitored by the formation of a ^{32}P-labeled covalent ligase-adenylate (AMP) complex. This assay, as is described later, takes advantage of the first step of the DNA ligase reaction mechanism, in which the enzyme interacts with ATP to form a covalent enzyme-AMP intermediate. Advantages of this assay include its specificity for ATP-dependent nucleotidyl transferases and ability to detect the integrity of the DNA ligase polypeptides because the labeled polypeptides are visualized by phosphorimaging analysis after separation by SDS-PAGE (Laemmli, 1970). However, this assay is not quantitative in crude extracts. Coomassie blue staining or quantitative immunoblotting after separation by SDS-PAGE are better ways to estimate the expression levels in crude extracts. Monoclonal antibodies for DNA ligase I and DNA ligase III are available commercially (GeneTex, San Antonio, TX).

1. Assay for formation of ^{32}P-labeled DNA ligase-AMP.

10× AMP Buffer: 0.6 M Tris-HCl (pH 8.0), 100 mM MgCl$_2$, 50 mM DTT, 0.5 mg/ml BSA. Aliquot and freeze at $-20°$.
Assay Mix: 20 μl 10× AMP Buffer, 1 μl of α-^{32}P ATP (3000 Ci/mmol and 10 μCi/μl, GE Healthcare), and 179 μl of dH$_2$O (for 20 assays).
Assay mix (10 μl) is added to each Eppendorf tube containing the appropriate amount of column fraction (usually 2 μl) and incubated at room temperature (RT) for 15 min. Reactions are stopped by the addition of 5 μl of SDS-sample buffer (Laemmli, 1970). After heating at 90° for

5 min, polypeptides are separated by SDS-PAGE. For molecular mass standards, 5 μl of ^{14}C-methylated proteins (5 nCi/μl, GE Healthcare) are run per well. The unincorporated α-^{32}P ATP can be collected in the bottom reservoir buffer by running the bromophenol blue dye front off the bottom of the gel. The gel is fixed for 10–15 min in 10% acetic acid, dried onto a piece of filter paper, exposed to phosphorimager screen, and then visualized with a PhosphorImager.

2. Purification of recombinant human DNA ligase I.

Full-length human DNA ligase I has been expressed in and purified from both insect (Wang *et al.*, 1994) and *E. coli* cells (Teraoka *et al.*, 1993). Although DNA ligase I purified from insect cells is heterogeneously phosphorylated, the bacterial and insect cell expressed enzymes have similar catalytic activities. Below is the protocol for purifying untagged full length DNA ligase I after expression in *E. coli* cells. The expression vector pET-24a-hLig1 was constructed from pTD-T7-hLig1 (Teraoka *et al.*, 1993). After transformation of *E. coli* BL21(DE3)RP cells (Stratagene, La Jolla, CA) with the expression plasmid and selection for ampicillin resistance, a single colony is picked into 5 ml 2× YT medium (16 g tryptone, 10 g yeast extract, 5 g NaCl in 1 liter of H_2O [pH 7.0] supplemented with ampicillin (100 μg/ml) and chloramphenicol [34 μg/ml]), and incubated overnight (O/N) at RT. This O/N culture is diluted 1/125 into the same media and incubated O/N at RT. A 100-ml aliquot of the O/N culture is added to 900 ml of the same medium and incubation is continued at RT. When the A_{600} is between 0.72 and 0.74, iso-propyl-thio-galactoside (IPTG) is added to a final concentration of 0.1 mM to induce DNA ligase I expression. After incubation for 18 h at 16°, cells are harvested by centrifugation in JLA8.1 rotor (Beckman) at 5000g for 10 min at 2° and then resuspended in buffer A (50 mM Tris-HCl [pH 7.5], 1 mM EDTA, 1 mM DTT, 10% glycerol, 50 mM NaCl, 1 mM Benzamidine, 1 mM PMSF, 1 μg/ml pepstatin, and 1 μg/ml Aprotinin). After incubation on ice for 10 min, 10% NP-40 (10% stock solution) is added to a final concentration of 1.0 % and the cells are lysed by 2 passes through a cell disruptor (Avestin Emulsiflex C5). After clarification of the lysate by centrifugation at 12,000 rpm in a JA-12 rotor (Beckman) for 30 min at 2°, the protein concentration of the lysate is determined by the Bradford assay (Bradford, 1976) and the level of DNA ligase I in the lysate is visualized by Coomassie blue staining after electrophoresis through a 7.5% SDS-polyacrylamide gel. Human DNA ligase I should be visible as a prominent band with a molecular mass of 125 kDa. The lysate containing approximately 450 mg of total protein is loaded onto a 40-ml phosphocellulose (Whatman, Florham Park, NJ) column pre-equilibrated with buffer A. After washing with 5 column vol of buffer A, bound proteins are eluted with a gradient (10 column vol.) from 50 to 750 mM NaCl in buffer A. The presence of DNA ligase I in the column fractions (10 ml) is detected by Coomassie blue staining after SDS-PAGE. Fractions containing 125 kDa DNA ligase I (which elutes at about 250 mM NaCl) are pooled and then diluted 5-fold with buffer A containing no NaCl. This dilution brings the salt concentration to approximately 50 mM for loading onto a 5-ml Q Sepharose column (GE Healthcare Piscataway, NJ) pre-equilibrated in buffer A. After washing with buffer A, bound proteins are eluted with an 80-ml gradient from 50 to 350 mM NaCl in buffer A. The presence of DNA ligase I in the column fractions (2 ml) is detected by Coomassie blue staining after SDS-PAGE. Fractions containing 125-kDa DNA ligase I are pooled and then loaded onto a Blue Sepharose CL-6B column (GE Healthcare) equilibrated with buffer A containing 150 mM NaCl. After washing, bound proteins are eluted with an 80 ml gradient from 0.15 to 1 M NaCl in buffer A. Fractions (2 ml) containing 125 kDa DNA ligase I are pooled, concentrated with an Amicon Ultra centrifugal filter device (Millipore, Billerica, MA) to less than 5 ml, and loaded onto a Superdex 200 (16/60) gel filtration column (GE Healthare) pre-equilibrated with buffer A containing 150 mM NaCl. Fractions (3 ml) containing homogeneous or nearly homogeneous 125 kDa DNA ligase I are pooled and then dialyzed against storage buffer (25 mM Tris-HCl [pH 7.5], 150 mM NaCl, 0.1 mM EDTA, 1 mM DTT). DNA ligase I is then concentrated to >20 mg/ml with an Amicon Ultra centrifugal filter device, aliquoted, flash frozen in liquid nitrogen, and stored at $-80°$. Approximately, 4 mg of nearly homogenous DNA ligase I is usually obtained from a 1 liter culture.

3. Purification of recombinant DNA ligase IIIβ.

The nuclear form of DNA ligase IIIα is unstable in the absence of its partner protein XRCC1 (Caldecott et al., 1995). This instability also appears to occur when DNA ligase IIIα is expressed in E. coli, limiting the yield of active purified protein even after the addition of a tag for affinity chromatography. By contrast, the germ cell-specific form DNA ligase IIIβ, in which a small positively charged peptide sequence replaces the C-terminal BRCT of DNA ligase IIIα (Mackey et al., 1997), has been purified from baculovirus-infected insect cells and E. coli cells in much higher quantities (Mackey et al., 1999). Below is the protocol for purifying hexahistidine (his)-tagged full length DNA ligase IIIβ after expression in E. coli cells. After transformation of the E. coli strain M15 (Promega, Madison, WI) with the pQE (Qiagen, Valencia, CA) derived expression plasmid pHis-LigIIIβ encoding his-tagged DNA ligase IIIβ (Mackey et al., 1999) and selection for resistance to ampicillin and kanamycin, a single colony is picked into 5-ml LB medium containing 100 μg/ml ampicillin and 50 μg/ml kanamycin and grown O/N. The O/N culture is diluted 1/50 in 1 liter of the same media and incubated at 37°. At an A_{600} of 0.6, IPTG is added to a final concentration of 1 mM, and incubation continued for 6 h. After collection by centrifugation, cells are resuspended in 50 ml of ice-cold lysis buffer (50 mM Tris-HCl [pH 7.5], 50 mM NaCl, 2% Nonidet P-40, 1 mM phenylmethanesulfonyl fluoride, 1 mM benzamidine-HCl, 10 mM 2-mercaptoethanol) and sonicated (Branson Sonifier 150, Danbury, CT) 5 times, each time at 18 watts for 30 s with intervals on ice. The lysate is clarified by centrifugation at 35,000 rpm for 30 min at 4° in a 45Ti rotor (Beckman) and then loaded onto a 35-ml P11 column (Whatman) pre-equilibrated with buffer B (50 mM Tris-HCl [pH 7.5], 10% glycerol, 50 mM NaCl, 10 mM 2-mercaptoethanol, 1 mM phenylmethanesulfonyl fluoride, 1 mM benzamidine-HCl). After washing with buffer B, bound proteins are eluted stepwise with buffer B containing 0.2 M and then 0.5 M NaCl. Eluted fractions are assayed for total amount of proteins by the method of Bradford (Bradford, 1976) and for DNA ligase IIIβ by detecting the formation of the ^{32}P-labeled 97 kDa enzyme-adenylate intermediate. Fractions containing DNA ligase IIIβ, which is present in the 0.5 M eluate, are pooled and, after the addition of imidazole to a final concentration of 10 mM, incubated by constant rotation at 4° for 3 h with nickel Ni-NTA beads (Qiagen, 10 mg of protein per ml of resin). After washing with buffer B containing 20 mM imidazole, bound proteins are batch-eluted from the beads with buffer B containing 250 mM imidazole and 50 mM EDTA. Approximately 2 mg of nearly homogenous DNA ligase IIIβ is usually obtained from a 1-l culture. After dialysis against buffer B, purified DNA ligase IIIβ is aliquoted, flash-frozen in liquid nitrogen, and stored at −80°.

4. Purification of DNA ligase IV/XRCC4.

Similar to nuclear DNA ligase IIIα, DNA ligase IV is unstable in the absence of its partner protein XRCC4 (Grawunder et al., 1997, 1998). The DNA ligase IV/XRCC4 complex has been purified from insect cells coinfected with two recombinant baculoviruses, one of which encodes DNA ligase IV and the other XRCC4 (Grawunder et al., 1997; Lee et al., 2000). We constructed a dual expression baculovirus from pFastBac-dual (Life Technologies) encoding full-length untagged XRCC4 and full-length DNA ligase IV with C-terminal hemagglutinin (HA) and his-tags (Chen et al., 2000). Below is the protocol for purifying the recombinant DNA ligase IV/XRCC4 complex from baculovirus-infected Sf9 cells. Although a significant fraction of DNA ligase IV molecules remain adenylated during purification (Chen et al., 2000; Robins and Lindahl, 1996), purification of the recombinant DNA ligase IV/XRCC4 complex is monitored by assaying for the formation of the labeled DNA ligase IV-adenylate complex. Sf9 insect cells (600 ml, $\sim 10^6$/ml) in suspension are infected with the dual expression baculovirus and then harvested 60 h after infection. The cell pellet is flash-frozen and stored at $-80°$. After thawing on ice, cells (10^9) are resuspended in 50 ml of buffer C (50 mM Tris-HCl [pH 7.5], 300 mM NaCl, 10% glycerol, 10 mM 2-mercaptoethanol, 1 mM PMSF, 1 mM benzamidine HCl, 1 μg/ml leupeptin, 2 μg/ml aprotinin, and 1 μg/ml pepstatin) and lysed by sonication for 30 s. The lysate is clarified by centrifugation at 40,000 rpm for 30 min in a Ti-45 rotor (Beckman) at $2°$. After the addition of imidazole to a final concentration of 10 mM, the lysate is incubated on a rotator for 3 h at $4°$ with nickel Ni-NTA beads (Qiagen, 10 mg protein per ml of resin) pre-equilibrated with buffer C containing 10 mM imidazole. The beads are poured into a column and then washed with buffer C containing 40 mM imidazole. DNA ligase IV is eluted from the column with buffer C containing 200 mM imidazole. Fractions from the 200 mM eluate are assayed for protein by the method of Bradford (Bradford, 1976). Protein-containing fractions are pooled and then dialyzed against buffer D (50 mM Tris-HCl [pH 7.5], 50 mM NaCl, 0.5 mM DTT, 1 mM EDTA 10% glycerol, 1 mM PMSF, 1 mM benzamidine HCl, 1 μg/ml leupeptin, 2 μg/ml aprotinin, and 1 μg/ml pepstatin) prior to loading onto a Resource Q ion exchange column (GE Healthcare). After washing, bound proteins are eluted with a gradient from 50 to 750 mM NaCl in buffer D. Eluted fractions are assayed for DNA ligase IV by detecting the formation of the labeled 100-kDa enzyme-adenylate intermediate. Fractions containing DNA ligase IV are pooled and loaded onto a Superdex 200 16/60 gel filtration column pre-equilibrated with buffer D containing 150 mM NaCl. Eluted fractions containing DNA ligase IV are pooled, dialyzed against buffer D, and then loaded onto a 1-ml Mono S ion exchange column (GE Healthcare). After washing, bound proteins are eluted with a gradient from 50 to 750 mM NaCl in buffer D. The 100 kDa DNA ligase IV and 50 kDa XRCC4, visualized by Coomassie blue staining after SDS-PAGE, co-elute from the Mono S column. Peak fractions of DNA ligase IV/XRCC4 are aliquoted, flash frozen, and stored at $-80°$. This protocol yields approximately 0.2 mg of nearly homogeneous DNA ligase IV/XRCC4 complex from 10^9 infected insect cells.

DNA Ligation Assays

Phosphodiester bond formation, which converts the $5'$ phosphate group at a nick into a form that is resistant to phosphatases, has been measured by the joining of $5'$ end labeled oligonucleotides, generating phosphatase-resistant radioactivity within a DNA single strand captured on a nitrocellulose filter. In many instances, these assays utilized substrates composed of complementary homo-polymers and -oligomers (Arrand et al., 1986; Robins and Lindahl, 1996; Tomkinson et al., 1991). Although differences in activity with oligo dT/poly dA, oligo dT/poly rA, and oligo rA/poly dT were used to distinguish between DNA ligases I, III, and IV (Robins and Lindahl, 1996; Tomkinson et al., 1991), the biological relevance of these differences was not clear. The recent determination of the structure of the catalytic domain of DNA ligase I complexed with nicked DNA (Pascal et al., 2004) has provided a molecular explanation for the inability of this enzyme to join oligo dT molecules annealed to a poly rA template and suggests that, despite the conservation of amino acid sequence, there are significant differences in the catalytic domains of the human DNA ligases.

The discovery that the catalytic domain of DNA ligase I and probably the catalytic domains of DNA ligases III and IV encircle nicked DNA (Pascal et al., 2004) raises the possibility that these enzymes may be able to slide along the DNA duplex. Intriguingly, both DNA ligase I and DNA ligase IV interact with ring-shaped proteins, proliferating cell nuclear antigen (Levin et al., 1997) and Ku (Chen et al., 2000; Nick McElhinny et al., 2000), respectively. DNA ligase I also interacts with RFC (Levin et al., 2004), the protein complex that loads PCNA rings onto DNA. This has prompted the design of DNA substrates to examine the functional interactions of DNA ligases with DNA sliding clamps.

Biotinylated Linear DNA Substrates with Blocked DNA Ends

This type of DNA substrate was developed to examine the interactions between RFC and PCNA during the loading of the PCNA clamp onto DNA (Gomes et al., 2000; Waga and Stillman, 1998). We have adapted it to probe the functional interactions among DNA ligase I, PCNA and RFC (Levin et al., 2004). The biotin tag and the blocked ends permit the capture of the DNA-protein complexes on beads, in particular free sliding complexes that are topologically linked to the DNA duplex. In addition, it is possible to utilize either the same or a very similar DNA substrate to measure DNA joining.

The 5' biotinylated 90-mer oligonucleotide, Bio-5–90 is similar to the one described by Waga and Stillman (Waga and Stillman, 1998), except that the sequences correspond to the nucleotide positions from 4881 to 4971 of the M13mp19 (Fermentas) single-stranded DNA. Bio15-1 (5'-TGAGGCGGTCAGTAT-3') and Bio15-2 (5'-AAGATAAAACA-GAGG–3') are complementary to Bio-5–90. 200 pmol of Bio15-1 is 5' end labeled with 150 μCi of γ-^{32}P ATP using 20 units of T4 Polynucleotide Kinase (New England Biolabs, Ipswich, MA). After purification on Micro Bio-Spin 30 column (Bio-Rad), labeled Bio15-1 and Bio15-2, are annealed with Bio5–90 to generate a partial duplex of 30 base pairs containing a single ligatable nick in the middle flanked by single-stranded regions of 30 nucleotides. The biotinylated linear DNA (1.6 pmol; the efficiency of the binding of biotin labeled DNA to Streptavidin agarose beads is approximately 60–70%) is incubated with Streptavidin-agarose beads (10 μl, Pierce Biotechnology, Rockford, IL) in PBS for 30 min at RT. The DNA beads are washed three times in the ligation buffer (50 mM Tris-HCl [pH 7.5], 10 mM MgCl$_2$, 1 mM dithiothreitol, 0.25 mg/ml BSA, 100 μM ATP, and 100 mM NaCl), and then incubated with either 2 pmol of replication protein A (RPA)/pmol DNA or 2 pmol of $E.\ coli$ single-stranded DNA-binding protein (SSB)/pmol DNA (Sigma) in the same buffer for 15 min at RT to generate a nicked DNA substrate with both ends blocked by either RPA or SSB. After incubation with DNA ligase in the presence or absence of its interacting partners, the beads are collected by centrifugation, and washed two times in the ligation buffer. To detect proteins specifically retained on the beads, the beads are resuspended in 10 μl SDS loading buffer and boiled for 3 min. After separation by SDS-PAGE, proteins are detected by immunoblotting. To quantitate DNA ligation, the beads are spun down and the reaction is terminated by adding 10 μl of 5× stop mix (50% glycerol, 1% SDS, 20 mM EDTA and 0.05% bromophenol blue). After heating at 95° for 3 min to denature DNA, a 2 μl aliquot is mixed with 2 μl of 2× denaturing–PAGE dye (95% formamide, 0.05% bromophenol and 0.05% xylene cyanol) prior to electrophoresis on a 7 M urea-12% polyacrylamide gel. After drying, the gel is exposed to a Storage Phosphor screen and subjected to Phosphor-Imager analysis (Molecular Dynamics, Sunnyvale, CA).

Linear DNA Substrates to Measure Intra- and Intermolecular DNA Joining

The breakage of both strands of the DNA duplex presents a difficult challenge for the cellular DNA repair pathways. One conserved pathway, non-homologous end joining, involves the bringing together of DNA ends followed by their ligation (Hefferin and Tomkinson, 2005). Many studies have used linear DNA fragments with cohesive ends generated by restriction endonucleases to detect and quantitate this type of repair in cell extracts and $in\ vivo$ (for example, Baumann and West, 1998; Boulton and Jackson, 1998). Although DNA ligases alone can directly join this type of DNA substrate, studies in cell extracts and $in\ vivo$ have indicated that DNA ligases act together with other factors in this repair pathway (Hefferin and Tomkinson, 2005).

The joining of linear DNA molecules with cohesive ends can occur either via an intra-molecular mechanism generating circular molecules or via an intermolecular mechanism, generating linear multimers with the length of the DNA substrate influencing the mode of ligation. With oligo-nucleotides of around 50 base pairs, intermolecular ligation is preferred presumably because of the difficulty in making circles with short molecules. In contrast, DNA molecules of about 400 base pairs are predominantly joined by intramolecular ligation. We have used this type of substrate to identify proteins that change the mode of ligation from intra- to intermo-lecular by acting as end-bridging factors (Chen *et al.*, 2000, 2001). The protocol for preparing the DNA substrate and assaying for intra- and intermolecular ligation is described later.

Linear duplex DNA fragments of the appropriate length, with 5′ and 3′ complementary single-stranded overhangs, are generated by restriction enzyme digestion of the circular plasmid. After electrophoresis through a 1% Tris-acetate agarose gel, the DNA fragments are detected by staining with ethidium bromide and purified with the QIAquick Gel Extraction Kit (Qiagen). Following extraction, the DNA fragment is incubated with Calf Intestinal Phosphatase (NEB) for 30 min at 37° and then purified using the QIAquick Enzymatic Reaction Clean Up protocol (Qiagen). The DNA fragment is end labeled by incubation with T4 Polynucleotide kinase (NEB) and γ^{32}P-ATP (3000 Ci/mmol) for 1 h at 37°, and then purified using the Micro BioSpin-30 Column (Bio-Rad) prior to storage at 4°.

DNA ligase, in combination with other repair factors, is incubated with the labeled DNA substrate (40 fmol) in 20 μl of reaction buffer (60 mM Tris (pH 7.5), 10 mM MgCl$_2$, 5 mM DTT, 1 mM ATP, and 50 μg/ml BSA) at 25° for 90 min. Reactions are stopped by the addition of 20 μl of the stop mix (100 mM EDTA, 1% SDS). The total reaction volume is brought up to 200 μl by the addition of TE buffer (pH 8.0) and then the DNA is purified by phenol/chloroform extraction and ethanol precipitation. The DNA precipitate is resuspended in 20 μl of DNA loading buffer (0.4% bromo-phenol blue, 0.4% xylene cyanol, 5% glycerol) and separated by electro-phoresis through a 0.8% Tris-acetate agarose gel at 85 v for 2 h. After drying, the gel is exposed to a Storage Phosphor screen and subjected to PhosphorImager analysis (Molecular Dynamics).

Ligation Assays Using Fluorescent DNA Substrates

The gel-based assays with radioactively labeled substrates described previously are not suitable for carrying out a large number of assays. This prompted us to develop a fluorescence-based assay that can be used for high throughput screening of large chemical libraries for DNA ligase inhibitors. In this assay, phosphodiester bond formation results in the phy-sical linkage of an oligonucleotide containing a fluorophore, Alexa Fluor 488 (AF488) to an oligonucleotide containing a quencher, Black Hole Quencher-1 (BHQ1). In Fig. 1, we show the design of DNA substrates to

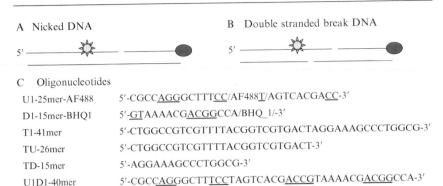

FIG. 1. Fluorescence-labeled ligation substrates. (A) Nicked DNA duplex are formed by annealing the up-stream primer U1–25mer-AF488, 5' phosphorylated down-stream primer D1–15mer-BHQ1 and the template T1–41mer. (B) Double stranded break DNA duplex with adhesive overhangs of 10 bases are formed by annealing the same fluorescence labeled primers with up-stream template TU-26mer and the 5' phosphorylated down-stream template TD-15mer. (C) Oligonucleotides used to form ligation substrates and for fluorescence detection. Base pairs that form hairpin structures are underlined. The position of the AF488 label is indicated with a star and the position of the BHQ1 label is indicated with a shaded oval.

quantitate nick ligation and the joining of linear duplex DNA molecules with short cohesive single-stranded ends. In the nicked DNA duplex substrate, the AF488 is 11 nucleotides from the 3' terminus of the nick and the BHQ1 is 15 nucleotides from the 5' terminus of the nick. Under these conditions and following ligation of both the nicked and double-stranded break substrates, the distance between the AF488 fluorescent donor and BHQ1 acceptor prevents significant quenching. To detect ligation, the oligonucleotide substrates and ligation products are denatured and then re-natured in the presence of a 20-fold excess of an oligonucleotide U1D1–40mer, that is identical to the ligated strand containing the fluorescent donor and acceptor except that it lacks these modifications. Under these conditions, the modified oligonucleotides are single-stranded and can adopt secondary structures. In the ligated oligonucleotide, the juxtaposition of the donor close to the acceptor results in static quenching, thus diminishing quantum emission from AF488. To optimize quenching efficiency, the DNA sequence of the ligated oligonucleotide is designed to form a stable hairpin structure thus causing direct contact between the donor and acceptor moieties in the single-stranded ligation product. Neither the AF488 labeled oligonucleotide nor the BHQ1 labeled oligonucleotide form hairpins alone. The fluorescent properties of the DNA substrates and the ligated products during the ligation assay are shown schematically in Fig. 2 (panels A to D).

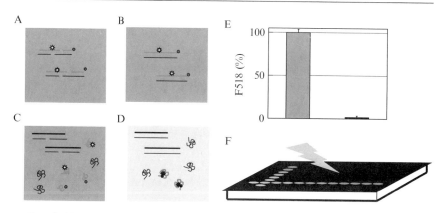

FIG. 2. Fluorescence-based ligation assay. (A) Fluorescence-labeled double stranded break DNA duplex without ligation shows 100% fluorescence intensity at 518 nm. (B) Ligated DNA duplex has the same fluorescence intensity, owing to the linear distance between the donor and the acceptor groups. (C) Because the single-stranded AF488-labeled up-stream primer and BHQ1-labeled down-stream primer are physically separate in solution, fluorescence measurement at 518 nm shows 100% intensity. (D) Under single-stranded conditions, the fluorescent donor, AF488, and the acceptor, BHQ1, are stacked within the ligated oligonucleotide, resulting in efficient intra-molecular quenching. (E) The un-ligated single stranded substrates, depicted in C, demonstrate 100% fluorescence emission (gray bar); whereas in the single-stranded ligation products, depicted in D, there is more than 95% quenching of the fluorescence (black bar). (F) For high throughput applications, the reaction is performed in 96-well format and quantitated using a multiplate fluorometer.

DNA ligase is incubated with 10 pmol of DNA substrate for 90 min at 25° in the ligation buffer described previously (total volume 30 μl). Reactions are diluted to 200 μl by adding 200 pmol of the unmodified competitor oligonucleotide in the annealing buffer (10 mM Tris-HCl [pH 8.0], 5 mM MgCl$_2$) and heated at 95° for 5 min in a Thermocycler (Bio-Rad). After cooling to room temperatures at a rate of 2° per minute, fluorescence at 518 nm is measured using a Cary Eclipse Multiplate Spectrofluorometer (Varian Inc.). Under these conditions, complete ligation by T4 DNA ligase reduces AF488 fluorescence by approximately 95% (Fig. 2, panel E). As illustrated in Fig. 2 (panel F), this assay can be adapted for high throughput screening. In addition, the fluorescence-based ligation assay is highly quantitative and can be used as an alternative to the electrophoresis-based assays.

Concluding Remarks

Differences in DNA substrate specificity and protein partners underlie the specific participation of the mammalian DNA ligases in different DNA transactions. With the exception of mitochondrial DNA ligase IIIα and the nuclear DNA ligase IIIα/XRCC1 complex, active recombinant versions of the human DNA ligases have been purified in sufficient quantities for biochemical and biophysical analyses. The recent determination of the structure of the catalytic domain of human DNA ligase I complexed with nicked DNA will not only stimulate further studies on the catalytic mechanism of DNA ligases but also the functional interactions between the DNA ligases and their partner proteins. At the present time, specific inhibitors of DNA ligases are not available. Since DNA joining is necessary to complete almost all DNA repair pathways, DNA ligase inhibitors may have clinical utility when used in combination with radiation or chemotherapy agents. We have developed a fluorescence-based ligation assay that is suitable for high throughput screening.

Acknowledgments

We thank former and present members of the Tomkinson laboratory for their contributions. Work in the Ellenberger, Tomkinson, and Wilson laboratories is supported by grants from the National Institutes for Health (PO1 CA92584 to A.E.T. and T.E., RO1 GM47251, RO1 GM57479 and RO1 ES12512 to A.E.T., and RO1 CA102428 to G.M.W.) Additional support for the Center for Fluorescence Spectroscopy was provided by P41 RR08119 from the National Institutes of Health.

References

Arrand, J. E., Willis, A. E., Goldsmith, I., and Lindahl, T. (1986). Different substrate specificities of the two DNA ligases of mammalian cells. *J. Biol. Chem.* **261**, 9079–9082.

Barnes, D. E., Johnston, L. H., Kodama, K., Tomkinson, A. E., Lasko, D. D., and Lindahl, T. (1990). Human DNA ligase I cDNA: Cloning and functional expression in Saccharomyces cerevisiae. *Proc. Natl. Acad. Sci. USA* **87**, 6679–6683.

Barnes, D. E., Tomkinson, A. E., Lehmann, A. R., Webster, A. D., and Lindahl, T. (1992). Mutations in the DNA ligase I gene of an individual with immunodeficiencies and cellular hypersensitivity to DNA-damaging agents. *Cell* **69**, 495–503.

Baumann, P., and West, S. C. (1998). DNA end-joining catalyzed by human cell-free extracts. *Proc. Natl. Acad. Sci. USA* **95**, 14066–14070.

Boulton, S. J., and Jackson, S. P. (1998). Components of the Ku-dependent non-homologous end-joining pathway are involved in telomeric length maintenance and telomeric silencing. *EMBO J.* **17**, 1819–1828.

Bradford, M. M. (1976). A rapid and sensitive method for the quantitation of microgram quantities of protein utilizing the principle of protein-dye binding. *Anal. Biochem.* **72**, 248–254.

Caldecott, K. W., McKeown, C. K., Tucker, J. D., Ljungquist, S., and Thompson, L. H. (1994). An interaction between the mammalian DNA repair protein XRCC1 and DNA ligase III. *Mol. Cell. Biol.* **14**, 68–76.

Caldecott, K. W., Tucker, J. D., Stanker, L. H., and Thompson, L. H. (1995). Characterization of the XRCC1-DNA ligase III complex *in vitro* and its absence from mutant hamster cells. *Nucleic Acids Res.* **23**, 4836–4843.

Chen, J., Tomkinson, A. E., Ramos, W., Mackey, Z. B., Danehower, S., Walter, C. A., Schultz, R. A., Besterman, J. M., and Husain, I. (1995). Mammalian DNA ligase III: Molecular cloning, chromosomal localization, and expression in spermatocytes undergoing meiotic recombination. *Mol. Cell. Biol.* **15,** 5412–5422.

Chen, L., Trujillo, K., Ramos, W., Sung, P., and Tomkinson, A. E. (2001). Promotion of Dnl4-catalyzed DNA end-joining by the Rad50/Mre11/Xrs2 and Hdf1/Hdf2 complexes. *Mol. Cell* **8,** 1105–1115.

Chen, L., Trujillo, K., Sung, P., and Tomkinson, A. E. (2000). Interactions of the DNA ligase IV-XRCC4 complex with DNA ends and the DNA-dependent protein kinase. *J. Biol. Chem.* **275,** 26196–26205.

Dimitriadis, E. K., Prasad, R., Vaske, M. K., Chen, L., Tomkinson, A. E., Lewis, M. S., and Wilson, S. H. (1998). Thermodynamics of human DNA ligase I trimerization and association with DNA polymerase beta. *J. Biol. Chem.* **273,** 20540–20550.

Gomes, X. V., Gary, S. L., and Burgers, P. M. (2000). Overproduction in *Escherichia coli* and characterization of yeast replication factor C lacking the ligase homology domain. *J. Biol. Chem.* **275,** 14541–14549.

Grawunder, U., Wilm, M., Wu, X., Kulesza, P., Wilson, T. E., Mann, M., and Lieber, M. R. (1997). Activity of DNA ligase IV stimulated by complex formation with XRCC4 protein in mammalian cells. *Nature* **388,** 492–495.

Grawunder, U., Zimmer, D., Kulesza, P., and Lieber, M. R. (1998). Requirement for an interaction of XRCC4 with DNA ligase IV for wild-type V(D)J recombination and DNA double-strand break repair *in vivo. J. Biol. Chem.* **273,** 24708–24714.

Hefferin, M. L., and Tomkinson, A. E. (2005). Mechanism of DNA double-strand break repair by non-homologous end joining. *DNA Repair (Amst.)* **4,** 639–648.

Laemmli, U. K. (1970). Cleavage of structural proteins during the assembly of the head of bacteriophage T4. *Nature* **227,** 680–685.

Lakshmipathy, U., and Campbell, C. (1999). The human DNA ligase III gene encodes nuclear and mitochondrial proteins. *Mol. Cell. Biol.* **19,** 3869–3876.

Lee, K. J., Huang, J., Takeda, Y., and Dynan, W. S. (2000). DNA ligase IV and XRCC4 form a stable mixed tetramer that functions synergistically with other repair factors in a cell-free end-joining system. *J. Biol. Chem.* **275,** 34787–34796.

Lehman, I. R. (1974). DNA ligase: Structure, mechanism, and function. *Science* **186,** 790–797.

Levin, D. S., Bai, W., Yao, N., O'Donnell, M., and Tomkinson, A. E. (1997). An interaction between DNA ligase I and proliferating cell nuclear antigen: Implications for Okazaki fragment synthesis and joining. *Proc. Natl. Acad. Sci. USA* **94,** 12863–12868.

Levin, D. S., Vijayakumar, S., Liu, X., Bermudez, V. P., Hurwitz, J., and Tomkinson, A. E. (2004). A conserved interaction between the replicative clamp loader and DNA ligase in eukaryotes: Implications for Okazaki fragment joining. *J. Biol. Chem.* **279,** 55196–55201.

Mackey, Z. B., Niedergang, C., Murcia, J. M., Leppard, J., Au, K., Chen, J., de Murcia, G., and Tomkinson, A. E. (1999). DNA ligase III is recruited to DNA strand breaks by a zinc finger motif homologous to that of poly(ADP-ribose) polymerase. Identification of two functionally distinct DNA binding regions within DNA ligase III. *J. Biol. Chem.* **274,** 21679–21687.

Mackey, Z. B., Ramos, W., Levin, D. S., Walter, C. A., McCarrey, J. R., and Tomkinson, A. E. (1997). An alternative splicing event which occurs in mouse pachytene spermatocytes generates a form of DNA ligase III with distinct biochemical properties that may function in meiotic recombination. *Mol. Cell. Biol.* **17,** 989–998.

Nick McElhinny, S. A., Snowden, C. M., McCarville, J., and Ramsden, D. A. (2000). Ku recruits the XRCC4-ligase IV complex to DNA ends. *Mol. Cell. Biol.* **20,** 2996–3003.

Pascal, J. M., O'Brien, P. J., Tomkinson, A. E., and Ellenberger, T. (2004). Human DNA ligase I completely encircles and partially unwinds nicked DNA. *Nature* **432**, 473–478.

Prasad, R., Singhal, R. K., Srivastava, D. K., Molina, J. T., Tomkinson, A. E., and Wilson, S. H. (1996). Specific interaction of DNA polymerase beta and DNA ligase I in a multiprotein base excision repair complex from bovine testis. *J. Biol. Chem.* **271**, 16000–16007.

Riballo, E., Critchlow, S. E., Teo, S. H., Doherty, A. J., Priestley, A., Broughton, B., Kysela, B., Beamish, H., Plowman, N., Arlett, C. F., Lehmann, A. R., Jackson, S. P., and Jeggo, P. A. (1999). Identification of a defect in DNA ligase IV in a radiosensitive leukaemia patient. *Curr. Biol.* **9**, 699–702.

Robins, P., and Lindahl, T. (1996). DNA ligase IV from HeLa cell nuclei. *J. Biol. Chem.* **271**, 24257–24261.

Shuman, S., and Schwer, B. (1995). RNA capping enzyme and DNA ligase: A superfamily of covalent nucleotidyl transferases. *Mol. Microbiol.* **17**, 405–410.

Soderhall, S., and Lindahl, T. (1973). Two DNA ligase activities from calf thymus. *Biochem. Biophys. Res. Commun.* **53**, 910–916.

Teraoka, H., Minami, H., Iijima, S., Tsukada, K., Koiwai, O., and Date, T. (1993). Expression of active human DNA ligase I in *Escherichia coli* cells that harbor a full-length DNA ligase I cDNA construct. *J. Biol. Chem.* **268**, 24156–24162.

Tomkinson, A. E., Roberts, E., Daly, G., Totty, N. F., and Lindahl, T. (1991). Three distinct DNA ligases in mammalian cells. *J. Biol. Chem.* **266**, 21728–21735.

Waga, S., and Stillman, B. (1998). Cyclin-dependent kinase inhibitor p21 modulates the DNA primer-template recognition complex. *Mol. Cell. Biol.* **18**, 4177–4187.

Wang, Y. C., Burkhart, W. A., Mackey, Z. B., Moyer, M. B., Ramos, W., Husain, I., Chen, J., Besterman, J. M., and Tomkinson, A. E. (1994). Mammalian DNA ligase II is highly homologous with vaccinia DNA ligase: Identification of the DNA ligase II active site for enzyme-adenylate formation. *J. Biol. Chem.* **269**, 31923–31928.

Wei, Y. F., Robins, P., Carter, K., Caldecott, K., Pappin, D. J., Yu, G. L., Wang, R. P., Shell, B. K., Nash, R. A., Schar, P., Barnes, D. E., Haseltine, W. A., and Lindahl, T. (1995). Molecular cloning and expression of human cDNAs encoding a novel DNA ligase IV and DNA ligase III, an enzyme active in DNA repair and recombination. *Mol. Cell. Biol.* **15**, 3206–3216.

[4] Enzymatic Mechanism of the WRN Helicase/Nuclease

By ROBERT M. BROSH, JR., PATRICIA L. OPRESKO, and VILHELM A. BOHR

Abstract

Werner syndrome (WS) is a premature aging disorder characterized by genomic instability and increased cancer risk (Martin, 1978). The *WRN* gene product defective in WS belongs to the RecQ family of DNA helicases (Yu *et al.*, 1996). Mutations in RecQ family members BLM and RecQ4 result in two other disorders associated with elevated chromosomal

METHODS IN ENZYMOLOGY, VOL. 409
Copyright 2006, Elsevier Inc. All rights reserved.

0076-6879/06 $35.00
DOI: 10.1016/S0076-6879(05)09004-X

instability and cancer, Bloom syndrome and Rothmund-Thomson syndrome, respectively (for review see Opresko *et al.*, 2004a). RecQ helicase mutants display defects in DNA replication, recombination, and repair, suggesting a role for RecQ helicases in maintaining genomic integrity.

The *WRN* gene encodes a 1432 amino acid protein that has several catalytic activities (Brosh and Bohr, 2002) (Fig. 1). WRN is a DNA-dependent ATPase and utilizes the energy from ATP hydrolysis to unwind double-stranded DNA. WRN is also a 3' to 5' exonuclease, consistent with the presence of three conserved exonuclease motifs homologous to the exonuclease domain of *Escherichia coli* DNA polymerase I and RNase D. Most recently, WRN (Machwe *et al.*, 2005) and other human RecQ helicases (Garcia *et al.*, 2004; Machwe *et al.*, 2005; Sharma *et al.*, 2005) have been reported to possess an intrinsic single-strand annealing activity. In addition to its catalytic activities, WRN interacts with a number of proteins involved in various aspects of DNA metabolism.

To understand the role of WRN in the maintenance of genome stability, a number of laboratories have undertaken a thorough characterization of its molecular and cellular functions. Here, we describe methods and approaches used for the functional and mechanistic analysis of WRN helicase or exonuclease activity. Protocols for measuring ATP hydrolysis, DNA binding, and catalytic unwinding or exonuclease activity of WRN protein are provided. Application of these procedures should enable the researcher to address fundamental questions regarding the biochemical properties of WRN or related helicases or nucleases, which would serve as a platform for further investigation of its molecular and cellular functions.

Expression and Purification of Recombinant WRN Protein

WRN Bacmid Construction and Baculovirus Infection

Recombinant WRN bacmid DNA was constructed as previously described (Gray *et al.*, 1997). Briefly, a *Sal*I-*Ssp*I fragment containing the full-length human *WRN* cDNA (nt 220–4745, accession number L76937) was subcloned into the *Sal*I-*Eco*RI cloning sites of pBlueBacHis2A (Invitrogen, Grand Island, NY), placing a hexahistidine tag at the N-terminus of the recombinant WRN protein. pBlueBacHis-WRN DNA was cotransfected with linearized wild-type *Autographa californica* nuclear polyhedrosis virus (AcNMPV) DNA into *Spodoptera frugiperda* cells (*Sf9* strain) using a cationic liposome-mediated transfection procedure (Invitrogen). Plaque-purified recombinant WRN virus was amplified in *Sf9* cells. For optimal infection and recombinant WRN expression, *Sf9* cells are typically infected with baculovirus at a multiplicity of infection of 10 for 72 h at 28°.

WRN Protein Purification

Recombinant WRN protein is purified as previously described (Opresko *et al.*, 2004b; Sharma *et al.*, 2004). Briefly, WRN purification is achieved via DEAE-Sepharose, Q Sepharose, and Ni-NTA resin chromatography (Orren *et al.*, 1999). Then eluents are loaded on an SP-Sepharose column (Amersham, Pharmacia, Piscataway, NJ) pre-equilibrated with SP-buffer (150 mM Tris [pH 8.0], 10% glycerol, and a protease inhibitors cocktail [Roche Molecular Biochemicals, Indianapolis, IN] plus 50 mM NaCl). After washes with 50 mM and 100 mM NaCl, the protein is eluted with 400 mM NaCl buffer. Protein concentration of the final fractions is determined by Bradford assay (Bio-Rad, Hercules, CA). Bovine serum albumin (BSA) is added to a final concentration of 100 μg/ml for stability during storage at $-80°$.

Characterization of WRN Helicase Activity

Directionality and Loading Properties of WRN Helicase

The directionality of movement of WRN helicase along single-stranded DNA (ssDNA) has been inferred from strand displacement assays using a linear partial duplex DNA substrate containing two radiolabeled oligonucleotides that are annealed to the very proximal opposite ends of a long ssDNA molecule (Matson *et al.*, 1994) (Fig. 1). Preferential release of one

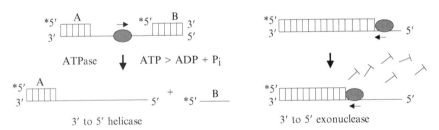

FIG. 1. WRN helicase and exonuclease activities. Shown is a cartoon representation of the three catalytic activities of the WRN protein: DNA-dependent ATPase, helicase, and exonuclease. WRN strand annealing activity is not depicted. WRN helicase directionality is defined as 3' to 5' with respect to the strand that the enzyme is inferred to be bound based on the preferential release of oligonucleotide B in the figure. WRN ATP hydrolysis is required for the dsDNA unwinding reaction. WRN exonuclease directionality is defined as 3' to 5' with respect to the direction that the single strand of the DNA molecule is degraded, as depicted in the figure.

of the two oligonucleotides on such a substrate has been used to define the 3' to 5' polarity of movement of WRN helicase along the bound ssDNA residing between the duplexes (Shen *et al.*, 1998). However, formal proof for unidirectional translocation of WRN helicase along a DNA molecule is warranted, and has been provided recently for the PcrA helicase (Dillingham *et al.*, 2000).

Although directional unwinding by WRN helicase can be inferred from the result of a strand displacement assay using a directionality DNA substrate, WRN does not require a pre-existing ssDNA tail of defined polarity for either loading or initiation of unwinding (Brosh *et al.*, 2002). This principle becomes important for a variety of DNA replication or repair intermediates (e.g., replication fork, 5' flap, D-loop, Holliday junction) that the helicase may act upon *in vivo*. In the subsequent helicase sections, protocols for preparing DNA substrates and characterizing WRN unwinding of various DNA structures are described.

WRN, like certain other DNA helicases, prefers a forked duplex compared to a duplex with a ssDNA tail for unwinding; however, this is not a universal property of all DNA helicases. A length of 10 nucleotides in either the 3'- or 5'-ssDNA tail provides an optimal forked duplex substrate for unwinding by WRN; however, DNA duplex substrates with shorter 3'-ssDNA tails can be unwound by WRN (Brosh *et al.*, 2002). The DNA substrate specificity of a helicase may be conferred by its DNA binding preference (Sun *et al.*, 1998) or by DNA structural elements that serve to occlude one of the single strands of the duplex from the central channel of the helicase (Kaplan, 2000). A useful tool for probing the importance of DNA structural elements for WRN helicase unwinding function is a DNA substrate with a specifically positioned steric block such as a biotin-streptavidin complex whose strong affinity and size can block a helicase when it is positioned on the single strand on which the enzyme translocates (Brosh *et al.*, 2002). However, certain DNA helicases have the ability to displace streptavidin from biotin-labeled oligonucleotides (Morris and Raney, 1999; Morris *et al.*, 2002), suggesting a role for removing proteins during cellular processes such as DNA repair. In the case of WRN, the helicase unwound a 3' ssDNA tailed duplex substrate with streptavidin bound to the end of the 3' ssDNA tail, suggesting that WRN does not require a free DNA end to unwind the duplex; however, WRN was completely blocked by streptavidin bound to the 3' ssDNA tail 6 nucleotides upstream of the single-stranded/double-stranded DNA junction. WRN efficiently unwound the forked duplex with streptavidin bound just upstream of the junction, suggesting that WRN recognizes elements of the fork structure to initiate unwinding (Brosh *et al.*, 2002).

Length Dependence for Unwinding by WRN Helicase

The length dependence for duplex DNA unwinding is an important property of DNA helicases (Matson *et al.*, 1994). By testing duplex DNA substrates of increasing lengths, information pertaining to the processivity of a DNA helicase can be determined. Whereas some helicases are very processive and can unwind thousands of base pairs of duplex DNA, other helicases such as WRN can only efficiently unwind fairly short duplexes (<50 bp) (Brosh *et al.*, 1999). The presence of an auxiliary factor can greatly alter the ability of a helicase to unwind duplex DNA, and in some cases enable a relatively non-processive helicase to act in a processive fashion and unwind very long duplexes. An example of a protein auxiliary factor for a DNA helicase is the human single-stranded DNA binding protein Replication protein A (RPA) (see section on *WRN Protein Interactions* and Table I).

TABLE I
SUMMARY OF WRN PROTEIN INTERACTIONS THAT ALTER WRN CATALYTIC ACTIVITIES

Protein binding partner	Effect on WRN activities	Reference
RPA	Stimulates WRN helicase activity on short and long duplex substrates	Brosh *et al.*, 1999; Doherty *et al.*, 2005; Shen *et al.*, 1998, 2003
Ku heterodimer	Stimulates WRN exonuclease activity on a variety of substrates	Cooper *et al.*, 2000; Li and Comai, 2000, 2001
p53	Inhibits WRN helicase and exonuclease activities	Brosh *et al.*, 2001; Sommers *et al.*, 2005; Yang *et al.*, 2002
DNA-PK kinase	Phosphorylation of WRN inhibits WRN helicase and exonuclease activities	Karmakar *et al.*, 2002; Yannone *et al.*, 2001
Bloom	Inhibits WRN exonuclease activity	von Kobbe *et al.*, 2002
TRF2	Stimulates WRN helicase activity on relatively short duplex substrates	Opresko *et al.*, 2002
PARP-1	Inhibits both WRN helicase and exonuclease activities	von Kobbe *et al.*, 2004
RAD52	Inhibits or stimulates WRN helicase activity dependent on the structure of the DNA-substrate	Bayton *et al.*, 2003
Mre11/Rad50/Nbs1	Promotes WRN helicase activity on short duplex substrates	Cheng *et al.*, 2004
APE1	Inhibits WRN helicase unwinding of base excision repair intermediates	Ahn *et al.*, 2004

WRN Unwinding of Alternate or Damaged DNA Structures

In addition to classical B-form DNA duplex substrates, WRN can unwind DNA:RNA hybrids but not RNA duplexes (Suzuki *et al.*, 1997). WRN can efficiently unwind alternate DNA structures including triplexes (Brosh *et al.*, 2001a) and tetraplexes (Fry and Loeb, 1999; Mohaghegh *et al.*, 2001), which can form spontaneously in sequences that are widely distributed throughout the human genome. Such alternate DNA structures that deviate from the canonical Watson-Crick base-pairing potentially interfere with cellular processes such as replication or transcription, and may give rise to genomic instability. At least some of the genomic instability detected in human RecQ disorders may be a consequence of the absence of the respective DNA helicase that acts to resolve these structures.

DNA damage may also hinder pathways of nucleic acid metabolism *in vivo*. Helicases are likely to encounter DNA lesions during the cellular processes of replication or DNA repair. Generally speaking, helicase inhibition by helix-distorting lesions is strand-specific (i.e., a DNA adduct inhibits DNA unwinding catalyzed by a helicase when the adduct is positioned in the strand that the helicase translocates upon, but inhibits to a significantly lesser degree when the adduct is on the opposite strand) (for review, see Villani and Tanguy, 2000). Stereochemistry and orientation of the adduct can impact the ability of the adduct to deter DNA unwinding, as demonstrated for the WRN helicase (Driscoll *et al.*, 2003). DNA substrates with site-specific DNA modifications can be constructed from oligonucleotides harboring covalent adducts essentially as described for standard unadducted duplex DNA substrates (see the section entitled *Preparation of Radiolabeled Duplex DNA Substrate*).

Helicase Assays

The next several sections for characterizing WRN helicase-catalyzed DNA unwinding will focus on the classic strand displacement assay using a radiolabeled DNA substrate and the kinetic fluorometric-based helicase assay. The protocols for radiometric helicase assays are relatively easy to implement and interpret, enabling the researcher to gain some useful insight to properties of the DNA substrate that influence WRN helicase activity and characterization of the enzyme's unwinding activity on DNA substrates that might be acted upon in DNA metabolism. The radiometric-based helicase assay has also been profitably utilized for the investigation of functional WRN protein interactions (Table I), coordination of WRN helicase and exonuclease activities (see later), and other applications. For more sophisticated kinetic analyses, a real-time fluorometric-based helicase assay can be used. A fluorometric kinetic helicase assay that is useful for mechanistic studies is described in the following section.

Preparation of Radiolabeled Duplex DNA Substrate

A procedure for preparing two- or three-stranded duplex DNA substrates is described here. Radiolabel the gel-purified oligonucleotide at its 5' end by adding 10 pmol of oligonucleotide (1 μl, 10 pmol/μl) to a microfuge tube containing 13 μl H$_2$O, 2 μl of 10× T4 Kinase buffer (New England Biolabs, Ipswich, MA), and 3 μl of [γ-^{32}P]ATP (>5000 Ci/mmol). Add 1 μl T4 Polynucleotide Kinase (New England Biolabs). Mix the contents and briefly pulse down in a microfuge. Incubate the reaction mixture at 37° for 60 min. Inactivate the kinase reaction by heating at 65° for 10 min. To remove the unincorporated [γ-^{32}P]ATP, load the 20 μl reaction mixture onto a pre-spun MicroSpin G-25 column. Centrifuge at 735g for 1 min, and collect the eluate.

For the annealing reaction, add 25 μl H$_2$O, 2.5 μl 1 M NaCl, and 25 pmol (2.5 μl, 10 pmol/μl) of complementary oligonucleotide to the tube containing 20 μl ^{32}P-labeled oligonucleotide kinase reaction. The ratio of ^{32}P-labeled oligonucleotide to complementary oligonucleotide (1:2.5) ensures that a greater majority of labeled oligonucleotide becomes annealed to complementary oligonucleotide. Incubate the mixture in a boiling water bath for 5 min and then briefly pulse down the tube's contents. Immediately, transfer the tube to a 65° water bath. Upon transfer, turn off the thermostat of the water bath and allow oligonucleotides to anneal upon cooling to room temperature over a period of 2–3 h. Alternatively, the mixture may be heated at 95° for 5 min in a water bath, and then allowed to cool to room temperature by switching off the thermostat.

For three-stranded duplex DNA substrates in which an upstream primer is annealed to the 3' single stranded tail of a duplex, add 50 pmol (2 μl, 25 pmol/μl) of upstream primer to the tube containing an annealed duplex DNA mixture (50 μl). Incubate the mixture in a 37° water bath for 60 min. After incubation, turn off the thermostat of the water bath and allow upstream oligonucleotide to anneal to duplex DNA substrate upon cooling to room temperature over a period of 3 h. To construct a synthetic replication fork structure with duplex leading and lagging strand arms, add 50 pmol (1 μl, 50 pmol/μl) of complementary leading and lagging strand primers to the tube containing annealed duplex DNA mixture (50 μl), incubate at 37° for 2 h, and allow to slowly cool to room temperature.

Store the mixture containing annealed duplex DNA substrate shielded at 4° until required. The final concentration of the DNA substrate is

200 fmol/μl and can be diluted to 10 fmol/μl in substrate dilution buffer (10 mM Tris-HCl [pH 8.0], 50 mM NaCl, 1 mM EDTA) for addition (1 μl, 10 fmol/μl) to helicase reaction mixtures prior to experiment.

For preparation of streptavidin-bound helicase substrates, specifically positioned biotinylated oligonucleotides can be purchased from a vendor such as Midland Certified Reagent Co. (Midland, TX) and the duplex DNA substrates can be prepared as described previously. 15 nM streptavidin is preincubated with the DNA substrate with all reaction components except WRN for 10 min at 37°. WRN can then be added to initiate the unwinding reaction (see *Radiometric Helicase Assay*).

Preparation of Four-Stranded Oligonucleotide-Based Holliday Junction

Since WRN helicase has the ability to catalyze branch fork migration (Constantinou *et al.*, 2000), a highly relevant substrate to test for WRN helicase activity is the model Holliday junction (HJ) structure. The protocol for preparing a synthetic four-stranded Holliday junction structure is a modified version of that used for preparing the more standard two-stranded duplex substrate. Synthetic HJ(X12) is made by annealing four 50-mer oligonucleotides (X12-1, X12-2, X12-3, and X12-4) whose sequences are listed in Mohaghegh *et al.* (2001). HJ substrates with a 5′ ^{32}P label on oligonucleotides X12-1, X12-2, X12-3, or X12-4 are designated HJ(X12-1), HJ(X12-2), HJ(X12-3), or HJ(X12-4), respectively.

Radiolabel the gel-purified oligonucleotide X12-1 at its 5′ end by adding 200 ng of oligonucleotide X12-1 (1 μl, 200 ng/μl) to a microfuge tube containing 4.5 μl H_2O, 1 μl of 10× T4 Kinase buffer, and 2.5 μl of [γ-^{32}P] ATP (>5000 Ci/mmol). Add 1 μl T4 Polynucleotide Kinase (10 U/μl). Mix the contents and briefly pulse down in a microfuge. Incubate the reaction mixture at 37° for 60 min. Inactivate the kinase reaction by adding 1 μl of 0.5 M EDTA and heating at 65° for 20 min. To remove the unincorporated [γ-^{32}P]ATP, load the 10 μl reaction mixture onto a pre-spun MicroSpin G-25 column. Centrifuge at 735g for 1 min, and collect the eluate. Remove 0.5 μl of the eluate, add to 49.5 μl TE, and save for determination of specific activity.

For the annealing reaction, add 1000 ng of oligonucleotides X12-2, X12-3, and X12-4 (1 μl, 1000 ng/μl) to the tube containing 9.5 μl of ^{32}P-labeled X12-1. Incubate the mixture at 90° for 5 min, 65° for 10 min, and then 37° for 10 min. After 10 min incubation, turn off the thermostat of water bath and allow oligonucleotides to anneal upon cooling to room temperature over a period of 2–3 h.

Pour a non-denaturing 10% polyacrylamide gel (1.5 mM thick, 40 cm long) with wells of 8-mm width. To the 12.5 μl annealed HJ(X12-1) mixture, add equal volume of 2× helicase load buffer. Carefully load annealing reaction mixture in helicase load buffer (2 lanes, 12.5 μl per lane) onto a nondenaturing 10% (19:1 acrylamide:bisacrylamide) vertical slab polyacrylamide gel mounted on the Sturdier Model SE400 unit (Hoefer Scientific Instruments). Electrophorese the samples at 125 V in 1× TBE for 4 h. Dismantle gel apparatus at the end of the run and place glass plate horizontally on the bench behind shield. Discard running buffer into ^{32}P liquid waste. Separate glass plates with a plastic wedge and allow gel to stick to one glass plate. Carefully blot off excess liquid surrounding gel with a paper towel and cover the gel bound to one glass plate with Saran Wrap. Expose the gel to X-ray film for ~2–5 min. Process by autoradiography. Carefully align the film with the gel and identify the desired band representing the four-stranded HJ structure. The four-stranded synthetic HJ will have migrated less than 2 cm into the gel and represents the slowest migrating species on the gel. Mark the desired band, and excise it from the rest of the gel using a razor blade. Re-expose gel lacking excised band to verify that radiolabeled band was removed. Carefully place the gel slice in a dialysis bag (Spectra/Por, MWCO: 6000–8000) that has been pre-equilibrated in 0.5× TBE. Add 0.5 ml 0.5× TBE to the bag containing gel slice. Clamp with weighted dialysis clips (Spectrum, Rancho Dominguez, CA). Place clamped dialysis bag in horizontal gel electrophoresis unit containing 0.5× TBE. Electrophorese the sample at 60 V for 4 h at room temperature. At the end of the 4 h, reverse the current for 1 min. Remove the radioactive liquid sample containing the HJ (X12-1) and store substrate shielded at 4° until needed.

Remove 1 μl of HJ(X12-1) sample and place in scintillation vial with 3 ml scintillation cocktail (Vial A). Place 1 μl of the diluted eluate of the ^{32}P-labeled X12-1 oligonucleotide in a scintillation vial that contains 3 ml of scintillation cocktail (Vial B). Count Vials A and B in a liquid scintillation counter using a ^{32}P program. Use the cpm value from Vial B to determine specific activity and calculate the concentration (ng/μl) of HJ-(X12-1) using the cpm value from Vial A. Convert the concentration of HJ (X12-1) to pmol/μl using the molecular weight of HJ(X12-1).

Preparation of Three-Stranded Oligonucleotide-Based D-Loop

Strand invasion D-loop structures occur as recombination intermediates and at the telomeric end, and the RecQ helicases have been implicated in the dissociation of alternate structures that occur during recombination and telomere metabolism (Opresko et al., 2004b; Saintigny et al., 2002). The WRN helicase and exonuclease dissociate D-loop structures by releasing the invading 3′ strand. A synthetic telomeric D-loop is prepared by anneal-ing three oligonucleotides to create a bubble with two 33-bp duplex arms, and a 33-bp melted region in which an invading strand is hybridized. The invading strand contains 15-bp of duplex DNA at the 5′ end (generated by a hairpin structure) followed by 33 nucleotides of single stranded DNA (see Fig. 2 for a schematic). The sequences of the oligonucleotides (INV, BT, and BB) used to generate a telomeric D-loop, in which the invading strand mimics the 3′ telomeric tail, are listed in Opresko *et al.* (2004b). Oligonucleotides BT and BB contain three phosphorothioated nucleotides

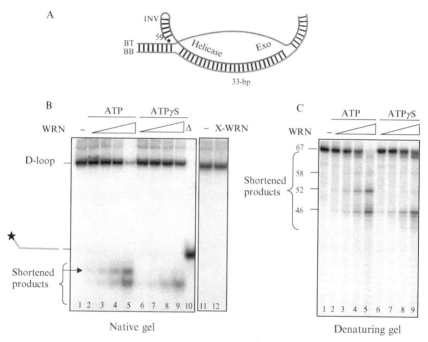

Fig. 2. WRN helicase and exonuclease cooperate to dissociate a telomeric D-loop. (A) Telomeric D-loop schematic. A bubble was formed by annealing the BT and BB strand. The INV strand with a 5′ duplex end (hairpin) was hybridized in the melted region. The 4 tandem telomeric repeats are highlighted. See (Opresko *et al.*, 2004b) and text for details. (B) and (C) Analysis of WRN helicase and exonuclease products from a model telomeric D-loop substrate. Wild type WRN protein (lanes 2–9) or the exonuclease dead mutant (X-WRN) (lane 12) was incubated with the D loop substrate containing a labeled INV strand (0.5 nM) for 15 min at 37°. Reaction aliquots were run on an 8% native gel (B) and on a 14% denaturing gel (C). The reactions contained 0.75, 1.5, 3, and 6 nM WRN with either 2 mM ATP (lanes 1–5) or 2 mM ATPγS (S, lanes 6–9), or 6 nM X-WRN with 2 mM ATP (lane 12). Δ, heat denatured substrate. Numbers in (C) indicate product length. The arrow in (B) indicates the products generated in the presence of active helicase, but that are absent when the helicase is inactive. (See color insert.)

at the 3' end to block digestion by the WRN exonuclease, which is normally active at the blunt ends of bubble substrates.

Radiolabel 10 pmol of gel-purified oligonucleotide INV at its 5' end in a 20 μl reaction using the procedure described in the section *Preparation of Radiolabeled Duplex DNA Substrate.* Annealing reactions (50 μl) are performed in a thermal cycler to allow for step-wise cooling, which enhances the proper alignment of the telomeric repeats. LiCl is used instead of NaCl to prevent G-quadruplex formation of the telomeric sequence. Add 27.5 μl H$_2$O, 2.5 μl 1 M LiCl, and 16 pmol (1.6 μl, 10 pmol/μl) of complementary oligonucleotide (BB) to a PCR tube containing 18 μl of ^{32}P-labeled INV oligonucleotide (9 pmol). Program a thermal cycler to incubate the mixture at 95° for 5 min, then cool step-wise to 60° by running 30 cycles of the program: $-1.2°$/cycle, 1 min/cycle. Pause the program, and add 20 pmol BT oligonucleotide (1.6 μl, 10 pmol/μl) that was preheated at 95° for 3 min. Mix by pipetting up and down. Restart the program and incubate at 60° for 1 h. Set the program to resume step-wise cooling to 25° : 30 cycles, $-1.2°$/cycle, 1 min/cycle. Store at 4°. The final substrate concentration is 180 fmol/μl and should migrate as a single species on a nondenaturing 8% polyacrylamide gel.

Radiometric Helicase Assay

Typically, to resolve the products of a helicase reaction with a radiolabeled DNA substrate, a nondenaturing polyacrylamide gel is prepared. The final percentage polyacrylamide of the native gel depends on the size of the intact duplex DNA substrate and unwound radiolabeled product of the helicase reaction. For example, resolution of an unwound 44-mer from a forked 19-bp duplex flanked by 25 nt 3' and 5' ssDNA arms can be achieved by electrophoresis at 180 V for 2 h on a 12% polyacrylamide gel. The percentage polyacrylamide of the gel and running time for electrophoresis can be altered to optimize resolution of the intact substrate from the released strand. For M13 partial duplex substrates, the percentage acrylamide is generally lowered to 6–8% to ensure that the M13 partial duplex migrates out of the well into the gel.

Reaction mixtures are typically set up in 20 μl aliquots. Add 1 μl of appropriately diluted helicase enzyme to a prechilled microfuge tube sitting on ice containing 4 μl of 5× reaction buffer (150 mM HEPES [pH 7.4], 25% glycerol, 200 mM KCl, 500 μg/ml bovine serum albumen [BSA], 5 mM MgCl$_2$), 10 fmol of forked duplex or selected DNA substrate (1 μl, 10 fmol/μl), 2 μl 10 mM ATP (or the specified nucleoside triphosphate, final concentration 1 mM nucleoside triphosphate), and 12 μl H$_2$O. The components of the reaction buffer (pH, buffering agent, salt, stabilizing agent [BSA or glycerol], nucleoside 5' triphosphate) should be optimized for the DNA helicase under investigation. For example, WRN helicase activity increases with Mg^{2+}:ATP ratios up to 1; however, greater Mg^{2+}:ATP ratios of up to 4 do not significantly further increase or decrease WRN unwinding (Choudhary *et al.*, 2004). An optimum concentration of ATP-Mg^{2+} for WRN helicase activity was determined to be 1 mM. Mn^{2+} and Ni^{2+} substituted for Mg^{2+} as a cofactor for WRN helicase, whereas Fe^{2+} or Cu^{2+} (10 μM) profoundly inhibited WRN unwinding in the presence of Mg^{2+}. Importantly, the reaction conditions for helicase catalyzed unwinding should be optimized for the DNA helicase under investigation.

After the reaction mixture is mixed, pulse down the tube contents briefly and incubate in a 37° (or desired temperature) water bath for 15 min (or the appropriate time). Quench reactions by adding 20 μl of 2× helicase load buffer (36 mM EDTA, 50% glycerol, 0.08% bromophenol blue, 0.08% xylene cyanol) containing a 10-fold molar excess of unlabeled oligonucleotide with the same sequence as the labeled strand. The presence of excess unlabeled oligonucleotide prevents reannealing of the unwound strand to its complementary strand.

Carefully load products of the helicase reactions onto nondenaturing 12% (19:1 acrylamide:bisacrylamide) polyacrylamide gels. Electrophorese the samples at 180 V in 1× TBE for 2 h or the appropriate time. Dismantle the gel apparatus at the end of the run and place the glass plate horizontally on the bench. Separate the glass plates with a plastic wedge and allow the gel to stick to one glass plate. Blot the excess liquid surrounding gel with a paper towel and cover the gel bound to one glass plate with Saran Wrap. Expose the gel to a phosphorimager screen for 5–15 h. Alternatively, expose the gel to X-ray film overnight at −20° and process by autoradiography the following day.

Visualize radiolabeled DNA species in polyacrylamide gel using a phosphorimager. An example of a phosphorimage of a native gel showing the WRN helicase products from a reaction containing a 5' flap DNA substrate is shown in Fig. 3. The identity of the reaction products can be confirmed by resolving control samples of the 1-, 2-, or 3-stranded radiolabeled DNA structures on the polyacrylamide gel. For a control reaction, set up a helicase reaction mixture in which the enzyme is omitted from the incubation. This serves as the "no enzyme" control. A second control is the heat-denatured DNA substrate control in which a "no enzyme" control tube is incubated at 95° for 5 min prior to loading. Quantitate helicase reaction products using the ImageQuant software (Molecular Dynamics, Sunnyvale, CA).

Percent helicase substrate unwound can be calculated using the following formula: percent unwinding = $100 \times (P/(S + P))$, where P is the product

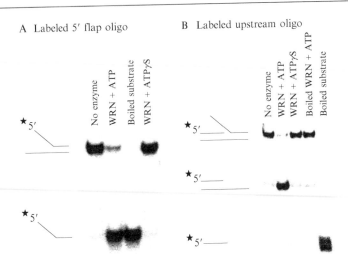

FIG. 3. WRN unwinds a 5′ ssDNA flap substrate. A 26 nt 5′ flap substrate (see Brosh *et al.*, 2002 and text for details) with a 5′ ³²P label on the 5′ flap oligonucleotide (A) or the 25-mer upstream oligonucleotide (B) was incubated with WRN in helicase reaction buffer (30 m*M* HEPES [pH 7.6], 5% glycerol, 40 m*M* KCl, 0.1 mg/ml BSA, 8 m*M* MgCl₂) containing 10 fmol of duplex DNA substrate (0.5 n*M* DNA substrate concentration) and 2 m*M* ATP where indicated. Helicase reactions were initiated by the addition of WRN and then incubated at 37° for 15 min. Reaction mixtures (20 μl) were quenched with 20 μl 2× helicase load buffer (36 m*M* EDTA, 50% glycerol, 0.08% bromophenol blue, 0.08% xylene cyanol) containing a 10-fold excess of unlabeled oligonucleotide with the same sequence as the labeled strand. The products of the helicase reactions were resolved on nondenaturing 12% (19:1 acrylamide: bisacrylamide) polyacrylamide gels. Radiolabeled DNA species were visualized using a phosphorimager. (A) lane 1, no enzyme control; lane 2, 3.8 n*M* WRN + ATP; lane 3, heat-denatured DNA substrate control; lane 4, 3.8 n*M* WRN + ATPγS. (B) lane 1, no enzyme control; lane 2, 3.8 n*M* WRN + ATP; lane 3, 3.8 n*M* WRN + ATPγS; lane 4, heat-denatured WRN protein control + ATP; lane 5, heat-denatured DNA substrate control.

and S is the residual substrate. Values of P and S are determined by subtracting background values in controls having no enzyme and heat-denatured substrate, respectively.

Real-Time Fluorometric Helicase Assay

To better understand WRN helicase mechanism and function, kinetic analyses of helicase-catalyzed unwinding of duplex DNA using fluorescence stopped-flow instrumentation is valuable since unlike classic radiometric assays, the data is collected continuously throughout the reaction in real time. The fluorometric assay uses the principle of fluorescence resonance energy transfer (FRET) to observe the unwinding of duplex DNA (Bjornson *et al.*, 1994). FRET occurs between a donor (e.g., fluoroscein) and acceptor (e.g., hexachlorofluoroscein) covalently attached to the complementary strands of the duplex substrate. Upon separation of the complementary strands, F and HF are no longer in close proximity, and the fluorescence emission from F excitation can be detected by a photosensor (Fig. 4A). Unwinding data from fluorometric assays can be substantiated using chemical quench flow kinetic analyses of data from radiometric helicase assays. Fast kinetics of helicase activity can be studied under two sets of conditions: (1) single-turnover, in which the DNA substrate is saturated with an excess of enzyme; (2) pre-steady state, in which the enzyme is saturated with an excess of DNA substrate. Results from single-turnover and pre-steady state kinetic analyses can be used to detect lag phase or initial burst kinetics of helicase-catalyzed DNA unwinding, respectively, under a given set of conditions. The kinetic mechanism for the formation of the active monomeric or multimeric helicase-DNA complex or the kinetic step size in sequential unwinding mechanisms can be determined from such analyses.

A fluorometric assay with the DNA substrate shown in Fig. 4A was used to study WRN helicase kinetics (Choudhary *et al.*, 2004). Prepare the DNA substrates as described previously for radiometric substrates with the exception that DNA substrates used for fluorometric assays are not 5'-^{32}P-labeled. Kinetic helicase assays are performed using an Applied Photophysics SX.18MV stopped-flow reaction analyzer (Applied Photophysics, Ltd., Leatherhead, UK). The instrument is equipped with a 150 W xenon lamp and the monochromator is set to a slit width of 1 mm. Fluorescein is excited at a wavelength of 492 nm and the fluorescence emission is monitored at wavelengths greater than 520 nm with the 51,300 cut-on filter from Oriel Corporation (Stratford, CT). Experiments are carried out in the two-syringe mode, where WRN and ATP are preincubated at 37° in one syringe for 1 min while the DNA substrate is preincubated at 37° in the second syringe. Each syringe contains 30 mM HEPES (pH 7.4), 5% glycerol, 40 mM KCl, 1 mM MgCl$_2$, and 100 ng/μl BSA. Concentrations of WRN, ATP, and DNA fork substrate in the syringe are double that of the indicated final concentration in the reaction. Mix equal volumes (60 μl) of sample from both syringes to initiate the reaction, which takes place at 37°. One thousand data points are collected from monitoring 20 μl of each kinetic time course reaction. For converting the output data from volts to percent unwinding, a time course with the same set up is performed, except

A

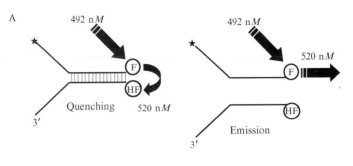

B WRN helicase kinetics

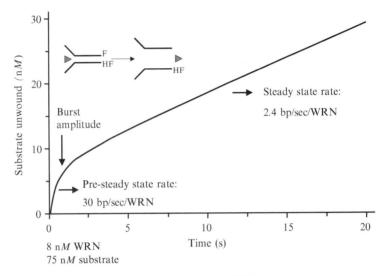

FIG. 4. Kinetic analyses of DNA unwinding by WRN helicase using a fluorometric stopped-flow assay. (A) Schematic representation of the fluorescent DNA substrate design. 19-bp forked duplex with two noncomplementary ssDNA tails of 26 nt (5′ tail) or 25 nt (3′ tail). Fluorescein (F) is covalently attached to the 3′ blunt end and hexachlorofluorescein (HF) is attached to the 5′ blunt end of the DNA substrate. The asterisk indicates the position of the 5′ ^{32}P label for the radiolabeled DNA substrate. When in close proximity, HF quenches the signal emitted from F upon excitation. Upon unwinding, the emission from F upon excitation is free to be detected because the HF is no longer in close proximity with F. (B) Analysis of WRN helicase kinetics under presteady state conditions (Choudhary *et al.*, 2004). Helicase reaction mixtures contained 75 nM forked DNA substrate and 1 mM ATP-Mg^{2+}. Kinetic time courses of WRN helicase activity were determined from the fluorometric assay at 8 nM WRN as shown.

with only the fluorescein oligonucleotide instead of the fluorescent forked duplex substrate. Data is then normalized by defining the voltage obtained with the fluorescein oligonucleotide as 100% unwinding.

Pre-steady state conditions can be established by using 75 nM DNA substrate and 4, 8, or 16 nM WRN with an optimal ATP-Mg^{2+} concentration of 1 mM (Fig. 4B). A 20-s time course can then be carried out using the fluorescence-based assay. Previously, it was shown that the resulting WRN unwinding kinetics curve displayed biphasic kinetics. The first phase is referred to as the pre-steady state phase, whereas the second phase is referred to as the steady state phase. For each level of WRN, the pre-steady state phase achieved an amplitude at a concentration of unwound DNA that was very close to the concentration of WRN enzyme in the reaction. The WRN:DNA substrate ratio of 1:1 suggests that a 19-bp duplex DNA substrate is unwound by a single WRN molecule under the reaction conditions tested. It was concluded from the kinetic analysis that WRN can function as a monomer to unwind the duplex DNA substrate (Choudhary *et al.*, 2004).

ATPase Assay

The assay described provides a facile thin layer chromatography (TLC) method to determine the kinetic parameters for WRN ATP hydrolysis. Like most DNA helicases, WRN ATPase activity is dependent on the presence of a DNA cofactor in the reaction. Presumably WRN binding to the DNA molecule induces a conformational change in the protein that is favorable for catalytic turnover of ATP to release ADP and the free phosphate ion. Hydrolysis of dATP or ATP, and to a lesser extent hydrolysis of dCTP or CTP, supports WRN-catalyzed DNA unwinding (Shen *et al.*, 1998).

Set up reaction mixtures in 20-μl aliquots for K_m determination and 30 μl-aliquots for k_{cat} determination. For simplicity, the 30-μl reaction volume will be described here. Add 2 μl of appropriately diluted helicase enzyme to a prewarmed microfuge tube containing 6 μl of 5× reaction buffer (150 mM HEPES [pH 7.4], 25% glycerol, 200 mM KCl, 500 μg/ml BSA, 5 mM MgCl$_2$), 3 μl of the desired DNA effector that has been appropriately diluted (e.g., M13 ssDNA circle, 0.1 mg/ml or 200 nM single-stranded or duplex oligonucleotide DNA molecules), 9.6 μl 2.5 mM [^{3}H]ATP (final concentration 0.8 mM) or [γ-^{32}P]ATP, and 9.4 μl H$_2$O. Alternatively, DNA effector can be omitted and replaced with 3 μl H$_2$O. The components of the reaction buffer (pH, buffering agent, salt, stabilizing agent [BSA or glycerol], nucleoside 5′ triphosphate) should be optimized for the DNA helicase under investigation. Mix the reaction mixture briefly, pulse down in a microfuge briefly, and incubate in a 37° (or desired temperature) water bath for a fixed time (e.g., 10 min) to determine K_m or a set of kinetic time points (e.g., 2, 4, 6, 8, 10 min) to determine k_{cat}.

At the end of incubation, carefully remove 5-μl aliquots from reaction mixture and add to 5-μl aliquots of ATPase stop solution (6.7 mM rADP, 6.7 mM rATP, 33 mM EDTA) that have been previously dispensed in labeled microfuge tubes. Mix and pulse down tube contents in microfuge. Repeat for all time points. To prepare the TLC sheet (Baker-flex Cellulose PEI sheet, Philipsburg, NJ [J. T. Baker]) for spotting, cut the 20-cm by 20-cm sheet in half with scissors. With a pencil and ruler, draw a line 1 cm from the bottom of a half sheet. Make small tick-marks along the line at 1-cm intervals with a pencil. These tick-marks will serve as position markers for spotting aliquots of quenched ATPase reaction mixtures.

Spot 2–3 μl of quenched reaction mixture samples on to appropriate position on TLC sheet. Proceed with next sample, spotting 2–3 μl each time. After an entire set of samples have been spotted consecutively, allow the TLC sheet to dry by air or with a blow dryer (low heat setting). Repeat the spotting for each sample until the entire 10 μl has been spotted. This will require 4–5 sample applications.

Once the TLC plate has dried from the final application of samples, carefully dip the TLC plate into a glass baking dish containing a thin layer of deionized distilled water. Allow buffer to wick up toward the 1-cm horizontally marked line on the TLC plate. Prior to the water-front reaching the 1-cm line, carefully place the TLC plate in the tank containing a thin layer of 1 M formic acid/0.8 M LiCl for the [^{3}H]ATP hydrolysis assay or 0.75 M KH$_2$PO$_4$ for the [γ^{32}]ATP hydrolysis assay. Allow the solvent front to reach 1 cm from the top of the TLC sheet for the [^{3}H]ATP assay. Carefully remove the TLC sheet from the tank and place the TLC sheet horizontally on the bench under a low intensity heat lamp. Do not over-dry the TLC plate as this will result in problems during the scraping step (samples will flake). If [γ^{32}]ATP was used, the radioactivity on the TLC plate can be directly visualized on a phosphorimager and quantitated using the ImageQuant software. The released ^{32}P$_i$ will have migrated significantly farther than [γ^{32}P]ATP. If [^{3}H]ATP was used, then use a handheld short-wavelength UV lamp to visualize ADP and ATP spots, which migrate to higher and lower positions, respectively, relative to the origin of spotting. Circle ADP and ATP spots with a pencil. Using a blunt-ended metal spatula, carefully scrape spots off and transfer with blunt-ended forceps to scintillation vials. Add 3 ml scintillant (Complete Counting Scintillation Cocktail 3a70B, Research Product International Corp., Mt. Prospect, IL) and count using a ^{3}H program. Subtract background radioactivity (counts obtained from ADP spots of control reactions lacking the helicase) from total radioactivity for each ADP spot to determine counts of ADP produced for each time reaction. Determine the specific activity of the ATP using the values of the ATP spots from the no enzyme control reactions and the calculated amount of ATP (4000 pmol in this case) in each spot. Using the specific activity term, convert the corrected ADP counts to pmol ADP. Initial rates of ATP hydrolysis (k_{cat}) can be determined by linear regression analyses of kinetic plots (pmol ADP produced versus time [min]). Experiments should be designed such that at a given helicase concentration, <20% of the substrate ATP is consumed in the reaction over the entire time course of the experiment.

Several properties of the WRN ATPase have been elucidated by the described method. The ATPase activity of WRN is stimulated to a much greater extent by single-stranded effectors than by double-stranded effectors (Orren et al., 1999). These results on the ATPase activity of WRN are consistent with the higher affinity of WRN for ssDNA than for double-stranded DNA (dsDNA). Secondly, longer ssDNA molecules are better effectors for WRN ATPase activity than shorter ssDNA tracts. This suggests that WRN may have the ability to translocate along long stretches of ssDNA without additional binding steps, resulting in maximal ATP hydrolysis. The k_{cat} for WRN ATP hydrolysis using circular ssDNA molecules, the optimal DNA effector for WRN, ranges from ~50–200 min^{-1} (Brosh et al., 1999; Orren et al., 1999). Interestingly, although RPA greatly enhances the processivity of WRN unwinding, the ssDNA binding protein does not affect WRN ATP hydrolysis (Brosh et al., 1999).

Gel Mobility Shift Assay to Measure DNA Binding

Although WRN has the ability to unwind a large variety of DNA substrates including 3′ tailed or forked duplexes, Holliday Junctions, triplexes, and tetraplexes, recombinant full-length WRN does not stably bind to a number of DNA substrates as detected by nitrocellulose filter binding or gel-shift assays; however, an important exception to that general rule is the synthetic 4-stranded Holliday junction structure, which WRN binds relatively well compared to other substrates (Constantinou et al., 2000). WRN also has been reported to bind bubble DNA structures (Shen and Loeb, 2000). Gel-shift assays using various duplex DNA structures and recombinant truncated fragments of WRN revealed that WRN contains three structure-specific DNA binding domains, one in the N-terminus and two in the C-terminus (RCQ and HRDC) (von Kobbe et al., 2003).

Described later is a gel mobility shift assay procedure that can be applied to measure HJ binding by WRN or more generally to other DNA helicase-substrate interactions.

Pour a nondenaturing 5% polyacrylamide gel (1.5-mM thick) with wells of 8-mm width. After the gel is polymerized (2–3 h), let it cool at 4°. Binding reaction mixtures are set up in 20-μl aliquots. Add 1 μl of appropriately diluted helicase enzyme to a prechilled microfuge tube sitting on ice containing 2 μl of 5× DNA binding reaction buffer (125 mM Tris-HCl [pH 7.5], 100 mM NaCl, 10 mM MgCl$_2$, 500 μg/ml BSA), 1 μl 10 mM dithiothreitol, 10 fmol of DNA duplex or HJ substrate (1 μl, 10 fmol/μl), 2 μl 10 mM ATPγS (or the specified nucleotide, final concentration 1 mM), and 13 μl H$_2$O. Note that nucleotide-dependent conformational changes in protein structure may influence DNA binding properties of the helicase. Therefore, binding mixtures containing ATPγS, ADP, or no nucleotide should be tested to address the effect of nucleotide on DNA binding.

Mix the reaction mixture briefly, pulse down in a microfuge briefly, and incubate in a 37° (or desired temperature) water bath for 15 min (or the appropriate time). At the end of the incubation, add 3 μl native loading buffer (74% glycerol, 0.01% xylene cyanol, 0.01% bromophenol blue). Carefully load DNA binding reaction mixtures onto nondenaturing 5% (19:1 acrylamide:bisacrylamide) polyacrylamide gels. To optimize resolution, the ratio of acrylamide: bisacrylamide used for the preparation of the native gel can be modified. A 37.5:1 ratio could be tried. Electrophorese the samples at 200 V in 1 × TBE (or 0.5× TBE) for 2 h at 4°. Dismantle gel apparatus and wrap plate in Saran Wrap. Expose the gel to a phosphorimager screen for 5–15 h. Alternatively, expose the gel to X-ray film overnight at −20° and process by autoradiography the following day. Shorter exposure times are preferable since DNA diffusion occurs more readily in lower percentage gels. Visualize radiolabeled DNA species in polyacrylamide gel using a phosphorimager. Quantitate fraction of DNA substrate bound by protein using the ImageQuant software (Molecular Dynamics).

The following formula, based on Scatchard analysis, can be used to analyze the data by Hill plot:

$$K_d = (1 - f)[Pt]/f; \tag{1}$$

$$\log[Pt] = \log(f/(1 - f)) + \log K_d \tag{2}$$

K_d is the apparent dissociation constant of the DNA-protein complex, [Pt] is the total concentration of helicase protein present in the reaction, and f is the ratio of the amount of bound DNA over the total amount of DNA present in the reaction. Plot the logarithm of [Pt] against the logarithm of f/(1-f); the y-intercept represents the logarithm of K_d.

To trap WRN protein-DNA complexes, it may be helpful to add a cross-linking agent. For example, incubate the protein-DNA complex mixture with 0.25% glutaraldehye (final concentration) for 10 min at 37°. Use triethanolamine in place of Tris as the buffering agent when glutaldehyde is used for cross-linking. Specific binding of the protein to the DNA substrate may be confirmed by the detection of a super-shifted complex when an antibody against the DNA binding helicase protein is incubated with the protein-DNA complex mixture; however, if the DNA binding region and the antibody recognition (epitope) region overlap, a super-shifted species may not be detected. A super-shifted complex of a helicase bound to its protein partner and the DNA substrate may be detected in some circumstances. A good example of this is the WRN-FEN-1 interaction in which it was shown that WRN recruits FEN-1 to a Holliday junction (Sharma *et al.*, 2004). See Fig. 5 for a representation of this result.

Exonuclease Assay

WRN is unique among the human RecQ helicases in that the protein also contains a 3′ to 5′ exonuclease domain, which resides in the protein's N-terminus. The following sections describe protocols for characterizing the WRN exonuclease-catalyzed stepwise removal of 5′ mononucleotides from the 3′ hydroxyl termini of various DNA substrates.

The standard radiometric exonuclease assays typically utilize a 5′ end-labeled DNA substrate to detect the degradation of DNA molecules. The radiometric exonuclease assay utilizes similar reaction parameters and protocols as described for the radiometric helicase assay, but differs primarily in the electrophoretic methods used to resolve and visualize the products. The WRN exonuclease activity is separable from the helicase activity. Therefore, protocols for assaying the exonuclease both independently and in concert with the helicase are described. Both approaches have been used to characterize the WRN exonuclease activity on substrates that mimic intermediates in DNA metabolism, and to investigate factors that modulate the WRN exonuclease activity (i.e. protein partners [Table I] or WRN posttranslational modification state).

Substrate Specificity of WRN Exonuclease

The WRN exonuclease cleaves a phosphodiester bond to produce a 5′-dNMP and resected DNA molecule, but will not cleave a thioester bond. The substrate preparation for the radiometric exonuclease assay is

WRN (73 nM)   −   +   +   −

FEN-1 (116 nM)   −   −   +   +

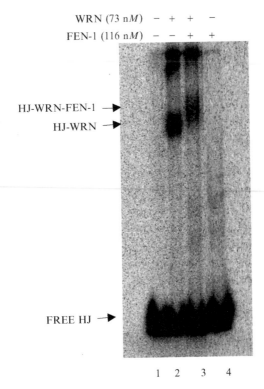

HJ-WRN-FEN-1 →

HJ-WRN →

FREE HJ →

1   2   3   4

FIG. 5. WRN recruits FEN-1 to the Holliday junction. WRN-HJ complex is super-shifted by FEN-1 (Sharma *et al.*, 2004). WRN (73 nM) and FEN-1 (116 nM) were incubated in HJ binding buffer (20 mM triethanolamine [pH 7.5], 2 mM MgCl$_2$, 0.1 mg/ml BSA, 1 mM dithiothreitol) with 25 fmol of HJ(X12-1) and 1 mM ATPγS in a total volume of 20 μl at 24° for 20 min. Protein-DNA complexes were fixed in the presence of 0.25% glutaraldehyde cross-linking agent for 10 min at 37°. Protein-DNA complexes were analyzed on nondenaturing 5% polyacrylamide 0.5× TBE gels and visualized by phosphorimager analysis.

identical to that used for the helicase assays, but the WRN exonuclease differs somewhat in the substrate requirements. The exonuclease initiates digestion at a 3' recessed terminus but not at a terminus with a protruding 3' single strand, and is largely inactive on single-stranded DNA. The WRN exonuclease does not degrade duplex molecules with blunt ends, but will initiate digestion at a duplex blunt end if the substrate also contains a junction or alternate structure such as a fork, bubble or hairpin (Shen and Loeb, 2000). WRN exonuclease is also active at the 3' hydroxyl group of a nicked or gapped DNA substrate, and can digest mismatched bases (Huang *et al.*, 2000). Analysis of the activity on molecules containing abnormal or modified bases indicated that the exonuclease is selectively blocked or inhibited by apurinic, 8-oxo-guanine, 8-oxoadenine, and cholesterol adducts, but is not blocked by other lesions such as uracil or hypoxanthine (Machwe et al., 2000). Such modifications are useful for probing the active site of the exonuclease, and the specific contacts between the enzyme and the DNA substrate that are required for activity.

Differential Effects of Zn^{2+} and Mg^{2+} on WRN Exonuclease Activity

As described in the helicase section, WRN catalytic activity is affected by the solution reaction conditions. We have observed that the extent of WRN exonuclease digestion is dependent on the concentration of free divalent cations in the reaction. Higher ratios of Mg^{2+}:ATP increased the extent of substrate degradation, and very little exonuclease activity was detected at equal molar ratios of Mg^{2+}:ATP (Choudhary et al., 2004 and our unpublished data). Results from WRN helicase and exonuclease assays demonstrated that in contrast to Mg^{2+}, Zn^{2+} dramatically stimulated WRN exonuclease activity on a forked duplex substrate in the presence of ATP but failed to serve as a cofactor for WRN-catalyzed unwinding (Choudhary et al., 2004). Direct comparison of WRN exonuclease activity (in the absence of ATP) further demonstrated that Zn^{2+} is clearly preferred as a metal cofactor for WRN exonuclease activity at a low concentration of metal (100 μM). Furthermore, Zn^{2+} strongly stimulated the exonuclease activity of a WRN exonuclease domain fragment, suggesting a Zn^{2+} binding site in the WRN exonuclease domain. Under conditions of DNA unwinding where both ATP and Mg^{2+} are present, low concentrations of Zn^{2+} stimulated the exonuclease activity of full-length WRN, suggesting that Zn^{2+} acts as a molecular switch converting WRN from a helicase to an exonuclease (Choudhary et al., 2004). These results emphasize the importance of solution reaction conditions for studying WRN helicase and nuclease activities.

Radiometric Exonuclease Assay

Typically the substrates are labeled at the 5' end so the shortened DNA molecule products are detected and measured rather than the released 5'-dNMP. Several factors influence the extent of digestion, resulting in a population of molecules that have been resected to varying lengths. The length of the products depends on the processivity of the enzyme, (i.e., the number of phophodiester bonds broken before the enzyme dissociates). Also, unless the reactions are performed under single-hit conditions, product length is determined by the number of encounters a single DNA molecule has with an active exonuclease protein prior to the reaction termination or loss of enzyme activity. Typically, the products of a radiometric exonuclease are resolved on a denaturing polyacrylamide gel, similar to sequencing gels, so the labeled strands of the substrate and products migrate as single strand oligonucleotides and not duplex molecules. The final percentage polyacrylamide of the denaturing gel is chosen to achieve optimal separation so that molecules that differ by a single nucleotide can be separated and distinguished. This depends on the lengths of the initial substrates and products. For example, higher percentage gels (20%) are better suited for resolving longer oligonucleotides. An example is shown for a 34-bp forked duplex with 15-mer single stranded tails in Fig. 6. The WRN exonuclease initiates degradation at the 3'OH of the blunt end, thereby shortening the 5'-end labeled top strand of the fork. These shortened products appear as a ladder on a denaturing gel (Fig. 6, lane 2). Separation of a 49-mer (the top strand of a 34-bp forked duplex) and the ladder of degraded products can be achieved by electrophoresis at 35 Watts for 2 h on a denaturing 14% polyacrylamide gel.

The reaction volumes (10 μl) are normally half the size of helicase assay reactions, since the products are electrophoresed on very thin denaturing gels, although the final enzyme and DNA concentrations are similar. To a microfuge tube on ice add 1 μl of the 10× Exonuclease Reaction buffer (40 mM Tris-HCl pH 8.0, 4 mM MgCl$_2$, 5 mM DTT, and 0.1 mg/ml BSA), 5 fmol of DNA substrate (0.5 μl, 10 fmol/μl), 1 μl of 20 mM ATP (final concentration 2 mM ATP), and 6.5 μl H$_2$O. As for the helicase assay, the components of the reaction buffer should be optimized for the DNA exonuclease under investigation. For example, the WRN exonuclease activity is enhanced at higher ratios of Mg^{2+}:ATP (see previously), and is increased at 37° compared to room temperature incubations. Initiate the reactions by adding 1 μl of the appropriate WRN protein dilution. A titration of various protein amounts should be conducted for each substrate tested. A typical range to test is 5 to 200 fmol of enzyme in a reaction with 5 fmol of DNA substrate.

After the addition of enzyme, mix well by pipetting up and down and then incubate in a 37° water bath (or tapered dry heat block) for 15 min (the time should be optimized). Terminate the reactions by adding 10 μl of formamide stop dye (80% formamide, 0.5× Tris borate, 0.1% bromophenol blue, and 0.1% xylene cyanol). Denature the samples by heating at 95° for 5 min, and then spin down the samples briefly. Next, load the samples on a denaturing 14–20% (19:1 acrylamide:bisacrylamide) polyacrylamide gel containing 7 M urea. Electrophorese the samples at 35 Watt in 1× TBE

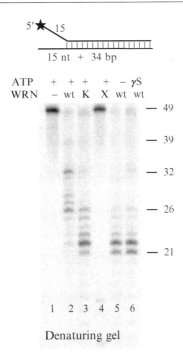

FIG. 6. WRN exonuclease digestion of forked duplex molecules. The reactions contained either 7.5 nM wild-type WRN (WT), the helicase-dead mutant K577M (K), or the exonuclease-dead mutant E84A (X), as indicated (see Opresko *et al.*, 2001 for details). The enzymes were incubated with the 34-bp forked duplex substrate (0.5 nM) under the standard reaction conditions except that either ATP was omitted or 2 mM ATPγS (γS) was substituted where indicated. Reactions were terminated with formamide stop dye after incubation for 15 min at 37° and were analyzed on 14% denaturing polyacrylamide gels. Values indicate the length of the products and the extent of degradation.

for 2 h or the appropriate time. Dismantle the gel apparatus and recover the gel as described for the helicase assay. The denaturing gel is thinner and more brittle than the native gels, so extra care should be taken in prying the plates apart. It is important to siliconize one plate prior to pouring the gel to prevent the gel from sticking to both plates. Cover the gel bound to one glass plate with Saran Wrap and place in a cassette with a phosphorimager screen. Expose for 5–15 h. X-ray film may also be used.

　　Visualization and quantitation of the denatured products is similar to that described for the helicase radiometric assay. A control reaction should be included which lacks enzyme (Fig. 6, lane 1), which serves as a marker for the intact substrate. A ladder of products should be visible that consists of different DNA fragment lengths, which migrate faster than the intact substrate (Fig. 6, lane 2). Each individual product can be quantitated. Determine the size of each product by comparing the migration with marker oligonucleotides to determine how many nucleotides were removed by the WRN exonuclease.

The proportion of unreacted substrate and products shortened to each length can be determined by dividing the amount of radioactivity in a given product band, by the total radioactivity quantitated in the lane. The following formula can be used to estimate the amount of nucleotides excised (fmols) that each product band represented: (Proportion of product length n) * (fmol total DNA molecules) * (initial substrate length – product length n). Values should be corrected for background in the no enzyme control lane, as described for the helicase assay.

Independent Analysis of WRN Exonuclease and Helicase Activities

Both WRN helicase and exonuclease activities can act on various DNA substrates simultaneously, including forked duplexes and bubbles. Thus, there is an interest in studying the WRN exonuclease in the absence of contributions from the helicase to simplify and clarify product analysis. For example, the helicase converts forked duplex substrates to single-stranded DNA, and the WRN exonuclease is largely inactive on single-stranded DNA. So the digestion of forked duplexes by the WRN exonuclease progresses further in the absence of helicase activity (Fig. 6) (Opresko et al., 2001). Several approaches are used to analyze the action of the exonuclease alone. The first is to modify the protein by either introducing a mutation that inactivates the helicase activity (K577M) (Fig. 6, lane 3) or by using a truncated protein (aa 1–368), which only contains the exonuclease domain. The latter will also result in the removal of domains that interact with DNA and protein partners. Another approach has been to modify the reaction conditions to favor the exonuclease. Since the helicase unwinding activity is coupled to ATP hydrolysis, the omission of ATP (Fig. 6, lane 5) or substitution with a nonhydrolyzable analog (ATPγS) (Fig. 6, lane 6), will also eliminate the helicase activity. A third approach is to use a substrate on which the helicase is inactive. A common substrate for the WRN exonuclease is a duplex molecule with a recessed 3′ OH and 5′ single-stranded tail. The WRN exonuclease initiates digestion at the 3′ OH; however, the helicase is unable to initiate unwinding of a duplex molecule with a 5′ single-stranded tail.

In turn, several approaches have been used to examine the helicase activity separately from the exonuclease. A mutation in the exonuclease domain (E84A) inactivates the exonuclease (Fig. 6, lane 4). In addition, truncated WRN fragments containing only the helicase domain actively unwind DNA substrates (Harrigan *et al.*, 2003). Reaction conditions that contain equal molar ATP and Mg^{2+} also appear to be less favorable for the exonuclease activity. The incubation of the reactions at room temperature instead of 37° will also decrease the exonuclease activity.

Combined Analysis of WRN Helicase and Exonuclease Activities

Although separable, the WRN helicase and exonuclease activities reside in the same polypeptide; therefore, it is also important to study how these activities contribute to the processing of a DNA substrate in a simultaneous or cooperative manner. For this purpose, it is important to empirically determine reaction conditions under which both the helicase and exonuclease are active. The following procedure describes a protocol for analyzing helicase and exonuclease products generated after the incubation of WRN protein with a telomeric D-loop substrate.

The procedure for constructing a telomeric D-loop is described earlier (see section on *Substrate Preparation*) (Fig. 2A). This substrate is not sufficiently unwound by the WRN helicase alone to displace the full length 67-mer 5′-end labeled invading strand (INV), since WRN is poorly processive alone. However, the exonuclease initiates degradation at the 3′ OH of the invading strand (INV), thereby shortening the duplex length so that it can now be completely unwound by the helicase. Once the strand is released, it is no longer degraded since the WRN exonuclease is inefficient on single-stranded DNA.

The reactions are carried out in 30 µl total volume. To a microfuge tube on ice add 3 µl of the 10× exonuclease reaction buffer (40 mM Tris-HCl [pH 8.0], 4 mM $MgCl_2$, 5 mM DTT, and 0.1 mg/ml BSA), 15 fmol of DNA substrate (1.5 µl, 10 fmol/µl), 3 µl of 20 mM ATP (final concentration 2 mM ATP), and 21.5 µl H_2O. Initiate the reactions by adding 1 µl of the appropriate WRN protein dilution. A titration of various protein amounts should be conducted for each substrate tested (a typical range is 7.5 to 300 fmol of enzyme with 15 fmol of DNA substrate).

After adding the enzyme, mix the reactions well and incubate in a 37° water bath (or tapered dry heat block) for 15 min (the time should be optimized). Add 10 µl of formamide stop dye to a microfuge tube. To terminate the reactions, remove a 10-µl aliquot from the reaction and add it to the tube containing 10 µl of formamide stop dye. To the remaining 20 µl, add 10 µl of 3× native stop buffer (50 mM EDTA, 40% glycerol, 0.9% SDS, 0.1% bromophenol blue and 0.1% xylene cyanol). The helicase stop buffer should also contain a 10-fold molar excess (100 fmol) of unlabeled competitor oligonucleotide that is identical to the labeled strand of the duplex. This prevents reannealing of the unwound ssDNA products. Process the aliquot terminated in formamide dye as described for the exonuclease assay and run on a 14% denaturing gel. Process the aliquot terminated in native stop dye as described for the helicase assay, and run on an 8% native gel. The native gel is required to visualize unwound products, whereas the exonuclease products are better resolved on a denaturing gel. The native gel shows WRN helicase products that migrate faster than the substrate heat-denatured marker (67-mer) (Fig. 2B, compare lanes 2–9 to lane 10). This is due to the action of the WRN exonuclease degrading the labeled INV strand, and the subsequent release of the shortened strands by the WRN helicase. The degradation products are better visualized on the denaturing gel (Fig. 2C). To further substantiate the helicase contribution to the observed products, perform control reactions in which ATP is substituted with ATPγS, to inactivate the helicase (Fig. 2B and 2C, lanes 6–9). In these reactions, the exonuclease degrades the labeled strand further until the duplex becomes thermally unstable and dissociates. The released longer products (indicated by the arrow) are no longer visible in the native gel (Fig. 2B, compare lanes 2–5 to lanes 6–9), indicating that these products depend on the helicase activity. The contribution of the exonuclease to the WRN generated products can be assessed by using an exonuclease-dead WRN mutant (X-WRN, E84A) (Fig. 2A, lane 12).

Effects of Post-Translational Modifications on WRN Catalytic Activities

Since human Ku heterodimer physically interacts with WRN and stimulates its exonuclease activity (Fig. 7) (Cooper *et al.*, 2000; Li and Comai, 2000) and the Ku-DNA-PKcs complex participates in double strand break repair (Featherstone and Jackson, 1999), it has been hypothesized that WRN plays a role in the DNA-PK damage response pathway. WRN was found to be a target for DNA-PKcs phosphorylation resulting in the down-regulation of WRN exonuclease activity (Karmakar *et al.*, 2002; Yannone *et al.*, 2001). Treatment of purified recombinant WRN protein with a Ser/Thr phosphatase enhances WRN exonuclease and helicase activities; moreover, the effect can be reversed by the rephosphorylation of WRN with DNA-PK. For a description of *in vitro* phosphorylation and phosphatase treatments of WRN, see Karmakar *et al.*, 2002. Tyrosine phosphorylation of WRN by the kinase c-Abl results in inhibition of both WRN

A

5′ *TGACGTGACGACGATCAGGGTACGTTCAGCAG³′

3′ ASTGCACTGCTGCTAGTCGCATGCAAGATCGTCGTCAGACGTGA₅′

B

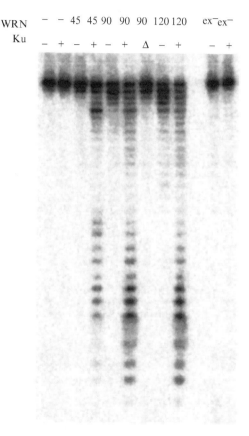

FIG. 7. The WRN exonuclease is stimulated by Ku86/70. (A) Schematic of the WRN exonuclease substrate containing a 5′ overhang. (B) Ku stimulation of WRN exonuclease at different concentrations of WRN (fmols); Ku was 64 fmols (10 ng) (see Cooper *et al.*, 2000 for details). The reactions were carried out in exonuclease reaction buffer for 1 h at 37°. Products were analyzed on 15% denaturing gels. Δ, heat inactivated Ku; ex-, exonuclease-dead WRN mutant.

exonuclease and helicase activities (Cheng *et al.*, 2003). Other WRN post-translational modifications including ATR/ATM phosphorylation (Pichierri *et al.*, 2003), SUMO-1 conjugation (Kawabe *et al.*, 2000), or acetylation (Blander *et al.*, 2002) may also affect WRN catalytic activities.

WRN Protein Interactions that Modulate WRN Catalytic Activity

The physical and functional interactions of WRN with human nuclear proteins implicated in DNA metabolic processes suggest that WRN plays a direct and coordinate role in processing DNA structures that arise during replication, recombination, and/or repair that might interfere with normal DNA transactions (for review, see Lee *et al.*, 2005; Ozgenc and Loeb, 2005). Although WRN modulates the catalytic functions of a number of other proteins (e.g., FEN-1, EXO-1, and DNA polymerase β), the focus of this section will be on WRN protein interactions that modulate the intrinsic catalytic activities of WRN (Table I). As stated previously, a major protein interaction of WRN is with the Ku heterodimer. Ku dramatically stimulates WRN exonuclease activity by its physical interaction with WRN (Fig. 7) (Cooper *et al.*, 2000; Li and Comai, 2000). Ku not only increases the extent of exonuclease digestion (Cooper *et al.*, 2000; Li and Comai, 2000), thereby increasing the percent of shorter products (Fig. 7), but also promotes WRN digestion of DNA molecules that are normally poor substrates for the WRN exonuclease (i.e. ssDNA, DNA duplexes blunt at both ends) (Li and Comai, 2001).

A major functional interaction for WRN helicase activity is with the single-stranded DNA binding protein RPA. RPA interacts with WRN (Brosh *et al.*, 1999; Shen *et al.*, 1998), BLM (Brosh *et al.*, 2000), and RECQ1 (Cui *et al.*, 2003) helicases and stimulates their DNA unwinding activities. In the presence of RPA (but not the heterologous ssDNA binding protein *E. coli* SSB), WRN can unwind DNA substrates with long duplex tracts such as a 257 bp (Fig. 8) or 849 bp M13 partial duplex (Brosh *et al.*, 1999). Although the M13 partial duplex substrate is not physiological, its value for examining the duplex length properties of WRN (and other helicases) and their functional protein interactions should not be underestimated. For a detailed description of the M13 partial duplex substrate preparation, see Brosh *et al.*, 1999. Since RPA has been shown to have roles in DNA replication, recombination, and repair, WRN is likely to function with the ssDNA binding protein during one or more of these processes. The large subunit of human RPA (RPA70) stimulates WRN helicase to the same extent as the RPA heterotrimer (Shen *et al.*, 2003). Recent evidence suggests that the physical interaction between RPA and WRN helicase

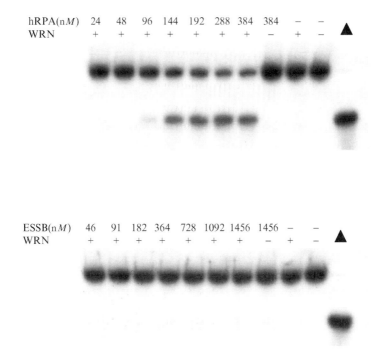

FIG. 8. RPA interacts with WRN and stimulates its helicase activity. RPA stimulates WRN to unwind a long (249 bp) M13 partial duplex DNA substrate whereas the heterologous *E. coli* single-stranded DNA binding protein ESSB fails to support the WRN helicase reaction (Brosh *et al.*, 1999).

plays an important role in the mechanism for RPA stimulation of WRN-catalyzed DNA unwinding (Doherty *et al.*, 2005).

In vitro studies show that p53 inhibits RPA-stimulated WRN helicase activity on an 849 bp M13 partial duplex substrate (Sommers *et al.*, 2005). p53 also inhibited WRN unwinding of short (19-bp) forked duplex substrate in the absence of RPA in a specific manner since a WRN helicase domain fragment or the related RECQ1 helicase was not inhibited by p53 (Sommers *et al.*, 2005). p53 also regulates WRN unwinding of replication intermediates (Sommers *et al.*, 2005) and a Holliday junction structure (Yang *et al.*, 2002). p53 also inhibits WRN exonuclease activity by its physical interaction with WRN (Brosh *et al.*, 2001b). Collectively, these results and cellular data suggest that regulation of WRN function by p53 is likely to play a role in genomic integrity surveillance, a vital function in the prevention of tumor progression.

Acknowledgments

We wish to recognize all the laboratories that developed various biochemical techniques for the study of DNA helicases and exonucleases that could be applied to the mechanistic analyses of WRN. We also thank those labs engaged in the study of RecQ helicases, particularly WRN, for their contribution to the development of protocols and reagents to study WRN catalytic functions. We wish to thank Joshua Sommers and Jason Piotrowski in The Laboratory of Molecular Gerontology, National Institute on Aging, NIH for critical reading of the manuscript and other members of the lab for helpful discussion.

References

Ahn, B., Harrigan, J., Indig, F. E., Wilson, D. M., 3rd, and Bohr, V. A. (2004). Regulation of WRN helicase activity in human base excision repair. *J. Biol. Chem.* **279**, 53465–53474.

Bayton, K., Otterlei, M., Bjoras, M., von Kobbe, C., Bohr, V. A., and Seeberg, E. (2003). WRN interacts physically and functionally with the recombination mediator protein RAD52. *J. Biol. Chem.* **278**, 36476–36486.

Bjornson, K. P., Amaratunga, M., Moore, K. J., and Lohman, T. M. (1994). Single-turnover kinetics of helicase-catalyzed DNA unwinding monitored continuously by fluorescence energy transfer. *Biochemistry* **33**, 14306–14316.

Blander, G., Zalle, N., Daniely, Y., Taplick, J., Gray, M. D., and Oren, M. (2002). DNA damage-induced translocation of the Werner helicase is regulated by acetylation. *J. Biol. Chem.* **277**, 50934–50940.

Brosh, R. M., Jr., Majumdar, A., Desai, S., Hickson, I. D., Bohr, V. A., and Seidman, M. M. (2001a). Unwinding of a DNA triple helix by the Werner and Bloom syndrome helicases. *J. Biol. Chem.* **276**, 3024–3030.

Brosh, R. M., Jr., Karmakar, P., Sommers, J. A., Yang, Q., Wang, X. W., Spillare, E. A., Harris, C. C., and Bohr, V. A. (2001b). p53 modulates the exonuclease activity of Werner syndrome protein. *J. Biol. Chem.* **276**, 35093–35102.

Brosh, R. M., Jr., Li, J. L., Kenny, M. K., Karow, J. K., Cooper, M. P., Kureekattil, R. P., Hickson, I. D., and Bohr, V. A. (2000). Replication protein A physically interacts with the Bloom's syndrome protein and stimulates its helicase activity. *J. Biol. Chem.* **275**, 23500–23508.

Brosh, R. M., Jr., Orren, D. K., Nehlin, J. O., Ravn, P. H., Kenny, M. K., Machwe, A., and Bohr, V. A. (1999). Functional and physical interaction between WRN helicase and human Replication protein A. *J. Biol. Chem.* **274**, 18341–18350.

Brosh, R. M., Jr., Waheed, J., and Sommers, J. A. (2002). Biochemical characterization of the DNA substrate specificity of Werner syndrome helicase. *J. Biol. Chem.* **277**, 23236–23245.

Brosh, R. M., Jr., and Bohr, V. A. (2002). Roles of the Werner syndrome protein in pathways required for maintenance of genome stability. *Exp. Gerontol.* **37**, 491–506.

Cheng, W. H., von Kobbe, C., Opresko, P. L., Arthur, L. M., Komatsu, K., Seidman, M. M., Carney, J. P., and Bohr, V. A. (2004). Linkage between Werner syndrome protein and the Mre11 complex via Nbs1. *J. Biol. Chem.* **279**, 21169–21176.

Cheng, W. H., von Kobbe, C., Opresko, P. L., Fields, K. M., Ren, J., Kufe, D., and Bohr, V. A. (2003). Werner syndrome protein phosphorylation by abl tyrosine kinase regulates its activity and distribution. *Mol. Cell. Biol.* **23**, 6385–6395.

Choudhary, S., Sommers, J. A., and Brosh, R. M., Jr. (2004). Biochemical and kinetic characterization of the DNA helicase and exonuclease activities of Werner syndrome protein. *J. Biol. Chem.* **279**, 34603–34613.

Constantinou, A., Tarsounas, M., Karow, J. K., Brosh, R. M., Jr., Bohr, V. A., Hickson, I. D., and West, S. C. (2000). Werner's syndrome protein (WRN) migrates Holliday junctions and co-localizes with RPA upon replication arrest. *EMBO Reports* **1**, 80–84.

Cooper, M. P., Machwe, A., Orren, D. K., Brosh, R. M., Jr., Ramsden, D., and Bohr, V. A. (2000). Ku complex interacts with and stimulates the Werner protein. *Genes Dev.* **14**, 907–912.

Cui, S., Klima, R., Ochem, A., Arosio, D., Falaschi, A., and Vindigni, A. (2003). Characterization of the DNA-unwinding activity of human RECQ1, a helicase specifically stimulated by human replication protein A. *J. Biol. Chem.* **278**, 1424–1432.

Dillingham, M. S., Wigley, D. B., and Webb, M. R. (2000). Demonstration of unidirectional single-stranded DNA translocation by PcrA helicase: Measurement of step size and translocation speed. *Biochemistry* **39**, 205–212.

Doherty, K. M., Sommers, J. A., Gray, M. D., Lee, J. W., von Kobbe, C., Thoma, N. H., Kureekattil, R. P., Kenny, M. K., and Brosh, R. M., Jr. (2005). Physical and functional mapping of the RPA interaction domain of the Werner and Bloom syndrome helicases. *J. Biol. Chem.* **280**, 29494–29505.

Driscoll, H. C., Matson, S. W., Sayer, J. M., Kroth, H., Jerina, D. M., and Brosh, R. M., Jr. (2003). Inhibition of Werner syndrome helicase activity by benzo[c]phenanthrene diol epoxide dA adducts in DNA is both strand- and stereoisomer-dependent. *J. Biol. Chem.* **278**, 41126–41135.

Featherstone, C., and Jackson, S. P. (1999). DNA double-strand break repair. *Curr. Biol.* **9**, R759–R761.

Fry, M., and Loeb, L. A. (1999). Human Werner syndrome DNA helicase unwinds tetrahelical structures of the fragile X syndrome repeat sequence d(CGG)n. *J. Biol. Chem.* **274**, 12797–12802.

Garcia, P. L., Liu, Y., Jiricny, J., West, S. C., and Janscak, P. (2004). Human RECQ5beta, a protein with DNA helicase and strand-annealing activities in a single polypeptide. *EMBO J.* **23**, 2882–2891.

Gray, M. D., Shen, J. C., Kamath-Loeb, A. S., Blank, A., Sopher, B. L., Martin, G. M., Oshima, J., and Loeb, L. A. (1997). The Werner syndrome protein is a DNA helicase. *Nat. Genet.* **17**, 100–103.

Harrigan, J. A., Opresko, P. L., von Kobbe, C., Kedar, P. S., Prasad, R., Wilson, S. H., and Bohr, V. A. (2003). The Werner syndrome protein stimulates DNA polymerase beta strand displacement synthesis via its helicase activity. *J. Biol. Chem.* **278**, 22686–22695.

Huang, S., Beresten, S., Li, B., Oshima, J., Ellis, N. A., and Campisi, J. (2000). Characterization of the human and mouse WRN 3′–>5′ exonuclease. *Nucleic. Acids. Res.* **28**, 2396–2405.

Kaplan, D. L. (2000). The 3′-tail of a forked-duplex sterically determines whether one or two DNA strands pass through the central channel of a replication-fork helicase. *J. Mol. Biol.* **301**, 285–299.

Karmakar, P., Piotrowski, J., Brosh, R. M., Jr., Sommers, J. A., Miller, S. P., Cheng, W. H., Snowden, C. M., Ramsden, D. A., and Bohr, V. A. (2002). Werner protein is a target of DNA-dependent protein kinase *in vivo* and *in vitro*, and its catalytic activities are regulated by phosphorylation. *J. Biol. Chem.* **277**, 18291–18302.

Kawabe, Y., Seki, M., Seki, T., Wang, W. S., Imamura, O., Furuichi, Y., Saitoh, H., and Enomoto, T. (2000). Covalent modification of the Werner's syndrome gene product with the ubiquitin-related protein, SUMO-1. *J. Biol. Chem.* **275**, 20963–20966.

Lee, J. W., Harrigan, J., Opresko, P. L., and Bohr, V. A. (2005). Pathways and functions of the Werner syndrome protein. *Mech. Ageing Dev.* **126**, 79–86.

Li, B., and Comai, L. (2000). Functional interaction between Ku and the Werner syndrome protein in DNA end processing. *J. Biol. Chem.* **275,** 28349–28352.

Li, B., and Comai, L. (2001). Requirements for the nucleolytic processing of dna ends by the Werner syndrome protein-Ku70/80 complex. *J. Biol. Chem.* **276,** 9896–9902.

Machwe, A., Ganunis, R., Bohr, V. A., and Orren, D. K. (2000). Selective blockage of the 3′–>5′ exonuclease activity of WRN protein by certain oxidative modifications and bulky lesions in DNA. *Nucleic. Acids. Res.* **28,** 2762–2770.

Machwe, A., Xiao, L., Groden, J., Matson, S. W., and Orren, D. K. (2005). RecQ family members combine strand pairing and unwinding activities to catalyze strand exchange. *J. Biol. Chem.* **280,** 23397–23407.

Martin, G. M. (1978). Genetic syndromes in man with potential relevance to the pathobiology of aging. *Birth Defects Orig. Artic. Ser.* **14,** 5–39.

Matson, S. W., Bean, D. W., and George, J. W. (1994). DNA helicases: Enzymes with essential roles in all aspects of DNA metabolism. *Bioessays* **16,** 13–22.

Mohaghegh, P., Karow, J. K., Brosh, R. M., Jr., Bohr, V. A., and Hickson, I. D. (2001). The Bloom's and Werner's syndrome proteins are DNA structure-specific helicases. *Nucleic. Acids. Res.* **29,** 2843–2849.

Morris, P. D., Byrd, A. K., Tackett, A. J., Cameron, C. E., Tanega, P., Ott, R., Fanning, E., and Raney, K. D. (2002). Hepatitis C virus NS3 and simian virus 40 T antigen helicases displace streptavidin from 5′-biotinylated oligonucleotides but not from 3′-biotinylated oligonucleotides: Evidence for directional bias in translocation on single-stranded DNA. *Biochemistry* **41,** 2372–2378.

Morris, P. D., and Raney, K. D. (1999). DNA helicases displace streptavidin from biotin-labeled oligonucleotides. *Biochemistry* **38,** 5164–5171.

Opresko, P. L., Cheng, W. H., and Bohr, V. A. (2004a). At the junction of RecQ helicase biochemistry and human disease. *J. Biol. Chem.* **279,** 18099–18102.

Opresko, P. L., Laine, J. P., Brosh, R. M., Jr., Seidman, M. M., and Bohr, V. A. (2001). Coordinate action of the helicase and 3′ to 5′ exonuclease of Werner syndrome protein. *J. Biol. Chem.* **276,** 44677–44687.

Opresko, P. L., Otterlei, M., Graakjaer, J., Bruheim, P., Dawut, L., Kolvraa, S., May, A., Seidman, M. M., and Bohr, V. A. (2004b). The Werner Syndrome helicase and exonuclease cooperate to resolve telomeric D loops in a manner regulated by TRF1 and TRF2. *Mol. Cell* **14,** 763–774.

Orren, D. K., Brosh, R. M., Jr., Nehlin, J. O., Machwe, A., Gray, M. D., and Bohr, V. A. (1999). Enzymatic and DNA binding properties of purified WRN protein: High affinity binding to single-stranded DNA but not to DNA damage induced by 4NQO. *Nucleic. Acids. Res.* **27,** 3557–3566.

Ozgenc, A., and Loeb, L. A. (2005). Current advances in unraveling the function of the Werner syndrome protein. *Mutat. Res.* **577,** 237–251.

Pichierri, P., Rosselli, F., and Franchitto, A. (2003). Werner's syndrome protein is phosphorylated in an ATR/ATM-dependent manner following replication arrest and DNA damage induced during the S phase of the cell cycle. *Oncogene* **22,** 1491–1500.

Saintigny, Y., Makienko, K., Swanson, C., Emond, M. J., and Monnat, R. J., Jr. (2002). Homologous recombination resolution defect in Werner syndrome. *Mol. Cell Biol.* **22,** 6971–6978.

Sharma, S., Otterlei, M., Sommers, J. A., Driscoll, H. C., Dianov, G. L., Kao, H. I., Bambara, R. A., and Brosh, R. M., Jr. (2004). WRN helicase and FEN-1 form a complex upon replication arrest and together process branch-migrating DNA structures associated with the replication fork. *Mol. Biol. Cell* **15,** 734–750.

Sharma, S., Sommers, J. A., Choudhary, S., Faulkner, J. K., Cui, S., Andreoli, L., Muzzolini, L., Vindigni, A., and Brosh, R. M., Jr. (2005). Biochemical analysis of the DNA unwinding and strand annealing activities catalyzed by human RECQ1. *J. Biol. Chem.* **280,** 28072–28084.

Shen, J. C., Gray, M. D., Oshima, J., and Loeb, L. A. (1998). Characterization of Werner syndrome protein DNA helicase activity: Directionality, substrate dependence and stimulation by replication protein A. *Nucleic. Acids. Res.* **26,** 2879–2885.

Shen, J. C., Lao, Y., Kamath-Loeb, A., Wold, M. S., and Loeb, L. A. (2003). The N-terminal domain of the large subunit of human replication protein A binds to Werner syndrome protein and stimulates helicase activity. *Mech. Ageing Dev.* **124,** 921–930.

Shen, J. C., and Loeb, L. A. (2000). Werner syndrome exonuclease catalyzes structure-dependent degradation of DNA. *Nucleic. Acids. Res.* **28,** 3260–3268.

Sommers, J. A., Sharma, S., Doherty, K. M., Karmakar, P., Yang, Q., Kenny, M. K., Harris, C. C., and Brosh, R. M., Jr. (2005). p53 modulates RPA-dependent and RPA-independent WRN helicase activity. *Cancer Res.* **65,** 1223–1233.

Sun, H., Karow, J. K., Hickson, I. D., and Maizels, N. (1998). The Bloom's syndrome helicase unwinds G4 DNA. *J. Biol. Chem.* **273,** 27587–27592.

Suzuki, N., Shimamoto, A., Imamura, O., Kuromitsu, J., Kitao, S., Goto, M., and Furuichi, Y. (1997). DNA helicase activity in Werner's syndrome gene product synthesized in a baculovirus system. *Nucleic. Acids. Res.* **25,** 2973–2978.

Villani, G., and Tanguy, L. G. (2000). Interactions of DNA helicases with damaged DNA: Possible biological consequences. *J. Biol. Chem.* **275,** 33185–33188.

von Kobbe, C., Harrigan, J. A., Schreiber, V., Stiegler, P., Piotrowski, J., Dawut, L., and Bohr, V. A. (2004). Poly(ADP-ribose) polymerase 1 regulates both the exonuclease and helicase activities of the Werner syndrome protein. *Nucleic. Acids. Res.* **32,** 4003–4014.

von Kobbe, C., Karmakar, P., Dawut, L., Opresko, P. L., Zeng, X., Brosh, R. M., Jr., Hickson, I. D., and Bohr, V. A. (2002). Colocalization, physical, and functional interaction between Werner and Bloom syndrome proteins. *J. Biol. Chem.* **277,** 22035–22044.

von Kobbe, C., Thoma, N. H., Czyzewski, B. K., Pavletich, N. P., and Bohr, V. A. (2003). Werner syndrome protein contains three structure specific DNA binding domains. *J. Biol. Chem.* **278,** 52997–53006.

Yang, Q., Zhang, R., Wang, X. W., Spillare, E. A., Linke, S. P., Subramanian, D., Griffith, J. D., Li, J. L., Hickson, I. D., Shen, J. C., Loeb, L. A., Mazur, S. J., Appella, E., Brosh, R. M., Jr., Karmakar, P., Bohr, V. A., and Harris, C. C. (2002). The processing of Holliday junctions by BLM and WRN helicases is regulated by p53. *J. Biol. Chem.* **277,** 31980–31987.

Yannone, S. M., Roy, S., Chan, D. W., Murphy, M. B., Huang, S., Campisi, J., and Chen, D. J. (2001). Werner syndrome protein is regulated and phosphorylated by DNA-dependent protein kinase. *J. Biol. Chem.* **276,** 38242–38248.

Yu, C. E., Oshima, J., Fu, Y. H., Wijsman, E. M., Hisama, F., Alisch, R., Matthews, S., Nakura, J., Miki, T., Ouais, S., Martin, G. M., Mulligan, J., and Schellenberg, G. D. (1996). Positional cloning of the Werner's syndrome gene. *Science* **272,** 258–262.

[5] Analysis of the DNA Unwinding Activity of RecQ Family Helicases

By CSANÁD Z. BACHRATI and IAN D. HICKSON

Abstract

The RecQ family of DNA helicases is highly conserved in evolution from bacteria to mammals. There are five human RecQ family members (RECQ1, BLM, WRN, RECQ4 and RECQ5), defects, three of which give rise to inherited human disorders. Mutations of *BLM* have been identified in patients with Bloom's syndrome, *WRN* has been shown to be mutated in Werner's syndrome, while mutations of *RECQ4* have been associated with at least a subset of cases of both Rothmund-Thomson syndrome and RAPADILINO. The most characteristic features of these diseases are a predisposition to the development of malignancies of different types (particularly in Bloom's syndrome), some aspects of premature aging (particularly in Werner's syndrome), and on the cellular level, genome instability. In order to gain understanding of the molecular defects underlying these diseases, many laboratories have focused their research on a study of the biochemical properties of human RecQ helicases, particularly those associated with disease, and of RecQ proteins from other organisms (e.g., Sgs1p of budding yeast, Rqh1p of fission yeast, and RecQ of *E.coli*). In this chapter, we summarize the assay systems that we employ to analyze the catalytic properties of the BLM helicase. We have successfully used these methods for the study of other RecQ and non-RecQ helicases, indicating that they are likely to be applicable to all helicases.

Introduction

Helicases are motor proteins that couple the hydrolysis of nucleoside triphosphate (usually ATP) to the breakage of the hydrogen bonds between the complementary bases of duplex DNA. This process is often referred to as DNA unwinding or DNA strand separation. DNA helicases function in all aspects of DNA metabolism where they are required to permanently or transiently unwind regions of the genome to permit, for example, initiation of DNA replication or transcription, or the execution of DNA repair. Because of their roles in a diverse range of processes, many

METHODS IN ENZYMOLOGY, VOL. 409
0076-6879/06 $35.00
DOI: 10.1016/S0076-6879(05)09005-1

helicases have become highly specialized to a particular step in a single key process (such as the *E. coli* DnaB helicase that unwinds the DNA ahead of the translocating replication fork), while others play roles in more than one cellular process (such as the XPB and XPD helicases that function in nucleotide excision repair and in transcription as components of TFIIH). Based on comparisons of primary sequence, DNA helicases have been classified into distinct superfamilies (Singleton and Wigley, 2002). Superfamily I and II represent the bulk of the known helicases. Representatives of these superfamilies contain a conserved helicase domain that comprises seven characteristic sequence motifs. In contrast, the superfamily III and IV helicases contain three and five of these signature motifs, respectively. In all cases, two of these motifs represent the so-called Walker A and B boxes that are essential for NTP binding and hydrolysis. Members of the RecQ family of helicases (Bachrati and Hickson, 2003) contain all seven of the signature motifs, as well as a number of other conserved stretches of sequence that distinguish them from other superfamily I and II members.

Preparation of Substrates

RecQ helicases have been shown to be able to be capable of unwinding a wide array of DNA duplex and partial duplex substrates (Bachrati and Hickson, 2003). Although these structures are designed to model many different putative *in vivo* DNA substrates for these helicases and may appear superficially to be very different, in fact they can be created using just a few standard procedures. Following initial preparation of these substrates, it is essential that they be purified to remove nonincorporated precursor molecules or the by-products of synthesis, which might serve as binding competitors.

Substrates Prepared from Annealed Oligonucleotides

The main advantage of oligonucleotide-based substrates is the ease with which they can be generated. Some of these substrates are simple DNA structures; however, many of them have been designed to model a particular *in vivo* DNA structure such as a replication fork or a Holliday junction recombination intermediate (Fig. 1). Nevertheless, one must be aware of the limitations of these model structures, as they may either have features not existing *in vivo* or lack features of the genuine *in vivo* substrate.

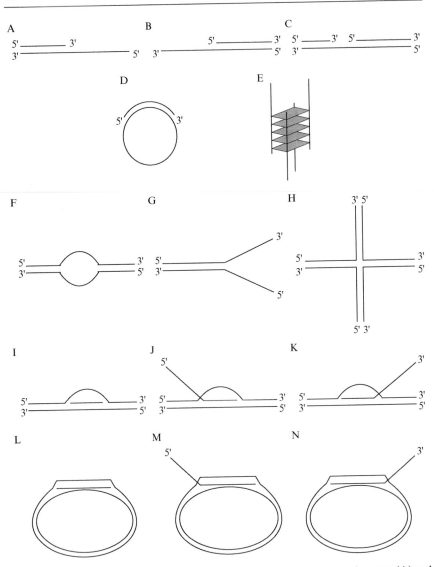

FIG. 1. Schematic representation of the most commonly used helicase substrates. (A) and (B) Partial duplexes with 5′ and 3′ protruding ends, respectively. (C) Partial duplex with a gap in the middle; commonly used to determine the orientation of helicase unwinding. (D) Partial duplex made by annealing a synthetic oligonucleotide to the complementary sequences of a single-stranded circular dna, such as φX174 or m13 bacteriophage. (E) G4 quadruplex DNA. (F) "Bubble." (G) Replication fork. (H) Synthetic Holliday junction. (I–K) Oligonucleotide-based, static D-loops. (L–N) Plasmid-based, mobile D-loops.

Detection of reaction products in helicase assays is generally based on the incorporation of radiolabel into one or more of the oligonucleotides that comprise the substrate. Generally speaking, only one of the oligonucleotides is labeled and its release from the intact substrate is monitored. The oligonucleotides must be of the highest quality and homogeneous in terms of length in order to obtain interpretable data. This is best achieved by PAGE purification, a detailed method which can be found in several standard laboratory manuals. Current Protocols in Molecular Biology (2005) is our preferred text.

Procedure

1. To generate 5' ^{32}P-labeled oligonucleotides, incubate 10 pmol of oligonucleotide with T4 polynucleotide kinase and [γ-^{32}P]ATP using the forward reaction protocol. If it is necessary to label the 3' end of the oligonucleotide, use terminal deoxynucleotidyl transferase and [α-^{32}P]-ddATP (N.B. an extra nucleotide will be added). Separate the unincorporated nucleotides from the oligonucleotide on a Sephadex G50-based microcentrifuge spin column (such as the mini Quick Spin Columns of Roche Diagnostics, Indianapolis, IN), keeping the elution volume low (<30 μl). Measure the radioactivity of the labeled oligonucleotide and determine its specific activity assuming an ~95% DNA recovery from the column.

2. In a screw-capped microcentrifuge tube on ice containing the labeled, purified oligonucleotide solution, add the remaining reaction components in the following order: 5 μl 100 mM Tris-HCl, pH 7.5; 5 μl 100 mM MgCl$_2$ and 30 pmoles each of the other oligonucleotides. Bring the volume up to 50 μl with ddH$_2$O.

3. It is not necessary to have any specialist equipment to set up the annealing of oligonucleotides; this can be done with "homemade" equipment. Heat up an almost full 2-liter glass beaker of water to 95°, and carefully place the beaker into a polystyrene box that can easily accommodate it. Place the tube with the annealing mixture into a polystyrene tube rack, and float this on the surface of the hot water. Close the polystyrene box with its lid; it will provide adequate insulation to allow the water to cool down slowly over several hours.

Sometimes it is necessary to anneal more than two oligonucleotides in a particular order. In this case, the incubation and the controlled cooling of the annealing mixture are most conveniently carried out in an ordinary PCR machine (thermal cycler) (Opresko *et al.*, 2004).

1. Program your PCR machine as follows: hold at 95° for 5 min; cool down to 60° at 1.2°/min; hold at 60° for 60 min; and cool down to 20° again at 1.2°/min. Some PCR machines cannot be programmed to such slow automatic cooling rate; stepwise cooling can work equally well.

2. In a 200-μl thin-walled PCR tube set up the mixture as stated previously, but containing only the first two of the multiple oligonucleotides to be annealed. Put into the PCR machine and start the program described in step 1. above. When the temperature reaches 60°, add the third oligonucleotide and continue the program.

At the completion of the annealing process continue to substrate purification.

Generation of Displacement Loop (D-loop) Substrates with RecA-Mediated Strand Invasion

In order to gain a more accurate picture of the function of helicases, one should consider using substrates *in vitro* that model the *in vivo* DNA structures as closely as possible. One such example is the use of plasmid-based displacement loop (D-loop) structures generated by RecA-mediated strand transfer, instead of oligonucleotide-based D-loops that have the same sequence. These D-loop substrates represent the initial stage of homologous recombination reactions. The plasmid-based D-loop substrates can be built according to the method of McIlwraith *et al.* (McIlwraith *et al.*, 2001). The invading radiolabeled oligonucleotide is coated with RecA molecules in the first step and then a supercoiled plasmid is added. RecA facilitates a homology search between the plasmid and the oligonucleotide and catalyzes the invasion of the homologous single-stranded sequences into the plasmid while displacing the complementary strand of the plasmid. The invasion step is reversible and if the reaction is allowed to proceed too long, RecA catalyzes the removal of the invading oligonucleotide from the plasmid. The reaction must, therefore, be stopped at the point where the maximal level of D-loop has been generated. This point is best determined experimentally.

Materials

10× RecA buffer
500 mM triethanolamine-HCl, pH 7.5
150 mM MgCl$_2$
10 mM DTT
1 mg/ml BSA.

Procedure

1. In a screw-capped microcentrifuge tube, mix the following:

    10.68 μl 10× RecA buffer
    12.0 μl 200 mM phosphocreatine (in 1× RecA buffer)
    1.2 μl 200 U/ml creatine phosphokinase (in 1× RecA buffer)
    2.4 μl 100 mM ATP
    360.0 pmol (nucleotides) radiolabeled oligonucleotide.

2. Bring up the volume with ddH$_2$O such that, when the further components of the reaction mix are added (see later), the final volume will be 120 μl.

3. Add 480 pmol RecA (using the 2 mg/ml RecA product supplied by New England BioLabs, Beverly, MA; this will be 9.08 μl), mix well and incubate at 37° for 5 min.

4. Add supercoiled plasmid DNA (ideally purified with two consecutive runs of ethidium-bromide saturated CsCl equilibrium gradient ultracentrifugation) to the final concentration of 300 μM (nucleotides), mix well and incubate at 37° for the optimal length of time determined previously.

5. Stop the reaction by the addition of 24 μl of 100 mM Tris-HCl, pH 7.5, 100 mM MgCl$_2$, 3% SDS, and 10 mg/ml proteinase K. Mix well and incubate at 37° for a further 30 min to allow deproteinization.

Proceed to substrate purification.

Generation of G4 Quadruplex Substrates

G-DNA structures have been shown to form readily *in vitro* in certain oligonucleotides that contain runs of guanines, and are stabilized by Hoogsteen hydrogen bonding (Sen and Gilbert, 1988). Depending on strand stoichiometry, strand orientation, and the presence of monovalent cations, G-DNAs can exist in different forms. In the presence of Na$^+$ ions the preferred configuration is G4 DNA, formed from four parallel strands, while in the presence of K$^+$ ions the planar structure is formed between two fold-back strands (G2'DNA) (Sen and Gilbert, 1990). The *in vivo* presence of G-DNAs has not yet been proven conclusively, although transient formation of G-DNAs has long been suggested at the rDNA locus, telomeres, in the immunoglobulin heavy chain switch regions, and at the c-Myc promoter (Arthanari and Bolton, 2001). Members of the RecQ family of helicases have been shown to be capable of unwinding not only B-form DNA, but also G4 quadruplex DNA (Huber *et al.*, 2002; Li *et al.*, 2001; Sun *et al.*, 1998).

G4 quadruplex DNA substrates are formed from unlabeled oligonucleotides and purified in that form. The purified substrate is stable in storage buffer for several months and aliquots of it can be labeled as required.

Procedure

1. Set up 100 μl of G4 oligonucleotide solution with oligonucleotide at 2 μg/μl in 1 *M* NaCl.
2. Heat denature at 100° for 5 min and then incubate at 60° for 48 h.
3. Gel purify using the procedure for nonradioactive substrates (see later).
4. Precipitate the substrate by the addition of 1/10 volume of 3 *M* Na-acetate, pH 5.2 and 1 volume of isopropanol at −20°. Collect the precipitate by centrifugation in an Eppendorf microcentrifuge (or equivalent) for 5 min. Discard the supernatant.
5. Dissolve the pellet in 100 μl of 100 m*M* NaCl, 10 m*M* Tris-HCl, pH 8, 1 m*M* EDTA.
6. Store at −20°.

Notes: Formation of G4 DNA is facilitated by high oligonucleotide and Na$^+$ concentrations. Even under these circumstances, there is an equilibrium between the G4 quadruplex form and the single-stranded oligonucleotide form of the substrate DNA. Though the G4 quadruplex is relatively stable, it is vital to carry out the electroelution step at 4° in order to keep the ratio between the G4 and single-stranded forms as high as possible.

Purification of Radioactive Substrates

In order to remove impurities that might adversely affect the helicase reaction, the substrate DNA must be gel purified. We have found that a good separation of full-length substrate and impurities can be achieved on 5% polyacrylamide gels. Moreover, because of the lower percentage of the gel, recovery of the DNA is more efficient.

Procedure

1. Add DNA gel loading dye to the product of the substrate generation reaction above and load into the wells of a 5% polyacrylamide TBE (Tris-Borate-EDTA) gel.

2. Run in 1× TBE buffer (90 m*M* Tris base, 90 m*M* boric acid, 2 m*M* EDTA, pH 8.0) at 4°. The running parameters depend on the size of gel used and the size of molecules to be separated; we usually run 12-cm long gels at 25 mA constant current for 60 min.

3. Remove one of the support plates and cover the gel in Saran Wrap.

4. Expose the gel to X-ray film for around 5 min in a dark room. Put markers on the X-ray film to allow alignment of the developed film and the gel. Develop the film.

5. Cut out the area of the film that corresponds to the radioactive substrate (there may be multiple bands present, the structurally intact substrate is usually the one with the slowest mobility), align the film with the gel, and use the film as a template to locate the area of the gel that contains the substrate. Cut the gel along the template "window" with a scalpel.

6. Lift the cut gel slice out of the gel and transfer it into a dialysis tube with 1× TBE for electroelution. We find that the electroelution cassettes of the Qbiogene (Irvine, CA) ProteoPlus kit are very handy for this purpose; the dialysis step can then be performed in the same chamber without the need to transfer the radioactive liquid into another dialysis bag.

7. Carry out the electroelution by running at constant voltage (usually 120 V for 2 h) in 1× TBE at 4°. At the end of electroelution, reverse the polarity of the current for 30 s. This step dissociates the DNA molecules from the inner wall of the dialysis tubing.

8. Rinse the inner wall of the dialysis tube several times with the liquid inside the chamber and transfer the liquid into a Slide-A-Lyzer dialysis cassette (Pierce Biotechnology, Rockford, IL) for dialysis.

9. Dialyze against 1000 ml of 10 mM MgCl$_2$ and 10 mM Tris-HCl, pH 7.5, for at least 2 h at 4°, changing the dialysis buffer every 20 min.

10. Transfer the dialyzed substrate into screw-cap microcentrifuge tubes and freeze in 100-μl aliquots.

11. Measure the radioactivity of the purified substrate and, using the specific activity of the labeled oligonucleotide, calculate the chemical concentration of the substrate.

Notes: Although plasmid-based D-loop molecules are large and best purified from agarose gels, we usually carry out the purification on 5% polyacrylamide gels, as described previously, since detection of the radioactive DNA molecules and handling of the thinner acrylamide gel is more straightforward. Also, we found that DNA recovery methods from agarose gels often give rise to impurities that inhibit the helicase reaction, or cause the D-loop molecules to dissociate during the purification process. It should also be noted that the supercoiled plasmid, which is added at a large molar excess to the reaction, copurifies with the radioactive D-loop. As RecQ helicases do not unwind supercoiled double-stranded plasmid DNA molecules, in most cases this is not a problem.

Purification of Nonradioactive Substrates

As the G4 quadruplex substrate is generated from nonlabeled oligonucleotides, detection of the DNA for purification purposes is achieved through UV shadowing.

Procedure

1. Run the prepared substrate on a 5% polyacrylamide gel, as described previously.
2. Cover a thin layer chromatography (TLC) plate, which contains a UV fluorescent dye for detection (such as the POLYGRAM CEL 300 PEI/ UV$_{254}$ of Macherey-Nagel, Germany) in Saran Wrap.
3. Remove one of the supporting plates from the gel and lay the gel onto the TLC plate. Remove the top supporting plate and cover the whole assembly in Saran Wrap. (N.B. the gel might have flipped during this, reversing the loading order.)
4. In the dark room, use a handheld UV lamp to expose the gel/TLC plate assembly briefly to short wavelength UV light (254 nm). As DNA molecules absorb UV light, the band in the gel where the DNA is present should make a shadow on the fluorescing TLC plate. Mark the position of the band with a marker pen. UV exposure must be for as brief a period as possible, in order to avoid damaging the DNA. If there are multiple bands on the same gel to be isolated, it is a good practice to cover the lanes that are not being processed immediately with a glass plate for protection.
5. Cut out the marked gel slice with a scalpel, and continue with electroelution and dialysis as described previously.

Analysis of Helicase Activity

The Assay System

Buffer Composition: RecQ helicases display two enzymatic activities: Mg^{2+} and DNA-dependent ATPase activity, and ATP-dependent helicase activity. Although the ATPase activity requires DNA as a cofactor, unwinding of the DNA is not a prerequisite for ATP hydrolysis. Hence, ATP hydrolysis can be uncoupled from DNA unwinding. The reaction buffer must, therefore, contain ATP and Mg^{2+}, as well as the DNA substrate. The optimal buffer composition must always be determined experimentally; the following guidelines might serve as a starting point:

1. The optimal pH for the reaction is generally around 7.5 to 8.0, which can be accomplished by the use of Tris-based buffers. Helicases that prefer neutral conditions are best analyzed in HEPES-based buffers.

2. Though the presence of monovalent cations is not an absolute requirement, we found that the optimal buffer contains 50–100 mM Na$^+$. K$^+$ ions would be more physiological; however, as the presence of K$^+$ ions influences the conformation of G-DNAs, they should be avoided if analysis of G4 quadruplex is carried out. Most protocols utilize Cl$^-$ anions, but several alternatives are available. We favor acetate-based buffers for the RecQ helicases.

3. The ATP molecules are complexed with Mg^{2+} ions in the reaction buffer. The optimal ATP/Mg^{2+} ratio is generally around 1 (Harmon and Kowalczykowski, 2001). Most helicases will utilize 1 mM ATP efficiently, but this should be tested experimentally.

4. The configuration of Holliday junction substrates is Mg^{2+} dependent. Unwinding of these substrates might be inhibited at high Mg^{2+} concentrations.

5. DTT should be included in the buffer to maintain the reduced state of the enzyme molecules, therefore preventing aggregation and precipitation.

6. The presence of BSA (100 μg/ml) in the reaction buffer is also protective.

In unwinding assays using a number of DNA helicases (BLM, WRN, UvrD, dmRecQ5β) we found optimal rates of unwinding using the following reaction buffer (1×):

66 mM Na-acetate
33 mM Tris-acetate, pH 7.8
1 mM ATP
1 mM Mg-acetate
100 μg/ml BSA
1 mM DTT.

A 10× concentrated buffer (excluding ATP and Mg-acetate) can be prepared in advance.

Substrate Enzyme Ratio, Enzyme Concentration: Many laboratories studying the biochemistry of helicases find that the concentration of enzyme needed for detectable unwinding exceeds the substrate concentration; in some cases by many fold. This is a phenomenon that might be explained in several ways. It is possible that due to suboptimal reaction circumstances only a fraction of the protein is active enzymatically. The enzyme might also have lost full activity during purification. Several helicases have been shown to be oligomeric in solution (Patel and Picha, 2000, and see references therein); the oligomeric state of the enzyme might decrease the effective enzyme concentration and specific activity. BLM (Cheok *et al.*, 2005; Machwe *et al.*, 2005), WRN (Machwe *et al.*, 2005),

RECQL1 (Sharma *et al.*, 2005) and RECQ5β (Garcia *et al.*, 2004) have been shown to possess DNA strand annealing activity, which works against the unwinding activity. It is also possible that several helicase molecules, or possibly helicase oligomers, simultaneously engage on one DNA strand, like beads on a thread.

No direct experimental data support any single explanation of the above phenomenon. Most likely, several features of helicase action all contribute to a reaction kinetic that must be considered as a single turnover reaction.

Helicase Reaction

For detailed enzymatic analysis of helicases the reaction circumstances (i.e., optimal substrate and enzyme concentrations) should be determined experimentally. Following the substrate preparation and purification guidelines detailed previously, the concentration of the purified substrate will generally be in the low nM range (0.5–5 nM). The radioactive detection method enables one to use as low as a 0.5 fmol of the substrate. Using a much higher amount of substrate might be disadvantageous as the high substrate concentration can inhibit unwinding.

The optimal enzyme concentration should be determined by titrating the enzyme preparation by serial dilutions. The dilution must be carried out in the same dilution buffer that the enzyme was dissolved in originally, to avoid changing any other component (such as monovalent cation concentration) that might influence unwinding activity. An example for such a titration experiment can be seen in Fig. 2.

Procedure 1

1. Keeping everything on ice, prepare a serial dilution of the enzyme such that 1 μl will contain the desired amount.
2. In a screw-capped microcentrifuge tube on ice, make up the required amount of master reaction mix as below:

   > 1 μl 10× helicase buffer (see composition above)
   > 0.1 μl 100 mM ATP
   > 1-x μl 10 mM Mg-acetate (the substrate already contains 10 mM Mg^{2+})
   > x μl substrate (0.5–1 fmol)
   > y μl ddH$_2$O to make the final volume 9 μl.

3. Mix well and dispense into screw-capped microcentrifuge tubes on ice, 9 μl each.
4. Place in a 37° water bath or other incubation device.

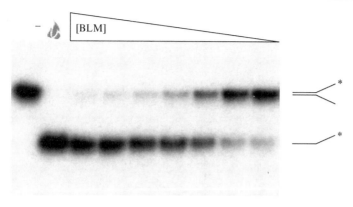

FIG. 2. An example of the concentration-dependence of unwinding. 25, 18.75, 15, 12.5, 10.7, 9.4, 8.3, and 0 fmol BLM was used to unwind approximately 2 fmol replication fork substrate in a 10-μl reaction volume for 45 min. '–' indicates untreated substrate; the flame symbol depicts heat-denatured substrate.

5. Start the helicase reaction by the addition of 1 μl diluted enzyme.
6. At the end of the incubation time, add 1.2 μl 10× loading dye (50% glycerol; 100 mM Tris-HCl, pH 7.5; 50 mM EDTA; 4% SDS; 0.1% bromo-phenol-blue; 0.1% xylene-cyanol), mix, and put the tubes back on ice until electrophoresis.

Procedure 2

A more quantitative picture of helicase activity can be obtained by following the progress of the unwinding reaction in a timecourse.

1. Prepare 9 reaction termination tubes on ice that each contain 1.2 μl 10× loading dye (see above).
2. In a screw-capped microcentrifuge tube set up reaction mixture for 10 samples on ice:

    10 μl 10× helicase buffer (see composition above)
    1 μl 100 mM ATP
    10-x μl 10 mM Mg-acetate
    x μl substrate (0.5–1 fmol)
    y μl ddH$_2$O to make the final volume 100 μl minus the amount of enzyme.

3. Mix and put into a 37° hotblock.
4. Add 10× the amount of enzyme that corresponds to the optimal dilution determined above in Procedure 1. Mix briefly.
5. Take a 10-μl sample; add it to the first pre-prepared reaction termination tube.

6. At 30 s, 1, 2, 3, 5, 7, 10, and 15 min take a further 10-μl sample as above.
7. Keep the collected samples on ice until electrophoresis.

It is possible to run simultaneously more than one timecourse since the addition of the enzyme, mixing, and putting the sample back on the hotblock takes approximately 10 s. In this case, the very first sample (timepoint 0) should be taken before the addition of the enzyme, and each sample manipulation should be staggered by 10 s.

The amount of reaction product often reaches a plateau in 5–7 min, and therefore taking samples up to a maximum of 15 min is usually sufficient. Because the progress curve changes more rapidly in the beginning of the reaction, as many samples as possible should be taken during the first 5 min; after that the initial period sampling can be less frequent. In some cases, it might be necessary to extend the reaction duration to longer than 15 min. In this case, taking more samples is more favorable than rearranging the schedule outlined previously.

An example of a timecourse experiment carried out as described previously is shown in Fig. 3.

Electrophoresis and Detection: The reaction product (i.e., the released single-stranded oligonucleotide) is separated from the intact substrate by polyacrylamide gel electrophoresis. For the analysis of oligonucleotide-based, "traditional" substrates (Holliday junction, replication fork, static D-loops, etc.) 10% polyacrylamide; for G4 substrates 15% polyacrylamide; for plasmid-based D-loops 4–20% gradient polyacrylamide TBE gels are recommended. The electrophoresis tank should be chilled during gel running to avoid heating the sample, which could cause "spontaneous" denaturation of the substrate. Using the gels recommended above, running the gel for 60 min at a constant current of 25 mA usually provides sufficient separation of substrate and product.

After electrophoresis, the gel should be transferred to a Whatman 3MM paper and dried. Detection and quantification is best carried out using phosphorimaging systems, such as a Molecular Dynamics Storm device, and ImageQuant software (Amersham Biosciences, UK).

Safety Issues

It is important to stress that the handling of radioisotopes is potentially hazardous, and precautions should be taken to avoid contamination of work areas and personnel. The quantities of ^{32}P used in the previous procedures is quite low, and therefore the operator can protect himself or herself quite easily by carrying out the manipulations behind a 1-cm thick

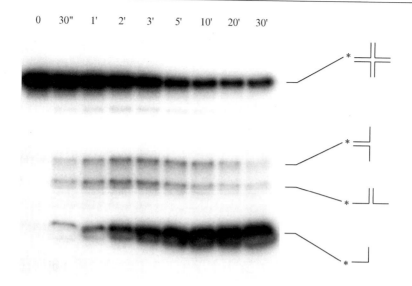

FIG. 3. An example of a timecourse experiment. The reaction contained 3.27 n*M* BLM and 56 p*M* Holliday junction substrate, and the reaction was monitored over a 30-min period. The side panel indicates the schematic structure of the products of unwinding.

Perspex screen of the type available from several commercial sources (e.g., Nalgene, Rochester, NY). It should also be noted that materials and equipment such as gel tanks inevitably become contaminated with low levels of ^{32}P. Safety rules vary from country to country and therefore it is important to consult with local radiation safety personnel to receive training and guidance in the safe handling of radioisotopes.

References

Arthanari, H., and Bolton, P. H. (2001). Functional and dysfunctional roles of quadruplex DNA in cells. *Chem. Biol.* **8,** 221–230.

Bachrati, C. Z., and Hickson, I. D. (2003). RecQ helicases: Suppressors of tumorigenesis and premature aging. *Biochem. J.* **374,** 577–606.

Cheok, C. F., Wu, L., Garcia, P. L., Janscak, P., and Hickson, I. D. (2005). The Bloom's syndrome helicase promotes the annealing of complementary single-stranded DNA. *Nucleic Acids Res.* **33,** 3932–3941.

Garcia, P. L., Liu, Y., Jiricny, J., West, S. C., and Janscak, P. (2004). Human RECQ5b, a protein with DNA helicase and strand-annealing activities in a single polypeptide. *EMBO J.* **23,** 2882–2891.

Harmon, F. G., and Kowalczykowski, S. C. (2001). Biochemical characterization of the DNA helicase activity of the Escherichia coli RecQ helicase. *J. Biol. Chem.* **276,** 232–243.

Huber, M. D., Lee, D. C., and Maizels, N. (2002). G4 DNA unwinding by BLM and
   Sgs1p: Substrate specificity and substrate-specific inhibition. *Nucleic Acids Res.* **30,**
   3954–3961.
Li, J. L., Harrison, R. J., Reszka, A. P., Brosh, R. M., Jr., Bohr, V. A., Neidle, S., and Hickson,
   I. D. (2001). Inhibition of the Bloom's and Werner's syndrome helicases by G-quadruplex
   interacting ligands. *Biochemistry (Mosc).* **40,** 15194–15202.
Machwe, A., Xiao, L., Groden, J., Matson, S. W., and Orren, D. K. (2005). RECQ family
   members combine strand pairing and unwinding activities to catalyze strand exchange.
   J. Biol. Chem. **280,** 23397–23407.
McIlwraith, M. J., Hall, D. R., Stasiak, A. Z., Stasiak, A., Wigley, D. B., and West, S. C.
   (2001). RadA protein from Archaeoglobus fulgidus forms rings, nucleoprotein filaments
   and catalyses homologous recombination. *Nucleic Acids Res.* **29,** 4509–4517.
Opresko, P. L., Otterlei, M., Graakjaer, J., Bruheim, P., Dawut, L., Kolvraa, S., May, A.,
   Seidman, M. M., and Bohr, V. A. (2004). The Werner syndrome helicase and exonuclease
   cooperate to resolve telomeric D loops in a manner regulated by TRF1 and TRF2. *Mol.
   Cell* **14,** 763–774.
Patel, S. S., and Picha, K. M. (2000). Structure and function of hexameric helicases. *Annu.
   Rev. Biochem.* **69,** 651–697.
Sen, D., and Gilbert, W. (1988). Formation of parallel four-stranded complexes by guanine-
   rich motifs in DNA and its implications for meiosis. *Nature* **334,** 364–366.
Sen, D., and Gilbert, W. (1990). A sodium-potassium switch in the formation of four-stranded
   G4-DNA. *Nature* **344,** 410–414.
Sharma, S., Sommers, J. A., Choudhary, S., Faulkner, J. K., Cui, S., Andreoli, L.,
   Muzzolini, L., Vindigni, A., and Brosh, R. M., Jr. (2005). Biochemical analysis of the
   DNA unwinding and strand annealing activities catalyzed by human RECQ1. *J. Biol.
   Chem.* **280,** 28072–28084.
Singleton, M. R., and Wigley, D. B. (2002). Modularity and specialization in superfamily 1 and
   2 helicases. *J. Bacteriol.* **184,** 1819–1826.
Sun, H., Karow, J. K., Hickson, I. D., and Maizels, N. (1998). The Bloom's syndrome helicase
   unwinds G4 DNA. *J. Biol. Chem.* **273,** 27587–27592.

Further Reading

Ausubel, F. M., Brent, R., Kingston, R. E., Moore, D. D., Seidman, J. G., Smith, J. A., and
   Struhl, K. (2005). Current Protocols in Molecular Biology. John Wiley & Sons. Current
   Protocols. Chanda, V. B. (Ed.).

[6] Characterization of Checkpoint Responses to DNA Damage in *Saccharomyces cerevisiae*: Basic Protocols

By RITU PABLA, VAIBHAV PAWAR, HONG ZHANG, and WOLFRAM SIEDE

Abstract

In spite of certain special features of its cell cycle, the yeast *Saccharomyces cerevisiae* has proved to be an excellent and widely used model to study eukaryotic checkpoint responses to DNA damage. This chapter primarily summarizes selected cytological methods that are useful for initial characterization of cell cycle responses. These can be useful in order to study mutants, conditions, or selected DNA damaging agents and experimental examples are given. We have also included protocols for flow-cytometric cell cycle analysis and for determination of Rad53 phosphorylation, a commonly used indicator of checkpoint activation.

General Introduction

The budding yeast *S. cerevisiae* represents the eukaryotic model system in which the checkpoint concepts were initially developed and many arguments can be made in favor of the continued use of *S. cerevisiae* as a model organism for checkpoint studies (Nyberg *et al.*, 2002; Weinert and Hartwell, 1988). If one studies cell cycle arrest following DNA-damaging treatments, cycle stages have to be distinguished and peculiar features of the budding yeast cell cycle need to be considered. For instance, there is no easily discernible G2 stage. The intranuclear mitotic spindle is assembled already during S-phase. Also, the degree of chromosome compaction during mitosis is insufficient to allow visualization of individual chromosomes precluding the definitive assignment of mitotic stages by conventional light microscopy. On the other hand, there are a number of morphological cell cycle landmarks available that are easily traceable with very little experience and experimental manipulation. Even such a complex methodology as fluorescence activated cell sorting (FACS) can be readily adapted, with additional calibration for the small cell size and the relatively low DNA content of yeast. In the following, we have compiled protocols for selected common methods of yeast cell cycle analysis. In general, we have assumed time course experiments that monitor cell cycle progression after initial checkpoint activation (e.g., by irradiation of synchronized cells). Individually, some of these protocols will only suffice for preliminary studies, and

METHODS IN ENZYMOLOGY, VOL. 409
0076-6879/06 $35.00
DOI: 10.1016/S0076-6879(05)09006-3

a combination of methods is often necessary to detect certain cell cycle transitions unequivocally. General yeast protocols can be found elsewhere (Amberg *et al.*, 2005; Guthrie and Fink, 1991).

Budding Analysis

Introduction

In *S. cerevisiae*, bud emergence is a landmark of early S-phase and transition through START that can easily be observed with a well-adjusted phase contrast microscope. Although uncoupling of bud emergence and S-phase initiation is known to occur in certain mutants, determination of the frequency of *small*-budded cells over time can serve as a fairly reliable initial measurement of S-phase entry of synchronized cells and of its possible delay. In response to DNA damage, such budding delay (e.g., as detected following UV radiation treatment) frequently does not last for more than one hour and is of unknown significance; its extent is also significantly influenced by the genetic background. In contrast to small-budded cells, the exact cell cycle stage of a *large*-budded yeast cell cannot be determined by light microscopy alone. Such cells could be in late S, G2/M, in various stages of M, or even in G1, if cells have yet to separate following mitosis. The last type of cells is still held together by the cell walls but can usually be separated by mild sonication. Many agents that introduce spindle damage (such as nocodazole) or DNA double-strand breaks (such as γ-irradiation, streptonigrin, or bleomycin and its derivatives) cause a prominent mitotic arrest response that can last for many hours. Accumulation of enlarged budding cells of typical dumbbell shape can easily be observed even if an initially asynchronous logarithmic-phase culture was used. Thus, determination of the fraction of large-budded cells can quickly provide useful preliminary information, for instance, if an unknown agent triggers checkpoint arrest or if a strain behaves abnormally towards a certain agent.

In the following protocol, UV-induced budding delay is determined as a measurement of G1/S arrest. Similar protocols can easily be established for a variety of different synchronization and treatment regimens.

Protocol

An early-logarithmic phase culture of a *MAT*a haploid yeast strain ($2 \times 10^6 - 1 \times 10^7$ cells/ml) is grown at 30° in YPD growth medium (2% peptone, 1% yeast extract, 2% dextrose; dextrose should be autoclaved separately as a 20% stock solution). Low-density cultures can be concentrated by centrifugation (1500*g*, 3 min) and resuspended in fresh YPD at not more than 2×10^7 cells/ml. This culture is synchronized in G1 by exposure to mating factor α (Sigma Chemicals, St. Louis, MO, or US Biological, Swampscott, MA). A stock solution is prepared by dissolving α-factor at 1 mg/ml in deionized water and freezing in aliquots at $-20°$. Typically, α-factor is added to the culture up to a final concentration of 10 μg/ml in two aliquots, separated by 75 min of incubation with shaking. The total time required for synchronization is strain dependent and amounts to 120 to 180 min. Synchronization with α-factor results in characteristically shaped, elongated cells ("shmoo" phenotype). We recommend to carefully monitor the duration of α-factor treatment since the bud will typically emerge at the elongated cell tip (see Fig. 1B) and may not be easy to identify if excessive "shmooing" occurs. For the same reason, we do not recommend the use of α-factor hypersensitive *bar1* mutant strains where this effect is difficult to control. If synchronization conditions have been firmly established, a starting population containing virtually no background of small-budded cells can be generated. Cells are harvested with a low-speed centrifuge (1500g, 3 min), washed in the same volume sterile deionized water, resuspended in 2–5 ml deionized water, and sonicated (once for 5–10 s at low setting). For this and other protocols listed later, this sonication step is critical. In a typical case, we found that sonication of a 5-ml cell sample at a low setting for 10 s will separate the vast majority of attached G1 cells (e.g., by use of a Fisher Scientific [Pittsburgh, Sonic Dismembrator 60, equipped with a microtip and set to "5"). Even a sonication period of 6×10 s with intermittent cooling on ice will not cause any lethality. The culture is expanded with deionized water and a titer of 2.5×10^7 cells/ml is adjusted. The cell suspension is transferred to a disposable plastic petri dish and a stir bar is added. A volume of 4–6 ml can be treated in a petri dish of 6-cm diameter, whereas treatment of 15-ml samples requires a petri dish of 10-cm diameter. The cell suspension is irradiated under constant stirring with a calibrated germicidal UV-C radiation source (254 nm). An aliquot of the cell suspension is kept untreated and serves as unirradiated control. Other protocols for UV treatment of yeast can be found in the literature (e.g., irradiation of cells on filters, on plates). It should be noted that irradiation of cells in suspension as described here requires a dose about 3-fold higher for the same biological effect than direct irradiation (e.g., on the surface of a YPD plate). Irradiated cells are collected by aspiration with an automatic pipetor, spun down as stated previously, and resuspended in fresh YPD. A culture density is chosen that is convenient for microscopic analysis, but no more than 1×10^7 cells/ml. The kinetics of cell cycle reentry appears to be dependent on culture density and a series of experiments should always be performed at the same density. Cells are incubated with shaking at 30°.

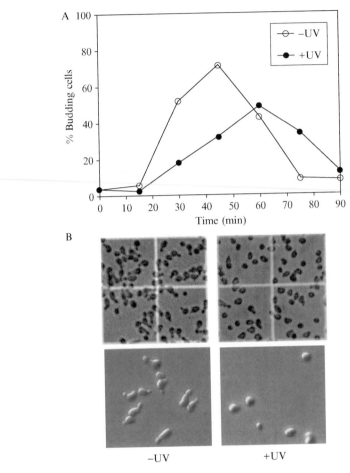

FIG. 1. Determination of budding delay. (A) Budding delay following UV radiation treatment (80 J/m^2). G1 cells of strain BY4741 were irradiated in suspension following release from α-factor arrest (closed circles). The fraction of small-budded cells was determined as a function of time and compared to the unirradiated control (open circles). (B) Left panels: A population of unirradiated yeast cells having entered S-phase with high synchronicity about 40 min after release from α-factor arrest. At this point, the majority of the cells have formed a small to medium-size bud. Right panels: At the same time, the UV treated portion of the culture shows only a low frequency of budding cells. Note the elongated cell shape due to α-factor treatment.

The fraction of small-budded cells among all cells is determined microscopically in a hemacytometer at a magnification level of at least 600× and plotted as a function of time after release from α-factor arrest (Fig. 1). At least 200 cells should be classified per time point with the aid of a mechanical multichannel cell counter. It should be noted that the fraction of budded cells diminishes following a maximum fraction of up to 80%, depending on the degree of synchronization (Fig. 1A). The shape of this portion of the curve will depend on the preset size required to classify a large bud as a separate cell. Typically, a bud that is larger than 1/3 of the mother cell is counted as a separate cell. Thus, there is an element of subjectivity in establishing budding curves. However, the initial burst of small buds can be determined quite accurately and analysis by different individuals will yield very similar numbers. The given example (Fig. 1A) shows a typical result for budding delay following UV treatment. Figure 1B illustrates the synchronous "burst" of small buds in an untreated culture typically found 20–40 min after release from α-factor arrest which is delayed in the UV-treated culture.

Microcolony Assay

Introduction

If an arrest phenomenon lasts for a significant amount of time in a majority of the treated cells, it is highly desirable to "trap" all cycling cells in a stage downstream of the arrest point and thus avoid confusion by preventing cell cycle reentry. This can be achieved by the use of chemical inhibitors or thermoconditional cell cycle mutations (see e.g., Gardner *et al.*, 1999). More generally applicable (but less specific) is the use of a microcolony assay. Basically, treated and untreated cells are placed on solid media and the number of cell "bodies" (separated cells or large buds) per colony is analyzed over time. Thus, cells that have progressed through the cell cycle can be easily distinguished from those that stay arrested. Small buds are difficult to identify in cells on plate and also, a discrimination of separated cells from large buds is impossible. Consequently, other methods need to be applied to define the arrest stage more precisely. The technique is only recommended for agents whose effects on cell cycle progression have been well established.

The following example illustrates the general technique by using the S-phase inhibitor hydroxyurea or the M-phase inhibitor nocodazole as a synchronizing agent and γ-irradiation or UV irradiation as the checkpoint trigger.

Protocol

A small volume of an early-logarithmic phase culture of haploid or diploid yeast cells is synchronized in S-phase by incubation in YPD with hydroxyurea (HU) (200 mM for about 150 min). Because of the required high concentration, preparation of a stock solution is not feasible. Add the powder to YPD at the working concentration and filter-sterilize. There is considerable overlap between checkpoint responses to HU and DNA damage; obviously, checkpoint mutants defective in HU-induced S-phase arrest cannot be synchronized with this method. A widely used inhibitor to synchronize cells in M-phase is the antimicrotubule drug nocodazole. The working concentration is 10 μg/ml by use of a 40 mg/ml stock in DMSO. With either inhibitor, the majority of cells will accumulate in a large-budded stage. A portion of the culture can then be treated with γ-irradiation (e.g., 70 Gy in a ^{137}Cs irradiator for a haploid strain). If available, a ^{60}Co-irradiator is highly recommended because of the higher dose rate and shorter irradiation period. Because of higher resistance, this type of irradiator is essential for treating diploid cells or nocodazole-synchronized haploid cells. Aliquots of cells of the irradiated and unirradiated culture are spun down (1500g, 3 min), washed free of the inhibitor with an equal volume of YPD, spun down again, and resuspended in YPD at a density convenient for observation (about 1×10^7 cells/ml). Following brief sonication, samples of the cell suspensions are streaked out on a YPD plate that has been prewarmed to 30°. In the case of UV irradiation, the cells can be treated directly on plate by irradiating plate sectors with the desired dose. The plate is incubated at 30°. Periodically, the cells on plate are examined microscopically (at least at 160× magnification). In each microscopic field, they are classified as being in the single-cell stage (should be very few), in the double-cell, or in the microcolony stage (i.e., containing more than two cell bodies) (Fig. 2). At least 200 cells/microcolonies should be classified for each time point. Care should be taken to count within regions of comparable cell density. Depending on the dose, the arrest can last for several hours (Fig. 2A–C). Microscopic examination after about 24 h of incubation permits an estimate of the fraction of surviving, macrocolony forming cells.

If desired, the number of cells per microcolony can be categorized in more detail (see e.g., Yamamoto et al., 1996). For instance, accumulation of cells in a 4-cell stage following initial arrest in a double-cell stage could indicate adaptation to the initial checkpoint triggering event followed by repeated arrest due to residual damage. Although the random analysis outlined here can provide satisfactory results, more accurate data can be generated if cells of a desired stage are micromanipulated and distributed along a grid on solid YPD. This ensures that identical populations of cells are analyzed at each time point and individual pedigrees can be established. See Lee et al. (1998) and Bennett et al. (2001) for examples.

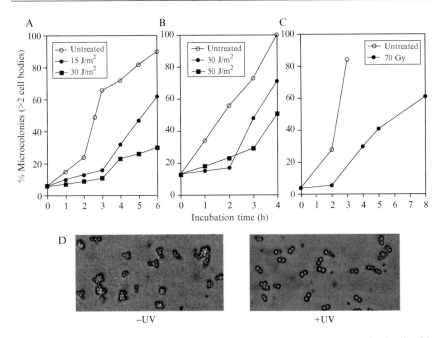

FIG. 2. Delayed formation of microcolonies following treatment of synchronized cells with UV (two doses) (A, B, D) or γ-irradiation, administered with a ^{137}Cs source (C). (A) Strain BY4741 was synchronized in S-phase with hydroxyurea and cells streaked out on a YPD plate. The fraction of microcolonies with more than 2 cell bodies was determined as a function of time, in unirradiated control (open circles) and irradiated samples (closed symbols). (B) Same as A, but cells were synchronized in M-phase with nocodazole. Note that the UV dose is not directly comparable to Fig. 1 where cells were treated in suspension and not on plate. (C) Same as A, but HU-synchronized cells were treated with γ-irradiation (70 Gy). (D) An example of microcolony formation in the control (left) as compared to predominant arrest in a two-cell body stage (G2/M arrest) in the treated sample (right). The pictures were taken 3 h after HU-synchronization and plating. (C) Adapted from Zhang and Siede (2004), with permission.

Visualization of the Mitotic Spindle

Introduction

In order to further classify yeast cells in the budded stage, the mitotic spindle can be visualized by indirect immunofluorescence using tubulin-specific antibodies. Although already formed during S-phase, the intranuclear rod-like spindle consisting of tubulin bundles undergoes a very characteristic elongation at the metaphase/anaphase transition (Fig. 3A). The following procedure has been adapted from D. Amberg's protocol, previously available

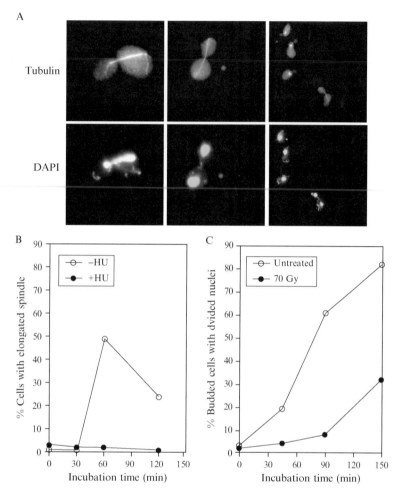

FIG. 3. Classification of large-budded yeast cells as pre- or post-mitotic. (A) Various examples of elongated mitotic and short premitotic spindles (upper row) compared to nuclear staining pattern with DAPI of the same cells (lower row). (B) Spindle elongation is inhibited by hydroxyurea. Haploid cells (strain BY4741) were released from G1 arrest into YPD medium with (closed circles) or without HU (open circles). (C) Delayed progression of cells through mitosis following treatment with γ-irradiation. Haploid cells (strain BY4741) that had been synchronized in S-phase with HU were treated with 70 Gy, resulting in mostly large-budded cells containing an undivided nucleus. Using nuclear staining with DAPI, premitotic and postmitotic cells (with divided nuclei) were distinguished. Adapted from Zhang and Siede (2004), with permission.

on the World Wide Web, a modification of a protocol from M. Rose. A few examples of stained cells are shown in Fig. 3A, refer to Kilmartin and Adams (1984) for further illustrations. The use of a strain containing GFP-tagged tubulin provides another simple way of spindle detection (Clarke *et al.*, 2001).

Spindle elongation is prevented by agents that inhibit replication, such as hydroxyurea. The example shown in Fig. 3B illustrates such a checkpoint response.

Materials

Teflon coated slides (e.g., Medical Packaging Corp., Camarillo, CA, Fisher Scientific Cat. No. NC9706026)

YOL1/34 Rat monoclonal antibody to yeast tubulin (Genetex Inc., San Antonio, TX)

Fluorescein-conjugated goat-anti-rat IgG (Pierce, Rockford, IL) 40 mM KPO_4 (pH 6.5)/500 μM $MgCl_2$, with or without 1.2 M sorbitol ("sorbitol buffer")

Zymolyase 100T (e.g., from U.S. Biological), dissolved at 10 mg/ml in sorbitol buffer and frozen in aliquots at $-80°$

Blocking solution: PBS (pH 7.4)/0.5% BSA/0.5% Tween 20 Polylysine (e.g., from Sigma, > 400,000 MW), dissolved as a 1% stock in water and frozen in aliquots at $-80°$

Mount: 100 mg p-phenylenediamine is dissolved in 10 ml PBS, pH is adjusted to above 8.0 with 0.5 M sodium carbonate buffer (pH 9.0), and the volume is brought to 100 ml with glycerol. From a 1-mg/ml stock solution in water, 4′,6-diamidino-2-phenylindole (DAPI) is added to 50 ng/ml. After mixing, the solution is stored at $-20°$. Solutions that have turned brown are discarded.

Protocol

Between 5×10^6 and 5×10^7 cells are harvested and resuspended in 5 ml 40 mM KPO_4 (pH 6.5)/500 μM $MgCl_2$. 0.5 ml 37% formaldehyde (best quality) is added, and cells are stored for at least 4 h at 4°. During a time course experiment, cells are kept in this stage until all samples have been collected. Cells can be kept overnight. The fixed cells are washed two times in the same buffer, once in the same buffer containing 1.2 M sorbitol, and gently resuspended in 0.5 ml sorbitol buffer. 30 μl Zymolyase 100T is added and cells are typically incubated between 10–30 min at 30°. Microscopically, cells should look transparent but intact, not misshapen or dark, which would indicate overdigestion. Cells are spun down (3 min, 1500g), carefully washed once with sorbitol buffer, very gently resuspended in 100–500-μl sorbitol buffer, and placed on ice. The wells of a teflon-coated slide are coated with 0.1% polylysine for 10 min at room temperature. It is recommended that all solutions that go on wells have been spun free of particle matter just before using (10 min, microfuge). Slides are incubated in a moist chamber. The wells are washed 4–5 times with water and dried. Coated slides can be prepared in advance and stored dust-free. Next, 20-μl cell suspension is spotted on the wells and incubated at room temperature for 10 min. Most of the liquid is aspirated while not letting the slides dry. Blocking solution is layered on top of the cells and incubated for 15 min at room temperature. The blocking step can be extended to overnight incubation. Next, cells are incubated with the primary tubulin antibody at a dilution of 1:300 in blocking solution for 1 h at room temperature. Cells are washed 4 × 5 min with blocking solution. Do not let dry at any point. Cells are incubated with secondary antibody at a dilution of 1:500 in blocking solution for 1 h at room temperature, then washed as before. Mount is applied to the slide next to the cells and a cover slip is carefully put down, thus overlaying the cells with the mount. Paper towels are used to clean up the excess mount that is squeezed out. The slide is cleaned and the edges of the cover slip are sealed with nail polish. When dry, slides can be stored at $-20°$. Cells are examined on a fluorescence microscope at high magnification with appropriate filter sets (e.g., Nikon B2A [EX 450–490, DM 505, BA 520] for fluorescein or Nikon DAPI [EX 360/40, DM 400, BA 460/50]). Note that the mount contains DAPI (4',6-diamidino-2-phenylindole) and by switching to a different filter set during fluorescence microscopy the status of the nuclear DNA and the mitotic spindle can be determined in the same cell (Fig. 3A). Refer to following section for simplified protocols if just nuclear staining is required.

Nuclear Staining with DAPI

Introduction

Further information on the cell cycle stage can be gained from the staining of nuclear DNA and large-budded yeast cells can be classified as being pre- or postmitotic. Cells in mitosis are characterized by an elongated nucleus in-between mother and daughter cell (Fig. 3B). Cells that have undergone anaphase are clearly identified by their divided nuclei. Nuclear staining is routinely done with the fluorescent dye 4',6-diamidino-2-phenylindole (DAPI). Cells can be stained with or without fixation and a variety of protocols will give excellent results (Amberg *et al.*, 2005). Without fixation, cells can be stained with 50-ng/ml DAPI in deionized water. However, a time course experiment will normally involve collection of many samples and thus fixation of cells before going on.

Since microscopic analysis may take time, we prefer to stain cells fixed and sealed in mounting medium, essentially as described previously. The example in Fig. 3C shows delayed progression through mitosis following γ-irradiation of S-phase cells, as detected by nuclear staining and determination of the fraction of budded cells with divided nuclei.

Protocol

See materials listed under *Materials* in the previous section. Between 5×10^6 and 5×10^7 cells are harvested and resuspended in 5-ml deionized water. 0.5 ml 37% formaldehyde is added and cells are stored for at least 4 h at 4°. Alternatively, cells can be fixed in ethanol as described later. During a time course experiment, cells are kept in this stage until all samples have been collected. Cells can be kept overnight. Next, the fixed cells are washed two times and resuspended in 0.5 ml deionized water. The wells of a teflon-coated slide are coated with 0.1% polylysine for 10 min at room temperature. Slides are incubated in a moist chamber. The wells are washed 4–5 times with water and dried. 20-μl cell suspension is spotted on the wells and incubated at room temperature for 10 min. Most of the liquid is aspirated while not letting the slides dry. Mount is applied to the slide next to the cells and a cover slip is carefully put down, thus overlaying the cells with the mount. Paper towels are used to clean up the excess mount that is squeezed out. The slide is cleaned and the edges of the cover slip are sealed with nail polish. When dry, slides can be stored at −20°. Cells are examined on a fluorescence microscope at high magnification with an appropriate filter set.

Flow Cytometry

Introduction

In this method, DNA of fixed cells is stained with a fluorescent dye like propidium iodide or Sytox Green (Molecular Probes, Inc., Eugene, OR) (Clarke *et al.*, 2001) and a histogram of cells sorted by DNA content is established by laser flow cytometry (Givan, 1992). With careful detector adjustments, the method is suitable for the analysis of cell cycle distributions within a yeast cell sample and satisfactory results can easily be obtained. Although simpler staining procedures can be found in the literature, we have achieved the most consistent results with the following protocol, adapted from Paulovich and Hartwell (1995). The protocol described later assumes the use of Becton-Dickinson, San Jose, CA FACScan® or FACSort® sorter with CELLquest® software. The manuals should be consulted extensively for details of using the instrument and its software.

Materials

FACS Analysis tubes (FALCON 2054, 12 × 75 mm)
50 m*M* Sodium citrate, pH 7.0
16 μg/ml Propidium iodide (PI) (e.g., from SIGMA), in 50 m*M* sodium citrate (pH 7.0). Make up in a large volume or as a 10× stock solution. RNase A, 10 mg/ml in STE (100 m*M* NaCl, 10 m*M* TRIS, 1 m*M* EDTA, pH 8.0), made DNase free by boiling for 10 min (Sambrook *et al.*, 1989)
Proteinase K, 10 mg/ml in water
Sonicator*Protocol*

About 4 × 10⁶ – 2 × 10⁷ cells are spun down in 1.5-ml microfuge tubes for a few seconds at 14,000g. Cells are resuspended in 1-ml deionized water and spun down again. The fluid is poured off and cells are resuspended in the remaining traces of water by vortexing (this prevents excessive clumping during the following fixation step). 1 ml of absolute ethanol is added and the sample is mixed well. In this stage, cells can be stored at 4°. During a time course experiment, cells are kept in this stage until all samples have been collected. When ready for analysis, the samples are vortexed extensively, cells are spun down for 1 min at 14,000g, washed with 1 ml of water, and spun down again. The cell pellet is resuspended with 1-ml 50 m*M* sodium citrate (pH 7.0) and transferred to appropriate FACS analysis tubes (this volume can be adjusted according to cell density). 8 μl of 10 mg/ml DNase-free RNase A is added and samples are incubated for 1 h at 50°. Without problems, we have stored cells overnight at 4° after this step. 25 μl of 10 mg/ml proteinase K is added and incubation at 50° is continued for another hour. 1 ml of 16 μg/ml propidium iodide in 50 m*M* sodium citrate (pH 7.0) is added. Protect from light! (Overnight storage of cells at 4° is possible at this stage.) Briefly sonicate all samples just before analysis. Filtration is usually not required. Vortex briefly just before mounting the tube and run each sample in the cell sorter set to low speed. The detector set-up is critical and may vary somewhat from experiment to experiment. It is suggested to adjust the settings with an unsynchronized logarithmic-phase cell sample, (e.g., taken from the same experiment before synchronization).

First, the forward-scatter threshold is set to 24. Second, a dot plot of propidium iodide fluorescence signal (FL2-A) versus side scatter (SSC-H) or forward scatter (FSC-H) is created. By manipulating detector voltage and amplifier gain, using initial values similar to those in Fig. 4A as approximate settings, a diagram resembling the one shown in Fig. 4B is created during real-time analysis. Two cell populations (G1, G2) that are distinguished by FL2-A signal intensity should be clearly visible.

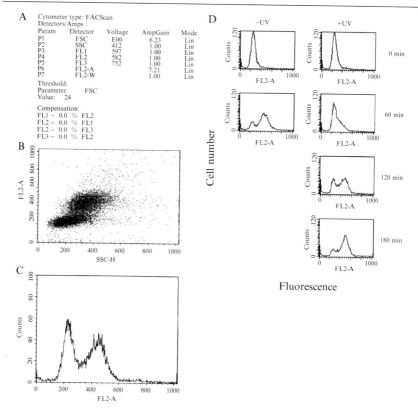

FIG. 4. FACS analysis of haploid yeast cells (BD FACScan). (A) Example of appropriate detector settings used for FACS of PI-stained yeast cells. (B) Dot-plot of side scatter (SSC-H) vs. PI fluorescence (FL2-A) of asynchronously dividing yeast cells. (C) Histogram of PI fluorescence of the same sample. (D) Cell cycle progression of a UV irradiated culture (haploid strain SX46A, 40 J/m^2, suspension) (right) following release from α–factor arrest as compared to untreated control (left). Adapted from Zhang and Siede (2004), with permission.

Lastly, the FL2-A histogram is created (Fig. 4C) and fine-tuned by adjusting FL2 voltage and FL2-A amplifier gain. At least 10,000 cells are analyzed for each time point. Gating is unnecessary with well-stained samples. Since yeast has a tendency to clog the fluidics of the sorter, the cleanup and shut-down procedures should be followed meticulously. The example in Fig. 4D shows the cell cycle response in response to UV irradiation in a population released from α-factor arrest. Following a short G1 arrest, a slow S phase and the beginning of G2/M arrest is evident. Although FACS analysis can be used to detect arrest stages successfully, treatment of cells with certain inhibitors such as nocodazole may negatively affect the histogram quality by broadening of peaks due to overstaining of cellular DNA.

Analysis of Rad53 Phosphorylation

Introduction

The evolutionary conserved Rad53 kinase of *S. cerevisiae* (CHK2 in mammalian cells) is essential for checkpoint activation following a wide variety of treatments that damage DNA or inhibit replication. Phosphorylation of Rad53 results in its activation and phosphorylation of many downstream target proteins involved in cell cycle arrest and transcriptional regulation of DNA repair. Rad53 is phosphorylated at multiple residues by the upstream acting Mec1 kinase and is also subject to autophosphorylation in *trans*, mediated by the Rad9 protein (Allen *et al.*, 1994; Friedberg *et al.*, 2005; Gilbert *et al.*, 2001; Lee *et al.*, 2003; Sanchez *et al.*, 1996; Sun *et al.*, 1996). Rad53 phosphorylation is easily detectable using a commercial antibody and conventional SDS-PAGE procedures. This provides a widely used robust assay to monitor checkpoint activation in response to numerous agents, under various growth conditions and in mutant backgrounds. Figure 5 shows Rad53 phosphorylation following treatment with the double-strand break inducing agent streptonigrin. The following protocol is adapted from Foiani *et al.* (1994). Note that the abundance of phosphorylated forms and the degree of shift in mobility of Rad53 are good indications but not proof of Rad53 activation. Protocols for estimating Rad53

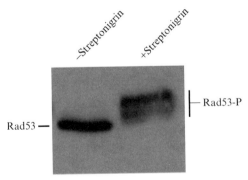

FIG. 5. Phosphorylation of Rad53 in an asynchronously dividing culture of strain SX46A treated with streptonigrin in YPD (3 h at 10 μl/ml).

activity based on autophosphorylation of membrane-bound renatured Rad53 are available elsewhere (Lee et al., 2003, 2004; Pellicioli et al., 1999).

Materials

0.5 mm detergent-washed zirconia/silica beads (Biospec Products Inc., Bartlesville, OK)

20% and 5% trichloroacetic acid (TCA)

1× SDS gel loading buffer: 50 mM Tris-HCl (pH 6.8), 100 mM dithiothreitol, 2% SDS, 0.1% bromophenol blue and 10% glycerol.

Vertical slab gel electrophoresis apparatus

Molecular weight protein markers

10× TBS (Tris-buffered saline): 0.2 M Tris base, 9% NaCl; adjust pH to 7.6.

Wash buffer 1× TBST: 1× TBS, 0.1% Tween-20 Blocking buffer and antibody dilution buffer: 1× TBS, 0.1% Tween-20 with 5% w/v nonfat dry milk.

Primary Rad53 antibody: goat polyclonal antibody raised against a C-terminal polypeptide derived from Rad53 (RAD53 yC-19 from Santa Cruz Biotechnology, Santa Cruz, CA)

Horseradish peroxidase (HRP)-conjugated anti-goat secondary antibody (e.g., from Santa Cruz Biotechnology).

Yeast Protein Preparation

Spin down $1\text{-}4 \times 10^8$ cells, wash once with deionized water, transfer to a 2-ml microfuge tube, and spin again (cell pellets can be frozen at $-80°$ at this point). Wash cell pellet with 500-μl 20% TCA, spin, and resuspend with 200 μl 20% TCA. Add glass or zirconia beads up to 2/3 volume and vortex with intermittent cooling on ice. The duration of vortexing required for lysis depends on the type of cells. While logarithmic phase haploid cells can be efficiently lysed within 5–10 min, stationary phase cells or diploid cells may require up to 30 min vortexing at $4°$, and microscopic monitoring of lysis is recommended. If many samples are to be processed, use of an adaptor that can be mounted in place of the conventional mixer head is a great convenience (e.g., TurboJet adaptor for Vortex Genie 2, Fisher Scientific, Pittsburgh, PA). Transfer supernatant into new 1.5-ml microfuge tube (conveniently done with gel loading tips), wash beads twice with 200 μl 5%TCA, and collect liquid into same tube. Spin for 10 min at 3000 rpm at room temperature. Discard supernatant and solubilize the precipitate by vortexing in 180 μl of PAGE-SDS-loading buffer (lacking dithiothreitol); this may take time. Add 20 μl of 1 M dithiothreitol and 100 μl of

1 *M* TRIS base (pH 9.0). Boil samples for 3 min, centrifuge the samples for 5 min at top speed, and save the supernatants.

Western Blotting

A wide variety of gel running and blotting conditions will give excellent results. We typically load 10–100-μl protein sample per lane of a large vertical slab gel (16 cm length, 1.5-mm thickness, 10% resolving gel). Samples are run at 150 V and transferred to nitrocellulose by semidry blotting. After transfer, wash nitrocellulose membrane with 1× TBST for 5 min. Incubate membrane with blocking buffer for 1 h at room temperature. Wash three times for 5 min each with 1× TBST. Incubate nitrocellulose membrane with primary Rad53 antibody at a 1:1000 dilution in antibody dilution buffer with gentle agitation at 4° overnight. Wash 3 times for 5 min each with TBST. Incubate membrane with HRP-conjugated secondary antigoat antibody for 1 h at 1:5000 dilution. Wash as stated previously. Use a chemoluminescence detection system according to manufacturer's instructions and expose to X-ray film.

Acknowledgments

We thank Gulnaz Bachlani and Larry Oakford for technical assistance. Work in the authors' laboratory is supported by grants CA87381 and ES11163 from the National Institutes of Health.

References

Allen, J. B., Zhou, Z., Siede, W., Friedberg, E. C., and Elledge, S. J. (1994). The *SAD1/RAD53* protein kinase controls multiple checkpoints and DNA damage induced transcription in yeast. *Genes Dev.* **8,** 2401–2415.

Amberg, D. C., Burke, D. J., and Strathern, J. N. (2005). "Methods in Yeast Genetics: A Cold Spring Harbor Laboratory Course Manual, 2005 Edition." Cold Spring Harbor Laboratory Press, Cold Spring Harbor, NY.

Bennett, C. B., Lewis, L. K., Karthikeyan, G., Lobachev, K. S., Jin, Y. H., Sterling, J. F., Snipe, J. R., and Resnick, M. A. (2001). Genes required for ionizing radiation resistance in yeast. *Nat. Genet.* **29,** 426–434.

Clarke, D. J., Segal, M., Jensen, S., and Reed, S. I. (2001). Mec1p regulates Pds1p levels in S phase: Complex coordination of DNA replication and mitosis. *Nature Cell Biol.* **3,** 619–627.

Foiani, M., Marini, F., Gamba, D., Lucchini, G., and Plevani, P. (1994). The B subunit of the DNA polymerase α-primase complex in *Saccharomyces cerevisiae* executes an essential function at the initial stage of DNA replication. *Mol. Cell. Biol.* **14,** 923–933.

Friedberg, E. C., Walker, G. C., Siede, W., Wood, R. D., Schultz, R. A., and Ellenberger, T. (2005). "DNA Repair and Mutagenesis." American Society of Microbiology Press, Washington, D.C.

Gardner, R., Putnam, C. W., and Weinert, T. (1999). *RAD53, DUN1* and *PDS1* define two parallel G2/M checkpoint pathways in yeast. *EMBO J.* **18**, 3173–3185.

Gilbert, C. S., Green, C. M., and Lowndes, N. F. (2001). Budding yeast Rad9 is an ATP-dependent Rad53 activating machine. *Mol. Cell* **8**, 129–136.

Givan, A. L. (1992). "Flow Cytometry - First Principles." Wiley-Liss, New York.

Guthrie, C., and Fink, G. F. (1991). "Guide to Yeast Genetics and Molecular Biology, Methods in Enzymol. 194." Academic Press, San Diego, CA.

Kilmartin, J. V., and Adams, A. E. M. (1984). Structural rearrangements of tubulin and actin during the cell cycle of the yeast Saccharomyces. *J. Cell Biol.* **98**, 922–933.

Lee, S.-J., Duong, J. K., and Stern, D. F. (2004). A Ddc2-Rad53 fusion protein can bypass the requirements for *RAD9* and *MRC1* in Rad53 activation. *Mol. Biol. Cell* **15**, 5443–5455.

Lee, S.-J., Schwartz, M. F., Duong, J. K., and Stern, D. F. (2003). Rad53 phosphorylation site clusters are important for Rad53 regulation and signaling. *Mol. Cell. Biol.* **23**, 6300–6314.

Lee, S. E., Moore, J. K., Holmes, A., Umezu, K., Kolodner, R. D., and Haber, J. E. (1998). *Saccharomyces* Ku70, Mre11/Rad50, and RPA proteins regulate adaptation to G2/M arrest after DNA damage. *Cell* **94**, 399–409.

Nyberg, K. A., Michelson, R. J., Putnam, C. W., and Weinert, T. A. (2002). Toward maintaining the genome: DNA damage and replication checkpoints. *Annu. Rev. Genet.* **36**, 617–656.

Paulovich, A. G., and Hartwell, L. H. (1995). A checkpoint regulates the rate of progression through S phase in *S. cerevisiae* in response to DNA damage. *Cell* **82**, 841–847.

Pellicioli, A., Lucca, C., Liberi, G., Marini, F., Lopes, M., Plevani, P., Romano, A., Di Fiore, P. P., and Foiani, M. (1999). Activation of Rad53 kinase in response to DNA damage and its effect in modulating phosphorylation of the lagging strand DNA polymerase. *EMBO J.* **18**, 6561–6572.

Sambrook, J., Fritsch, E. F., and Maniatis, T. (1989). "Molecular Cloning - A Laboratory Manual." Cold Spring Harbor Laboratory Press, Cold Spring Harbor, NY.

Sanchez, Y., Desany, B. A., Jones, W. J., Liu, Q., Wang, B., and Elledge, S. J. (1996). Regulation of *RAD53* by the *ATM*-like kinases *MEC1* and *TEL1* in yeast cell cycle checkpoint pathways. *Science* **271**, 357–360.

Sun, Z., Fay, D. S., Marini, F., Foiani, M., and Stern, D. F. (1996). Spk1/Rad53 is regulated by Mec1-dependent protein phosphorylation in DNA replication and damage checkpoint pathways. *Genes Dev.* **10**, 395–406.

Weinert, T. A., and Hartwell, L. H. (1988). The *RAD9* gene controls the cell cycle response to DNA damage in *Saccharomyces cerevisiae*. *Science* **241**, 317–322.

Yamamoto, A., Guacci, V., and Koshland, D. (1996). Pds1p, an inhibitor of anaphase in budding yeast, plays a critical role in the APC and checkpoint pathway(s). *J. Cell Biol.* **133**, 99–110.

Zhang, H., and Siede, W. (2004). Analysis of the budding yeast *Saccharomyces cerevisiae* cell cycle by morphological criteria and flow cytometry. *In* "Cell Cycle Checkpoint Protocols" (H. B. Lieberman, ed.), Vol. 241, pp. 77–91. Humana Press, Totowa, NJ.

[7] Recruitment of ATR-ATRIP, Rad17, and 9-1-1 Complexes to DNA Damage

By XIAOHONG HELENA YANG and LEE ZOU

Abstract

The ATR (ataxia-telangiectasia mutated and rad3-related)-ATRIP (ATR-interacting protein) kinase complex plays a central role in the checkpoint responses to a variety of types of DNA damage, especially those interfering with DNA replication. The checkpoint-signaling pathway activated by ATR-ATRIP regulates and coordinates cell-cycle progression, DNA replication, DNA repair, and many other cellular processes critical for genomic stability. Upon DNA damage or DNA replication interference, ATR-ATRIP and two of its key regulators, the Rad17 and the 9-1-1 complexes, are localized to sites of DNA damage and stalled replication forks. Recent biochemical and cell biological studies have revealed that RPA-coated single-stranded DNA, a common structure generated at sites of DNA damage and stalled replication forks, plays crucial roles in the recruitment of ATR-ATRIP, Rad17, and 9-1-1 complexes. The recruitment of ATR-ATRIP and its regulators to DNA damage is a key step for the recognition of DNA damage by the checkpoint, and is likely important for the regulation of ATR activity and/or function in response to DNA damage. The methods used to characterize the DNA association of ATR-ATRIP, Rad17, and 9-1-1 complexes have laid a foundation for further biochemical studies, which may ultimately lead us to understand the molecular mechanisms by which ATR-ATRIP monitors and protects genomic integrity.

Introduction

The ATR-mediated checkpoint pathway plays a crucial role in the maintenance of genomic stability. In response to DNA damage or DNA replication stress, many proteins involved in checkpoint signaling or its downstream cellular processes, such as Chk1 and Rad17, are phosphorylated in an ATR-ATRIP-dependent manner (Liu *et al.*, 2000; Zou *et al.*, 2002). Concomitant with the phosphorylation of ATR substrates, ATR and ATRIP are localized to discrete damage-induced nuclear foci, suggesting that the activity and/or function of the ATR-ATRIP kinase complex are stimulated at sites of DNA damage (Cortez *et al.*, 2001; Tibbetts *et al.*, 2000). The damage-induced phosphorylation of most ATR substrates also

METHODS IN ENZYMOLOGY, VOL. 409   0076-6879/06 $35.00

requires the Rad17 and the Rad9-Rad1-Hus1 (9-1-1) complexes, two evolutionarily conserved protein complexes important for the regulation of ATR-ATRIP after DNA damage. The Rad17 complex is a replication factor C (RFC)-like complex in which the large RFC subunit Rfc1 is replaced by Rad17, whereas the 9-1-1 complex is a ring-shape heterotrimer structurally resembling PCNA (Lindsey-Boltz *et al.*, 2001; Shiomi *et al.*, 2002; Venclovas and Thelen, 2000). During DNA replication, RFC specifically recognizes the 3' ends of primers annealed to DNA template. In an ATP-dependent manner, RFC recruits PCNA to DNA and enables it to encircle DNA (Waga and Stillman, 1994). The recruitment of PCNA is greatly important for the structure, function, and regulation of DNA replication forks. The similarities between the Rad17 complex and RFC, as well as those between the 9-1-1 complex and PCNA, strongly suggested that the Rad17 and 9-1-1 complexes can associate with DNA in a structure-specific manner, and that they might function as DNA damage sensors for the ATR checkpoint pathway.

To address how ATR-ATRIP, Rad17, and 9-1-1 complexes are recruited to sites of DNA damage, a number of *in vivo* and *in vitro* methods have been developed using several different model organisms. In budding yeast, a single double-stranded DNA break (DSB) can be generated in cells using the HO endonuclease. The localization of the yeast homologues of ATR, ATRIP, Rad17, and Hus1 to the HO-induced DSBs have been examined using fluorescently tagged proteins and chromatin immunoprecipitation (ChIP) (Kondo *et al.*, 2001; Lisby *et al.*, 2004; Melo *et al.*, 2001). The binding of ATR-ATRIP, Rad17, 9-1-1 complexes to chromatin has been extensively studied *in vitro* using Xenopus egg extracts (Costanzo *et al.*, 2003; Guo *et al.*, 2000; Hekmat-Nejad *et al.*, 2000; You *et al.*, 2002), which can efficiently support rapid replication of various DNA templates. In human cells, the association of ATR-ATRIP, Rad17, and 9-1-1 complexes with chromatin have been analyzed using an extract fractionation method (Dart *et al.*, 2004; Post *et al.*, 2003; Zou *et al.*, 2002). Furthermore, recombinant human ATR-ATRIP, Rad17, and 9-1-1 complexes have been reconstituted, purified, and characterized *in vitro* (Bermudez *et al.*, 2003; Ellison and Stillman, 2003; Unsal-Kacmaz and Sancar, 2004; Unsal-Kacmaz *et al.*, 2002; Zou and Elledge, 2003; Zou *et al.*, 2003). A mechanistic picture of ATR activation has begun to emerge from the studies using the previously stated methods. Although Rad17 and 9-1-1 complexes are important for the function of ATR-ATRIP, the ATR-ATRIP kinase complex is localized to sites of DNA damage independently of these two regulators. The Rad17 complex probably directly recognizes specific DNA structures induced by damage, and it recruits 9-1-1 complexes onto DNA in a damage-stimulated manner. Importantly, recent studies revealed that

RPA-coated single-stranded DNA (ssDNA), a common structure generated at sites of DNA damage and stalled replication forks, plays critical roles in the recruitment of ATR-ATRIP, Rad17, and 9-1-1 complexes (Ball *et al.*, 2005; Dart *et al.*, 2004; Ellison and Stillman, 2003; Itakura *et al.*, 2004; Zou and Elledge, 2003; Zou *et al.*, 2003). These findings have provided a platform upon which new biochemical assays can be developed to address how ATR is regulated on damaged DNA. In this chapter, we will discuss the methods used to characterize the recruitment of human ATR-ATRIP, Rad17, and 9-1-1 complexes to DNA damage.

Chromatin Association of ATR, Rad17, Rad9, and RPA

To examine how ATR, Rad17, and Rad9 associate with DNA in human cells, we adapted a chromatin fractionation protocol used to characterize proteins involved in DNA replication (Mendez and Stillman, 2000). In this protocol, cells are first permeablized with buffer containing detergent. Nuclei are subsequently isolated by low-speed centrifugation in sucrose-containing buffer and lysed in solution with no salt. The insoluble fractions derived from the lysed nuclei are significantly enriched for proteins associating with chromatin. These proteins can be released from the chromatin-enriched pellets by nuclease treatments.

To introduce DNA damage or replication stress to cells, 293T or Hela cells were treated with 10–20 Gy of ionizing radiation (IR), 20–80 J/m^2 of ultraviolet light (UV), or 1 mM hydroxyurea (HU). At various time points after cells were treated, approximately 3×10^6 cells were harvested and briefly washed with cold 1× PBS buffer. Cells were then sedimented and resuspended in 200 μl of solution A (10 mM Hepes [pH 7.9], 10 mM KCl, 1.5 mM MgCl$_2$, 0.34 M sucrose, 10% glycerol, 1 mM DTT, 1 mM Na$_2$VO$_3$, protease inhibitors), and Triton X-100 was subsequently added to a final concentration of 0.1% to permeablize the cells. After a 5-min incubation on ice, samples were spun at low speed (1300g for 4 min) to separate nuclei (fraction P1) and cytoplasmic proteins (fraction S1) (Fig. 1A). The isolated nuclei were washed once with 200 μl of Solution A and lysed in 200 μl of Solution B (3 mM EDTA, 0.2 mM EGTA, 1 mM DTT). After a 10-min incubation on ice, samples were spun again (1700g for 4 min) to separate soluble nuclear proteins (fraction S2) and chromatin-enriched pellets (fraction P2) (Fig. 1A). The chromatin-enriched pellets were then washed once using Solution B and recovered by high-speed centrifugation (10,000g for 1 min). It was estimated that the chromatin-enriched fractions contained approximately 5–10% of the total proteins in whole-cell extracts.

Using the previously stated method, ATR, Rad17, and Rad9 were detected in the chromatin-enriched fractions, such as the known

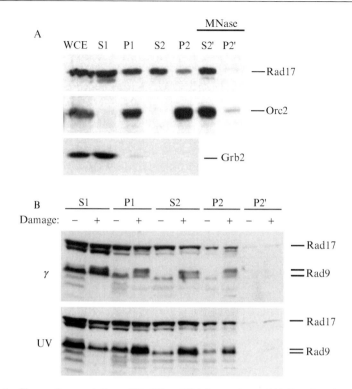

FIG. 1. Chromatin association of Rad17 and 9-1-1 complexes. (A) Fractionation of 293T cell extracts. Orc2 and Grb2 are markers for chromatin-bound and cytoplasmic proteins, respectively. (WCE) whole-cell extracts; (S1) cytoplasmic proteins; (P1) intact nuclei; (S2) soluble nuclear proteins; (P2) chromatin-enriched sediment; (S2′ and P2′) soluble fraction and sediment from micrococcal nuclease-treated nuclei, respectively; (MNase) micrococcal nuclease. (B) DNA damage-induced recruitment of 9-1-1 complexes to chromatin. 293T cells were untreated or treated with γ or UV radiation, and were fractionated as in (A). The levels of Rad17 and Rad9 in various fractions were analyzed by immunoblotting (Zou *et al.*, 2002).

chromatin-bound protein Orc2 (Fig. 1A). To demonstrate that the presence of these proteins in the chromatin-enriched fractions was due to their association with chromatin, we used micrococcal nuclease (MNase) to cleave chromatin into nuclesomes, which did not sediment under the centrifugal condition previously stated. To digest chromatin with MNase, isolated nuclei (P1) were resuspended in Solution A containing 1 mM CaCl$_2$ and 50 units of MNase (Sigma, St. Louis, MO). MNase digestion was carried out at 37° for 2 min, and was terminated by 2 mM EGTA. The MNase-treated nuclei were subsequently recovered and lysed in Solution B as previously stated. After MNase treatment, ATR, Rad17, and Rad9 were detected in the soluble nuclear fractions (fraction P1′) but not in insoluble pellets (fraction P2′) (Fig. 1A), confirming that they indeed associate with chromatin rather than other insoluble cellular structures.

Using this chromatin fractionation method, we found that ATR, Rad17, and Rad9 associated with chromatin in asynchronously growing undamaged human cells (Zou et al., 2002). Other studies using synchronously growing cells showed that Rad17 associates with chromatin throughout the cell cycle (Post et al., 2003), whereas ATR preferentially associates with chromatin during S phase (Dart et al., 2004). A Rad17 mutant lacking a functional ATP-binding motif (K132E) failed to associate with chromatin efficiently (Garg et al., 2004; Zou et al., 2002). After DNA damage, the amounts of Rad9 on chromatin increased significantly, indicating that 9-1-1 complexes were recruited onto chromatin in a damage-induced manner (Fig. 1B) (Zou et al., 2002). The recruitment of Rad9 onto chromatin is dependent upon Rad17, suggesting that 9-1-1 complexes are loaded onto damaged DNA by the Rad17 complex (Zou et al., 2002). Furthermore, elevated levels of chromatin-bound RPA, an ssDNA-binding protein complex, were detected after UV treatment, which implies that UV-induced DNA damage leads to increased amounts of RPA-coated ssDNA (Zou and Elledge, 2003). In contrast to Rad9 and RPA, the amounts of Rad17 and ATR associating with chromatin did not change substantially after DNA damage (Fig. 1B) (Post et al., 2003; Zou et al., 2002). It is possible that the amounts of Rad17 and ATR recruited to DNA damage do not significantly affect the overall levels of these proteins on chromatin, or that Rad17 and ATR redistribute to sites of DNA damage from other regions on chromatin. Thus, this chromatin fractionation method is useful for the analysis of recruitment of some (e.g., 9-1-1 and RPA), but not all, checkpoint proteins to DNA damage.

Purification of ATR, ATRIP, and Reconstitution of the ATR-ATRIP Complex

The large size of ATR (>300 kDa) has made its purification challenging. So far only Flag-tagged and GST-tagged ATR overexpressed in human cells, but not native ATR, has been successfully purified (Hall-Jackson et al., 1999; Unsal-Kacmaz et al., 2002; Zou and Elledge, 2003). The Flag-ATR transiently expressed in 293T cells was purified using anti-Flag M2 affinity gel (Sigma). To obtain Flag-ATR with high purity, it is necessary to extensively wash the ATR-bound M2 beads with buffer containing 1 M NaCl. Highly purified Flag-ATR can be eluted from the beads with buffer containing 200 μg/ml Flag peptide (Sigma). The Flag-ATR purified using this condition was devoid of ATRIP as shown by silver staining and immunoblotting. When the ATR-bound beads were washed with buffer containing only 150 mM NaCl, substantial but substoichiometric amounts of ATRIP were found to associate with Flag-ATR. Several other proteins, including RPA, were also found to associate with Flag-ATR under this condition.

Unlike ATR, recombinant ATRIP can be readily expressed in insect cells infected with baculovirus (Zou and Elledge, 2003). Flag-ATRIP was purified from insect cell extracts using anti-Flag M2 affinity gel. To reconstitute the ATR-ATRIP complex *in vitro*, we combined purified Flag-ATR and Flag-ATRIP at a 1:1 ratio and incubated them together for 1 h at room temperature. Although significant fractions of Flag-ATR and Flag-ATRIP formed complexes under this condition, free forms of these proteins remained detectable.

Recruitment of ATRIP and ATR-ATRIP to RPA-Coated
  Single-Stranded DNA

As revealed by the studies using budding yeast, increased amounts of ssDNA are generated at stalled replication forks and other checkpoint-activating DNA lesions, such as DSBs and aberrant telomeres. To assess whether ATR and ATRIP can directly associate with ssDNA or protein-DNA complexes containing ssDNA, we developed an *in vitro* protein-DNA binding assay using biotinylated DNA, purified ATR, ATRIP, and other proteins. In this assay, the biotinylated DNA, together with the proteins bound to it, can be retrieved from binding reactions with streptavidin-coated beads. The proteins present in the protein-DNA complexes can be detected by immunoblotting and analyzed by other functional assays. Thus, this method not only allows one to analyze the interactions between proteins and defined DNA structures, but also provides an effective means to isolate the protein-DNA complexes for further biochemical analysis.

To generate ssDNA for the *in vitro* binding assay, synthetic DNA oligomers (75 or 100-nucleotide) were biotinylated at either 3' or 5' end and attached to streptavidin-coated magnetic beads (Dynal, Oslo, Norway). Typically 100 pmol of biotinylated DNA oligomers were incubated with 100 μl of Dynal beads for 30 min at room temperature. The ssDNA-bound beads were subsequently recovered and unbound DNA oligomers were washed away. To generate double-stranded DNA (dsDNA) or other primed-ssDNA structures, biotinylated DNA oligomers were first annealed to their complementary DNA, which were not labeled with biotin. To ensure that most of the biotinylated DNA was incorporated to the desired structures, unlabeled DNA was present in 2 ~4-fold molar excess relative to the biotinylated DNA. The resultant DNA structures were then attached to streptavidin-coated beads, and the excess unlabeled DNA was removed by washing. Using ssDNA generated as stated previously, we found that ATRIP does not associate with ssDNA by itself (the binding conditions are described later). Purified ATR, on the other hand, bound weakly to ssDNA. However, ATR did not exhibit a preference for ssDNA over dsDNA. Based on these results, we suggested that neither ATR nor ATRIP can specifically recognize naked ssDNA.

The correlation between ssDNA generation and checkpoint activation in yeast prompted us to investigate whether certain protein-DNA complexes assembled on ssDNA are involved in recruiting and/or activating ATR-ATRIP. To examine whether ATR and ATRIP can associate with RPA-coated ssDNA, we pre-assembled the RPA-ssDNA complexes *in vitro* using recombinant RPA complexes purified from *E. coli* and biotinylated DNA oligomers. Various amounts of RPA (1–8 pmol) were incubated with 5 pmol of ssDNA bound to Dynal beads. The reactions were carried out in 500 μl of binding buffer (10 mM Tris-Cl [pH 7.5], 100 mM NaCl, 10 μg/ml BSA, 0.01% NP40, and 10% glycerol) for 30 min at room temperature. After the formation of RPA-ssDNA complexes, approximately 0.5 pmol of purified ATR or ATRIP were added to the reactions. The reactions were continued for another 30 min at room temperature to allow complex formation. Finally, the protein-DNA complexes formed on the biotinylated ssDNA were retrieved after 2–3 washes with binding buffer.

Using this *in vitro* binding assay, we found that RPA significantly stimulates the binding of ATRIP to ssDNA (Fig. 2). In contrast to RPA, SSB, an ssDNA-binding protein from *E. coli*, does not stimulate the binding of ATRIP to ssDNA. In the presence of RPA, ATRIP preferentially binds to ssDNA versus dsDNA, and the binding of ATRIP to RPA-ssDNA is length-dependent. These findings suggest that ATRIP can directly recognize the RPA-ssDNA complexes formed at DNA damage. Furthermore, the length dependence of the binding of ATRIP to RPA-ssDNA may enable ATR-ATRIP to distinguish the long stretches of ssDNA induced by DNA damage from those present during normal DNA metabolism.

The binding of ATR to ssDNA, unlike that of ATRIP, is not stimulated by RPA. When the ATR-ATRIP complexes reconstituted *in vitro* were tested using the binding assay stated previously, the binding of both ATR and ATRIP to ssDNA was stimulated by RPA. Thus, our *in vitro* analyses suggest that the ATR-ATRIP complex is sufficient to recognize ssDNA through the interaction between ATRIP and RPA-ssDNA. Consistent with this model, RPA is required for the recruitment of ATRIP and its yeast homologue Ddc2 to DSBs and stalled replication forks *in vivo* (Lucca *et al.*, 2004; Nakada *et al.*, 2004; Zou and Elledge, 2003). Together these findings strongly argue that the recognition of RPA-ssDNA by ATRIP plays a critical role in the recruitment of ATR-ATRIP to DNA damage. It should

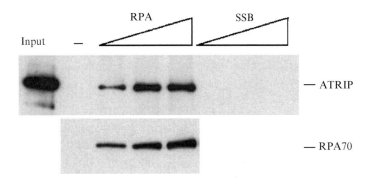

Fig. 2. Recruitment of ATRIP to ssDNA by RPA. The biotinylated 75-nucleotide oligomer (5 pmol) was attached to streptavidin-coated beads and incubated with various amounts of purified RPA or SSB (0, 1, 2.5, or 8 pmol). After a 30-min incubation, ATRIP (0.5 pmol) was added to the reactions and incubated for 30 min. The ssDNA oligomer was retrieved with streptavidin beads, and the unbound proteins were washed away. The ATRIP and RPA associated with ssDNA was detected by immunoblotting with the corresponding antibodies. Input, total input of ATRIP (Zou and Elledge, 2003).

be noted that although ATR-ATRIP is sufficient to bind to RPA-ssDNA, this association might be regulated by additional factors and/or protein modifications *in vivo*. The *in vitro* binding assay described here provides a useful tool to reveal and characterize the novel mechanisms involved in ATR-ATRIP regulation.

Purification of Rad17 and 9-1-1 Complexes

To determine how Rad17 and 9-1-1 complexes associate with DNA and how they regulate ATR-ATRIP on DNA, we and others sought to purify these protein complexes and characterize them *in vitro* (Lindsey-Boltz *et al.*, 2001; Shiomi *et al*,. 2002; Zou *et al.*, 2003). Recombinant Rad17 complexes consisting of Flag-Rad17, His_6-Rfc3, and untagged Rfc2, 4, and 5 have been reconstituted in insect cells and affinity purified (Lindsey-Boltz *et al.*, 2001; Zou *et al.*, 2003). To purify Rad17 complexes, approximately 1×10^8 insect cells (Sf9) were simultaneously infected with baculoviruses expressing Flag-Rad17, His6-Rfc3, and Rfc2, 4, and 5 (each at MOI of 5). Forty-eight hours after infection, cells were lysed in 5 ml of lysis buffer (10 mM Tris-Cl [pH7.5], 500 mM NaCl, 2 mM MgOAc, 0.05% NP40, 10% glycerol, 50 mM NaF, 0.1 mM NaVO3, 2 mM benzamidine, 1.4 μg/ml pepstatin A, 1 μg/ml aprotinin, 2 μg/ml leupeptin, and 0.2 mM PMSF). Rad17 complexes were first bound to Ni-NTA agarose (Invitrogen, Grand Island, NY), washed with lysis buffer containing 25 mM imidazole, and eluted with 150 mM imidazole. The partially purified Rad17 complexes were then further purified using M2 anti-Flag affinity gel and eluted with 200 μg/ml Flag peptides. Recombinant Rad17 complexes containing His$_6$-Flag-Rad17 have also been purified using similar strategies (Bermudez et al., 2003). Furthermore, native Rad17 complexes have been affinity purified using antibodies against Rfc2 and Rad17 (Ellison and Stillman, 2003).

Recombinant 9-1-1 complexes have also been reconstituted in insect cells and affinity purified (Lindsey-Boltz et al., 2001; Shiomi et al., 2002; Zou et al., 2003). Baculoviruses expressing Rad9-Flag, Rad1, and Hus1 were used to co-infect insect cells at MOI of 5. Reconstituted 9-1-1 complexes were bound to M2 anti-Flag affinity gel, washed with lysis buffer containing 300 mM NaCl, and eluted with 200 μg/ml Flag peptides. The 9-1-1 complexes purified from insect cells appear to contain hyper-phosphorylated Rad9. While studying the in vitro loading of 9-1-1 complexes onto DNA by the Rad17 complex, we found that Rad17 complex bound efficiently to various DNA structures containing ssDNA regions in the presence of RPA (see later). Furthermore, a fraction of the Flag-purified 9-1-1 complexes bound to the RPA-coated DNA structures even in the absence of Rad17 complex. We reasoned that this fraction of 9-1-1 complexes might be associated with the Rad17 complexes and/or other unknown factors from insect cells so that it bypasses the requirement of human Rad17 complex in vitro. To overcome this problem, a biotinylated "gapped DNA" structure (see later) coated with RPA was used to deplete the contaminating insect Rad17 complexes from the 9-1-1 preparations. After this purification step, 9-1-1 complexes were recruited to gapped DNA only in the presence of both RPA and Rad17 complexes (Fig. 3), recapitulating the RPA- and Rad17-dependent recruitment of 9-1-1 in vivo. Recombinant 9-1-1 complexes have also been reconstituted in E. coli and purified using Talon beads, phosphocellulose chromatography, and glycerol gradients (Ellison and Stillman, 2003).

Recruitment of 9-1-1 Complexes to Primed ssDNA by RPA and Rad17 Complexes

In vivo studies using budding yeast revealed that the recruitment of 9-1-1 complexes to DSBs requires both the Rad17 complex and RPA (Melo et al., 2001; Zou et al., 2003). To understand the molecular mechanisms of recruitment of 9-1-1 complexes to DNA damage, we sought to recapitulate this process in vitro using purified RPA, Rad17, 9-1-1 complexes, and defined DNA structures.

Many types of DNA damage known to activate the ATR checkpoint, such as resected DSBs and stalled replication forks, associate with junctions of single-/double-stranded DNA. To recreate this damage-induced

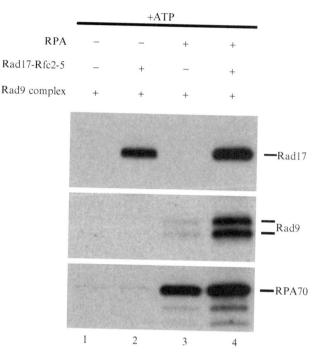

FIG. 3. A Rad17- and RPA-dependent recruitment of 9-1-1 complexes *in vitro*. The 9-1-1 complex purified from insect cells was precleared with beads carrying gapped DNA coated with RPA. The precleared Rad9 complex was then incubated with the Rad17 complex, RPA, and the gapped DNA template attached to streptavidin beads as indicated. The Rad17, Rad9, and RPA70 bound to the gapped DNA were detected by immunoblotting with the respective antibodies (Zou *et al.*, 2003).

DNA structure, biotinylated ssDNA oligomers (100-nucleotide) were annealed to circular M13 ssDNA, producing "primed ssDNA" with single-/double-stranded DNA junctions. The primed ssDNA structures can be retrieved using streptavidin-coated Dynal beads after annealing, and the excess unlabeled M13 ssDNA can be washed away. Since the biotinylated ends of primers are attached to streptavidin beads, the resultant primed ssDNA structures contain either free 3′ or 5′ single/double-stranded DNA junctions. When two or more different primers are simultaneously annealed to M13 ssDNA, ssDNA gaps between the primers can be generated. Using this method, we created a "gapped DNA" structure with two 100-nuleotide primers and a 200-nucleotide gap in between.

To examine the recruitment of 9-1-1 complexes to various primed ssDNA, purified 9-1-1 complexes (0.5 pmol) were incubated with Dynal beads coated with gapped DNA or primed ssDNA (5 pmol) in the presence or absence of RPA and Rad17 complexes (0.5 pmol). The binding reactions were carried out in 500 μl of binding buffer (40 mM Tris-Cl [pH7.8], 150 mM NaCl, 10 mM MgCl$_2$, 100 μM ATP, 1 mM DTT, 100 μg/ml BSA, 0.01% NP40, 10% glycerol) for 30 min at room temperature. The resultant protein-DNA complexes were washed 3 times with binding buffer containing 250 mM NaCl but no ATP. Using this *in vitro* binding assay, we found that 9-1-1 complexes were efficiently recruited to gapped DNA and primed ssDNA only in the presence of both RPA and Rad17 complexes (Fig. 3). The recruitment of 9-1-1 complexes was not observed in the absence of ATP or in the presence of ATR-γ-S, suggesting that this is a process requiring ATP hydrolysis (Zou *et al.*, 2003). Furthermore, 9-1-1 complexes were recruited to primed ssDNA but not ssDNA oligomers, indicating that junctions of single-/double-stranded DNA are important for this process (Zou *et al.*, 2003). Thus, these findings confirm several predications from previous *in vivo* and *in vitro* studies, supporting that the assay system can be used to reveal unknown mechanisms involved in the recruitment of 9-1-1.

Using the *in vitro* binding assay stated previously, we found that 9-1-1 complexes can be recruited to DNA structures with either 5' or 3' single/double-stranded DNA junctions, and that they bind more efficiently to gapped DNA than primed ssDNA. These results suggest that the DNA-structure specificity of the Rad17 complex is different from that of RFC. While RFC only uses the 3' single-/double-stranded DNA junctions to load PCNA onto DNA, the Rad17 complex can use both 5' and 3' junctions to recruit 9-1-1, and it exhibits a preference for ssDNA gaps (Zou *et al.*, 2003). The *in vitro* recruitment of 9-1-1 complexes to DNA has also been reconstituted by three other laboratories (Bermudez *et al.*, 2003; Ellison and Stillman, 2003; Majka and Burgers, 2003). Together, these studies provide possible explanations of how the Rad17 complex recognizes different types of single-/double-stranded DNA junctions at DNA damage and stalled replication forks.

Conclusion

The assays described here have enabled us to reveal the important roles of RPA in the recruitment of ATR-ATRIP, Rad17, and 9-1-1 complexes to DNA damage. These assays have also helped us to define some of the DNA structures recognized by the checkpoint sensors. The findings of the studies using these assays have provided important clues for future *in vivo* characterizations of the checkpoint. More important, these assays have laid a foundation upon which further biochemical assays can be developed to determine how ATR-ATRIP is regulated by its regulators on DNA and how DNA damage signals are relayed to downstream effectors. Tractable biochemical systems for checkpoint activation and checkpoint signaling, together with cell biological and genetic approaches, may ultimately lead us to understand how the checkpoint orchestrates the cellular processes critical for genomic integrity.

Acknowledgments

We are indebted to Drs. Stephen Elledge (Harvard Medical School) and Bruce Stillman (Cold Spring Harbor Laboratory), in whose laboratories most of the methods described in this chapter were developed. L. Z. is supported in part by a Smith Family New Investigator Award from the Medical Foundation.

References

Ball, H. L., Myers, J. S., and Cortez, D. (2005). ATRIP binding to replication protein A-single-stranded DNA promotes ATR-ATRIP localization but is dispensable for Chk1 phosphorylation. *Mol. Biol. Cell* **16,** 2372–2381.

Bermudez, V. P., Lindsey-Boltz, L. A., Cesare, A. J., Maniwa, Y., Griffith, J. D., Hurwitz, J., and Sancar, A. (2003). Loading of the human 9-1-1 checkpoint complex onto DNA by the checkpoint clamp loader hRad17-replication factor C complex *in vitro. Proc. Natl. Acad. Sci. USA* **100,** 1633–1638.

Cortez, D., Guntuku, S., Qin, J., and Elledge, S. J. (2001). ATR and ATRIP: Partners in checkpoint signaling. *Science* **294,** 1713–1716.

Costanzo, V., Shechter, D., Lupardus, P. J., Cimprich, K. A., Gottesman, M., and Gautier, J. (2003). An ATR- and Cdc7-dependent DNA damage checkpoint that inhibits initiation of DNA replication. *Mol. Cell* **11,** 203–213.

Dart, D. A., Adams, K. E., Akerman, I., and Lakin, N. D. (2004). Recruitment of the cell cycle checkpoint kinase ATR to chromatin during S-phase. *J. Biol. Chem.* **279,** 16433–16440.

Ellison, V., and Stillman, B. (2003). Biochemical characterization of DNA damage checkpoint complexes: Clamp loader and clamp complexes with specificity for 5′ recessed DNA. *PLoS Biol.* **1,** E33.

Garg, R., Callens, S., Lim, D. S., Canman, C. E., Kastan, M. B., and Xu, B. (2004). Chromatin association of rad17 is required for an ataxia telangiectasia and rad-related kinase-mediated S-phase checkpoint in response to low-dose ultraviolet radiation. *Mol. Cancer Res.* **2,** 362–369.

Guo, Z., Kumagai, A., Wang, S. X., and Dunphy, W. G. (2000). Requirement for Atr in phosphorylation of Chk1 and cell cycle regulation in response to DNA replication blocks and UV-damaged DNA in Xenopus egg extracts. *Genes Dev.* **14,** 2745–2756.

Hall-Jackson, C. A., Cross, D. A., Morrice, N., and Smythe, C. (1999). ATR is a caffeine-sensitive, DNA-activated protein kinase with a substrate specificity distinct from DNA-PK. *Oncogene* **18,** 6707–6713.

Hekmat-Nejad, M., You, Z., Yee, M. C., Newport, J. W., and Cimprich, K. A. (2000). Xenopus ATR is a replication-dependent chromatin-binding protein required for the DNA replication checkpoint. *Curr. Biol.* **10,** 1565–1573.

Itakura, E., Takai, K. K., Umeda, K., Kimura, M., Ohsumi, M., Tamai, K., and Matsuura, A. (2004). Amino-terminal domain of ATRIP contributes to intranuclear relocation of the ATR-ATRIP complex following DNA damage. *FEBS Lett.* **577,** 289–293.

Kondo, T., Wakayama, T., Naiki, T., Matsumoto, K., and Sugimoto, K. (2001). Recruitment of Mec1 and Ddc1 checkpoint proteins to double-strand breaks through distinct mechanisms. *Science* **294**, 867–870.

Lindsey-Boltz, L. A., Bermudez, V. P., Hurwitz, J., and Sancar, A. (2001). Purification and characterization of human DNA damage checkpoint Rad complexes. *Proc. Natl. Acad. Sci. USA* **98**, 11236–11241.

Lisby, M., Barlow, J. H., Burgess, R. C., and Rothstein, R. (2004). Choreography of the DNA damage response: Spatiotemporal relationships among checkpoint and repair proteins. *Cell* **118**, 699–713.

Liu, Q., Guntuku, S., Cui, X. S., Matsuoka, S., Cortez, D., Tamai, K., Luo, G., Carattini-Rivera, S., DeMayo, F., Bradley, A., Donehower, L. A., and Elledge, S. J. (2000). Chk1 is an essential kinase that is regulated by Atr and required for the G(2)/M DNA damage checkpoint. *Genes Dev.* **14**, 1448–1459.

Lucca, C., Vanoli, F., Cotta-Ramusino, C., Pellicioli, A., Liberi, G., Haber, J., and Foiani, M. (2004). Checkpoint-mediated control of replisome-fork association and signalling in response to replication pausing. *Oncogene* **23**, 1206–1213.

Majka, J., and Burgers, P. M. (2003). Yeast Rad17/Mec3/Ddc1: A sliding clamp for the DNA damage checkpoint. *Proc. Natl. Acad. Sci. USA* **100**, 2249–2254.

Melo, J. A., Cohen, J., and Toczyski, D. P. (2001). Two checkpoint complexes are independently recruited to sites of DNA damage *in vivo*. *Genes Dev.* **15**, 2809–2821.

Mendez, J., and Stillman, B. (2000). Chromatin association of human origin recognition complex, cdc6, and minichromosome maintenance proteins during the cell cycle: Assembly of prereplication complexes in late mitosis. *Mol. Cell. Biol.* **20**, 8602–8612.

Nakada, D., Hirano, Y., and Sugimoto, K. (2004). Requirement of the Mre11 complex and exonuclease 1 for activation of the Mec1 signaling pathway. *Mol. Cell. Biol.* **24**, 10016–10025.

Post, S. M., Tomkinson, A. E., and Lee, E. Y. (2003). The human checkpoint Rad protein Rad17 is chromatin-associated throughout the cell cycle, localizes to DNA replication sites, and interacts with DNA polymerase epsilon. *Nucleic Acids Res.* **31**, 5568–5575.

Shiomi, Y., Shinozaki, A., Nakada, D., Sugimoto, K., Usukura, J., Obuse, C., and Tsurimoto, T. (2002). Clamp and clamp loader structures of the human checkpoint protein complexes, Rad9-1-1 and Rad17-RFC. *Genes Cells* **7**, 861–868.

Tibbetts, R. S., Cortez, D., Brumbaugh, K. M., Scully, R., Livingston, D., Elledge, S. J., and Abraham, R. T. (2000). Functional interactions between BRCA1 and the checkpoint kinase ATR during genotoxic stress. *Genes Dev.* **14**, 2989–3002.

Unsal-Kacmaz, K., Makhov, A. M., Griffith, J. D., and Sancar, A. (2002). Preferential binding of ATR protein to UV-damaged DNA. *Proc. Natl. Acad. Sci. USA* **99**, 6673–6678.

Unsal-Kacmaz, K., and Sancar, A. (2004). Quaternary structure of ATR and effects of ATRIP and replication protein A on its DNA binding and kinase activities. *Mol. Cell. Biol.* **24**, 1292–1300.

Venclovas, C., and Thelen, M. P. (2000). Structure-based predictions of Rad1, Rad9, Hus1 and Rad17 participation in sliding clamp and clamp-loading complexes. *Nucleic Acids Res.* **28**, 2481–2493.

Waga, S., and Stillman, B. (1994). Anatomy of a DNA replication fork revealed by reconstitution of SV40 DNA replication *in vitro*. *Nature* **369**, 207–212.

You, Z., Kong, L., and Newport, J. (2002). The role of single-stranded DNA and polymerase alpha in establishing the ATR, Hus1 DNA replication checkpoint. *J. Biol. Chem.* **277**, 27088–27093.

Zou, L., Cortez, D., and Elledge, S. J. (2002). Regulation of ATR substrate selection by Rad17-dependent loading of Rad9 complexes onto chromatin. *Genes Dev.* **16**, 198–208.

Zou, L., and Elledge, S. J. (2003). Sensing DNA damage through ATRIP recognition of RPA-ssDNA complexes. *Science* **300,** 1542–1548.

Zou, L., Liu, D., and Elledge, S. J. (2003). Replication protein A-mediated recruitment and activation of Rad17 complexes. *Proc. Natl. Acad. Sci. USA* **100,** 13827–13832.

[8] Multiple Approaches to Study *S. cerevisiae* Rad9, a Prototypical Checkpoint Protein

By AISLING M. O'SHAUGHNESSY, MURIEL GRENON, CHRIS GILBERT, GERALDINE W.-L. TOH, CATHERINE M. GREEN, and NOEL F. LOWNDES

Abstract

The *Saccharomyces cerevisiae RAD9* checkpoint gene is the prototypical checkpoint gene and is required for efficient checkpoint regulation in late G1, S, and at the G2/M cell cycle transition following DNA damage. Rad9 is required for the activation of Rad53 after damage and has been proposed to have roles in lesion recognition as well as DNA repair and the maintenance of genome stability. Here we describe methodology suitable for the study of G1, intra-S, and G2/M checkpoints in budding yeast, the analysis of Rad9/Rad53 phospho-forms, the biochemical analysis of Rad9 and Rad53, the fractionation of soluble and chromatin associated proteins, including Rad9, and the live cell imaging of GFP tagged Rad9.

Introduction

Damage to genomic DNA caused by ionizing radiation, ultraviolet radiation, as well as other environmental mutagens is a constant problem for eukaryotic cells. However cells have evolved sophisticated surveillance mechanisms, termed cell cycle checkpoints, to monitor for the presence of DNA damage (Hartwell and Weinert, 1989). The checkpoint response is a largely intranuclear signal transduction cascade that once activated, induces cell cycle arrest or slow-down at key cell cycle transitions or phases (reviewed in Nyberg *et al.*, 2002), induces a DNA damage-specific transcriptional program, regulates specific DNA repair mechanisms (Aboussekhra *et al.*, 1996; de la Torre-Ruiz *et al.*, 1998) and, presumably if all else fails apoptosis (Yamaki *et al.*, 2001). Full checkpoint activation in budding yeast is dependent on Mec1/Tel1 PIKKs (phosphoinositide 3-kinase-like kinases), as well as downstream effector kinases, such as Rad53 and Chk1 (de la Torre-Ruiz *et al.*, 1998; Sanchez *et al.*, 1999). In addition, other important regulators of the checkpoint response include Rad9 and the

METHODS IN ENZYMOLOGY, VOL. 409
0076-6879/06 $35.00
DOI: 10.1016/S0076-6879(05)09008-7

members of the Rad24 epistasis group: *RAD24, RAD17, MEC3*, and *DDC1*. *RAD24* encodes a protein homologous to all five subunits of the "sliding clamp loader," Replication Factor C (RFC) is found in a complex with Rfc2, 3, 4, and 5. Rad17, Mec3, and Ddc1 form a heterotrimeric sliding clamp homologous to replicative sliding clamp, Proliferating Cell Nuclear Antigen (PCNA) (Green *et al.*, 2000; Majka and Burgers, 2003).

The *S. cerevisiae RAD9* checkpoint gene is the prototypical checkpoint gene being required for efficient checkpoint regulation in late G1 (G1/S boundary) (Siede *et al.*, 1993, 1994) intra-S (Paulovich *et al.*, 1997), and the G2/M cell cycle transition (Hartwell, 1989) following DNA damage. Here we describe assays used to study activation of these DNA damage checkpoints. Further study showed that Rad9 is phosphorylated during S/G2/M phases of the cell cycle in undamaged cells. These forms of Rad9 are collectively termed hypophosphorylated Rad9 and their function is currently unknown. Following damage, Rad9 is hyperphosphorylated in a Mec1/Tel1-dependent manner (Emili, 1998; Vialard *et al.*, 1998) and this event is often used as an early marker of checkpoint activation. Cell cycle and hyperphosphorylated Rad9 can be resolved using specific polyacrylamide gel running conditions that will be presented. Hyperphosphorylated Rad9 has been shown to mediate activation of the downstream kinase, Rad53 (Vialard *et al.*, 1998). Rad9 exists in two distinct soluble complexes in cell extracts (Gilbert *et al.*, 2003). The larger complex (>850 kDa), found in undamaged cells contains hypophosphorylated Rad9 associated with chaperone proteins, whereas the smaller complex (560 kDa), which forms after damage, contains the hyperphosphorylated Rad9 and the Rad53 kinase (Gilbert *et al.*, 2001). Purifying the Rad9 complex to homogeneity allows the determination of the components of both the larger and smaller complexes of Rad9, while gel filtration and glycerol gradient sedimentation allows both the size and shape of the Rad9 complex to be determined. We have previously described the detailed methodology used for the purification of the Rad9 complexes and the hydrodynamic analysis of Rad9 and other checkpoint proteins (Gilbert *et al.*, 2003; Green and Lowndes, 2004; Green *et al.*, 2000). A role for Rad9 in the activation of Rad53 after damage was addressed using the Rad53 release and catalysis assay described in this chapter. These experiments support a model in which hyperphosphorylated Rad9 serves primarily as a scaffold protein and recruits incompletely phosphorylated Rad53 molecules, bringing them into close proximity. This results in an increase in Rad53 local concentration and facilitates *in trans* autophosphorylation of Rad53, which presumably results in full activation of Rad53 (Gilbert *et al.*, 2001; van den Bosch and Lowndes, 2004).

Numerous checkpoint proteins have been shown to interact with damaged DNA (Lisby, 2004; Melo *et al.*, 2001). Our data suggests that at least some forms of Rad9 are recruited to sites of DNA damage (unpublished data). Here, we described two techniques, live cell imaging of a Rad9 GFP fusion protein and fractionation of chromatin-associated proteins, which we have optimized to successfully study Rad9 recruitment to damaged chromatin.

Cell Cycle Checkpoint Analysis

One of the consequences of the DNA damage checkpoint being activated is the temporary arrest or slow-down of the cell cycle preventing cells from replicating and segregating damaged DNA (Hartwell and Weinert, 1989). This delay occurs at key cell cycle transitions that are tightly regulated in all eukaryotic cells: in late G1 phase to prevent cells from committing to a new cell cycle (G1 checkpoint) and, at the end of the G2 phase to prevent mitotic entry in the presence of DNA damage (G2/M checkpoint). In addition, S phase is slowed down in presence of DNA damage (intra-S phase checkpoint) to allow complete and correct DNA replication.

In both of the widely used yeast model systems, *S. cerevisiae* and *Schizosaccharomyces pombe*, cell cycle perturbation is easy to detect by simply observing cells by microscopy (Reed *et al.*, 1995), since yeast cells show specific cell morphology relating to different stages of the cell cycle (Fig. 1). Therefore, methodologies to study cell cycle delays induced by DNA damage are relatively simple, requiring only simple microscopic observation of cells, either unstained or suitably stained (e.g., DAPI staining of DNA) using transmission or fluorescence microscopy. In addition, DNA content is easily assessed using fluorescence-activated cell sorting (FACS) analysis to follow S-phase progression. For the experiments described later, cell samples were fixed with formaldehyde for microscopic analysis or ethanol for FACS analysis. DAPI staining and FACS analysis were performed as previously described (Haase and Lew, 1997; Jacobs *et al.*, 1988; Moreno *et al.*, 1991).

S. cerevisiae divides by budding (see Fig. 1A). An exponential growing culture is roughly composed of one-third of unbudded G1 cells, one-third of very small budded cells, which have just passed START (the point of irreversible commitment into the next cell cycle and equivalent to the restriction point in mammals) and entered S phase, and one-third of large budded cells, which range from late S, through G2 and into M phase. Note that in *S. cerevisiae*, the G2/M transition is not as well defined as in fission yeast or higher cells. Events classically occurring in mitosis are less obvious or differently regulated. For example chromosomal condensation is not

A   *S. cerevisiae* cell cycle:

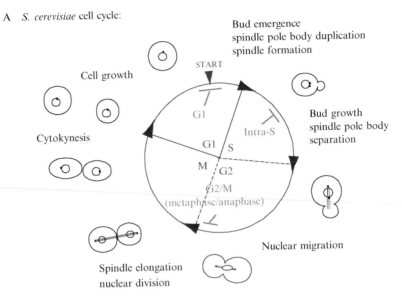

Bud emergence
spindle pole body duplication
spindle formation

Cell growth

START

Cytokynesis

G1

Intra-S

G1   S

M   G2

G2/M
(metaphase/anaphase)

Bud growth
spindle pole body
separation

Nuclear migration

Spindle elongation
nuclear division

B   Specific cell morphologies:

Normal cell cycle forms:

Unbudded   Small budded   G2/M

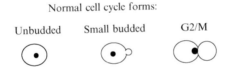

Synchronised cells:

Alpha factor   Nocodazole

G2/M checkpoint arrest:

Fig. 1. The cell cycle of budding yeast. (A) The budding yeast cell cycle showing the points of action of the three principal checkpoint arrests. (B) Specific cell morphologies observed in the normal cell cycle, after synchronization in G1 (α factor) and G2/M (nocodazole) and at the G2/M checkpoint. (See color insert.)

readily detectable and spindle pole formation is initiated during DNA replication to facilitate budding and nuclear migration. Therefore, the G2/M checkpoint is more accurately described as a metaphase/anaphase checkpoint, and it is this transition that is regulated by the DNA damage checkpoint. At this checkpoint, cells arrest as large budded cells with the nucleus in the bud neck and, as cell growth continues during the arrest, daughter cells attain a similar size to mother cells (Fig. 1B). In response to DNA damage, *S. cerevisiae* also arrests at the end of the G1 phase, before START. In addition, S phase progresses significantly slower in presence of DNA damage and this can be detected by FACS analysis.

It is possible to follow cell cycle delay induced by DNA damage in an asynchronous population of exponentially dividing cells. This assay provides a reliable and easy method to monitor a general checkpoint defect in cells that are not physiologically perturbed in any way other than the damage treatment itself. The assay measures the accumulation of cells at the G2/M transition, which is visible two hours after irradiation. DAPI staining can be used to count large budded cells with a single nucleus at the bud neck (G2/M cells). However, to more specifically assess checkpoint activation in individual phases of the cell cycle, cells should be synchronized in specific stages of the cell cycle. Furthermore, lesions that are generated in one phase of the cell cycle, but are not detected in that phase because of a checkpoint defect, can still activate the following checkpoint. Therefore, some defects cannot be detected using a minimally perturbing assay on asynchronous cells (Grenon *et al.*, 2001).

The ability to synchronize cells at a certain stage of the cell cycle and, once arrested, to treat them with a DNA damaging agent followed by their immediate release from the arrest, allows independent assessment of the integrity of the G1, intra-S, and G2 checkpoints, by specifically analyzing the relevant cell cycle transition. Regardless of whether asynchronously growing or synchronized cells are used, it is essential to perform the experiment on genuinely cycling cells. Therefore, dilution of saturated overnight cultures with fresh media should not be used as a significant and variable proportion of the cells in the population will be noncycling. Cell cultures should be appropriately inoculated the evening before to be at exponential phase the following day when the experiment is to be started. Cell cultures should be at a concentration of 5×10^6 cells/ml (i.e., early to mid-logarithmic phase in rich medium) with a minimum budding index of 70%. Synchronization methods in G1 and G2 often rely on the use of alpha factor (Breeden, 1997) and nocodazole respectively. Alpha factor (α factor) is a physiological yeast peptide, which arrests *MATa* cells in G1 before START (i.e., as unbudded cells; Fig. 1B). Nocodazole is a microtubule inhibitor, which arrests cells as large budded cells in late G2 with the nucleus in the center of mother cells, (i.e., just prior to the metaphase/anaphase checkpoint arrest point; Fig. 1B). The concentration of each of these components must be optimized by prior titration for each strain. Note that: (1) The G1 arrest can be maintained for longer (from 120 to 180 min once arrested) than the G2 arrest induced by nocodazole (about 60 to 80 min once arrested). (2) The highest degree of synchrony is achieved by performing all experiments in rich media and releasing cells as soon as the full arrest point has been reached. (3) The genetic dependency of checkpoint activation can be different depending on whether cells are grown in minimal or rich media. For example, the involvement of Tel1 in hyperphosphorylation of Rad9 is less detectable when experiments are performed in minimal media (Jorge Vialard, personal communication).

Both the G1 and intra-S checkpoints can be assessed in a single experiment. In this experimental protocol, cells synchronized in G1 (>95% unbudded cells, 0% small budded cells) are released after irradiation into medium without α factor, but containing nocodazole, to arrest in G2/M all those cells that successfully complete S phase and preventing them from entering a new cell cycle. Cell cycle progression is monitored by both budding index, to measure exit from G1 therefore allowing quantification of the G1 checkpoint activation (Fig. 2A), and FACS analysis of DNA content, to measure S-phase progression and thereby quantify activation of the intra-S phase checkpoint in response to DNA damage (Fig. 2B).

To specifically assay the induction of the G2/M checkpoint, corresponding to metaphase/anaphase arrest in budding yeast, the reciprocal experiment to the one described previously should be performed. In this protocol, cells synchronized in G2 using nocodazole (>80% large budded cells, 0% small budded cells) are released after irradiation into medium without nocodazole, but containing α factor, thus preventing cells that have successfully completed mitosis from cycling further by arresting them in G1. Cell cycle progression can be monitored by budding index, as the quantification of large budded cells will measure the extent of time cells spend in G2 and M, by FACS analysis of DNA content, to measure the time required for cells to accumulate 1C DNA content cells and, finally, by DAPI staining of cells, which will provide a more accurate assessment of cell cycle stage, particularly during mitosis (metaphase, as well as early and late anaphase cells can be distinguished by DAPI staining). Precise determination of mitotic arrest can be of importance since the G2/M checkpoint not only controls the metaphase/anaphase transition but also the exit of mitosis (Sanchez et al., 1999). Different quantification methods can be found in the literature, including the quantification of G2/M cells (nucleus in the bud neck, see Fig. 3B) and the counting of bi-nucleated cells (Longhese et al., 1997).

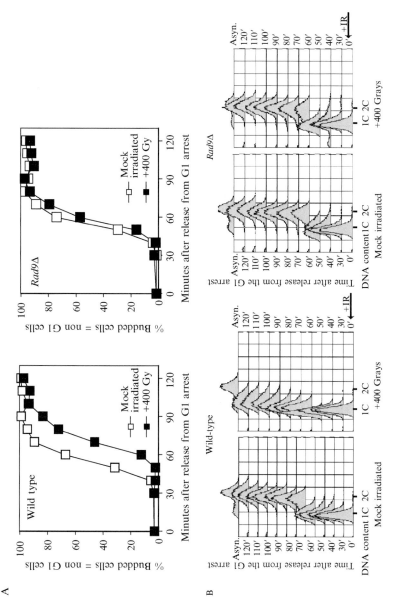

FIG. 2. Assaying the G1 and intra-S checkpoints in budding yeast. (A) The G1 checkpoint. Budding index analysis reveals the G1 checkpoint response of wild type and *rad9Δ* cells after ionizing radiation (400 Gy). (B) The intra-S checkpoint activation. FACS analysis reveals the intra-S checkpoint response of wild type and *rad9Δ* cells after ionizing radiation (400 Gy). Note that this experiment was done using centrifugation.

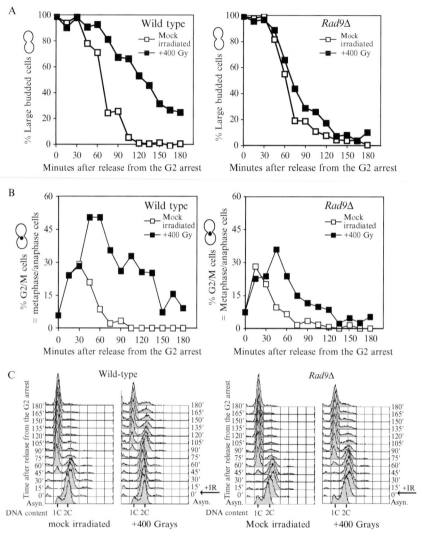

FIG. 3. Assaying the G2/M checkpoint in budding yeast. (A) Quantification of the G2/M checkpoint by large budded cells analysis in wild type and *rad9Δ* cells after ionizing radiation (400 Gy). (B) Quantification of the G2/M checkpoint by G2/M cells (DAPI) analysis in wild type and *rad9Δ* cells after ionizing radiation (400 Gy). (C) Quantification of the G2/M checkpoint by FACS analysis in wild type and *rad9Δ* cells after ionizing radiation (400 Gy). Note that this experiment was done using centrifugation.

1. Asynchronous cell cycle checkpoint experiment.

Cells are grown exponentially in YPD to a density of 2×10^6 cells/ml. Cells are collected and treated with DNA damaging agents or mock treated. Cells are returned to 30° and formaldehyde fixed samples are taken every hour (for 8 h) to determine the budding index and the percentage of G2/M cells.

2. G1 and intra-S checkpoint.

Cells are grown exponentially in YPD to a density of 5×10^6 cells/ml. Cells are arrested by addition of α factor (5 μg/ml final concentration is typical for wild-type strain, W303 but the optimum concentration should be determined for each batch of alpha factor and each strain). For W303, α factor arrest should take between 90 to 105 min. Once arrested, as judged by the absence of budded cells (in practice a budding index of <5% but more importantly, the absence of small budded cells serves as a good guideline) the culture is divided into two. One-half of the culture is mock treated while the other half is treated with the desired dose or concentration of damaging agent. The cells are then washed once with prewarmed saline (0.9% NaCl), once with prewarmed YPD and resuspended in YPD containing nocodazole at 5 μg/ml (Stock 5 mg/ml in DMSO). Time 0 is defined as the time when cells are put back in YPD during the first wash. Cells can be washed by centrifugation or using a multiple filtration station attached to a vacuum unit (type Biorad, technique described in Amon, 2002), which allows for faster and more efficient washes giving a higher degree of synchrony. Samples should be harvested every 10 min for 2 h, for budding analysis and FACS analysis in order not to miss S-phase progression (shift from 1C to 2C DNA content). The first sample is taken as quickly as possible. Typical budding index and FACS profiles are shown in Fig. 2A and B that quantify G1 and intra-S checkpoint activation for cells treated with ionizing radiation (400 Gy).

3. G2 checkpoint.

Cells are grown exponentially in YPD to a density of 5×10^6 cells/ml. Cells are arrested by addition of nocodazole at a final concentration of 5 μg/ml. Once arrested as judged by large budded cells (>80%) and the absence of small budded cells (it takes between 90 min and 105 min for wild type W303), cultures are divided into two. One half of the culture is mock treated and the other half is treated with the appropriate damaging agent. After treatment, cells are washed once with saline, once with YPD (as stated previously), and resuspended in YPD containing alpha factor at 5 μg/ml. Time 0 is again defined as the time when cells are resuspended in YPD during the first wash. Budding and FACS samples are taken every 15 min for 3 h. The budding index, proportion of G2/M checkpoint cells obtained by DAPI staining and FACS profiles after treatment with 400 Gy γ-irradiation are shown in Fig. 3A, B, and C.

Analysis of Rad9 Phospho-Forms by Western Blotting

As mentioned earlier, Rad9 is phosphorylated during the S/G2/M phases of the cell cycle in undamaged cells and these forms are collectively referred to as the hypophosphorylated Rad9. Although *RAD9* encodes a protein that is 1309 amino acids with a molecular weight of 148 kDa, Western blot analysis of cell lysates from asynchronously growing W303 cells display a broad band of polypeptides with mobilities ranging in size from 180 to 220 kDa (Fig. 4). Cells arrested in G1 with α factor (~ 2 μg/ml for 150 min) contain only the faster migrating forms (~ 180 kDa) (Fig 4). However cultures arrested in early S phase with hydroxyurea, (0.2 M for 120 min) which blocks the ribonucleotide reductase, exhibit both the faster and several slower migrating forms of Rad9 (Fig. 4). Cells arrested in G2/M with nocodazole (5 μg/ml for 120 min) display an increased accumulation of the slower migrating forms and the disappearance of the faster migrating forms (Fig. 4). Following damage of asynchronously growing W303 cells, there is a pronounced change in the Rad9 mobility profile. The slower migrating forms of Rad9 present before irradiation disappear while an even slower migrating form, termed hyperphosphorylated Rad9, appears (Fig. 4). Hyperphosphorylated Rad9 cannot be detected in whole cell extracts from mock treated cells. Good quality Rad9 Westerns can be achieved using trichloroacetic acid for whole cell extractions (Vialard *et al.*, 1998).

These Rad9 phospho-forms are best resolved on a 6.5% SDS-PAGE gel containing an acrylamide to bis-acrylamide ratio of 80:1. This ratio results in less cross-links and improves the resolution of phospho-forms

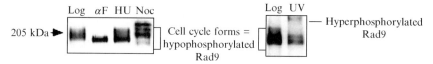

FIG. 4. Phospho-forms of Rad9 present in normally proliferating cells and after DNA damage. Left panel, Western blot analysis of Rad9 in asynchronous, G1, S, and G2-arrested cells (hypophosphorylated Rad9). Right panel, Western blot analysis of Rad9 in asynchronous, exponentially growing and UV treated (60 J/m^2) cells (hyperphosphorylated Rad9). Figure adapted from Vialard *et al.*, 1998 with permission from *EMBO Journal.*

of Rad9, in particular the cell cycle forms that are not otherwise readily detectable. Transfer can be either semi-dry or wet onto a nitrocellulose membrane (Hybond C, Amersham, Bucks, UK). The rabbit polyclonal antibody (NLO5) raised against the N-terminal region of Rad9 has been previously described (Vialard *et al.*, 1998). Note that the NLO5 rabbit polyclonal serum detects Rad9 more weakly as it becomes increasingly phosphorylated after damage. In brief the Western is performed as follows: the membrane is incubated overnight at room temperature with a 1/10,000 dilution of NLO5 in PBS containing 0.02% TWEEN20 and 0.5% milk proteins. Preincubation of the filter with PBS and milk proteins should not be performed. The filter is then washed with PBS containing 0.02% TWEEN20 by successive, rapid (<10 sec), small volume (20–30 ml) washes with a total of 400 ml PBS + TWEEN20 (0.02%) at room temperature. Secondary horseradish peroxidase, conjugated anti-rabbit antibody (Amersham) at 1/20,000 dilution in PBS/0.02% TWEEN20 is incubated with the membrane for 45 min at room temperature. The blot is revealed by chemiluminescence (Pierce, Rockford, IL) and exposed for approximately 30 s to 1 min. Short exposures are required to observe the different cell cycle forms. We and others have also detected Rad9 in Western blots by N-terminally tagging Rad9 with the hemagglutinin (HA) epitope (Naiki *et al.*, 2004; Schwartz *et al.*, 2002; Soulier and Lowndes, 1999), which can be detected by the commercially available mouse 12CA5 monoclonal antibody. This method has the advantage that Rad9 phospho-forms are detected as easily as nonphosphorylated Rad9, which is not the case for the NLO5 antibody. If the previously described running conditions are used, the different phospho-forms of Rad9 can be detected using this antibody.

Rad53 Release and Catalysis Assays

The strain used in these assays contains N-terminally tagged Rad9 with a single HA (hemagglutinin) epitope and 10 histidine residues (i.e., 1HA-10His). This strain is herein referred to as HH-RAD9 (Gilbert *et al.*, 2001; Vialard *et al.*, 1998). These assays are performed with partially purified HH-Rad9 complexes and specifically address the activation of Rad53 by hyperphosphorylated Rad9 that occurs in the soluble hyperphosphorylated Rad9-Rad53 complex. Firstly, in the Rad53 release assay, *in vivo* Rad53, which is partially activated by Mec1, is further phosphorylated by *in trans* autophosphorylation and then released from hyperphosphorylated Rad9 in an ATP dependent fashion. Secondly, in the Rad9 catalysis assay, hyperphosphorylated Rad9 is used to catalyse the *in trans* autophosphorylation of de-phosphorylated, bacterially produced recombinant Rad53.

1. The Rad53 release assay.

100 μl of crude whole cell extract (approximately 1 to 1.5 mg of total protein as determined by Bradford protein assay) from irradiated or non-irradiated *HH-RAD9* cells is bound to 10 μl of 12CA5 protein G beads (equilibrated with 1× lysis buffer: 100 mM potassium acetate [KAc] 50 mM HEPES/KOH pH 7.5, 10% glycerol, 0.5 mM EDTA, 4 mM β-mercaptoethanol, 0.5% NP40, 1× protease inhibitors [see recipe later], 1× phosphatase inhibitors [see recipe later]). Note: slight variations of this recipe will be used throughout. The beads are incubated with extract on a rotary shaker at 4° for 3 h, centrifuged at 3000 rpm for 1 min, and washed five times with the same lysis buffer. As much of the buffer as possible is removed using thin gel loader tips (Eppendorf, G. Kisker, Steinfurt, Germany) and resuspended in 1× kinase buffer (25 mM Hepes pH 7.5, 5 mM EGTA, 15 mm MgCl$_2$, 15 mM KAc, 1 mM ATP, 1× protease, and 1× phosphatase inhibitors). This is incubated for 30 min at 25°. Centrifuge and carefully remove the supernatant using the thin gel loader tips. The supernatants and boiled beads are analyzed by Western blotting.

2. Catalysis assay.

This assay is used to investigate Rad9's ability to directly catalyze Rad53 *in trans* autophosphorylation. It is performed in essentially the same way as the release assay except that 25–100 ng of recombinant de-phosphorylated Rad53 is added to the Rad9 beads prior to incubation with 1 mM ATP.

To obtain recombinant Rad53, a plasmid containing full length Rad53 amino-terminally epitope tagged with six histidine residues (H-Rad53) is expressed in *E. coli* to an OD$_{600}$ of 0.2 units. Expression of H-Rad53 is induced in a 400–800 ml culture, using 1 mM IPTG at 37° until the OD$_{600}$ reaches 0.6 units. Recombinant H-Rad53 is purified to homogeneity using Ni^{2+}-NTA (Qiagen, Valencia, CA) affinity chromatography. A 1 ml Ni^{2+} resin is used and elution conditions are as per manufacturer's instructions. Dephosphorylation of H-Rad53 is achieved by treating it (0.05 mg) with 4 units of lambda protein phosphatase (NEB) and incubating for 30 min at 30°. The reaction is stopped by adding 20-fold of 1× lysis buffer also containing 10 mM sodium vanadate and 50 mM sodium fluoride, as recommended by the manufacturer. The dephosphorylated H-Rad53 is repurified by binding for 1 h at 4° to Ni-NTA beads, which have been equilibrated with 1× lysis buffer containing 1× protease inhibitors (100× stock in ethanol: 25 μg/ml leupeptin, 125 μg/ml pepstatin A, 20 μg/ml PMSF, 30 mg/ml benzamidine, 125 μg/ml antipain, 80 μg/ml chymostatin) and 1× phosphatase inhibitors (50× stock in H$_2$O: 2 mg/ml NaF, 10 mg/ml

β-glycerophosphate, 2 mg/ml Na_3VO_4, 20 mg/ml EGTA, 100 mg/ml sodium pyrophosphate; in H_2O). The beads are washed three times with the same lysis buffer also containing 10 mM imidazole, poured into a mini column (1 ml syringe as stated previously) and eluted in 50 μl fractions with lysis buffer containing 200 mM imidazole. Fractions are checked on a 10% Coomassie-stained acrylamide gel. Following the catalysis incubation, any remaining Rad53 from the cell extracts is separated from the H-Rad53 by diluting the reaction 10-fold in 8 M urea, 0.01 M Na_2HP0_4, and 0.1 M Tris/HCl [pH8] (as for Rad53 release assay, previously). The histidine-tagged H-Rad53 is bound to pre-equilibrated Ni-NTA agarose beads as stated previously, washed three times with 1 ml of 1× lysis buffer, and boiled with 1× Laemmli buffer for analysis by Western blotting.

Fractionation of Soluble and Chromatin-Associated Proteins

This technique allows the detection of proteins specifically recruited to undamaged or damaged chromatin and is an adaptation of a previously published protocol (Fig. 5) (Liang and Stillman, 1997). Using this technique Rad9 can be detected on the chromatin after ionizing radiation (unpublished data).

Cells (1×10^8) are harvested by centrifugation (2000 rpm), washed with cold PBS (5 min), and the cell pellet is frozen on dry ice. Cell pellets are resuspended in 1.5 ml of Cell Suspension Buffer (CSB: 50 mM HEPES-KOH [pH 7.5], 150 mM NaCl, 0.8 M sorbitol, 10 mM DTT) in a 2-ml Eppendorf and incubated for 10 min at 30° with gentle shaking, followed by centrifugation for 5 min at 2000 rpm. The pellet is resuspended in 1.5 ml Spheroplasting Buffer (SB: 50 mM HEPES-KOH [pH7.5], 10 mM DTT, 0.8 M sorbitol) or equivalent amount so that A_{600} of 10 μl in 1 ml water is between 0.2–0.3. 20 μl of 20 mg/ml Zymolyase 100T (ICN) is added and incubated at 30° with mixing. The cells are spheroplasted for approximately 10 min at 30° until A_{600} has fallen to 1/10th of starting value. Alternatively one can check microscopically by mixing 5 μl of cells with 5 μl 10% SDS. Lysis should be >90%. Spheroplasts are pelleted by centrifugation in a cooled microfuge at 4°, 2000 rpm for 2 min. The size of the pellet is approximately 100–150 μl. Spheroplasts are washed in 1 ml of ice-cold SB without DTT, then 1 ml ice-cold Spheroblast Wash Buffer (SWB: 100 mM KCl, 50 mM HEPES-KOH [pH7.5], 2.5 mM $MgCl_2$, 0.6 M sorbitol). To obtain optimal yield of cells and DNA, and to avoid difficulty in repelleting the spheroplasts, it is best not to resuspend the pellet but rather add the wash buffer over the pellet and place the Eppendorf at 180° position (opposite side) relative to its position in the last spin. Spheroplasts are washed by spinning pellet from one side of the tube to the other in a

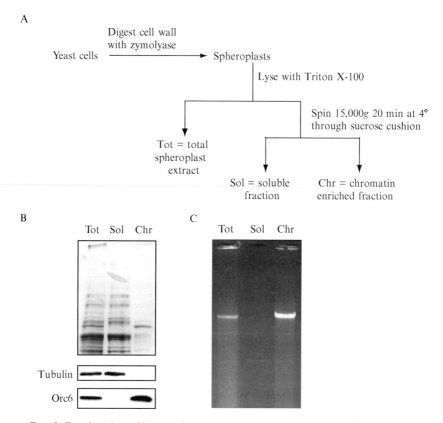

FIG. 5. Fractionation of chromatin bound and soluble proteins. (A) Schematic of the fractionation of total yeast spheroplast extract (Tot) and soluble (Sol) and chromatin-enriched (Chr) fractions. (B) Electrophoretic and Western blot analysis of fractionated protein. (C) Electrophoretic analysis of fractionated DNA. (See color insert.)

fixed-angle microfuge at 4°, 2000 rpm for 2 min. Resuspend washed pellets in ice-cold Extraction Buffer with Triton X-100 (ExB: 100 mM KCl, 50 mM HEPES-KOH [pH 7.5], 2.5 mM MgCl$_2$, 1× protease, and 1× phosphatase inhibitors, 1/5 volume of a 10% Triton X-100 solution) to a total volume of approximately 300 μl (scale up as necessary). Pipette with a cut-off P1000 tip until no clumps remain, then split the sample into two; 1/4 for total fraction and 3/4 for the soluble and chromatin-enriched fractions. To lyse spheroplasts, 1/3 volume of ExB plus inhibitors with Triton X-100, to 0.5% final concentration are added. Spheroplasts are left on ice for approximately 5 min, mixing occasionally by gentle inversion. Complete lysis is verified microscopically.

For the soluble/chromatin-enriched fraction, underlay the lysate with 50 μl of 30% sucrose, by using a cut off tip and adding the sucrose to the bottom of the tube. The lysate is spun at full speed for 20 min at 4° in a cooled microfuge. The supernatant (Soluble) is removed first, followed by the sucrose layer. The supernatant is snap frozen on dry ice. The pellet (Chr) is washed with 50 μl of extraction buffer (EB) without Triton X-100 and spun for 5 min at 10,000 rpm. Finally, the pellet is resuspended in 50–100 μl of EB plus inhibitors and $3\times$ Laemmli buffer is added, which is then boiled for 3 min. This is sonicated to fragment chromatin and release proteins. Sonication time can vary depending on the starting amount of cells and the efficiency of spheroplasting. However as the chromatin becomes fragmented, the size of the insoluble pellet decreases in size, following brief centrifugation. Successful fractionation is confirmed by analysis of total protein and DNA, in addition to probing with antibodies specific to soluble or chromatin bound proteins, for example, against alpha-tubulin and Orc6 (Diffley *et al.*, 1994; Liang and Stillman, 1997) (Fig. 5B). Fractions are then analyzed by Western blotting with antibodies to the relevant proteins of interest.

Live Cell Imaging

Live cell fluorescence microscopy is a powerful technique that is now widely used in studies of gene function and that has many advantages over indirect immunofluorescence. The most important advantage is that the localization and dynamics of proteins can be studied in real time in living cells. Thus the sequence of recruitment of checkpoint and repair proteins following DNA damage, for example, after double-strand breaks (DSBs) can be studied in great detail (Lisby *et al.*, 2004).

1. Construction of GFP fusion proteins.

Addition of GFP or any other tag to a protein of interest can be achieved most easily at either the N- or C-terminus. However there is no rule for choosing either terminus, as this can be both protein and tag dependent. In general, it is best to avoid, where possible, important domains located in the immediate N- or C-termini of the protein of interest but functionality of each tagged protein must be determined empirically for any tag used. We have tagged Rad9 at the genomic locus at the C-terminus using an integrative plasmid pJK1 (Melo *et al.*, 2001) which contains the carboxyl terminal of Rad9 fused to GFP (a kind gift from D. Toczyski). Other methods available to tag any gene of interest with GFP include the adaptamer-mediated PCR method (Reid *et al.*, 2002) and the PCR-based technique that allows for a one-step tagging of chromosomal genes at the

C-terminus as previously described (Longtine *et al.*, 1998; Wach *et al.*, 1997). It is also possible to tag a gene with GFP at the N-terminus as described by Kohlwein's group (Prein *et al.*, 2000).

To be useful, GFP fusion proteins must retain the properties of the untagged protein when expressed at similar levels. It is therefore essential to test functionality of the protein in as many assays as possible. For Rad9-GFP we have tested this in four assays: 1) survival after IR; 2) level of expression of the tagged protein; 3) Rad9 and Rad53 phosphorylation after IR; and 4) cell cycle delay after IR (using the protocol described in Fig. 3A,B).

3. Live cell fluorescence microscopy of Rad9.

Previously Rad9 has been reported to form very weak foci after damage that are difficult to detect above the diffuse background nuclear fluorescence (Melo *et al.*, 2001). The following protocol facilitates the detection of Rad9 foci after ionizing radiation. Background fluorescence can serve as a problem when it comes to visualizing foci formation after damage. Strains defective in the *ADE1* and *ADE2* genes accumulate a bright fluorescent pigment in the media, due to an intermediate in the adenine biosynthesis pathway, phosphoribosyl aminoimidazole (Ishiguro, 1989). It is therefore necessary to remove rich or mininal media prior to visualization (see later). For cells that are grown in minimum media, it is not necessary to wash the cells. However, the conditioned minimum media is replaced with fresh media prior to visualization as some byproducts of metabolism cause high background fluorescence. Cells that are grown in rich media are harvested by brief (30 sec) centrifugation in a microfuge at 12,000 rpm, media is removed and cells washed at least five times in prewarmed saline. Cells should be maintained at 30° during this period using a heating block to avoid temperature shock of the cells that can cause abnormal or "squashed" nuclear morphology. Following washing, the cells are resuspended in approximately 10 μl of prewarmed saline (0.9%).

A 2% solution of low melting point agar is used for mounting the cells onto prewarmed slide cover slips. Approximately 2 μl of the cells are transferred onto a coverslip, which is placed on a heating block at 30°, on top of which 2 μl of agar is layered. The prewarmed slide is immediately placed on top and a slight pressure applied.

4. Microscopic imaging.

A microscope enclosed in a temperature-controlled box is used (it is important that the stage and objectives are all at 30°) and is turned on for at least 60 min prior to imaging. For imaging, it is only possible to process two strains in a single experiment because of time limitations during the

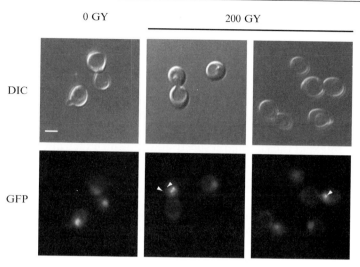

FIG. 6. Rad9 foci formation after IR. Representative images of Rad9-GFP foci in asynchronously growing WT cells before and after 200 Gy ionizing radiation. Rad9-GFP foci are indicated by arrow heads. Scale bar is 5 μM.

processing (washing) and imaging of cells. Approximately 10 min is required to image enough cells to score ($\sim$150) for each time point.

For the data shown in Fig. 6, live cell images were captured using an Orca-ER camera (Hamamatsu, Houston, TX) mounted on a Zeiss Axiovert S100 microscope (Carl Zeiss, UK) enclosed in a temperature-controlled box at 30°, with a PlanApo 63 × 1.4 NA objective with a Piezo Focus Drive (Physika Instrumente, Heidelberg, Germany). Excitation and emission filter wheels (Sutter, Novato, CA), transmission and fluorescence illumination shutters (Uniblitz, Rochester, NY) are controlled by a PC running AQM software (Kinetic Imaging, Nottingham, UK). For each field of cells, DIC and fluorescence images are taken in 15 Z-positions at 0.3-μM intervals. Fluorescence illumination is with an HBO 100 lamp, using either a single GFP filter set (excitation 470/40 nm, dichroic 440/80, emission 525/50, Omega Optical, Brattleboro, VT) or a dual GFP/RFP dichroic with separate excitation/emission filters (excitation 485/15 and 575/25, dichroic 490/575, emission 525/50 and 615/45, Omega Optical). Exposure times for Rad9 are typically 300 ms.

Summary

We have described various techniques that have been successfully used to study the function of DNA damage checkpoint proteins, in particular

Rad9. These include experiments designed to specifically investigate the G1, the intra-S, and G2/M phase checkpoints; Western blotting to resolve the different phospho-forms of Rad9, Rad53 release, and catalysis assays; chromatin fractionation; and finally live cell fluorescence microscopy. These and similar approaches in many laboratories worldwide have resulted in significant mechanistic understanding of the DNA-damage checkpoint pathway. However, a technical challenge for the future will be to devise novel methods for addressing the function of checkpoint proteins at, or close to, lesions in the context of the chromatin substrate with which they interact.

Acknowledgments

We thank Jorge Vialard for advice and members of the Genome Stability Laboratory for their critical review of the manuscript. Support from the Higher Education Authorities' Programme for Research in Third Level Institutions (PRTLI3) and from the Health Research Board (HRB) is gratefully acknowledged.

References

Aboussekhra, A., Vialard, J. E., Morrison, D. E., de la Torre-Ruiz, M. A., Cernakova, L., Fabre, F., and Lowndes, N. F. (1996). A novel role for the budding yeast RAD9 checkpoint gene in DNA damage-dependent transcription. *EMBO J.* **15,** 3912–3922.

Amon, A. (2002). Syncrhonisation procedures. *Methods Enzymol.* **351,** 457–467.

Breeden, L. L. (1997). Alpha-factor synchronization of budding yeast. *Methods Enzymol.* **283,** 332–341.

de la Torre-Ruiz, M. A., Green, C. M., and Lowndes, N. F. (1998). RAD9 and RAD24 define two additive, interacting branches of the DNA damage checkpoint pathway in budding yeast normally required for Rad53 modification and activation. *EMBO J.* **17,** 2687–2698.

Diffley, J. F., Cocker, J. H., Dowell, S. J., and Rowley, A. (1994). Two steps in the assembly of complexes at yeast replication origins *in vivo. Cell* **78,** 303–316.

Emili, A. (1998). MEC1-dependent phosphorylation of Rad9p in response to DNA damage. *Mol. Cell.* **2,** 183–189.

Gilbert, C. S., Green, C. M., and Lowndes, N. F. (2001). Budding yeast Rad9 is an ATP-dependent Rad53 activating machine. *Mol. Cell* **8,** 129–136.

Gilbert, C. S., van den Bosch, M., Green, C. M., Vialard, J. E., Grenon, M., Erdjument-Bromage, H., Tempst, P., and Lowndes, N. F. (2003). The budding yeast Rad9 checkpoint complex: Chaperone proteins are required for its function. *EMBO Rep.* **4,** 953–958.

Green, C. M., Erdjument-Bromage, H., Tempst, P., and Lowndes, N. F. (2000). A novel Rad24 checkpoint protein complex closely related to replication factor C. *Curr. Biol.* **10,** 39–42.

Green, C. M., and Lowndes, N. F. (2004). Purification and analysis of checkpoint protein complexes from Saccharomyces cerevisiae. *Methods Mol. Biol.* **280,** 291–306.

Grenon, M., Gilbert, C., and Lowndes, N. F. (2001). Checkpoint activation in response to double-strand breaks requires the Mre11/Rad50/Xrs2 complex. *Nat. Cell. Biol.* **3,** 844–847.

Haase, S. B., and Lew, D. J. (1997). Flow cytometric analysis of DNA content in budding yeast. *Methods Enzymol.* **283,** 322–332.

Hartwell, L. H., and Weinert, T. A. (1989). Checkpoints: Controls that ensure the order of cell cycle events. *Science* **246,** 629–634.

Ishiguro, J. (1989). An abnormal cell division cycle in an AIR carboxylase-deficient mutant of the fission yeast Schizosaccharomyces pombe. *Curr. Genet.* **15,** 71–74.

Jacobs, C. W., Adams, A. E., Szaniszlo, P. J., and Pringle, J. R. (1988). Functions of microtubules in the Saccharomyces cerevisiae cell cycle. *J. Cell. Biol.* **107,** 1409–1426.

Liang, C., and Stillman, B. (1997). Persistent initiation of DNA replication and chromatin-bound MCM proteins during the cell cycle in cdc6 mutants. *Genes Dev.* **11,** 3375–3386.

Lisby, M., Barlow, J. H., Burgess, R. C., and Rothstein, R. (2004). Choreography of the DNA damage response: Spatiotemporal relationships among checkpoint and repair proteins. *Cell* **118,** 699–713.

Longhese, M. P., Paciotti, V., Fraschini, R., Zaccarini, R., Plevani, P., and Lucchini, G. (1997). The novel DNA damage checkpoint protein ddc1p is phosphorylated periodically during the cell cycle and in response to DNA damage in budding yeast. *EMBO J.* **16,** 5216–5226.

Longtine, M. S., McKenzie, A., 3rd, Demarini, D. J., Shah, N. G., Wach, A., Brachat, A., Philippsen, P., and Pringle, J. R. (1998). Additional modules for versatile and economical PCR-based gene deletion and modification in Saccharomyces cerevisiae. *Yeast* **14,** 953–961.

Majka, J., and Burgers, P. M. (2003). Yeast Rad17/Mec3/Ddc1: A sliding clamp for the DNA damage checkpoint. *Proc. Natl. Acad. Sci. USA* **100,** 2249–2254.

Melo, J. A., Cohen, J., and Toczyski, D. P. (2001). Two checkpoint complexes are independently recruited to sites of DNA damage *in vivo. Genes Dev.* **15,** 2809–2821.

Moreno, S., Klar, A., and Nurse, P. (1991). Molecular genetic analysis of fission yeast Schizosaccharomyces pombe. *Methods Enzymol.* **194,** 795–823.

Naiki, T., Wakayama, T., Nakada, D., Matsumoto, K., and Sugimoto, K. (2004). Association of Rad9 with double-strand breaks through a Mec1-dependent mechanism. *Mol. Cell. Biol.* **24,** 3277–3285.

Nyberg, K. A., Michelson, R. J., Putnam, C. W., and Weinert, T. A. (2002). Toward maintaining the genome: DNA damage and replication checkpoints. *Annu. Rev. Genet.* **36,** 617–656.

Paulovich, A. G., Margulies, R. U., Garvik, B. M., and Hartwell, L. H. (1997). RAD9, RAD17, and RAD24 are required for S phase regulation in Saccharomyces cerevisiae in response to DNA damage. *Genetics* **145,** 45–62.

Prein, B., Natter, K., and Kohlwein, S. D. (2000). A novel strategy for constructing N-terminal chromosomal fusions to green fluorescent protein in the yeast Saccharomyces cerevisiae. *FEBS Lett.* **485,** 29–34.

Reed, S. I., Hutchison, C. J., and Macneill, S. (1995). Cell Cycle Control. Oxford University Press.

Reid, R. J., Lisby, M., and Rothstein, R. (2002). Cloning-free genome alterations in Saccharomyces cerevisiae using adaptamer-mediated PCR. *Methods Enzymol.* **350,** 258–277.

Sanchez, Y., Bachant, J., Wang, H., Hu, F., Liu, D., Tetzlaff, M., and Elledge, S. J. (1999). Control of the DNA damage checkpoint by chk1 and rad53 protein kinases through distinct mechanisms. *Science* **286,** 1166–1171.

Schwartz, M. F., Duong, J. K., Sun, Z., Morrow, J. S., Pradhan, D., and Stern, D. F. (2002). Rad9 phosphorylation sites couple Rad53 to the Saccharomyces cerevisiae DNA damage checkpoint. *Mol. Cell.* **9,** 1055–1065.

Siede, W., Friedberg, A. S., Dianova, I., and Friedberg, E. C. (1994). Characterization of G1 checkpoint control in the yeast Saccharomyces cerevisiae following exposure to DNA-damaging agents. *Genetics* **138,** 271–281.

Siede, W., Friedberg, A. S., and Friedberg, E. C. (1993). RAD9-dependent G1 arrest defines a second checkpoint for damaged DNA in the cell cycle of Saccharomyces cerevisiae. *Proc. Natl. Acad. Sci. USA* **90**, 7985–7989.

Soulier, J., and Lowndes, N. F. (1999). The BRCT domain of the S. cerevisiae checkpoint protein Rad9 mediates a Rad9-Rad9 interaction after DNA damage. *Curr. Biol.* **9**, 551–554.

van den Bosch, M., and Lowndes, N. F. (2004). Remodelling the Rad9 checkpoint complex: Preparing Rad53 for action. *Cell Cycle* **3**, 119–122.

Vialard, J. E., Gilbert, C. S., Green, C. M., and Lowndes, N. F. (1998). The budding yeast Rad9 checkpoint protein is subjected to Mec1/Tel1-dependent hyperphosphorylation and interacts with Rad53 after DNA damage. *EMBO J.* **17**, 5679–5688.

Wach, A., Brachat, A., Alberti-Segui, C., Rebischung, C., and Philippsen, P. (1997). Heterologous HIS3 marker and GFP reporter modules for PCR-targeting in Saccharomyces cerevisiae. *Yeast* **13**, 1065–1075.

Yamaki, M., Umehara, T., Chimura, T., and Horikoshi, M. (2001). Cell death with predominant apoptotic features in Saccharomyces cerevisiae mediated by deletion of the histone chaperone ASF1/CIA1. *Genes Cells* **6**, 1043–1054.

[9] Methods for Studying Adaptation to the DNA Damage Checkpoint in Yeast

By DAVID P. TOCZYSKI

Abstract

When yeast are faced with irreparable DNA damage, they will first arrest in G2/M, via the DNA damage checkpoint pathway, but will subsequently adapt to that arrest and resume division. Here, we summarize assays that we have used to examine checkpoint adaptation. Specifically, we discuss the merits of inducing DNA damage with ionizing radiation (IR) and IR-mimetic drugs, HO, and the *cdc13-1* mutation. We also discuss readouts that we have used to visualize adaptation.

Introduction

Cells alter their physiology in response to DNA damage by activating a signal transduction cascade referred to as the DNA damage checkpoint pathway (Melo and Toczyski, 2002; Nyberg *et al.*, 2002). This response affects several aspects of cell physiology, including transcription, repair, and cell cycle progression. These responses are thought, collectively, to aid the cell in repairing such damage and to avoid any permanent repercussions from it. Depending upon the nature of the initial insult, most cells are

METHODS IN ENZYMOLOGY, VOL. 409
0076-6879/06 $35.00
DOI: 10.1016/S0076-6879(05)09009-9

able to repair the lesion in question quite efficiently, although mutations or chromosomal rearrangements sometimes result as a consequence.

Occasionally, cells are unable to repair DNA damage during the arrest. When this occurs, cells are left with the options of either remaining arrested or temporarily ignoring the DNA damage and continuing cell division (Sandell and Zakian, 1993; Toczyski et al., 1997). This process is referred to as adaptation, and is conceptually analogous to adaptation seen in other signal transduction pathways, such as mammalian vision. The repercussions of adapting to irreparable DNA damage are variable. In the budding yeast *Saccharomyces cerevisiae*, it has been shown that cells are often able to repair DNA damage, such as double-stranded breaks (DSBs), in subsequent cell cycles (Galgoczy and Toczyski, 2001; Malkova et al., 1996; Sandell and Zakian, 1993; Toczyski et al., 1997). In addition, some studies suggest that adaptation may actually be required for some repair pathways, possibly due to these pathways being less efficient in metaphase (Galgoczy and Toczyski, 2001). Alternatively, a cell may lose the damaged chromosome altogether after checkpoint adaptation. The result of this will depend upon whether the damaged chromosome is essential. There are several situations in which the damaged chromosome may not be essential. For example, diploids are able to grow with only a single copy of one chromosome (monosomic diploids) with varying efficiencies, depending upon the chromosome in question. Diploids with a single copy of chromosome III, for example, will retain a division rate comparable to that of a euploid. This may be due in part to the fact that chromosome III encodes fewer genes than most of the other chromosomes. A haploid with one irreparably broken sister chromosome may also "survive," in that one of the two daughter cells formed after adaptation may receive the intact sister. Even the haploid daughter cell lacking a chromosome altogether could survive if it is able to mate in the subsequent G1. While each of these scenarios may be preferable to death for a yeast cell, metazoans need to be more careful about allowing a damaged cell to continue proliferation. While the process of adaptation has not been well characterized in mammalian cells, the choice between adaptation and apoptosis is likely to be strongly dependent on cell type.

Designing Adaptation Assays

Assays for checkpoint adaptation can vary in two ways: in the type of damage induced and in the readout for cell cycle progression. Importantly, the DNA damage that is induced to assay checkpoint adaptation must be irreparable. In yeast, this can be accomplished with either a DSB or by introducing damage-inducing temperature-sensitive mutations. The merits of each of these will be discussed. Several different readouts for checkpoint adaptation can be used. Cells can be either examined directly to determine whether they have progressed through the cell cycle (by microcolony assay) or scored for viability. Viability assays can only be used under special conditions, where the damage being induced is irreparable, but not lethal (Sandell and Zakian, 1993). Under these conditions, adaptation is required for viability, and thus adaptation-defective strains show lower plating efficiency (Toczyski *et al.*, 1997). In addition to these direct readouts, an adaptation defect can be inferred based upon phenotypes that have been observed in well characterized adaptation mutants, such as a decrease in chromosome loss rates. However, care must be taken to ensure that the mutants being examined do not exhibit other pleotrophic phenotypes that alter the interpretation of these assays.

Ensuring that DSBs Are Irreparable

Budding yeast repair DSBs most efficiently using homologous recombination (Krogh and Symington, 2004). While yeast are able to use other repair mechanisms, such as non-homologous endjoining (NHEJ) or direct ligation, these can be inefficient depending upon the method used to induce the DSB. During homologous recombination (HR), a DSB is repaired by using DNA replication off a perfect or near perfect template. Templates usually come in one of three forms: sister chromatids, homologues, or sequences at distinct loci that match the sequence at the DSB site. Sister chromatids are ideal templates in that the sequence is, by definition, identical. However, the ability to use a sister chromatid requires that the DSB occurs after the locus has been replicated. Thus, this option is not available in G1 or part of S phase. Repair off of a homologous chromosome is also efficient in budding yeast. When given a choice between a sister chromatid and a homologue that matches the break site perfectly, budding yeast will use the homologue only 10% of the time (Kadyk and Hartwell, 1992). When a DSB is introduced into a G1 haploid cell in which there exists neither a homologue nor a sister, the break typically cannot be repaired, resulting in lethality. The exception to this is situations in which the break site is flanked by direct repeats (which is rare) or situations in which a near perfect match to the break site occurs elsewhere in the genome. The most well characterized example of the latter is the repair of the enzymatically-induced break at the MAT locus on chromosome III (see later), which is efficiently repaired by homologous recombination using either of two template sequences at two different loci, HML and

HMR (Herskowitz *et al.*, 1992). In summary, the study of adaptation to an irreparable DSB must be designed in a system in which none of these repair mechanisms occurs with an appreciable frequency.

IR-Induced DSBs

DSBs are generally introduced by one of two methods: endonucleases or exposure to DNA damaging conditions. The latter is typically either ionizing irradiation (IR) or IR mimetic drugs, such as bleomycin or its derivatives, zeocin or phleomycin. The dose of IR required to induce one DSB per haploid yeast genome is quite high, as compared to mammalian cells. This is likely because the yeast genome is two orders of magnitude smaller and presents a smaller target. For this reason, gamma ray sources (which typically deliver a lower flux) are not as practical as X-ray machines when studying yeast. However, X-ray producing machines are not widely available. Zeocin, in contrast, is easy to use and can be purchased in pre-aliquotted sterile vials (Invitrogen, Grand Island, NY).

One can render IR or bleomycin-induced DSBs irreparable in either of two ways. First, haploid MATa cells can be arrested in G1 with alpha factor. When treated with IR, or zeocin, these cells are effectively repair-deficient (Kadyk and Hartwell, 1992). Alternatively, one can use cells that are mutated in genes required for HR. This is best done by deletion of the *RAD52* gene (Krogh and Symington, 2004). The use of a *rad52Δ* has several advantages. First, diploids can also be examined, since HR templated by a homolog similarly requires Rad52. Second, cells do not need to be synchronized in G1, which simplifies experiments and removes the requirement for MATa cells. Finally, we have found that the use of a *rad52Δ* yields cleaner data, likely owing to the difficulty in generating a culture in which 100% of cells are G1 arrested. For this reason, we have used *rad52Δ* strains almost exclusively when examining adaptation to IR (Galgoczy and Toczyski, 2001). The use of a *rad52Δ* also has several disadvantages. First is the simple fact that the deletion must be made in all strains to be examined. Second is the fact that *rad52Δ* strains are slightly sick, a problem which could be exacerbated in some mutant backgrounds. Finally, further deletions or epitope tagging cannot be carried out in *rad52Δ* strains, since gene targeting requires homologous recombination. This can be circumvented by: temporarily covering the *rad52Δ* with a *RAD52* plasmid; deleting *RAD52* as the last step in strain construction; or crossing other alleles to be used into the *rad52Δ* strain. It must be remembered, however, that *RAD52* is required for meiosis, and thus homozygous mutants cannot be used in crosses.

HO-Induced DSBs

The use of an endonuclease to generate a DSB allows for a much more precise examination of damage and circumvents many of the problems encountered when using IR. HO and SceI are two homing endonucleases that have been employed in budding yeast. Of these two, HO has been much more widely exploited (Herskowitz and Jensen, 1991). The use of HO as a regulatable damaging agent was greatly facilitated by placing it under the control of the *GAL1* promoter. Since site specific endonucleases recognize their targeting site on both sister chromatids, breaks can be induced in G1, G2, or asynchronous cultures without repair by sister chromatid recombination. In wild-type cells, HO recognizes a single target in the yeast genome at the MAT locus. This DSB is used by yeast to induce switching from one mating type to another by undertaking gene conversion of the HO-cut MAT site using either HML or HMR as a template (Herskowitz *et al.*,, 1992). While the HML and HMR loci also encode the HO recognition site, these sequences are not available for cutting by HO in a wild-type cell due to their chromatin structure.

If the endogenous MAT locus is to be used as the HO target, one can block the repair of the break by either deleting both HML and HMR or by disabling recombinational repair by deleting *RAD52* (Krogh and Symington, 2004). *HMLΔ/HMRΔ* strains have been used extensively by the Haber laboratory to examine adaptation to an HO break at the MAT locus on chromosome III (Lee *et al.*, 1998). Alternatively, one can use HO to target a sequence other than the endogenous site at the MAT locus (Sandell and Zakian, 1993). To do this, one must integrate a short sequence that targets the HO endonuclease elsewhere in the genome. When this is done, HML/HMR are unable to efficiently perform template repair due to the small size of the HO targeting sequence, and thus the DSB is not repairable. When a site other than MAT is used, the endogenous MAT site must be eliminated or made uncuttable, otherwise both the additional HO site and the MAT locus will be cleaved. To do this, one can employ an allele of the MAT locus that is not recognized by HO, called *MAT-inc*. Alternatively, the MAT locus itself can be deleted (Sandell and Zakian, 1993). However, MAT deletion has the disadvantage that *MATΔ* strains cannot be readily crossed, although we have developed a labor-intensive method for doing so (Toczyski *et al.*, 1997).

Adaptation is typically monitored using microcolony assays. These are performed by allowing cells to initially arrest at the DNA damage checkpoint, plating, and examining cells microscopically with time. Cells are scored on the plate for the number of cell bodies. Cell bodies are defined as the spheres that comprise yeast morphology, such that a large budded

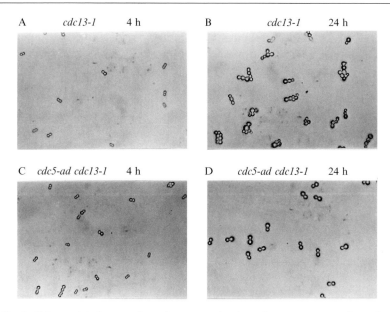

FIG. 1. Primary data from a microcolony assay. A microcolony assay was performed on an adaptation proficient *CDC5 cdc13-1* strain (panels A and B) and an adaptation-deficient *cdc5-ad cdc13-1* strain (panels C and D) at 32°. At 4 h, the 2 strains are both arrested as large budded cells (with 2 cell bodies). By 24 h, the *CDC5* strain has adapted, such that just 1 of the 17 cells in panel B (upper right) has only 2 cell bodies. Typically, cells are scored as having undergone adaptation if they contain more than 2 cell bodies. (Toczyski *et al.*, 1997, with permission from Elsevier.)

cell has two cell bodies, whereas the same cell will have three cell bodies after cytokinesis if one of the two daughter cells buds (Fig. 1). The term "cell body" is used, since one cannot determine easily whether a four-cell body microcolony represents two large budded cells, or one large budded cell adjacent to two unbudded cells. Either way, one can say definitively that cells with more than two cell bodies have re-initiated budding after the checkpoint arrest. Unpublished experiments from our laboratory examining lacI-GFP arrays in microcolonies grown on microscope slides have confirmed that several rounds of replication and segregation occur in microcolonies formed after adaptation to a *cdc13*-induced checkpoint arrest. In theory, one could use assays that monitor specific mitotic events, such as sister chromatid separation, as a measure of adaptation; however, there is an inherent problem with this. Adaptation occurs 6 to 10 h after damage induction, yet it takes only 10 min for a cell to proceed from the metaphase arrest into the next cell cycle. Because of this, one cannot simply examine cells after 8 h, since metaphase cells that are observed could represent either cells that have not yet arrested or cells that had adapted and completed another cycle. Preliminary data suggest that it may be possible to use alpha factor to trap cells after adaptation, but we have not explored this fully. Overexpression of a non-degradable B type cyclin can also be used to trap cells in anaphase following adaptation (Kaye *et al.*, 2004).

Our laboratory has extensively employed the use of a *MATΔ* strain expressing the HO endonuclease under the control of the *GAL1* promoter. The *GAL1* promoter is repressed by glucose in addition to being activated by galactose. Because of this, cells need to be starved of glucose before induction. Cultures are grown overnight in rich medium at 30° with 2% raffinose as a carbon source. Raffinose should be filter sterilized, and not autoclaved, as there is some breakdown of raffinose after autoclaving. Glycerol, lactate, and sucrose have also been employed for this purpose. We find that our strains grow better on raffinose than on glycerol and that sucrose, while inexpensive and an efficient carbon source, represses the *GAL1* promoter somewhat. Galactose is added to the overnight raffinose cultures to a final concentration of 2% and cells are incubated at 30° for 2 h. After sonicating, 5000 to 10,000 cells are plated on rich plates with 2% raffinose and 2% galactose plates. Sonication is done for 7 sec with a Fisher 550 probe sonicator on setting 2.5. The importance of this initial incubation in liquid is that it allows cells time to synchronize at the checkpoint arrest before plating. This is critical, since cells receiving a break after the arrest point, in metaphase, would continue through an additional cell cycle and be indistinguishable from cells that had first arrested at the checkpoint and then adapted. Cells can be scored at the indicated timepoint, or plates can be stored for up to 24 h at 4° before scoring. Adaptation typically starts after 6 h and is complete by 10 h (Fig. 2).

Using Disomic Strains

In strains in which one can induce irreparable damage that is not inherently lethal, adaptation-proficient cells are viable, whereas adaptation-deficient cells remain permanently arrested at the checkpoint and are thus inviable (Fig. 3) (Toczyski *et al.*, 1997). Irreparable, nonlethal DNA damage can be generated by inducing a DSB in a chromosome present in two copies in the cell. There are two advantages to examining adaptation under such a condition. The first is theoretical. If one activates a checkpoint by unraveling telomeres or inducing a DSB in an essential chromosome, then one runs the risk that the observed delay is due not only to the DNA

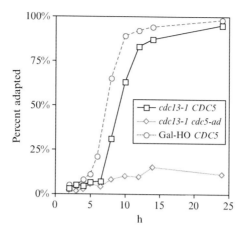

F<small>IG</small>. 2. Adaptation time courses. DNA damage is induced by induction of an HO break or by raising a *cdc13-1* strain to the nonpermissive temperature of 32°. Cells with three or more cell bodies are scored as having undergone adaptation. The exact timing of adaptation can depend upon many factors, including the media type, the carbon source, the temperature, and amount of damage present. Unlike the *cdc13-1 CDC5* cells, *cdc13-1 cdc5-ad* cells do not adapt.

damage checkpoint, but also to the loss of gene products encoded by loci adjacent to the damage site. This problem is especially acute during the study of adaptation, since this process is typically studied 6–10 h after induction of DNA damage. Since broken ends are processed into ssDNA at a rate of several kilobases per hour, genes as far as 10–20 kilobases from the break can be inactivated by the time adaptation has occurred (Lee *et al.*, 2000). The second advantage in examining irreparable damage in nonessential chromosomes is more practical: it allows for macroscopic assays for adaptation, which we will discuss later.

To generate a strain with a nonessential chromosome, one must either make a haploid strain with an extra copy of one chromosome, or a diploid strain in which one chromosome can be lost. We have worked almost exclusively with the former, as we have found that a single break in a diploid provides a significantly weaker checkpoint response. Specifically, we have worked with a haploid strain generated by the Zakian laboratory, which harbors an extra copy of chromosome VII containing a site for the HO endonuclease near its telomere (strain 19–20) (Sandell and Zakian, 1993). Disomic strains are generated by a somewhat labor intensive process in which karyogamy-defective cells are mated, resulting in partial genome

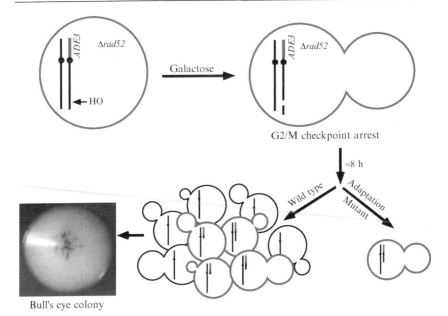

FIG. 3. The bull's eye assay for adaptation. A disomic strain containing an additional, and thus nonessential, copy of chromosome VII encoding the sole functional copy of the *ADE3* gene and a site for the HO endonuclease is shown. The illustrated strain (strain 19–21), is deleted for the MAT locus, such that the only copy of the HO site is on chromosome VII. The *RAD52* gene is also deleted, to prevent repair of the broken chromosome by homologous recombination off of the homolog. On induction of the HO endonuclease by galactose, the HO site on chromosome VII is cut and the cell arrests, via the checkpoint, in G2/M as a large budded cell (i.e., with 2 cell bodies). An adaptation-proficient strain will eventually continue division with the broken chromosome. The broken chromosome will continue to be degraded until it is lost, generating *ade3* progeny that do not accumulate the red pigment. This will form a "bull's eye" colony. In contrast, an adaptation mutant will remain arrested in G2/M. (Toczyski *et al.*, 1997; Sandell and Zakian, 1993; with permission from Elsevier.) (See color insert.)

transfer (Dutcher, 1981). The mating products are selected/screened for the transfer of the chromosome of interest, which is observed in a small fraction of the resultant cells. Strain 19–20 harbors one auxotrophic marker (*lys5* and *aro2*) on each copy of chromosome VII. One must maintain selection for both of these markers (by growing cells in the absence of lysine and tyrosine) when growing these strains, as one copy of chromosome VII can be lost and cells that have lost this copy will harbor a growth advantage over their progenitors. When HO is induced with galactose, the break that is induced is repaired off the second copy of chromosome VII in approximately 75% of the cells (Sandell and Zakian, 1993; Toczyski *et al.*, 1997). When this occurs, chromosome VII (and almost all of its markers) will be retained. To make this easy to follow, 19–20 was constructed such that it is *ade2*, and the only functional copy of *ADE3* is on the chromosome VII homolog with the HO cut site. *ade2 ADE3* strains are red, whereas *ade2 ade3* strains are white. Thus, even after an HO break, the 75% of cells that repair this HO break will retain the *ADE3* gene on the cut chromosome, and the strain will remain red.

The remaining 25% of cells adapt to the break. The adapting cell will passage the cut chromosome through several divisions before losing (or repairing) the break (Sandell and Zakian, 1993). This will often generate a colony that starts out as red (because it transiently passages the *ADE3*-containing cut chromosome) and eventually ends up white. The appearance of such a colony is that of a bull's eye, and these colonies can be taken as indicative of checkpoint adaptation (Fig. 3). To perform a bull's eye assay, cells are transferred from synthetic glucose -lysine -tyrosine media, to synthetic raffinose -lysine -tyrosine media and grown overnight. We have found that our disomic strains do not grow well in synthetic raffinose lysine -tyrosine liquid media, but grow somewhat better on solid media of the same composition. Cells are then induced with galactose for 6 h by either transferring cells to synthetic raffinose -lysine -tyrosine liquid media or by patching onto a synthetic raffinose -lysine -tyrosine plate. For bull's eye assays, cells are sonicated and plated on rich media with glucose. Plates lacking glucose, or containing supplemented adenine, produce a less distinct red pigmentation. Cells that do not cut the HO site within the 6 h incubation in galactose can be identified based on their retention of the *URA3* marker adjacent to the HO site. These colonies are ignored.

In the preceding assay, the broken chromosome is not truly irreparable. However, since the efficiency of repair is only 75%, the presence or absence of the 25% bull's eyes colonies allows one to assay a defect in adaptation. While this analysis is useful in that it provides a quantitative readout for checkpoint activation, it is relatively time consuming and not conducive to performing high throughput genetics. For the purposes of screening, a patch assay for sensitivity to galactose provides a rapid assay for checkpoint adaptation. However, this assay requires that the strains to be assayed be deleted for *RAD52*. If homologous recombination is disabled by deleting the *RAD52* gene (to generate strain 19–21), all cells must adapt to the galactose-induced break. Thus, an adaptation-defective *rad52* strain will arrest permanently at an HO break and will therefore be galactose sensitive. In this assay, strains are patched onto synthetic glucose -lysine -tyrosine plates and replica plated to synthetic raffinose -lysine -tyrosine plates after overnight growth. After 24 h on raffinose, this plate is replica plated to synthetic raffinose media (as a control) and synthetic raffinose plus galactose media (Fig. 4).

Extra care must be taken to maintain selection for both chromosomes in all but the last step of this experiment, since chromosome loss rates are 1000× higher in *rad52* strains. Because of this, 1% of a population of *rad52* cells will have lost the chromosome VII homolog with the HO site in a growing population. While these cells will not divide under selection on lysine-tyrosine media, they will remain viable. Obviously, selection for the two chromosomes must be released when cells are replica plated to synthetic galactose media in the final step of the assay, since adapting cells will lose the cut chromosome after the HO break in a *rad52* background. The upshot of this is that some small percentage of cells does not receive the HO break at the time of galactose induction because the

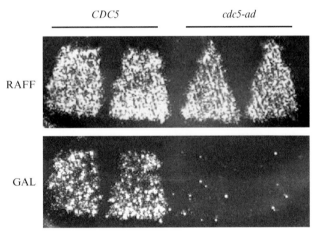

Fig. 4. Patch assay for adaptation. This assay is convenient and conducive to screening, but requires that the strain be disomic and *rad52Δ* (as in Figs. 3 or 5). Strains are first patched onto glucose synthetic plates lacking lysine and tyrosine to select for both chromosomes. These plates are then replica plated to raffinose plates lacking lysine and tyrosine, in order to relieve repression of the Gal promoter. These plates are in turn replica plated to either a complete synthetic plate with 2% raffinose (as a control) or a complete synthetic plate with 2% raffinose and 2% galactose. In the illustrated experiment, an adaptation-proficient strain *rad52 CDC5* has been patched as 2 trapezoids, whereas an adaptation-defective strain, *rad52 cdc5-ad*, has been patched as 2 triangles. It can be clearly seen that the adaptation-defective strain (right) is unable to grow on galactose (below), but grows well on raffinose (above). The colonies from the *cdc5-ad* strain that do come up on the galactose plate do not appear to be revertants, but are instead cells that had lost the chromosome containing the HO site before the final replica plating.

cells do not contain the chromosome with the HO site. These cells effectively bypass the assay entirely and grow on synthetic galactose plates. For this reason, care must be taken to replica plate cells to the final galactose plate quite lightly, since these cells will be viable even in the tightest of adaptation mutants. Fortunately, this background is reduced in adaptation-defective mutants, since these mutants have significantly lower spontaneous chromosome loss rates than do adaptation-proficient strains (see later).

Adaptation Assays Based on Chromosome Loss

The observation that both spontaneous and damage-induced chromosome loss rates are lower in adaptation mutants than in adaptation-proficient cells provides for a highly specific assay for adaptation mutants. We isolated over a half dozen adaptation-defective mutants and all displayed this phenotype ([Galgoczy and Toczyski, 2001] and data not shown). Chromosome loss events that are initiated by DNA damage are preceded by a checkpoint arrest. Only after adaptation to that arrest can the cell divide and lose the damaged chromosome. Adaptation-defective mutants affected in genes that function in DNA repair, such as the KU genes (Lee *et al.*, 1998), may have pleotrophic phenotypes including an increase in chromosome loss. Therefore, assays based upon indirect measures of adaptation must be used with care. While sectoring assays for chromosome loss are not quantitative, they are visual and macroscopic, and thus conducive to screening. Moreover, they are quite specific; mutations in a large number of pathways can lead to increases in the chromosome loss rate, whereas relatively few mutations decrease the rate of chromosome loss.

Only chromosome loss events induced by DNA damage require the DNA damage checkpoint adaptation pathway (Galgoczy and Toczyski, 2001). The 1000-fold elevated chromosome loss rate observed in *rad52* mutants is thought to be due to intrinsic DNA damage that is repaired in a *RAD52*-dependent manner in wild-type strains. Using the previously mentioned disomic *rad52* strain (19–21), one can observe chromosome loss in a whole colony by plating cells on rich glucose medium and examining sectoring. It is important to note that HO induction is not being used in this assay, and identical results can be obtained by using strains that lack an HO site entirely. When the chromosome with the *ADE3* gene is lost during the growth of a colony, the resulting cells appears as a white sector. Due to the high chromosome loss rates, *rad52* mutants produce highly sectored colonies (Fig. 5). In contrast, an adaptation-defective isogenic strain (*cdc5-ad*)

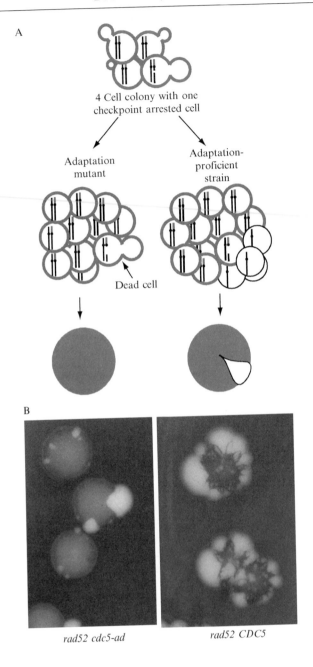

A

4 Cell colony with one
checkpoint arrested cell

Adaptation
mutant

Adaptation-
proficient
strain

Dead cell

B

rad52 cdc5-ad   rad52 CDC5

produces very few sectors (Galgoczy and Toczyski, 2001). Importantly, this strain is not inviable in the absence of the chromosome and thus a low level of chromosome loss can still be observed.

Adaptation to *cdc13*-Induced Damage

As an alternative to DSB induction, temperature-sensitive mutations can be employed to induce damage in the population. The best characterized of these is the *cdc13-1* mutant. Cdc13 associates with telomeres and is required to maintain the integrity of telomeric DNA. When *cdc13-1* mutants are brought to the nonpermissive temperature, they accumulate ssDNA at their telomeres and arrest in G2/M. This arrest depends upon the checkpoint machinery and, by all checkpoint assays, is analogous to the arrest induced by DSBs. As with a DSB, cells adapt to a *cdc13*-mediated checkpoint arrest, and this occurs with kinetics similar to that seen for an HO break or X-rays (Figs. 1 and 2) (Toczyski *et al.*, 1997). The *cdc13-1* allele has an unusually low maximum permissive temperature; *cdc13-1* cells are unable to form colonies at temperatures above 25°. Moreover, *cdc13-1* mutants appear to lose Cdc13 function progressively over a wide temperature (unpublished data). Because of this, *cdc13-1* mutants can be examined at different temperatures so as to generate cells with progressively more damage. Checkpoint adaptation is most easily observed at an intermediate temperature of 32°. At this temperature, almost all cells have a first cycle arrest in G2/M and adapt after 8 h. Thus, adaptation can be easily assayed. In contrast, at 36° there is a much more penetrant checkpoint arrest (Fig. 6).

For this assay, cells are grown overnight in rich media at 23°. A small culture grown to not more than 10^7 cell/ml is placed in a 32° water bath for 2.0 h. Cells are removed from the water bath only momentarily to allow

---

FIG. 5. Sectoring assay for adaptation-dependent chromosome loss. Disomic strains grown in the absence of selection for the 2 chromosomes will spontaneously lose either chromosome. If a visual marker is present on one of the disomic chromosomes, the portion of a colony that arises from a cell that has lost that chromosome will appear as a sector in the colony. The spontaneous rate of chromosome loss in *rad52* mutants is elevated by more than 1000-fold, and the observed loss events appear to be preceded by an initial checkpoint arrest, followed by adaptation to that arrest. Thus, adaptation mutants, such as *cdc5-ad*, produce fewer sectors. Cells that would have gone on to be scored as chromosome loss events will instead remain permanently checkpoint arrested. This phenomenon is cartooned in (A). (B) shows examples of such sectored colonies. (Galgoczy and Toczyski, 2001, with permission from the publisher (ASM). (See color insert.)

		Cell bodies							
		1	2	3	4	5	6	≤25	>25
CDC5 cdc13-1	32°	1 ± 1	7 ± 1	38 ± 5	23 ± 3	12 ± 2	8 ± 3	11 ± 4	0 ± 0
	36°	1 ± 1	54 ± 4	38 ± 5	4 ± 1	2 ± 1	1 ± 0	0 ± 0	0 ± 0
cdc5-ad cdc13-1	32°	4 ± 3	85 ± 5	2 ± 1	9 ± 1	0 ± 0	0 ± 0	0 ± 0	0 ± 0
	36°	6 ± 4	92 ± 4	0 ± 0	3 ± 2	0 ± 0	0 ± 0	0 ± 0	0 ± 0

FIG. 6. Adaptation kinetics of the *cdc13-1* induced checkpoint arrest is temperature dependent. *cdc13-1* cells were arrested at either 32° or 36° and microcolonies were examined after 24 h to determine the number of cells with different numbers of cell bodies. 32° is 5° above the maximum permissive temperature for *cdc13-1* mutants. At 36°, only half of the cells adapted, whereas at 32° 95% of cells adapted. Moreover, the adapted cells underwent more divisions at 32°. Higher temperatures could either generate more damage in *cdc13-1* mutants, or the process of adaptation could itself be temperature-sensitive.

sonication. 5000 to 10,000 cells are plated onto prewarmed rich plates and incubated at 32°. Microcolonies are scored over a 24 h time course. If scoring is not possible immediately, plates can be transferred to 4° for up to one day before scoring. When this is done, care must be taken not to stack plates at 4° to avoid the possibility that cells will have enough time at the permissive temperature to regain Cdc13 activity and recover from the arrest. Cells should not be left at 4° longer than 24 h or lysis will occur in some cells.

Conclusions

The methods described here were designed primarily for the genetic dissection of checkpoint adaptation. While a great deal of information can be gleaned from these assays, further work must be performed to generate better methods to explore the biochemistry of this phenomenon. There will be two hurdles to this. First, more synchronous induction of DNA damage would be helpful. HO induction is not particularly synchronous, and *cdc13-1* mutants only generate damage in S phase, and thus synchronized cells would need to be employed. Second, better methods for blocking cells in G1 following adaptation must be investigated. While the methods we have employed (overexpression of nondegradable Clb2) worked well for the examination of chromosome segregation, these methods may have unforeseen effects on cell cycle biology, which makes their use less optimal for examining the mechanisms of adaptation itself.

References

Dutcher, S. K. (1981). Internuclear transfer of genetic information in kar1-1/KAR1 heterokaryons in Saccharomyces cerevisiae. *Mol. Cell. Biol.* **1**, 245–253.

Galgoczy, D. J., and Toczyski, D. P. (2001). Checkpoint adaptation precedes spontaneous and damage-induced genomic instability in yeast. *Mol. Cell. Biol.* **21**, 1710–1718.

Herskowitz, I., and Jensen, R. E. (1991). Putting the HO gene to work: Practical uses for mating-type switching. *Methods Enzymol.* **194**, 132–146.

Herskowitz, I., Rine, J., and Strathern, J. (1992). Mating-type determination and mating type interconversion in Saccharomyces cerevisiae. *In* "The Molecular and Cellular Biology of the Yeast Saccharomyces" (E. W. Jones, J. R. Pringle, and J. R. Broach, eds.), Vol. 2, pp. 583–656. Cold Spring Harbor Press, Cold Spring Harbor, NY.

Kadyk, L. C., and Hartwell, L. H. (1992). Sister chromatids are preferred over homologs as substrates for recombinational repair in Saccharomyces cerevisiae. *Genetics* **132**, 387–402.

Kaye, J. A., Melo, J. A., Cheung, S. K., Vaze, M. B., Haber, J. E., and Toczyski, D. P. (2004). DNA breaks promote genomic instability by impeding proper chromosome segregation. *Curr. Biol.* **14**, 2096–2106.

Krogh, B. O., and Symington, L. S. (2004). Recombination proteins in yeast. *Annu. Rev. Genet.* **38**, 233–271.

Lee, S. E., Moore, J. K., Holmes, A., Umezu, K., Kolodner, R. D., and Haber, J. E. (1998). Saccharomyces Ku70, mre11/rad50 and RPA proteins regulate adaptation to G2/M arrest after DNA damage. *Cell* **94**, 399–409.

Lee, S. E., Pellicioli, A., Demeter, J., Vaze, M. P., Gasch, A. P., Malkova, A., Brown, P. O., Botstein, D., Stearns, T., Foiani, M., and Haber, J. E. (2000). Arrest, adaptation, and recovery following a chromosome double-strand break in Saccharomyces cerevisiae. *Cold Spring Harb. Symp. Quant. Biol.* **65**, 303–314.

Malkova, A., Ivanov, E. L., and Haber, J. E. (1996). Double-strand break repair in the absence of RAD51 in yeast: A possible role for break-induced DNA replication. *Proc. Natl. Acad. Sci. USA* **93**, 7131–7136.

Melo, J., and Toczyski, D. (2002). A unified view of the DNA-damage checkpoint. *Curr. Opin. Cell. Biol.* **14**, 237–245.

Nyberg, K. A., Michelson, R. J., Putnam, C. W., and Weinert, T. A. (2002). Toward maintaining the genome: DNA damage and replication checkpoints. *Annu. Rev. Genet.* **36**, 617–656.

Sandell, L. L., and Zakian, V. A. (1993). Loss of a yeast telomere: Arrest, recovery, and chromosome loss. *Cell* **75**, 729–739.

Toczyski, D. P., Galgoczy, D. J., and Hartwell, L. H. (1997). CDC5 and CKII control adaptation to the yeast DNA damage checkpoint. *Cell* **90**, 1097–1106.

[10] DNA Damage-Induced Phosphorylation of Rad55 Protein as a Sentinel for DNA Damage Checkpoint Activation in S. cerevisiae

By Vladimir I. Bashkirov, Kristina Herzberg, Edwin Haghnazari, Alexey S. Vlasenko, and Wolf-Dietrich Heyer

Abstract

Rad55 protein is one of two Rad51 paralogs in the budding yeast *Saccharomyces cerevisiae* and forms a stable heterodimer with Rad57, the other Rad51 paralog. The Rad55-Rad57 heterodimer functions in homologous recombination during the assembly of the Rad51-ssDNA filament, which is central for homology search and DNA strand exchange. Previously, we identified Rad55 protein as a terminal target of the DNA damage checkpoints, which coordinate the cellular response to genotoxic stress. Rad55 protein phosphorylation is signaled by a significant electrophoretic shift and occurs in response to a wide range of genotoxic stress. Here, we map the phosphorylation site leading to the electrophoretic shift and show that Rad55 protein is a *bona fide* direct *in vivo* substrate of the central DNA damage checkpoint kinase Mec1, the budding yeast equivalent of human ATM/ATR. We provide protocols to monitor the Rad55 phosphorylation status *in vivo* and assay Rad55-Rad57 phosphorylation *in vitro* using purified substrate with the Mec1 and Rad53 checkpoint kinases.

Introduction

DNA damage checkpoints coordinate the cellular response to genotoxic stress and regulate a multitude of effector pathways to ensure survival and genomic stability (reviewed in Melo and Toczyski, 2002; Nyberg *et al.*, 2002). Central to the DNA damage checkpoints is a web of protein kinases. In the budding yeast *S. cerevisiae*, the PI3-kinase-like protein kinase Mec1 (equivalent to human ATR) plays a pivotal role. Mec1 (and to a lesser extent the yeast ATM homolog Tel1) control the activity of other kinases, including Rad53, Dun1, and Chk1 (reviewed in Zhou and Elledge, 2000). In addition, Mec1 (and Tel1) are likely to have additional direct downstream target proteins that are phosphorylated in response to DNA damage (Melo and Toczyski, 2002; Nyberg *et al.*, 2002; Zhou and Elledge, 2000). The identification of such target sites was aided by the definition

METHODS IN ENZYMOLOGY, VOL. 409
Copyright 2006, Elsevier Inc. All rights reserved.
0076-6879/06 $35.00
DOI: 10.1016/S0076-6879(05)09010-5

of a minimal consensus preferred by the Mec1/Tel1 (ATR/ATM) kinases consisting of the S/TQ motif (Kim *et al.*, 1999). DNA damage checkpoint activation in *S. cerevisiae* is typically monitored by the activation of Rad53 kinase, which leads to a noticeable electrophoretic mobility change accompanied by activation of its kinase activity (Pellicioli *et al.*, 1999; Sanchez *et al.*, 1996; Sun *et al.*, 1996).

Homologous recombination is a ubiquitous pathway for the repair of DNA damage affecting both strands of the DNA duplex, including double-stranded breaks (DSBs) and interstrand crosslinks (reviewed in Krogh and Symington, 2004). Central to recombination is assembly of the Rad51 ssDNA filament, which performs homology search and DNA strand exchange. Assembly of the Rad51 filament on ssDNA coated by the ssDNA-binding protein RPA requires the function of mediator proteins, Rad52 and the Rad55-Rad57 heterodimer in *S. cerevisiae*. The *RAD55* and *RAD57* genes of *S. cerevisiae* encode the two Rad51 paralogs of budding yeast with a molecular weight of 46.3 and 52.2 kDa, respectively (Sung, 1997). Both proteins share the RecA core with Rad51 but have unique N- and C-terminal extensions. While the Rad55-Rad57 heterodimer exhibits ATPase activity and binds to DNA, it is devoid of the DNA strand exchange activity, which is pivotal for Rad51 and RecA function (Sung, 1997).

Based on the precedence of the bacterial SOS system, direct regulation of DNA repair, and specifically homologous recombination, by the DNA damage checkpoints has been discussed (Carr, 2002; Zhou and Elledge, 2000). In a systematic survey, we established that the Rad55 protein is a terminal substrate of the DNA damage checkpoints and specifically phosphorylated after induction of genotoxic stress (Bashkirov *et al.*, 2000). Here, we describe methods to monitor the Rad55 phosphorylation status *in vivo* (Protocol 1; Figs. 1, 2) and the development of a new budding yeast overexpression vector for the purification of heterodimeric proteins (see Methods; Fig. 3), which allows efficient purification of the Rad55-Rad57 heterodimer (Protocol 2; Fig. 4) that was used as a substrate in *in vitro* kinase assays with the Mec1 and Rad53 DNA damage checkpoint kinases (Protocols 3 and 4; Fig. 5).

Methods

Analysis of Rad55 Phosphorylation Status in Response to DNA Damage

In a survey of *S. cerevisiae* homologous recombination proteins to identify DNA damage checkpoint phosphorylation targets, we found a DNA damage-induced electrophoretic shift shown to be caused by DNA damage checkpoint-induced phosphorylation of the Rad55 protein (Fig. 1;

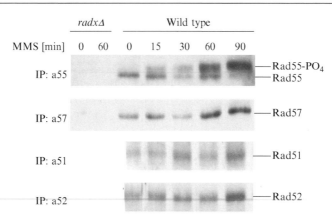

FIG. 1. Identification of Rad55 phosphorylation by electrophoretic shift. Electrophoretic mobility of Rad51, Rad52, Rad55 and Rad57 proteins upon DNA damage. Wild type (WDHY485 *MATα leu2-3,112 trp1-289 ura3-52 his7-2 lys1-1*), *rad55* (WDHY836: same as WDHY485 and *rad55-Δ::KanMX*), or *rad57* (WDHY838: same as WDHY485 and *rad57-Δ::KanMX*) cells were treated with 0.075% MMS for the indicated times, and Rad51, Rad52, Rad55, and Rad57 proteins were immunoprecipitated with the corresponding affinity-purified antibodies raised in rabbit. Immunoblot detection employed the corresponding protein-specific affinity-purified antibodies developed in rats. Note that only Rad55 protein underwent a detectable electrophoretic mobility shift.

Bashkirov *et al.*, 2000). No shift was detected for the Rad51, Rad52, Rad54, and Rad57 proteins (Fig. 1 and data not shown). However, this finding does not exclude that these proteins are phosphorylated constitutively or in response to DNA damage, because not all phosphorylation events lead to a shift in the electrophoretic profile. The DNA damage-induced electrophoretic shift of Rad55 provides a convenient assay to monitor activation of the DNA damage checkpoints (Protocol 1).

Using Protocol 1, we showed that the Rad55 phosphorylation leading to the electrophoretic shift was dependent on the central Mec1 checkpoint kinase (Fig. 2A; Bashkirov *et al.*, 2000). A deletion/substitution strategy identified serine 378 (S378), which resides in the SQ consensus motif for PI3-kinase-like kinases, as the likely Mec1 phosphorylation site (Fig. 2B, C).

Protocol 1: Immunoprecipitation and Detection of Rad55-Rad57 Heterodimer

Reagents. Methyl methanesulfonate (MMS; Kodak, Rochester, NY), PMSF (serine protease inhibitor; Roche, Indianapolis, IN), NaF and Na_3VO_4 (phosphatase inhibitors; ACROS, Geel, Belgium), affinity-purified polyclonal antibodies against Rad55 protein developed in rabbits and rats,

Protein G Sepharose 4F Fast Flow (GE Healthcare, Piscataway, NJ), anti-rat IgG/HRP conjugate (Dako, Carpinteria, CA), ECL chemiluminescence detection reagents (GE Healthcare).

Buffers. IP-A buffer: 20 mM Tris HCl (pH 7.5), 100 mM NaCl, 1 mM EDTA; IP-B buffer: 100 mM Tris HCl (pH 7.5), 100 mM NaCl, 0.4% Triton X-100; IP-C buffer: 50 mM Tris HCl (pH 7.5), 100 mM NaCl, 0.2% Triton X-100.

NOTE: IP-A, -B, and -C buffers were also supplemented with phosphatase inhibitors: 30 mM NaF and 0.1 mM Na$_3$VO$_4$ (final concentrations) and the protease inhibitor PMSF (1 mM final concentration except IP-C buffer with 0.2 mM); transfer buffer: 25 mM Tris base, 192 mM glycine, 1% SDS, 20% methanol; SDS loading buffer (2×): 125 mM Tris-HCl (pH 6.8), 4% SDS, 20% glycerol, 10% β-mercaptoethanol, 0.006% bromophenolblue.

Procedure

Cells are grown to mid-log phase (OD$_{600}$ 1.0 for haploid or OD$_{600}$ 2.0 for diploid strains), and MMS is added to 0.1%, when necessary to induce the DNA damage checkpoint response. At least 100 OD units of cells per immunoprecipitation (IP) are collected by centrifugation (3500 rpm, standard rotor F45–24–11, Eppendorf centrifuge 5415), washed once in IP-A buffer, frozen in liquid nitrogen, and stored at −70°. Cells are resuspended in 800 μl of IP-A buffer and transferred to a plastic tube with screw cap (Sarstedt, No. 72.694.005, Newton, NC) containing 300 μl of glass beads (0.5 mm, Biospec Products, Inc., Bartlesville, OK). Cell are disrupted in a FastPrep machine (FP120, Bio101, Savant) at setting "4" six times for 45 sec each resulting in 60–90% broken cells. The tube is placed on ice for 2 min after every other cycle of disruption. After the final disruption, the cell lysate is cleared by centrifugation at 13,000 rpm for 10 min in a microcentrifuge at 4°. The cleared protein extract is transferred to a plastic tube (5 ml polystyrene round-bottom tube, Falcon type 2054), and an equal volume of IP-B buffer is added. Affinity-purified rabbit polyclonal antibodies (0.2–0.3 μg) against Rad55 protein are added, and the tube is rocked at 4° for 2–3 h. To collect the immuno-complexes, 20–25 μl bed volume of Protein G Sepharose 4F Fast Flow suspension, prewashed in IP-C buffer, is added, and the tube is rocked at 4° for 1 h. Protein G Sepharose beads bound to the immunocomplexes are collected by centrifugation at 1800–2000 rpm for 2 min (F45–24–11, Eppendorf centrifuge 5415). The beads are gently resuspended in IP-C buffer and transferred to a fresh microcentrifuge tube. All steps are performed at 4°. The beads are washed four times with 1 ml of IP-C buffer. After the final centrifugation step, the supernatant

A

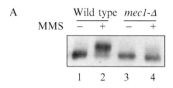

B

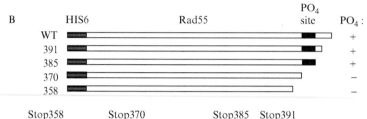

Stop358   Stop370   Stop385 Stop391

357 LKPDTANIASFPTLSTSSSSCSQVFNNIDSNDNPLPNAEGKEEIIYDSEG 406

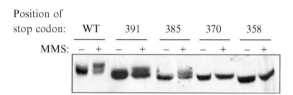

C

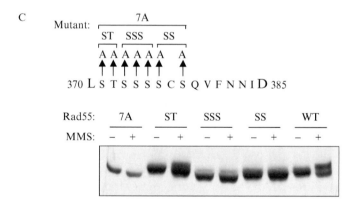

Fig. 2. *In vivo* analysis of Rad55 phosphorylation control. (A) The Rad55 electrophoretic mobility shift is controlled by Mec1 kinase. Rad55 phosphorylation status was analyzed as described for Fig. 1 in wild type (DES460 1n *MATa can1–100 ade2–1 trp1–1 leu2–3,112 his3–11,15 ura3–1 TRP::GAP-RNR1*) and *mec1-Δ* (DES459 1n *MATa can1–100 ade2–1 trp1–1 leu2–3,112 his3–11,15 ura3–1 TRP::GAP-RNR1 mec1-Δ::HIS3*) cells. (B) Deletion analysis to map the phosphorylation site causing the electrophoretic mobility shift. Upper panel: schematic illustration of the C-terminal region of Rad55 protein with the truncations analyzed and their effect on the DNA damage-induced electrophoretic mobility shift. Middle panel: amino acid sequence of the Rad55 C-terminus. Truncations were constructed by introducing

is removed completely, and 20 μl of 2× SDS loading buffer is added. After mixing, the suspension is heated at 100° for 10 min with intermittent mixing. The beads are collected at 13,000 rpm for 1 min (standard rotor F45–24–11, Eppendorf centrifuge 5415), and the supernatant is analyzed by

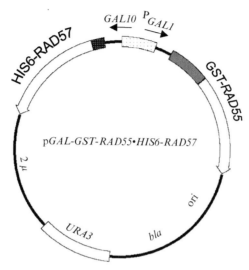

FIG. 3. Vector for overexpressing heterodimeric proteins in *S. cerevisiae*. Schematic representation of p*GAL-GST-RAD55·HIS6-RAD57* for overexpression of the Rad55-Rad57 complex under control of the inducible bidirectional *GAL1–10* promoter. Rad57 was expressed as an N-terminal His6-fusion and Rad55 as an N-terminal GST-fusion. The protein tags are shown filled in. *Pfu* DNA polymerase-generated PCR products containing the *RAD55* ORF with *Bam*HI ends were cloned into the *Bam*HI site of pJN58 to generate *GST-RAD55* under the control of the *GAL1* promoter. The *HIS6-RAD57* module with *Xho*I ends was PCR-amplified and inserted into the *Xba*I site of *pGAL-GST-RAD55*, placing it under the control of the *GAL10* promoter.

---

stop codons at the amino acid positions indicated by the enlarged characters using site-directed mutagenesis of p*GAL-GST-RAD57·HIS6-RAD55*. Lower panel: *In vivo* analysis of Rad55 phosphorylation status in cells expressing the variants shown in the upper panel. Cells were treated with 0.1% MMS for 2 h after 3 h of galactose-induced protein overexpression. Analysis was performed as described in (A). (C) Fine mapping Rad55 protein phosphorylation site by amino acid substitution analysis. Upper panel: amino acid sequence of Rad55 region responsible for Rad55 protein mobility shift, as mapped in (B). Arrows indicate the substitution of serine (S) and threonine (T) residues to alanine (A). ST mutant is S371A, T372A; SSS mutant-S373A, S374A, S375A, SS mutant -S376A, S378A, and 7A mutant has all 7 S/Ts substituted with A. Lower panel: phosphorylation status of Rad55 mutant proteins. Protein expression and immunoblotting were performed as in (B).

A   B

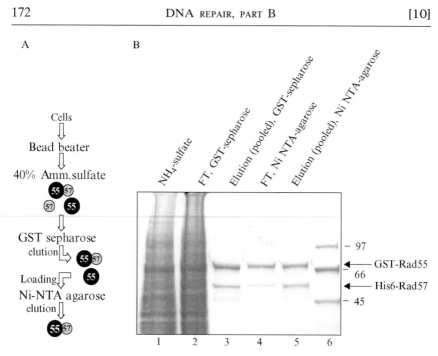

FIG. 4. Purification of Rad55-Rad57 heterodimer from undamaged and damaged *S. cerevisiae* cells. (A) Schematic illustration of purification strategy employing two consecutive affinity-chromatography steps on glutathione-S-sepharose beads and Ni-NTA agarose. (B) Summary of Rad55-Rad57 purification showing a Coomassie-stained 8% SDS-PAGE gel with fractions from a Rad55-Rad57 purification. Loaded were 40 μg (lanes 1, 2) or $\sim$2 μg (lanes 3–5) of total protein. Lane 6: molecular weight markers (kDa). FT: flow through.

9% SDS-PAGE to resolve the precipitated proteins. After electrotransfer of proteins from the gel to a nitrocellulose membrane (30 min at 500 mA in transfer buffer), Rad55 protein is detected with rat anti-Rad55 antibodies. Anti-rat IgG/HRP conjugate is used at a 1:6000 dilution as a secondary antibody. The ECL chemiluminescence system (GE Healthcare) is employed for visualization of the Rad55 protein.

Comments to Protocol 1

The IP technique described previously works best for detection of proteins with low abundance. It provides a more specific signal and generates lower background than direct Western blot analysis of cleared lysates. The latter method often leads to the problem of "protein band masking" due to poor resolution of proteins in one dimension. A smaller number of cells in the IP procedure will result in increased signal-to-noise ratios during detection by immunoblot. The protein of interest may have a mobility close to the IgG heavy or light chains, which may mask the specific signal after staining with secondary antibodies. The use of antibodies from different animal sources for IP (rabbit) and for detection (rat) overcomes this problem. The secondary anti-rat IgG antibodies conjugated with HRP do not cross-react with the heavy chain of rabbit IgGs used in IP and present on the blot. Note that it is critical to affinity-purify antibodies using the original antigen, typically a His6-tagged protein fragment expressed in *E. coli.* The protocol by Pringle *et al.* (1991) works well in our laboratory. Usually, ~300–400 μl of immune serum are used to absorb antibodies on the nitrocellulose strip containing 100–200 μg of His-tagged protein. This provides enough affinity-purified antibodies for about 10–20 immunoprecipitations. The amount of antibodies used for IP and immunostaining have to be optimized in each case.

In vitro *Phosphorylation of Rad55-Rad57 Protein by the Checkpoint Kinases*

To provide evidence that S378 of Rad55 protein is a direct Mec1 kinase site, we devised a strategy to efficiently purify functional Rad55-Rad57 heterodimer using a vector with an inducible, bidirectional promoter for overexpression of heterodimeric proteins in *S. cerevisiae* (Figs. 3, 4).

This vector is further discussed later (Vector System to Express Heterodimeric Proteins in *S. cerevisiae*) and has been useful not only in the purification of Rad55-Rad57 heterodimer (see later) but also for the Mus81-Mms4 heterodimer (K. Ehmsen and WDH, unpublished result). Using purified Rad55-Rad57 heterodimer (see Purification of Rad55-Rad57 Heterodimer from *S. cerevisiae;* Fig. 4, Protocol 2) and the purified wild-type and kinase-deficient (kd) DNA damage checkpoint kinases Mec1 and Rad53 (see *In vitro* Kinase Assays; Protocols 3 and 4), we show that Rad55-S378 is a *bona fide* direct Mec1 phosphorylation or recognition site, as wild-type Rad55-Rad57 but not mutant heterodimer that eliminates the SQ consensus motif can be phosphorylated *in vitro* by Mec1 kinase (Fig. 5A). The Mec1-kd mutant protein served as a control to demonstrate that phosphorylation is Mec1-dependent (Fig. 5A). Rad55-Rad57 is also directly phosphorylated by the Rad53 kinase *in vitro* (Fig. 5B).

Vector System to Express Heterodimeric Proteins in S. cerevisiae

Purification of heteromultimeric proteins poses the problem of expressing the individual subunits at equal levels to maximize formation of functional multimers. Moreover, multimer formation is often found to help in the stability and solubility of the individual protein subunits. Expression

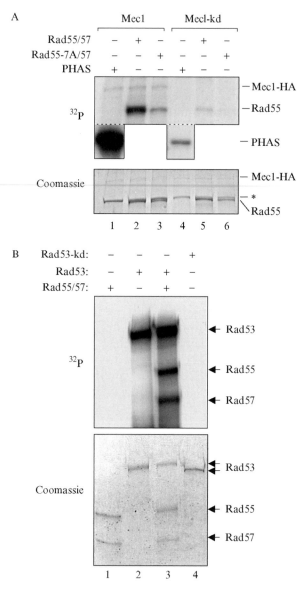

FIG. 5. *In vitro* kinase assays. (A) Rad55·Rad57 is phosphorylated by Mec1 kinase *in vitro*. Immunoprecipitated Mec1-HA and Mec1-kd-HA kinase with 1 μg of wild type GST-Rad55·HIS6·Rad57 or mutant GST·Rad55–7A·HIS6·Rad57 protein were used as a substrate. PHAS-1 peptide (1 μg) served as a control for Mec1 activity. The star denotes a band that coprecipitates with Mec1 kinase and possibly represents its Ddc2 DNA binding subunit. (B) Rad55·Rad57 is phosphorylated by Rad53 kinase *in vitro*. *In vitro* kinase assays were

of protein in the cognate host offers the advantage of cognate post-translational modifications and, in organisms such as yeasts, the opportunity for functional analysis by complementation of a chromosomal mutation. We designed an *S. cerevisiae–E. coli* shuttle vector for the expression of two genes from the strong, inducible, and bidirectional *GAL1–10* promoter (Fig. 3). The vector is based on pJN58 (Nelson *et al.*, 1996) and features the *S. cerevisiae URA3* gene and the 2 μ plasmid origin for selection and replication in yeast and the β-lactamase gene (bla) for selection in *E. coli*. The bidirectional *GAL1–10* promoter controls *GST-RAD55* and *HIS6-RAD57* fusion genes, which allows a consecutive dual affinity chromatography purification strategy that efficiently selects for the heterodimer product (see Fig. 4, Protocol 2). This promoter construct is strongly induced by the addition of galactose but is not completely repressed with glucose as a carbon source. This feature is useful in complementation studies, where we found better complementation on glucose-containing media than on galactose-containing media (not shown). This is likely due to negative effects associated with overexpression of Rad55-Rad57 heterodimer. The p*GAL-HIS6-RAD57-GST-RAD55* vector efficiently complemented the MMS sensitivity of a *rad55Δ rad57Δ* strain (data not shown; VIB and WDH, manuscript submitted), suggesting that the dually tagged heterodimer retained its biological function. This construct was used for protein purification (see later Protocol 2).

Purification of Rad55-Rad57 Heterodimer from *S. cerevisiae*

Protocol 2: Purification of Rad55-Rad57 Heterodimers

Reagents. Sodium lactate (Sigma, St. Louis, MO), glycerol (FisherBiotech, Pittsbugh, PA enzyme grade), glucose (Sigma), galactose (Sigma), Difco yeast nitrogen base w/o amino acids (BD, Palo Alto, CA), MMS (Kodak), protease inhibitors PMSF, leupeptin, pepstatin A (Roche) and benzamidin (Sigma), Glutathione Sepharose 4B (GE Healthcare), reduced glutathione (Sigma), imidazole (FisherBiotech, enzyme grade).

Buffers. Buffer A: 20 mM Tris HCl (pH 7.5), 1 mM EDTA, 1 M NaCl, 10% glycerol; buffer B: 10 mM Na$_2$HPO$_4$, 1.8 mM KH$_2$PO$_4$, 2.7 mM KCl, 1 M NaCl (pH 7.3); buffer C: 50 mM NaH$_2$PO$_4$ (pH 8.0), 1 M NaCl, 10% glycerol;

---

performed with affinity chromatography-purified GST-Rad53 kinase (1 μg) and purified GST-Rad55·HIS6-Rad57 heterodimer (2 μg) as a substrate. In (A) and (B) the upper panel shows the autoradiogram and the lower panel shows the Coomassie-stained gel (4–16% Tris-glycine gradient gel in A and 8% gel in B). Rad53-kd does not show autophosphorylation and is devoid of the electrophoretic shift caused by extensive autophosphorylation.

storage buffer: 20 mM Tris HCl (pH 7.5), 0.5 M NaCl, 0.1 mM EDTA, 1 mM DTT, 10% glycerol.

Procedure

Twelve liters of basic medium (6.7 g/l yeast nitrogen base w/o amino acids, 0.87 g/l drop out mixture w/o uracil, and 2% sodium lactate) with 3% glycerol are inoculated at OD$_{600}$ = 0.15–0.2 using an overnight preculture (grown in basic medium + 2% glucose + 3% glycerol) of the protease deficient strain BJ5464 (*MAT*α *ura3–52 trp1 leu2-Δ1 his3-Δ200 pep4::HIS3 prb1-Δ1.6R can1*) carrying the plasmid p*GAL-GST-RAD55·HIS6-RAD57* (Fig. 3). Cells are grown in an incubator-shaker at 25° for ~20 h to reach an OD$_{600}$ of 2.0. Protein expression is induced by the addition of galactose (to 2%) and growth is continued for 4.5 h. Cells are harvested by centrifugation, washed with Buffer A, and frozen in liquid nitrogen. Cells (60–70 g) are resuspended in 150 ml of Buffer A containing four protease inhibitors (1 mM PMSF, 2 μM leupeptin, 1 μM pepstatin A, 1 mM benzamidin) and disrupted using glass beads (350 g of 0.5 mm glass beads; BioSpec Products, Inc., Bartlesville, OK). The disruption chamber is loaded with the cell suspension and filled to the top with the same buffer, closed, and cells are disrupted in a Bead-Beater (BioSpec Products, Inc. Model 1107900) using eight 30-sec cycles with 2 min cooling intervals in between (on ice). Typically, more than 80% of the cells are disrupted under these conditions as monitored by microscopy. The suspension is withdrawn and glass beads are washed with an additional 100 ml of Buffer A. The combined suspensions are centrifuged in an ultracentrifuge (50.2 Ti rotor at 40,000 rpm for 45 min at 4°) to result in a total protein cell lysate. Proteins are precipitated with 40% ammonium sulfate in an ice bath and the precipitate is harvested by centrifugation in a Beckman JA-20 rotor at 10,000 rpm for 30 min at 4°. The precipitate is resuspended in 150 ml of Buffer B containing the four protease inhibitors. After additional centrifugation at 10,000 rpm for 10 min at 4° to remove insoluble proteins, the supernatant is collected and used for subsequent chromatography. GST-affinity chromatography is performed by applying the resuspended ammonium sulfate precipitate on a 4-ml Glutathione Sepharose 4B column that was pre-equilibrated with Buffer B containing the four protease inhibitors at a flow rate of 0.5 ml/min. The column is washed with Buffer B containing protease inhibitors until the A280 returns to baseline. The GST-tagged proteins are eluted with 40 ml Buffer A containing protease inhibitors and 20 mM of free reduced glutathione. Fractions containing GST-Rad55·His6-Rad57 hetero-dimer, as determined by 10% SDS-PAGE, are pooled and dialyzed overnight at 4° against Buffer C containing 0.2 mM PMSF, 2 μM leupeptin, 1 μM pepstatin A, and 1 mM benzamidin in a Slide-A-Lyzer Dialysis Cassette (Pierce, Rockford, IL). Metal binding-affinity chromatography is performed by applying the dialyzed fraction at a flow rate of 0.1 ml/min onto a 2 ml Ni-NTA-agarose column that was pre-equilibrated with 20 ml Buffer C containing 0.2 mM PMSF, 2 μM leupeptin, 1 μM pepstatin A, and 1 mM benzamidin. The column is washed with the same buffer and the bound protein is eluted with 20 ml of Buffer C containing 0.5 M NaCl, 0.1 mM PMSF, and 250 mM imidazole. After analysis of the fractions by 10% SDS-PAGE, the fractions containing stoichiometric Rad55-Rad57 complexes (as judged by Coomassie staining) are pooled and dialyzed against Buffer C containing 0.5 M NaCl and 0.1 mM PMSF to eliminate the imidazole. The protein solution is concentrated using Slide-A-Lyzer Cassette and Concentrating Solution (Pierce). Finally, after rinsing the cassette, dialysis is performed against Storage Buffer. The solution of purified Rad55-Rad57 heterodimer is aliquoted in plastic tubes, flash-frozen in liquid nitrogen, and stored at −70°. The final yield is typically 200–300 μg.

Comments to Protocol 2

We found that 4.5 h of galactose-inducible expression is optimal. At longer incubation times the His6-Rad57 subunit was destabilized. To purify the phosphorylated heterodimer, we added MMS up to 0.1% for an additional 2 h of expression at 25°. Incubation at 25° instead of the standard 30° was important to improve the yield of correctly folded heterodimer, and hence increased the amount of protein bound to the affinity columns. High salt concentrations are employed throughout the purification and were found to be important to keep the heterodimer soluble. Electrophoretic analysis from fractions during the purification protocol shows that the fraction eluting from the first affinity column (GST) contains an excess of the GST-Rad55 subunit (Fig. 4B). This excess is efficiently removed by the second affinity step that selects for the heterodimer (Fig. 4B). To prevent protein inactivation by frequent thawing-freezing cycles the protein is aliquoted in small batches and kept at −70°. Several thawing-freezing cycles usually did not affect the protein to serve as the substrate for kinase assays, although it may affect biochemical activity.

In Vitro Kinase Assays

Protocol 3: Pull-Down of GST-Tagged Rad53 Kinase

This pull-down protocol is provided as a small scale alternative to chromatographically purifying large amounts of Rad53 kinase described before (Bashkirov *et al.*, 2003). The purity of the material obtained in the pull-down procedure is equivalent to that in the chromatographic procedure using glutathione-affinity chromatography (data not shown).

Reagents. Raffinose (Sigma), glycerol (FisherBiotech, enzyme grade), glucose (Sigma), galactose (Sigma), Difco yeast nitrogen base w/o amino acids (BD), MMS (Kodak), protease inhibitors PMSF, leupeptin, pepstatin A (Roche) and benzamidin (Sigma), NaF and Na_3VO_4 (phosphatase inhibitors; ACROS), Glutathione Sepharose 4B (GE Healthcare).

Buffers. IP-A buffer: 20 mM Tris HCl (pH 7.5), 100 mM NaCl, 1 mM EDTA; IP-B buffer: 100 mM Tris HCl (pH 7.5), 100 mM NaCl, 0.4% Triton X-100; IP-C buffer: 50 mM Tris HCl (pH 7.5), 100 mM NaCl, 0.2% Triton X-100; IP-A, -B, and -C buffers are supplemented with phosphatase inhibitors (30 mM NaF, 0.1 mM Na_3VO_4) as well as protease inhibitors (1 mM PMSF, 2 μM leupeptin, 1 μM pepstatin A, 1 mM benzamidin [final concentrations]). Kinase buffer: 50 mM Tris HCl (pH 7.5), 10 mM $MgCl_2$, 10 mM $MnCl_2$, 1 mM dithiothreitol.

Procedure

The 200–300 ml cells of a *rad53Δ* strain (DES453 1n *MATa can1–100 ade2–1 trp1–1 leu2–3,112 his3–11,15 ura3–1 TRP::GAP-RNR1 rad53-Δ:: HIS3*) bearing p*GAL-GST-RAD53* (wild type or kinase-deficient mutant *rad53-K227A*) are grown to mid-log phase (OD_{600} = 1.0 for haploid or 2.0 for diploid strains) in SD-ura + 2% raffinose. Galactose is added to cultures to a final concentration of 2% from a 20% stock. Incubation of the cultures is continued at 30° in a shaker-incubator for 3 h to allow overexpression of the GST-tagged Rad53 proteins. Then, MMS is added to 0.1% and incubation is continued for an additional 2 h to activate Rad53 kinase. The cells are collected at 3500 rpm in an Allegra 6 centrifuge (Beckman, Fullerton, CA) using a GH-3.8 rotor and washed in 10 ml of IP-A buffer. The cell pellet is frozen in liquid nitrogen and stored at −70°. The resultant 200–250 OD units of cells are resuspended in 1 ml of IP-A buffer containing 1 mM PMSF and transferred to a 2 ml screw cap plastic tube (Sarstedt, No. 72.694.005) containing 700 mg (approx. 0.5 ml volume) glass beads. In case more than 250 OD units of cells are used, they are resuspended in 2 ml of buffer and distributed into two tubes before disruption. Typically 200–250 OD units provide enough protein kinase for eight *in vitro* kinase reactions. Cells are disrupted with glass beads in a FastPrep machine (FP120, Bio101, Savant) at setting "4" eight times for 45 sec each until 60 to 90% of the cells are broken. The tube is put on ice for 2 min after every two cycles of disruption. After the final disruption, the tubes are put on ice for 5 min and then centrifuged at 13,000 rpm for 10 min in an Eppendorf microcentrifuge (F-45–24–11 rotor) at 4° to clear the cell lysate. The cleared protein extract is transferred to a fresh plastic tube (5-ml polystyrene round-bottom tubes from Falcon type 2054) and an equal volume of

IP-B buffer is added. Then 160 μl of 50% (vol:vol) suspension of Glutathione Sepharose 4B (prewashed three times with 1 ml of cold IP-C buffer) is added and the tubes are incubated with rocking at 4° for 2.5–3 hours. The Glutathione Sepharose beads are collected at 1800–2000 rpm for 2 min (Allegra 6, GH-3.8 rotor). The supernatant is removed and the beads are resuspended in 1 ml of IP-C buffer, and transferred to a fresh Eppendorf tube. All steps are performed on ice. The beads are washed four times with 1 ml of IP-C buffer (2 min at 2000 rpm in an Eppendorf microcentrifuge). Next, the beads are washed once in 1 ml of the kinase buffer, then resuspended in 800 μl of the same buffer, and redistributed in eight Eppendorf tubes (100 μl each), to be used for eight *in vitro* reactions. Finally, the beads are collected by centrifugation (2 min, 2000 rpm), and excess buffer is removed. To the resulting 10 μl (bed volume) of beads carrying the GST-Rad53 kinase the substrate and additional components of the *in vitro* kinase reaction are added to perform the assay (see Protocol 4).

Comments to Protocol 3

Several important issues should be considered when choosing the method of isolation of checkpoint protein kinases for *in vitro* phosphorylation assays. The expression system used to produce the kinase is relevant, because noncognate systems may lack the mechanism to induce the kinase after genotoxic stress. For example, expression of Rad53 kinase in *E. coli* results in damage-independent autophosphorylation of Rad53 kinase, which appears more excessive than the natural damage-induced phosphorylation based on the electrophoretic mobility shift profile (Gilbert *et al.*, 2001). It was established that transphosphorylation of Rad53 kinase by the upstream Mec1/Tel1 kinases at [S/T]Q sites is an important step toward fully activated Rad53 kinase (Lee *et al.*, 2003), which is absent when expressing yeast kinases in bacteria. Using immunoprecipitates as a source of the kinase (and to a lesser degree using chromatographically purified protein kinase) poses the challenge of demonstrating that phosphorylation of a substrate is performed by the kinase under study. Checkpoint kinases interact with each other, and may be the part of the same protein complex, as shown for the transient Rad53 and Dun1 association (Bashkirov *et al.*, 2003). Thus, the IP of one kinase may result in coimmunoprecipitation of a partner kinase, resulting in potential difficulties interpreting the *in vitro* kinase assay results. This obstacle is overcome by including a kinase-deficient mutant kinase as a negative control to exclude the possibility of phosphorylation by a contaminating kinase. The use of phosphatase inhibitors during the isolation procedure is particularly recommended when working with kinases that are activated by phosphorylation.

Protocol 4: In Vitro *Kinase Assays with Rad53 and*
  Mec1 Kinases Reagents

[γ-^{32}P]-ATP (>7000 Ci/mmol) (GE Healthcare), ATP (GE Health-care), PMSF (Roche), 0.2 M stock of glutathione (reduced form) pH 7.5 (Sigma), murine monoclonal anti-HA antibodies (HA.11, BAbCO or Covance, Princeton, NJ), Protein A Sepharose CL-4B (GE Healthcare), PHAS-1 (Stratagene, La Jolla, CA).

Buffers. Rad53 kinase buffer: 50 mM Tris HCl (pH 7.5), 10 mM MgCl$_2$, 10 mM MnCl$_2$, 1 mM dithiothreitol; Mec1 kinase buffer: 10 mM HEPES (pH 7.4), 50 mM NaCl, 10 mM MnCl$_2$, 1 mM DTT; transfer buffer: 25 mM Tris base, 192 mM glycine, 20% (v/v) methanol, 0.1% (w/v) SDS; 5× SDS loading buffer: 313 mM Tris-HCl (pH 6.8), 10% SDS, 50% glycerol, 25% β-mercaptoethanol, 0.015% bromophenolblue.

Procedure

In vitro kinase assays with GST-Rad53 kinase (wild type or kinase-deficient Rad53-K227A), and purified GST-Rad55·His6-Rad57 as a substrate are performed in 25 μl of kinase buffer containing 0.25 mM ATP, 1.5 μCi of [γ-^{32}P]-ATP (>7000 Ci/mmol), Rad53 kinase (1 μg chromatographically purified [Bashkirov *et al.*, 2003] or the specified amount from the Rad53 pulldown [see protocol 3]) and 2 μg of purified Rad55-Rad57 heterodimer. When the pull-down GST-Rad53 kinase is used, the reactions are supplemented with 10 mM glutathione pH 8.0 (reduced form) to release the kinase from the Glutathione Sepharose beads.

Mec1 kinase assays employ immunoprecipitated HA-tagged Mec1, wild type, or kinase-deficient mutant. Cell growth, galactose induction of protein expression, and immunoprecipitation of Mec1-HA are performed using anti-HA antibodies (HA.11, BAbCO) and Protein A Sepharose CL-4B (GE), as described in Mallory *et al.* (2003). About 20 μl (bed volume) of beads with Mec1-HA prewashed in freshly prepared kinase buffer supplemented with 0.1 mM PMSF are used for one reaction. Usually, expression of Mec1-HA in 500 ml culture yielded enough protein for five reactions. Master mix (9 μl) consisting of 20 μCi [γ-^{32}P] ATP, 40 μM ATP, and 6 μl of kinase buffer and 1 μl (1 μg) of kinase substrate (PHAS-1 or purified Rad55-Rad57 heterodimer) are added to the beads.

All kinase reaction mixes are assembled on ice and the reaction is initiated by transferring the tubes to a 30° water bath for 30 min and reactions are mixed every five min. The reactions with Mec1-HA give better results with longer incubation times, up to 60 min. Upon completion, the tubes are transferred to ice, and 7 μl of 5× SDS loading buffer is added. After denaturing in a heating block at 100° for 5 min, reaction mixes (or supernatant after collecting the Sepharose beads) are analyzed by PAGE. A 4–16% gradient gel allows simultaneous resolution of the small PHAS peptide and larger proteins; for Rad55-Rad57 alone an 8% gel is appropriate. After Coomassie staining, gels are dried, photographed, and exposed to X-ray film or phosphorimager.

Comments to Protocol 4

Alternatively, after SDS-PAGE the proteins are transferred to nitrocellulose or PVDF Immobilon-P (Millipore, Bedford, MA) membranes in transfer buffer for 30–45 min at 500 mA, and blots are exposed to X-ray film or phosphorimager screen. This alternative method gives less background although it is not well suited for peptides (like PHAS-1) and small proteins used as kinase substrates, because of the large differences in optimal time required for protein transfer of small and large proteins. For control visualization of proteins on the blot we employ either immunostaining with specific antibodies or direct reversible staining of proteins on the membrane by BLOT-FastStain (GenoTech, St. Louis, MO).

Conclusions and Perspectives

We have identified that the Rad55-Rad57 heterodimer is a *bona fide in vivo* substrate of the Mec1 DNA damage checkpoint kinase and is specifically phosphorylated after induction of genotoxic stress. In this chapter, methods are detailed that allow monitoring of Rad55 phosphorylation *in vivo* and *in vitro* showing DNA damage-induced phosphorylation by Mec1 in the C-terminal region of Rad55 likely involving the S378Q motif. This provides an opportunity to monitor DNA damage checkpoint activation at the level of the top kinase, Mec1. We have devised a vector that allows expression of heterodimeric proteins from a single plasmid with the strongly inducible, bidirectional *GAL1–10* promoter in *S. cerevisiae*. This vector proved useful in the purification of Rad55-Rad57 and should aid in the purification of other heterodimeric protein from budding yeast. The elaboration of a robust Rad55-Rad57 purification protocol will allow a detailed biochemical analysis of the function of Rad55-Rad57 heterodimer in recombinational DNA repair and its regulation by DNA damage checkpoint kinases.

Acknowledgments

We thank Drs. S. Elledge, F. Fabre, and B. Jones for kindly providing yeast strains and Dr. J. Nelson for sharing pJN58. We are grateful to T. Doty for her careful proofreading. This work was supported by an NIH grant (CA92276) to W.D.H.

References

Bashkirov, V. I., Bashkirova, E. V., Haghnazari, E., and Heyer, W. D. (2003). Direct kinase-to-kinase signaling mediated by the FHA phosphoprotein recognition domain of the Dun1 DNA damage checkpoint kinase. *Mol. Cell. Biol.* **23,** 1441–1452.

Bashkirov, V. I., King, J. S., Bashkirova, E. V., Schmuckli-Maurer, J., and Heyer, W. D. (2000). DNA repair protein Rad55 is a terminal substrate of the DNA damage checkpoints. *Mol. Cell. Biol.* **20,** 4393–4404.

Carr, A. M. (2002). DNA structure dependent checkpoints as regulators of DNA repair. *DNA Repair* **1,** 983–994.

Gilbert, C. S., Green, C. M., and Lowndes, N. F. (2001). Budding yeast Rad9 is an ATP-dependent Rad53 activating machine. *Mol. Cell* **8,** 129–136.

Kim, S. T., Lim, D. S., Canman, C. E., and Kastan, M. B. (1999). Substrate specificities and identification of putative substrates of ATM kinase family members. *J. Biol. Chem.* **274,** 37538–37543.

Krogh, B. O., and Symington, L. S. (2004). Recombination proteins in yeast. *Annu. Rev. Genet.* **38,** 233–271.

Lee, S.-J., Schwartz, M. F., Duong, J. K., and Stern, D. F. (2003). Rad53 phosphorylation site clusters are important for Rad53 regulation and signaling. *Mol. Cell. Biol.* **23,** 6300–6314.

Mallory, J. C., Bashkirov, V. I., Trujillo, K. M., Solinger, J. A., Dominska, M., Sung, P., Heyer, W. D., and Petes, T. D. (2003). Amino acid changes in Xrs2p, Dun1p, and Rfa2p that remove the preferred targets of the ATM family of protein kinases do not affect DNA repair or telomere length in *Saccharomyces cerevisiae*. *DNA Repair* **2,** 1041–1064.

Melo, J., and Toczyski, D. (2002). A unified view of the DNA-damage checkpoint. *Curr. Opin. Cell Biol.* **14,** 237–245.

Nelson, J. R., Lawrence, C. W., and Hinkle, D. C. (1996). Thymine-thymine dimer bypass by yeast DNA polymerase zeta. *Science* **272,** 1646–1649.

Nyberg, K. A., Michelson, R. J., Putnam, C. W., and Weinert, T. A. (2002). Toward maintaining the genome: DNA damage and replication checkpoints. *Annu. Review Genet.* **36,** 617–656.

Pellicioli, A., Lucca, C., Liberi, G., Marini, F., Lopes, M., Plevani, P., Romano, A., Di Fiore, P. P., and Foiani, M. (1999). Activation of Rad53 kinase in response to DNA damage and its effect in modulating phosphorylation of the lagging strand DNA polymerase. *EMBO J.* **18,** 6561–6572.

Pringle, J. R., Adams, A. E. M., Drubin, D. G., and Haarer, B. K. (1991). Immunofluorescence methods for yeast. *Methods Enzymol.* **194,** 565–602.

Sanchez, Y., Desany, B. A., Jones, W. J., Liu, Q. H., Wang, B., and Elledge, S. J. (1996). Regulation of *RAD53* by the ATM-like kinases *MEC1* and *TEL1* in yeast cell cycle checkpoint pathways. *Science* **271,** 357–360.

Sun, Z. X., Fay, D. S., Marini, F., Foiani, M., and Stern, D. F. (1996). Spk1/Rad53 is regulated by Mec1-dependent protein phosphorylation in DNA replication and damage checkpoint pathways. *Genes Dev.* **10,** 395–406.

Sung, P. (1997). Yeast Rad55 and Rad57 proteins form a heterodimer that functions with replication protein A to promote DNA strand exchange by Rad51 recombinase. *Genes Devel.* **11,** 1111–1121.

Zhou, B. B. S., and Elledge, S. J. (2000). The DNA damage response: Putting checkpoints in perspective. *Nature* **408,** 433–439.

[11]   Methods for Studying Mutagenesis and Checkpoints in *Schizosaccharomyces pombe*

By MIHOKO KAI, LORENA TARICANI, and TERESA S.-F. WANG

Abstract

Mutations in genome caretaker genes can induce genomic instability, which are potentially early events in tumorigenesis. Cells have evolved biological processes to cope with the genomic insults. One is a multifaceted response, termed checkpoint, which is a network of signaling pathways to coordinate cell cycle transition with DNA repair, activation of transcriptional programs, and induction of tolerance of the genomic perturbations. When genomic perturbations are beyond repair, checkpoint responses can also induce apoptosis or senescence to eliminate those deleterious damaged cells. Fission yeast, *Schizosaccharomyces pombe* (*S. pombe*) has served as a valuable model organism for studies of the checkpoint signaling pathways. In this chapter, we describe methods used to analyze mutagenesis and recombinational repair induced by genomic perturbations, and methods used to detect the checkpoint responses to replication stress and DNA damage in fission yeast cells. In the first section, we present methods used to analyze the mutation rate, mutation spectra, and recombinational repair in fission yeast when replication is perturbed by either genotoxic agents or mutations in genomic caretaker gene such as DNA replication genes. In the second section, we describe methods used to examine checkpoint activation in response to chromosome replication stress and DNA damage. In the final section, we comment on how checkpoint activation regulates mutagenic synthesis by a translesion DNA polymerase in generating a mutator phenotype of small sequence alterations in cells, and how a checkpoint kinase appropriately regulates an endonuclease complex to either prevent or allow deletion of genomic sequences and recombinational repair when fission yeast cells experience genomic perturbation in order to avoid deleterious mutations and maintain cell growth.

Introduction

Our genome constantly faces many types of spontaneous errors, innate obstacles, and acquired lesions during the cell cycle. S-phase is the most genetically vulnerable period of the cell cycle; thus, maintaining genome integrity during S-phase is of paramount importance. It has been proposed

METHODS IN ENZYMOLOGY, VOL. 409
0076-6879/06 $35.00
DOI: 10.1016/S0076-6879(05)09011-7

that mutations in certain genome caretaker genes have the potential to induce mutations in other genes causing mutator phenotypes in cells and compromise genomic integrity, which might associate with many human diseases (Loeb, 1991, 1998). This notion has been validated by the findings of germline mutations in some of the mismatch repair genes that lead to a genetic predisposition for the hereditary nonpolyposis colon cancer (HNPCC) and sporadic tumors (Fishel and Kolodner, 1995; Fishel *et al.*, 1993; Kolodner, 1995). Here, we describe methods of using fission yeast as the model organism to analyze mutagenesis and checkpoint activation in cells.

Mutagenesis Studies

A fission yeast mutator assay is developed based on the premise that cells with a mutated nonfunctional $ura4^+$ gene are able to grow on 5-fluoroorotic acid (5-FOA) medium (Boeke *et al.*, 1987). Thus, $ura4^+$ gene at its endogenous chromosomal locus is used as the read-out for the mutator assay to estimate the mutation rate and mutation spectra in the replication stressed fission yeast cells (Kai and Wang, 2003; Liu *et al.*, 1999). An outline of the mutator assay is schematically illustrated in Fig. 1A.

Mutagenic DNA structures can also be processed by recombination events. An intrachromosomal recombination assay is developed to measure mitotic recombination frequency. This method uses an integrated plasmid carrying the heteroallelic ade6⁻ duplications with $ura4^+$ marker gene to measure conversion type or deletion type mutation frequencies in fission yeast. The strategy is outlined in Fig. 1B (Doe *et al.*, 2004; Kai *et al.*, 2005; Osman *et al.*, 2000; Schuchert and Kohli, 1988).

Protocol

Mutator Assay

Strains for the mutator analysis must have $ura4^+$ gene at its genomic locus. Strains are first grown on minimal medium as described in (Moreno *et al.*, 1991) lacking uracil to ensure each strain being $ura4^+$ and expressing the $ura4^+$ gene. For analysis of spontaneous mutations, the $ura4^+$ cells are plated on minimal medium containing uracil to allow accumulation of mutations. For thermosensitive replication mutants, cells are incubated at the semipermissive temperature on the minimal medium containing uracil to accrue mutations. The viability of the temperature sensitive strain used for the analysis must be greater than 80% at its semipermissive temperature. For subsequent median analysis, odd numbers of individual colonies,

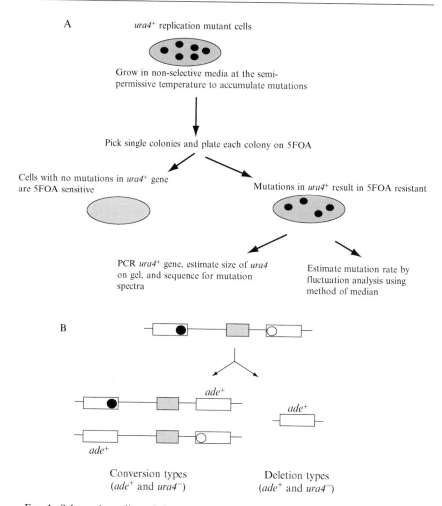

Fig. 1. Schematic outline of the mutator assay and recombination assay. (A) Forward mutation assay for identification of replication mutator by using *ura4+* gene expression as readout. See text for description. (B) Recombination assay substrate and products. Deletion type of recombination yields cells of *ade+* and *ura4−* genotype allowing cells to grow on a medium without adenine but containing uracil. Conversion type of recombination in cells yields cells of *ade+* and *ura4+* genotype, which allows cells to grow on a medium without both adenine and uracil. (Reprinted with permission from Kai *et al.*, 2005.)

at least 11 to 15 colonies, are picked for the analysis. Colonies are suspended in sterile water and plated on minimal medium agar containing 1 mg/ml of 5-FOA. Each assay should be repeated at least three times (a total of at least 33 to 45 independent colonies should be picked for the analysis). Colonies resistant to 5-FOA are scored after incubating at 25° for 3–4 days. Mutation rates on *ura4+* are calculated by fluctuation analysis using the method of the median as described in Lea and Coulson (1949) by using the *Analyze-it* Microsoft Excel (Microsoft, Redmond, WA) program. Experiments of mutation rate analysis are performed as described in Kai and Wang (2003). Significant levels compared between strains are determined by using the Mann-Whitney rank test and all comparisons should have a "significant" *p* level of $p < 0.005$.

For mutation spectra analysis, genomic DNA samples are isolated from 5-FOA^r colonies and the mutant *ura4* genes are PCR-amplified and analyzed by 1.5% agarose gel electrophoresis as described in (Kai and Wang, 2003). The *ura4*$^−$ PCR products are then gel purified and sequenced to determine the mutation types generated. At least 100 5-FOA^r colonies from each strain should be analyzed for the size of *ura4*-PCR product by agarose gel electrophoresis and 20 *ura4* PCR products should be sequenced.

Intrachromosomal Recombination Assay

Mitotic recombination frequencies are determined using strains harboring an intrachromosomal recombination substrate consisting of a nontandem direct repeat of *ade6*$^−$ heteroalleles flanking a functional *ura4+* gene as illustrated in Fig. 1B. Frequencies of spontaneous recombinants are determined by fluctuation analysis as described in (Kai *et al.*, 2005; Osman *et al.*, 2000). At least seven independent colonies are used for the assay for each strain. Each assay should be repeated at least three times. To determine recombination frequencies, single colonies are picked and cultured in YES media (0.5% [w/v] oxiod yeast extract, 3% [w/v] glucose plus 50–250 mg/liter adenine, histidine, leucine, uracil, and lysine hydrochloride) (Moreno *et al.*, 1991). Cells are first cultured in YES media at appropriate temperature for two to three days and then diluted to plate onto media without adenine (*ade*$^−$ plates) to select for *ade+* cells to score deletion type of recombinants. The recombination frequency of wild-type fission yeast is approximately $1–3 \times 10^{-3}$. The *ade+* recombinants are then replicated onto minimal media as described in Moreno *et al.* (1991) lacking both adenine and uracil to score conversion type of recombinants.

Median recombination frequencies of each strain are determined from three independent assays. The average recombination frequencies and percentage of conversion-types are determined from these medians, according to Lea and Coulson (1949) by *Analyze-it*, the Microsoft Excel program and significant levels between the strains are estimated by using the Mann-Whitney rank test. All comparisons should have a "significant" *p* level of $p < 0.005$.

Identification of Replication Mutator

By using the mutator assay described previously (Fig. 1A), thermosensitive replication mutant strains that display a mutation frequency 2-fold higher than the wild-type cells at their respective semipermissive temperature are considered to have a mutator phenotype and referred to as a replication mutator. By this criterion, three mutant alleles each of *polα* and *polδ*, one of each of two *polδ* small subunits (*cdc1-7* and *cdc27-k3*), a mutant allele of DNA ligase (*cdc17-K42*), and a deletion mutant of *S. pombe rad2*⁺ gene (homologue of human Fen1 and budding yeast *RAD27*) are replication mutators, whereas a mutation allele in *mcm2* (*cdc19-P1*), *mcm10* (*cdc23-M36*), *polε* (*cdc20-M10*), and *orc1* (*orp1*) are not replication mutators. The semipermissive growth temperature of each replication mutator is empirically defined. The extent of elevated mutation rate in the replication mutators relative to the wild-type cells are summarized in Fig. 2A and an example of *ura4* gene size in selective replication mutators relative to the wild-type *ura4*⁺ gene size are shown in Fig. 2B.

Checkpoint Responses to Replication Stalling and DNA Damage

The main function of checkpoint is to preserve genome integrity when cells experience genomic perturbation (Bartek *et al.*, 2004; Carr, 2002a,b; Melo and Toczyski, 2002; Nyberg *et al.*, 2002; Osborn *et al.*, 2002). The fission yeast checkpoint response pathways are schematically outlined in Fig. 3. Signals of genomic perturbations are sensed and transduced by a group of checkpoint protein complexes (Rad3-Rad26 complex, Rad17-Rfc2–5 complex, Rad9-Rad1-Hus1 complex, and Rad4/Cut5) to two effector kinases, Cds1 and Chk1. Genomic stress or damage during early S-phase activates replication checkpoint (S-phase checkpoint) kinase Cds1 to stabilize replication forks. Thus, Cds1 activation is often used to monitor checkpoint response to early S-phase perturbation by its kinase activity with myelin basic protein (MBP) as substrate (Fig. 4A). Genomic perturbations in late S-phase and G2 phase activate damage checkpoint (G2-M phase) kinase Chk1. Activation of Chk1 kinase regulates G2 to mitosis cell cycle transition. Chk1 activation is estimated by the presence of a phosphorylated form of influenza hemagglutinin (HA) epitope tagged Chk1 protein, which exhibits as a slower migrating HA-tagged Chk1 protein in SDS gels compared to the unphosphorylated HA-tagged Chk1 protein when detected by Western blot (Fig. 4B) (Walworth and Bernards, 1996). Furthermore, in response to genomic perturbation, checkpoint clamp associates with the perturbed genomic sites in a checkpoint clamp loader (Rad17-Rfc2–5 complex)-dependent manner (Kondo *et al.*, 2001; Melo *et al.*, 2001). Thus, chromatin association of checkpoint clamp components, Rad9, Rad1, and Hus1 has also been

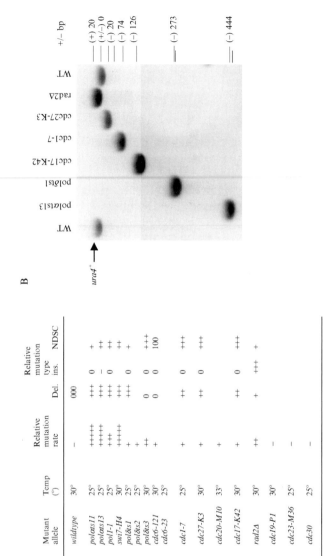

A

Mutation protein	Mutant allele	Temp (°)	Relative mutation rate	Relative mutation type Del.	ins.	NDSC
Wildtype	wildtype	30°	−	000		
Polα	polαts11	25°	++++	+++	0	+
	polαts13	25°	++++	+++	−	++
	pol1-1	25°	+++	+++	0	++
	swi7-H4	30°	++++	+++		++
Polδ	polδts1	25°	+	+++	0	+
	polδts2	25°	+			
	polδts3	30°	++	0	0	+++
	cdc6-121	30°		0	0	100
	cdc6-23	25°	+			
Subunits of Polδ	cdc1-7	25°	+	++	0	+++
	cdc27-K3	30°	+	++	0	+++
Polε	cdc20-M10	33°	+			
Ligase	cdc17-K42	30°	+	++	0	+++
FEN-1	rad2Δ	30°	++	+	+++	+
MCM2	cdc19-P1	30°	−			
MCM10	cdc23-M36	25°	−			
Orp1	cdc30	25°	−			

Del., deletion; Ins., insertion; NDSC, no detectable size change (point mutations and single base frameshifts.

Strains that had a mutation rate at least 5-fold relative to wildtype cells were designated as having a mutator phenotype. The relative mutation rate (++++) is expressed as having >100-fold higher than the wildtype mutation rate, which was 2.89×10^{-8} per generation. All mutation assays were performed at the semipermissive temperature for each strain or at 30° for the non-thermosensitive strains.

B

Lanes: WT, rad2Δ, cdc27-K3, cdc1-7, cdc17-K42, polδts1, polαts13, WT

ura4⁺ →

+/− bp
(+) 20
(+/−) 0
(−) 20
(−) 74
(−) 126
(−) 273
(−) 444

used to demonstrate checkpoint activation (Griffiths *et al.*, 2000; Kai and Wang, 2003; Kai *et al.*, 2001).

Protocol

Assay of Cds1 Kinase Activity

Cds1 kinase assay is based on the method described in Lindsay *et al.* (1998). Cells harboring the c-myc-epitope tagged $cds1^+$ at its endogenous

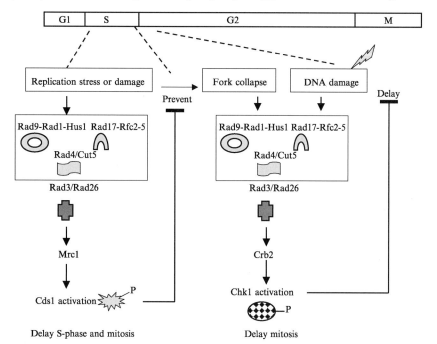

FIG. 3. Checkpoint responses in *S. pombe*. Signals of genomic perturbation and DNA damage in early S-phase and during S-phase sensed by checkpoint sensors and transducers activate Cds1 kinase to enforce S-phase delay and mitotic entry and stabilize replication forks. Signals of DNA damage in late S-phase and G2 sensed by checkpoint sensors and transducers activate Chk1 kinase to delay cell cycle transition.

FIG. 2. Replication mutators. See text for detailed description. (A) Replication mutators identified by using the mutator assay described in Fig. 1. (Adapted and modified with permission from Liu *et al.*, 1999.) (B) Deletion mutation spectrum of representative replication mutators. PCR products from 5-FOA resistant cells are analyzed on 1.5% agarose gel. The wild-type $ura4^+$ PCR product is 1.406-bp long and marked as having 0-bp alteration. The difference in size between the wild-type and replication mutators' *ura4* PCR products (+/− bp) is indicated to the right of the gel, with + indicating larger and − indicating shorter than the wild-type product. (Reprinted with permission from Liu *et al.*, 1999.)

chromosomal locus are used for the analysis. To induce S-phase perturbation, wild-type cells containing $cds1^+$:myc are treated with 12 mM hydroxyurea (HU) to induce early S-phase arrest. Replication mutants having the $cds1^+$:myc at its genomic locus are incubated in the semipermissive temperature. Cell lysates are prepared from the S-phase perturbed cells by glass bead disruption in HB buffer (25 mM Tris-HCl, pH 7.5, 15 mM MgCl$_2$, 15 mM EGTA, 1 mM DTT, and Complete Protease Inhibitor EDTA-Free Tablet [Roche Molecular Biochemical, Indianapolis, IN]) with the addition of 1% TX-100 plus 150 mM NaCl. Cds1 protein is immunoprecipitated from 1 mg of soluble protein in 300 μl volume, using Protein G plus Protein A agarose (Calbiochem, La Jolla, CA) prebound with anti-myc monoclonal antibody (9E10). Immunocomplexes are collected by centrifugation and washed four times with 1 ml of HB buffer described previously and further washed once with kinase buffer (10 mM Hepes pH 7.5, 75 mM KCl, 5 mM MgCl$_2$, 0.5 mM EDTA, 1 mM DDT). Twenty microliters (50% slurry) of the immunocomplex containing beads are incubated with 10 μl of 2$\times$ kinase buffer (5 μCi of γ^{32}P-ATP, 1 μl 2 mM ATP, 5 μl of myelin basic protein [MBP] [1mg/ml stock]). One-third of the sample is removed and analyzed by Western blotting as a protein loading control. Reaction mixtures after incubating at 30° for 15 min are terminated by boiling in an equal volume of 2$\times$ SDS sample buffer, and samples are fractionated on 15% SDS-gel. Gel is then dried and exposed on phosphorimager screen or X-ray films.

Assay of Chk1 Activation

Strains with triple-HA epitope tagged $chk1^+$ allele ($chk1^+$:HA) at the $chk1^+$ endogenous chromosomal locus are used for the anlaysis (Walworth and Bernards, 1996). Four hundred-ml of $chk1^+$:HA cells are grown at 30° in YES medium (Moreno et al., 1991) to midexponential phase (5 × 10^6 cells/ml). Camptothecin (CPT) is then added to one-half of the culture (200 ml) to a final concentration of 20–40 μM and incubated for 2 h at 30° to induce genomic damage. Cells are incubated on ice for 5 min and harvested by centrifugation at 3000 rpm for 5 min at 4°. Cell pellets are first washed with ice-cold stop buffer (150 mM NaCL, 50 mM NaF, 10 mM EDTA, 1 mM NaN$_3$, pH 8) and then spun down at 3000 rpm for 5 min and drained. Cell pellets are resuspended in 1 ml of freshly made LT300 buffer (25 mM Tris-HCl pH 7.4, 1 mM EDTA, 10% glycerol, 300 mM NaCL, 50 mM NaF, 1 mM Na3VO4, 0.1% Nonidet P-40, 10 mM β-mercaptoethanol, 1 mM phenylmethylsulfonyl fluoride, supplemented with complete protease inhibitors without EDTA from Roche Molecular Biochemicals) and lysed by using glass beads and a FastPrep (Bio 101) vortexing machine. Cell lysates are centrifuged for 10 min at 13,000 rpm and the supernatant is collected for protein determination using Bio-Rad Protein

Assay (Bio-Rad, Hercules, CA). Five mg of soluble protein is added to 100 μl of anti-HA (3F10) antibody affinity matrix (Roche Molecular Biochemicals) and incubated at 4° on a rotator for 3–5 h. Alternatively, 4 μl of Anti-HA High Affinity (3F10) matrix (Roche Molecular Biochemicals) can be added to 5 mg of soluble protein and incubated at 4° on a rotator for 3 h. Then 100 μl of Protein-G Plus-agarose suspension (Cal BioChem) are added and the immunoprecipitations are incubated for 1 h, washed three times with LT300 lysis buffer, and then resuspended in 30 μl of 2× SDS loading buffer. Immunoprecipitates are then fractionated on freshly prepared 8% SDS-PAGE, electroblotted onto a PVDF membrane (Biorad) for Western blot. The mobility of Chk1:HA protein is detected by mouse anti-HA (12CA5) at 1:500 dilution (Roche Molecular Biochemicals). Immunoreactive protein bands are detected by Horse Radish Peroxidase (HRP-) conjugated secondary goat anti-mouse IgG antibody (1:10,000) (New England BioLabs, Ipswich, MA) and the luminol-based ECL detection kit (Perkin-Elmer, Wellesley, MA).

Examples of assays of Cds1 kinase and Chk1 activation are shown in Fig. 4.

Chromatin Fractionation Assay

Logarithmically growing cells (5×10^8 cells) expressing an epitope-tagged checkpoint protein (for example myc-tagged checkpoint protein) are harvested in 1 mM sodium azide by centrifugation and washed sequentially with 25 ml of STOP buffer (0.9% NaCl, 1 mM NaN$_3$, 50 mM NaF, 10 mM EDTA), 25 ml double distilled water, and 10 ml of 1.2 M sorbitol. To prepare spheroplast, cells are first resuspended in 1.125 ml of CB1 (50 mM sodium citrate, 40 mM EDTA, 1.2 M sorbitol) and then added to 125 μl of CB1 containing 10 mg of lysis enzymes (Sigma, L2265, St. Louis, MO), 10 μg Zymolyase-20T (ICN), and 2.5 μl of β-mercaptoethanol. Digestion of cells was monitored by removing 2 μl of cell sample and adding an equal volume of 10 % (w/v) SDS and examining by microscope. When approximately 90% of cells are spheroplasts after zymolyase treatment, the digestions are terminated by adding an equal volume of ice-cold 1.2 M sorbitol. Spheroplasts are then harvested by centrifugation at low speed of 290g for 4 min, and washed twice with 1.2 ml of 1.2 M sorbitol. The spheroplasts are resuspended in 425 μl of 1.2 M sorbitol, frozen in liquid nitrogen, and stored at –80°. Samples can be subsequently thawed on ice and lysed by the addition of 50 μl of 10× lysis buffer (500 mM potassium acetate, 20 mM MgCl$_2$, 200 mM Hepes pH 7.9). To the lysates, a Complete Protease Inhibitor EDTA-Free Tablet (Roche Molecular Biochemical) and 20 μl of 25% Triton X-100 (TX-100) are added and incubated on ice for 10 min. Samples

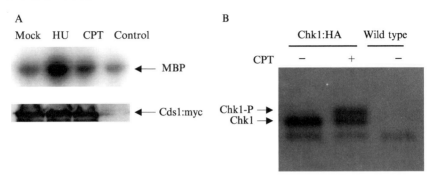

FIG. 4. Activation of Cds1 and Chk1 kinases. See text for detailed description. (A) Activation of Cds1 kinase. Wild-type cells are incubated in either 12 mM hydroxyurea or 30 μM CPT for 4 h. Cell extracts are prepared and Cds1 kinase assay are performed with myelin basic protein (MBP) as substrate as described in text. Control lane shows Cds1 kinase assay with anti-myc immunoprecipitates from cells containing *cds1*[+] without the myc-epitope tag. Bottom panel is the loading control as described in text. (Reprinted with permission from Kai *et al.*, 2005.) (B) Activation of Chk1. Chk1 activation induced by CPT damage shown as a slower migration form of phosphorylated Chk1 protein, see text for detailed description.

of 50 μl are removed and boiled in an equal volume of 2× SDS sample loading buffer and used for analysis of the total protein fraction (designated as Total). Extracts are then fractionated into soluble and pellet fractions by centrifugation at 12,000g for 15 min in a bench-top microcentrifuge. Supernatant is removed and boiled in equal volume of 2× SDS sample buffer (designated as Sup). Insoluble chromatin enriched pellet fraction is washed once with the lysis buffer without TX-100 and digested with DNaseI (100 U, Stratagene, La Jolla, CA) in the presence of 5 mM MgSO$_4$ and protease inhibitors described previously on ice for 30 min. The DNaseI digested chromatin enriched fraction was centrifuged for 5 min at 14,000g. Supernatant of this chromatin-enriched fraction is designated as the fraction that contains the chromatin associated proteins (designated as Chr), while the pellet was designated as the cellular scaffold and debris. Forty μg of Total and 5 vol equivalents of Chr fraction are fractionated on SDS-gel, followed by Western blotting with anti-myc monoclonal antibody (9E10).

Checkpoint Regulation of Mutagenesis

The primary mutator phenotypes exhibited in replication mutators are: deletion of genomic sequence flanked by short direct sequence repeats and small sequence alterations shown as base substitutions and single base frameshifts (Liu *et al.*, 1999). Replication mutators at their semipermissive growth conditions are under replication stress; however, these cells need to maintain continuous cell growth. In response to the genomic stress, checkpoint clamp is loaded onto chromatin in a checkpoint clamp-loader Rad17-Rfc2–5 dependent manner and replication checkpoint kinase Cds1 is mildly activated in the replication mutators (Kai and Wang, 2003; Kondo *et al.*, 2001; Melo *et al.*, 2001). The chromatin associated checkpoint clamp subsequently recruits translesion DNA polymerase such as polymerase κ (Polκ/DinB) onto chromatin to perform mutagenic synthesis, which generates point mutations and single base frameshifts. This perhaps is a way of how replication mutators tolerate the replication stress to facilitate replication fork progression in order to maintain the mutant's growth (Kai and Wang, 2003). In contrast, S-phase stress imposed by depletion of dNTP pool with HU treatment robustly activates replication checkpoint kinase Cds1 (Boddy *et al.*, 1998; Lindsay *et al.*, 1998). The robustly activated Cds1 kinase maintains the stability of replication fork and allows recovery of replication when HU is removed. Thus, HU treatment does not induce small sequence alteration type of mutator phenotype.

An endonuclease complex, Mus81-Eme1, which constitutively associates with chromatin, is able to generate deletion of genomic sequences in stressed replication mutators (Kai *et al.*, 2005). In wild-type cells, when S-phase is arrested by HU treatment, Cds1 kinase is robustly activated. Mus81 protein undergoes a Cds1-dependent phosphorylation and dissociates from chromatin. This is one method in which Cds1 kinase prevents Mus81 from cleaving replication fork and causing deleterious gross genomic rearrangement in S-phase stressed cells. This negative mode of Cds1 kinase regulation of Mus81-Eme1 also prevents gene conversion type of recombination described in the previous protocol and outlined in Fig. 1B (Kai *et al.*, 2005). In replication mutators, Mus81 protein undergoes limited Cds1-dependent phosphorylation, which allows Mus81-Eme1 to remain chromatin associated to generate deletion of nonessential genomic sequences, thereby allowing the replication mutator to tolerate the S-phase perturbation and maintain continuous cell growth (Kai *et al.*, 2005).

References

Bartek, J., Lukas, C., and Lukas, J. (2004). Checking on DNA damage in S phase. *Nat. Rev. Mol. Cell. Biol.* **5,** 792–804.

Boddy, M. N., Furnari, B., Mondesert, O., and Russell, P. (1998). Replication checkpoint enforced by kinase Cds1 and Chk1. *Science* **280,** 909–912.

Boeke, J. D., Trueheart, J., Natsoulis, G., and Fink, G. R. (1987). 5-Fluoroorotic acid a selective agent in yeast molecular genetics. *Methods Enzymol.* **154,** 164–175.

Carr, A. M. (2002a). Checking that replication breakdown is not terminal. *Science* **297,** 557–558.

Carr, A. M. (2002b). DNA structure dependent checkpoints as regulators of DNA repair. *DNA Repair (Amst.)* **1,** 983–994.

Doe, C. L., Osman, F., Dixon, J., and Whitby, M. C. (2004). DNA repair by a Rad22-Mus81-dependent pathway that is independent of Rhp51. *Nucleic Acids Res.* **32,** 5570–5581.

Fishel, R., and Kolodner, R. D. (1995). Identification of mismatch repair genes and their role in the development of cancer. *Curr. Opin. Genet. Dev.* **5**, 382–395.

Fishel, R., Lescoe, M. K., Rao, M. R. S., Copeland, N. G., Jenkins, N. A., Garber, J., Kane, M., and Kolodner, R. (1993). The human mutator gene homolog *MSH2* and its association with hereditary nonopolyposis colon cancer. *Cell* **75**, 1027–1038.

Griffiths, D., Uchiyama, M., Nurse, P., and Wang, T. S.-F. (2000). A novel allele of the chromatin-bound fission yeast checkpoint protein Rad17 separates the DNA structure checkpoints. *J. Cell Sci.* **113**, 1075–1088.

Kai, M., Boddy, M. N., Russell, P., and Wang, T. S.-F. (2005). Replication checkpoint kinase Cds1 regulates Mus81 to preserve genome integrity during replication stress. *Genes Dev.* **19**, 919–932.

Kai, M., Tanaka, H., and Wang, T. S.-F. (2001). Fission yeast Rad17 binds to chromatin in response to replication arrest or DNA damage. *Mol. Cell. Biol.* **21**, 3289–3301.

Kai, M., and Wang, T. S.-F. (2003). Checkpoint activation regulates mutagenic translesion synthesis. *Genes Dev.* **1**, 64–76.

Kolodner, R. D. (1995). Mismatch repair: Mechanisms and relationship to cancer susceptibility. *Trends Biochem. Sci.* **20**, 397–401.

Kondo, T., Wakayama, T., Naiki, T., Matsumoto, K., and Sugimoto, K. (2001). Recruitment of mec1 and ddc1 checkpoint proteins to double-strand breaks through distinct mechanisms. *Science* **294**, 867–870.

Lea, D. E., and Coulson, C. A. (1949). The distribution of the numbers of mutants in bacterial populations. *J. Genetics* **49**, 264–285.

Lindsay, H. D., Griffiths, D. J. F., Edwards, R., Murray, J. M., Christensen, P. U., Walworth, N., and Carr, A. M. (1998). S-phase specific activation of Cds1 kinase defines a subpathway of the checkpoint response in *S. pombe*. *Genes Dev.* **12**, 382–395.

Liu, V. F., Bhaumik, D., and Wang, T. S.-F. (1999). Mutator phenotype induced by aberrant replication. *Mol. Cell. Biol.* **19**, 1126–1135.

Loeb, L. A. (1991). Mutator phenotype may be required for multistage carcinogenesis. *Cancer Res.* **51**, 3075–3079.

Loeb, L. A. (1998). Cancer cells exhibit a mutator phenotype. *In* "Advances in Cancer Research" (G. F. Vande Woude and G. Klein, eds.), Vol. 72, pp. 25–56. Academic Press, San Diego.

Melo, J., and Toczyski, D. (2002). A unified view of the DNA-damage checkpoint. *Curr. Opin. Cell Biol.* **14**, 237–245.

Melo, J. A., Cohen, J., and Toczyski, D. P. (2001). Two checkpoint complexes are independently recruited to sites of DNA damage *in vivo*. *Genes Dev.* **15**, 2809–2821.

Moreno, S., Klar, A., and Nurse, P. (1991). Molecular genetic analysis of fission yeast *Schizosaccharomyces pombe*. *Methods Enzymol.* **194**, 795–823.

Nyberg, K. A., Michelson, R. J., Putnam, C. W., and Weinert, T. A. (2002). Toward maintaining the genome: DNA damage and replication checkpoints. *Annu. Rev. Genet.* **36**, 617–656.

Osborn, A. J., Elledge, S. J., and Zou, L. (2002). Checking on the fork: The DNA-replication stress-response pathway. *Trends Cell Biol.* **12**, 509–516.

Osman, F., Adriance, M., and McCready, S. (2000). The genetic control of spontaneous and UV-induced mitotic intrachromosomal recombination in the fission yeast Schizosaccharomyces pombe. *Curr. Genet.* **38**, 113–125.

Schuchert, P., and Kohli, J. (1988). The *ade6-M26* mutation of *Schizosaccharomyces pombe* increases the frequency of crossing over. *Genetics* **119**, 507–515.

Walworth, N. C., and Bernards, R. (1996). Rad-dependent response of the chk1-encoded protein kinase at the DNA damage checkpoint. *Science* **271**, 353–356.

[12] Methods for Determining Spontaneous Mutation Rates

By PATRICIA L. FOSTER

Abstract

Spontaneous mutations arise as a result of cellular processes that act upon or damage DNA. Accurate determination of spontaneous mutation rates can contribute to our understanding of these processes and the enzymatic pathways that deal with them. The methods that are used to calculate mutation rates are based on the model for the expansion of mutant clones originally described by Luria and Delbrück (1943) and extended by Lea and Coulson (1949). The accurate determination of mutation rates depends on understanding the strengths and limitations of these methods and how to optimize a fluctuation assay for a given method. This chapter describes the proper design of a fluctuation assay, several of the methods used to calculate mutation rates, and ways to evaluate the results statistically.

Introduction

Spontaneous mutations are mutations that occur in the absence of exogenous agents. They may be due to errors made by DNA polymerases during replication or repair, errors made during recombination, the movement of genetic elements, or spontaneously occurring DNA damage. The rate at which spontaneous mutations occur can yield useful information about cellular processes. For example, the occurrence of specific classes of mutations in different mutant backgrounds has been used to deduce the importance of various DNA repair pathways (Miller, 1996).

The mutation rate is the expected number of mutations that a cell will sustain during its lifetime. The mutant fraction or frequency is the proportion of cells in a population that are mutant.[1] Although mutant frequencies can be adequate indicators of the rate at which mutations are induced by DNA damaging agents, they are inadequate indicators of spontaneous mutation rates. This is because the population of mutants is composed of clones, each of which arose from a cell that sustained a mutation. The size of a given mutant clone will depend on when during the growth of the

---

[1] A mutation is a heritable change in the genetic material; a mutant is an individual that carries a mutation.

METHODS IN ENZYMOLOGY, VOL. 409   0076-6879/06 $35.00
DOI: 10.1016/S0076-6879(05)09012-9

population the mutation occurred. This is the fundamental property of spontaneous mutation that was exploited in the famous Luria and Delbrück fluctuation test (Luria and Delbrück, 1943). Among replicate cultures, the distribution of the numbers of mutations that were sustained is Poisson, but the distribution of the numbers of mutants that resulted is far from Poisson, and is usually referred to as the Luria-Delbrück distribution.

There are two basic methods to determine mutation rates: mutant accumulation and fluctuation analysis. These two methods are described later, with emphasis on fluctuation analysis.

Terminology

The definitions of the terms used in this chapter are given in Table I. It is important to distinguish between m, the mean number of mutations that occur during the growth of a culture, and μ = the mutation rate, which is the mean number of mutations that occur during the lifetime of a cell. Almost all methods to calculate mutation rates start by determining m and

TABLE I
TERMS USED

Term	Definition
m	Number of mutations per culture
μ	Mutation rate; probability of mutation per cell per division or generation
N	Number of cells
N_0	Initial number of cells in a culture = the inoculum
Nt	Final number of cells in a culture
r	Observed number of mutants in a culture
$\tilde{r}$	Median number of mutants in a culture
f	Mutant fraction or frequency = r/N
V	Volume of a culture
C	Number of cultures in experiment
p_0	Proportion of cultures without mutants
z	Dilution factor or fraction of a culture plated
p_r	Proportion of cultures with r mutants
c_r	Number of cultures with r mutants
Pr	Proportion of cultures with r or more mutants
Q_1	Value of r at 25% of the ranked series of r
Q_2	Value of r at 50% of the ranked series of r = the median
Q_3	Value of r at 75% of the ranked series of r
σ	Standard deviation
CL	Confidence limit = $(1\text{-}\alpha) \times 100$
α	Level of statistical significance, usually 0.05 or 0.01

then obtain μ by dividing m by some measure of the number of cell-lifetimes at risk for mutation, usually N_t. It is also important to distinguish between the number of mutations per culture, m, and the number of mutants per culture, r. r divided by N_t is the mutant fraction or mutant frequency, f.

The Lea-Coulson Model

The methods for calculating mutation rates discussed later are dependent on the model of expansion of mutant clones originally described by Luria and Delbrück (Luria and Delbrück, 1943) and extended by Lea and Coulson (Lea and Coulson, 1949). It is usually called the Lea-Coulson model and has the following assumptions.

1. The cells are growing exponentially.
2. The probability of mutation is not influenced by previous mutational events.
3. The probability of mutation is constant per cell-lifetime.
4. The growth rates of mutants and nonmutants are the same.
5. The proportion of mutants is always small.
6. Reverse mutations are negligible.
7. Cell death is negligible.
8. All mutants are detected.
9. No mutants arise after selection is imposed.

In addition, for assays using batch cultures, the following assumptions apply

10. The initial number of cells is negligible compared to the final number of cells.
11. The probability of mutation per cell-lifetime does not vary during the growth of the culture.

Departures from the Lea-Coulson model affect the distribution of mutant numbers and impact the mutation rate calculation. In general, most of the model's requirements can be met with proper experimental protocols, but some departures reflect real biological phenomenon.

Mutant Accumulation

When a population is growing exponentially, the appearance of new mutants plus the proliferation of preexisting mutants results in a constant increase in the mutant fraction each generation. The mutation rate is this increase, and can be determined by measuring the change in the mutant

Generation = k	Cells = N	New mutants	Growth of pre-existing mutants	Total mutants = r	Mutant fraction = r/N	Change per generation = μ
4	$N_t/16$	0		0	0	
3	$N_t/8$	1		1	$8/N_t$	$8/N_t$
2	$N_t/4$	2	2	4	$16/N_t$	$8/N_t$
1	$N_t/2$	4	4	4	12	$24/N_t$	$8/N_t$
0	N_t	8	8	8	8	32	$32/N_t$	$8/N_t$

FIG. 1. An illustration of the constant increase in the mutant fraction after a population reaches a size sufficiently large so that the accumulation of mutants is simply a function of population size. Luria's conventions are followed (Luria, 1951): k = the generation numbered backwards from 0; N = the number of cells present at each generation; N_t = the final number of cells in the population; μ = the mutation rate per cell (assuming a synchronous population). At each generation there are $N_t/2^k$ individuals that produce $\mu N_t/2^k$ new mutations, which will produce a total of μN_t mutant progeny by the last generation.

fraction over time (Fig. 1). However, an important caveat is that the population must be of sufficient size so that the probability that mutations occur each generation is essentially unity; otherwise, the chance occurrence of mutations will dominate the population (Luria, 1951). For batch cultures, the population must be large enough so that the average number of mutations per culture, m, is much greater than one. For a continuously dividing cell population, μ can be calculated by Eq. (1) (Drake, 1970):

$$\mu = \left(\frac{r_2}{N_2} - \frac{r_1}{N_1}\right) \div \left[\ln\left(\frac{N_2}{N_1}\right)\right] = (f_2 - f_1) \div (\ln N_2 - \ln N_1) \qquad (1)$$

Although conceptually simple, there are important technical difficulties that limit the use of this method. In general, by the time the population reaches a sufficient size, some mutations have already occurred, and these produce clones of mutants that make it impossible to accurately measure the accumulation of new mutants. Thus, to measure mutant accumulation a large population with few mutants must be generated. One way to do this is by generating and testing a number of populations and using the ones with a low mutant fraction. This is convenient if the population can be stabilized and used repeatedly for experiments, for example stocks of bacteriophage or viruses. Alternatively, a population can be purged of pre-existing mutants if the mutational target gives a phenotype that can be selected both for and against (i.e., a counterselectable marker). Selection against the mutants can be used to purge the population, and then selection for the mutants can be used to measure mutation rates. For example, Lac$^+$ bacteria with a temperature sensitive *galE* mutation are selected against by lactose at the nonpermissive temperature, but selected for by lactose at the permissive temperature (Reddy and Gowrishankar, 1997). In mammalian cells, mutations in the hypoxanthine-guanine phosphoribosyl transferase (*hprt*) locus make cells sensitive to HAT medium (which contains hypoxantine, aminopterin, and thymidine) but resistant to 6-thioguanine (Glaab and Tindall, 1997). A few other counterselectable markers are available (Reyrat *et al.*, 1998). Another possibility is to use a mutational target that allows mutants to be eliminated by cell sorting. For example, mutants that allow green fluorescent protein (GFP) to be produced can be eliminated by fluorescence-activated cell sorting (FACS); new mutants can then be detected by flow cytometry (Bachl *et al.*, 1999).

Mutant accumulation has been extensively used to measure mutation rates in chemostats (Kubitschek and Bendigkeit, 1964; Novick and Szilard, 1950). Cell number, N, is constant in a chemostat, so mutant accumulation is a function of the growth rate, λ, and the mutation rate per cell per generation μ. Thus:

$$\mu = \frac{1}{N\lambda} \frac{(r_2 - r_1)}{(t_2 - t_1)} \tag{2}$$

Where t_1 and t_2 are the times at which the numbers of mutants, r_1 and r_2, are measured.

Fluctuation Analysis

Experimental Design

A normal fluctuation test begins by inoculating a small number of cells into a large number of parallel cultures. The cultures are allowed to grow, usually to saturation, and then each culture is plated on a selective medium that allows the mutants to produce colonies. The total number of cells is determined by plating appropriate dilutions of a few cultures on nonselective medium. The distribution of the numbers of mutants among the parallel cultures is used to calculate the mutation rate. This basic design, invented by Luria and Delbrück (1943), can be used for single-celled microorganisms, cultured cells, bacteriophage, and viruses. So that individual mutants can be counted, a solid medium is usually used for selection, but the p_0 method (see later) also can be used with liquid medium.

The goal of designing a fluctuation assay is to maximize the precision with which the mutation rate is estimated. Precision is a measure of reproducibility, not accuracy (accuracy is how well the resulting estimate reflects the actual mutation rate, and that will depend on how well the underlying assumptions reflect reality). The important design parameters are: m, the number of mutations per culture; r, the number of mutants per culture; N_0, the initial number of cells; N_t, the final number of cells; V, the culture volume; and, C, the number of parallel cultures. The first step is to determine a preliminary r and m by plating aliquots from a few parallel cultures on the selective medium. A preliminary m is calculated from the mutant numbers obtained (e.g., using Method 2 or 3; described later) and then the other parameters adjusted so that the final m is within a useful range. The value of m will determine which methods can be used to calculate the mutation rate. None of the methods are reliable if m is less than 0.3 unless a prohibitive number of cultures are used. However, if m is above 15 some of the methods are not valid (Rosche and Foster, 2000). Obviously, the final m also has to be small enough so that the number of colonies on the selective plates is countable. However, if there are a few outlier high counts, they can be truncated at 150 with little loss of precision (Asteris and Sarkar, 1996; Jones et al., 1999).

The desired m is achieved by adjusting N_t, the final number of cells, either by manipulating the cell density or the culture volume. When using defined medium, cell density can be adjusted by limiting the carbon source. However, using other required growth factors, such as vitamins or amino acids, is not recommended because nonrequiring mutants will be selected and cell physiology may change. For example, in tryptophan-limited chemostats, bacterial mutation rates become time-dependent instead of generation-dependent (Kubitschek and Bendigkeit, 1964).

The desired N_t can be achieved by adjusting the culture volume. It is usually considered necessary to plate all the cells from each culture on the selective medium because sampling, or low plating efficiency (which is the same thing), increases the proportion of cultures with small numbers of mutants and narrows the distribution (Crane et al., 1996; Stewart et al., 1990). But this requirement restricts the volume of culture that can be used without concentrating the cells (which can be tedious with many cultures). However, if several mutant phenotypes are to be assayed in the same cultures, sampling is unavoidable. In addition, because a large culture contains more "information" than a small culture, it is better to plate a small aliquot from a large culture than all of a small culture if a proper correction can be applied (Jones et al., 1999). Some, but not all, of the methods for calculating mutation rates discussed later are amenable to such corrections.

The validity of the mutation rate calculation requires that N_t be the same in each culture. Usually, but not always, this can be accomplished by growing cells to saturation. If achieving a uniform N_t is a problem, the cell number in each culture can be monitored before mutant selection by measuring the optical density or by counting cells microscopically (e.g., using a Petroff-Hausser chamber). Because there is currently no valid method to correct for different N_t's, deviant cultures must be eliminated from the analysis.

The initial inoculum, N_0, must contain no preexisting mutants and must be small relative to N_t. Most of the methods to calculate the mutation rate are valid if N_0 is at least 1/1000 of N_t (Sarkar et al., 1992), but this may not ensure that N_0 contains no mutants. A reasonable rule of thumb is that N_0 roughly equals $(N_t/m) \times 10^{-5}$. The best way to ensure uniformity is to grow a starter culture in the same medium that will be used for the fluctuation assay, dilute these cells to the appropriate density in a large volume of fresh medium, and then distribute aliquots into individual cultures tubes for nonselective growth.

The precision of the estimate of m depends on C, the number of parallel cultures. Most experiments have 10 to 100 parallel cultures, with about 40 being most common. There is little gain in precision if C is larger unless m is less than about 0.3 (Jones et al., 1999; Rosche and Foster, 2000).

The fluctuation test was originally designed to test whether mutations occur before or after selection is imposed (Luria and Delbrück, 1943). If the selection is lethal, then the only mutants that appear must have pre-existed. However, if the selection is not lethal, for example reversion of an auxotrophy or utilization of a carbon source, mutants can arise both before and after selection has been applied. Postplating mutants can arise because the cells are proliferating on the selective medium (i.e., the selection is not stringent), or they can result from mutations that occur in nongrowing cells (adaptive mutations). In either case, the distribution of mutant numbers will be a combination of the Luria-Delbrück and Poisson, and the m estimated for preplating mutations will be inflated by the postplating mutations (Cairns et al., 1988). If nonmutant cells grow on the selective medium because of contaminating nutrients, one solution is to add an excess of scavenger cells that cannot mutate (because, for example, they have a deletion of the relevant gene) to consume the contaminants (Cairns and Foster, 1991). If the nonmutant cells grow because the selection is not stringent, the time it takes for a mutant to form a colony on the selective medium can be determined and then mutant colonies can be counted at the earliest possible time after plating.

Analyzing the Results of Fluctuation Assays

Fluctuation assays give the distribution of the numbers of mutants per culture, r, which is used to calculated m, the mean or most likely number of mutations per culture. m is not itself a particularly interesting parameter since it depends on the cell density and the volume of the culture. However, it is mathematically tractable and yields the mutation rate when divided by some measure of the number of cells. Although the distribution of mutants is not Poisson, the distribution of mutations is, so m is a Poisson parameter. There are many methods to calculate m (often called estimators) but they are all based on the theoretical distribution of mutant clone-sizes described by Luria and Delbrück (Luria and Delbrück, 1943) and Lea and Coulson (Lea and Coulson, 1949). Each method has its advantages and disadvantages, and the choice of method depends on the particular conditions of the experiment and the mathematical sophistication and persistence of the user. The MSS maximum likelihood method (Method 5) is the gold standard because it utilizes all of the results of an experiment and is valid over the entire range of mutation rates. Of the less complicated methods, the Lea-Coulson method of the median (Method 2) and the Jones median estimator (Method 3) are reliable when mutation rates are low to moderate, and the p_0 method (Method 1) can be used when mutation rates are very low ($m \leq 1$). Drake's Formula (Method 4) is particularly useful when comparing data reported as mutant frequencies instead of mutation rates. Methods 6 and 7 can be useful when not all the requirements of the clone-size distribution are met. No method using the mean number of mutants is valid, and none are given here. To see how these various methods behave with real data, see Rosche and Foster (2000).

Method 1: The p_0 Method

The distribution of the number of mutations that occur during the growth of parallel cultures has a Poisson distribution. If there are no mutants, there were no mutations, and so the mean number of mutations can be calculated from p_0, the proportion of cultures with no mutants (Luria and Delbrück, 1943). The 0th term of the Poisson distribution is:

$$p_0 = e^{-m} \tag{3}$$

So m is:

$$m = -\ln p_0 \tag{4}$$

Although simple, the p_0 method is limited. Its range of usefulness is $0.7 \geq p_0 \geq 0.1$ ($0.3 \leq m \leq 2.3$) and it performs best when p_0 is about 0.3. The p_0 method is inefficient (i.e., requires more cultures for the same precision) compared to other methods (Koziol, 1991; Rosche and Foster, 2000). In addition, p_0 is sensitive to several biologically relevant factors that complicate fluctuation analysis. Phenotypic lag (the delay in expression of a phenotype), poor plating efficiency, and selection against mutants all inflate p_0 and result in an erroneously low m. However, if all cells (not just mutants) have a plating efficiency of less than one, a correction factor can be applied to m. The same correction can be applied if only a fraction, of each culture is plated (Jones, 1993; Stewart et al., 1990). The actual m is calculated from the observed m using Eq. (5) where z is either the plating efficiency or the fraction plated[2]:

$$m_{act} = m_{obs} \frac{z - 1}{z \ln(z)} \tag{5}$$

Method 2: Lea-Coulson Median Estimator

This method is based on the observation that for $4 \leq m \leq 15$, a plot of the probability that a culture contains r or fewer mutants, versus $r/m - \ln(m)$ gives a skewed curve about a median of 1.24 (Lea and Coulson, 1949). Rearranging gives the following transcendental equation relating m to the median, $\tilde{r}$:

$$\frac{\tilde{r}}{m} - \ln(m) - 1.24 = 0 \tag{6}$$

that can be solved easily by iteration (an example of how to use a spread sheet to solve this equation is given in Fig. 2). The Lea and Coulson method of the median is easy to apply and remarkably accurate in computer simulations (Asteris and Sarkar, 1996; Stewart, 1994) and with real data (Rosche and Foster, 2000). It performs well over the range $2.5 \leq \tilde{r} \leq 60$ ($1.5 \leq m \leq 15$) (Rosche and Foster, 2000). Because it uses the median, it is relatively insensitive to deviations that affect either end of the distribution, especially if $\tilde{r}$ is relatively large. However, because little of the information obtained from the fluctuation test is used, the method is relatively inefficient (Rosche and Foster, 2000).

Method 3: The Jones Median Estimator

This estimator is based on the theoretical dilution of the experimental cultures that would be necessary to produce a distribution with a median of 0.5 (Jones et al., 1994). The basic equation is:

$$m = \frac{\tilde{r} - \ln(2)}{\ln(\tilde{r}) - \ln[\ln(2)]} = \frac{\tilde{r} - 0.693}{\ln(\tilde{r}) + 0.367} \tag{7}$$

[2] When $z = 1$, $\frac{z-1}{z\ln(z)} = 1$ from l'Hôpital's rule, and $m_{act} = m_{obs}$.

	A
1	[insert median here]
2	1.24
3	[try various values of *m* here]
4	= (A1/A3)-LN(A3)-(A2)

FIG. 2. Spreadsheet method to solve transcendental equations by iteration. The example is the Lea-Coulson median estimator. Method: insert the experimentally determined median in A1; try various values of *m* at A3 to get the value at A4 close to 0.

Two advantages of the Jones estimator are that it is explicit and that it accommodates dilutions. If z = the fraction of the culture that is plated or the plating efficiency, and $\tilde{r}_{obs}$ is the observed median, then (Jones *et al.*, 1994):

$$m = \frac{\left(\frac{\tilde{r}_{obs}}{z}\right) - 0.693}{\ln\left(\frac{\tilde{r}_{obs}}{z}\right) + 0.367} \tag{8}$$

In computer simulations over the range $3 \leq \tilde{r} \leq 40$ ($1.5 \leq m \leq 10$) the Jones estimator proved to be as reliable, and more efficient, than the Lea and Coulson median estimator (Jones *et al.*, 1994). The Jones estimator also performs well with real data (Rosche and Foster, 2000).

Method 4: Drake's Formula

Drake's formula (Drake, 1991) is an easy way to calculate mutation rates from mutant frequencies, and is especially useful in comparing data from different sources. Because it uses frequencies, Drake's formula gives the mutation rate, μ, instead of m (with $\mu = {}^{m}/_{N_t}$). Starting from Eq. (1) stated previously, Drake sets N_1 to be $1/\mu$, the population size at which the probability of mutation approaches unity. Assuming that no mutations occur before the population reaches this size, f_1 is zero; f_2 is the final mutant frequency, f; and, N_2 is the final population size, N_t. This gives:

$$u = f \div \ln(uN_t) \tag{9}$$

that can be solved for μ by iteration. Drake's formula is based on the same assumption discussed previously for Eq. (1) (i.e. that mutations occur only during the deterministic period of mutant accumulation). Using the median frequency (if available) instead of the mean minimizes the influence of jackpots (Drake, 1991). Since $\mu = m/N_t$, Drake's formula can be rearranged into the same form as Lea and Coulson's formula, Eq. (6):

$$\frac{\tilde{r}}{m} - \ln(m) = 0 \tag{10}$$

When $m < 4$, estimates of m obtained with Eq. (10) are significantly higher than those obtained with Eq. (6), but asymptotically approach those obtained with Eq. (6) as m becomes larger. If $m \geq 30$, the differences are trivial (Rosche and Foster, 2000).

Method 5: The MSS-maximum Likelihood Method

Sarkar et al. (1992) described a recursive algorithm[3] based on the Lea-Coulson generating function (Lea and Coulson, 1949) that efficiently computes the Luria-Delbrück distribution for a given value of m. Known as the MSS algorithm, it is:

$$p_0 = e^{-m}; \quad p_r = \frac{m}{r} \sum_{i=0}^{r-1} \frac{p_i}{(r-i+1)} \tag{11}$$

Note that the equation to calculate p_0 is the same as Eq. (3) and the proportion of cultures with each of the other possible values of r is given by the equation on the right. The algorithm is recursive, meaning that at a given m, the proportions of cultures with 0, 1, 2, 3, etc. mutants are $p_0 = e^{-m}$; $p_1 = m \times \frac{p_0}{2}$; $p_2 = \frac{m}{2} \times \left(\frac{p_0}{3} + \frac{p_1}{2}\right)$; $p_3 = \frac{m}{3} \times \left(\frac{p_0}{4} + \frac{p_1}{3} + \frac{p_2}{2}\right)$; etc. for all possible values or r.

This algorithm can be used to estimate m from experimental results using a maximum likelihood function, the formula for which is (Ma et al., 1992):

$$f(r|m) = \prod_{i=1}^{c} f(r_i|m) \tag{12}$$

where $f(r|m) = p_r$ from Eq. (11) and C is the number of cultures. The procedure is to start with a trial m (obtained from Eq. [6], for example) and use Eq. (11) to calculate the probability, p_r, of obtaining each possible r from 0 to the maximum value obtained (even if a given r was not obtained in the experiment it has to be included in the recursive equation). As mentioned previously, for most experiments, values of $r > 150$ can be lumped

---

[3] The algorithm itself was described earlier (Gurland, 1958, 1963) but was independently derived by Sarkar and coworkers and applied to fluctuation analysis (Ma et al., 1992; Sarkar et al., 1992).

into one category (Asteris and Sarkar, 1996). The likelihood function, Eq. (12), is the product of these calculated p_r's for r obtained in the experiment. The easiest way to do this calculation is to arrange the mutant counts in order and count the number of cultures that had each r mutants, c_r. Then the product of the p_r's is:

$$(p_0)^{c_0} \times (p_1)^{c_1} \times (p_2)^{c_2} \times (p_3)^{c_3} \times (p_4)^{c_4} \ldots \tag{13}$$

where each p_r is from Eq. (11) [alternatively, $c_r \ln(p_r)$ for each r can be added]. Note that values of r that were not obtained in the experiment give a value of 1 in Eq. (13) and thus do not have to be included. The procedure is repeated with different m's over a small range until a m that maximizes Eq. (13) is found.

As mentioned previously, the MSS-maximum likelihood method is the best method currently available to estimate m. It uses all the results from a fluctuation experiment and is valid over the entire range of values of m. In addition, computer simulations have shown it to behave in a manner that allows statistical evaluation (Stewart, 1994). A comparison with other methods using real data can be found in Rosche and Foster (Rosche and Foster, 2000)

Method 6: Accumulation of Clones

Luria (1951) pointed out that P_r, the proportion of cultures with r or more mutants, approaches $2m/r$ during the deterministic portion of growth (i.e., when m is 1 or greater). Formally:

$$P_r = \sum_{i=r}^{i=N_t} p_i = \frac{2m}{r} \tag{14}$$

Taking logarithms gives:

$$\ln(P_r) = -\ln(r) + \ln(2m) \tag{15}$$

A plot of $\ln(P_r)$ versus $\ln(r)$ will yield a straight line with a slope of -1 and an intercept [where $\ln(r) = 0$] equal to $\ln(2m)$. Dividing P_r at the intercept by 2 gives m.

Method 7: The Quartiles Method

The median is the central (50%) quartile of a distribution. More of the data from a fluctuation assay can be incorporated in the calculation of m if the upper (75%) and lower (25%) quartiles are also used. By regressing m versus the theoretical values of r at the quartiles, Koch (1982) derived the following empirical equations:

$$m_1 = 1.7335 + 0.4474\,Q_1 - 0.002755(Q_1)^2 \tag{16}$$

$$m_2 = 1.1580 + 0.2730 \, Q_2 - 0.000761 (Q_2)^2 \tag{17}$$

$$m_3 = 0.6658 + 0.1497 \, Q_3 - 0.0001387 (Q_3)^2 \tag{18}$$

where Q_1, Q_2, and Q_3 are the values of r at 25%, 50%, and 75% of the ranked series of observed r's. For a perfect Luria-Delbrück distribution, the three m's should be equal. These equations are valid over the range $2 \leq m \leq 14$; Koch also gives graphs that can be used up to values of $m = 120$ (Koch, 1982).

Calculating the Mutation Rate

The mutation rate, μ, is the mean number of mutations, m, normalized to some measure of the number of cells at risk for mutation. Three such measures are used, each of which is based on different assumptions about the underlying mutational process. If the probability of mutation is constant over the cell cycle, then m should be divided by the number of cell divisions that have taken place. Since the final number of cells in a culture, N_t, arose from $N_t - 1$ divisions, the mutation rate is (Luria and Delbrück, 1943):

$$u = \frac{m}{(N_t - 1)} \approx \frac{m}{N_t} \tag{19}$$

The same calculation applies if mutations are assumed to occur at or during division (Armitage, 1952). If mutations are assumed to occur at the beginning of the cell cycle (i.e., shortly after division), then m should be divided by the total number of cells that ever existed in the culture, which is $2N_t$ (because N_t cells had $N_t/2$ parents, $N_t/4$ grandparents, etc., and the sum of the series is $2N_t$). Thus, the mutation rate is (Armitage, 1952):

$$u = \frac{m}{2N_t} \tag{20}$$

However, cells usually are not growing synchronously, and in an asynchronous population there are an average of $N/\ln(2)$ cells during one generation period. Thus, the total number of divisions during the growth of a culture is $N_t/\ln(2)$ and the mutation rate is (Armitage, 1952):

$$u = m \frac{\ln(2)}{N_t} = 0.6932 \frac{m}{N_t} \tag{21}$$

For the same m, these three equations will give mutation rates that differ by 1: 0.5: 0.693. It is best to use one consistently and to describe which one was used so that readers can compare results obtained with different methods.

Statistical Methods to Evaluate Mutation Rates

The estimates of m or μ obtained from fluctuation tests are neither normally distributed nor unbiased; therefore, no matter how many times a fluctuation experiment is repeated, it is not valid to take the mean and standard deviation of the results (Asteris and Sarkar, 1996; Jones et al., 1994; Stewart, 1994). There are two approaches that allow reasonable confidence limits to be placed around estimates of mutation rates. The first approach is to put confidence limits around the parameter used to calculate m and calculate new m's using these values; the new m's will be estimates for the confidence limits of m (Wierdl et al., 1996). This approach is valid only for parameters that have defined distributions, such as p_0 and the median. The second approach is to find a transforming function that gives m a normal distribution; this has been successful only for the MSS maximum likelihood method (Method 5). Once confidence limits are obtained for m, these can be divided by N_t (or $2N_t$ or $N_t/\ln2$) to estimate the confidence limits for μ, the mutation rate. Of course, this procedure ignores the variance of the determination of N_t, (which is approximately N_t). Although nontrivial, it is probably justifiable to ignore this variance as long as N_t is determined accurately.[4]

Confidence Limits for p_0. Because a culture either has mutants or it does not, p_0 can be considered a binomial parameter with $p_0 = p$ and $(1 - p_0) = q$ (Lea and Coulson, 1949). For a sample population of n, the standard deviation of p is

$$\sigma_p = \sqrt{\frac{pq}{n-1}} \qquad (22)$$

But for most fluctuation assays, using the binomial is inappropriate and gives meaningless intervals. There are several more widely applicable methods to calculate the CLs for a proportion; the following one uses the F statistic (Zar, 1984):

---

[4] It is a little-appreciated fact that the expected value of the ratio of two variables is not the ratio of their expected values (i.e., not the ratio of their means). Furthermore, the calculation of the variance of a ratio is fairly complicated (e.g. see Rice, 1995). However, if the denominator is larger than the numerator, the variance of the ratio will be smaller than the variance of the numerator, and thus no great harm should be done by ignoring the variance of the denominator.

$$CL_{Upper} = \frac{(np + 1)F_{df1}}{nq + (np + 1)F_{df1}} \tag{23}$$

$$CL_{Lower} = \frac{np}{np + (nq + 1)F_{df2}} \tag{24}$$

where F is evaluated at the desired α (α = level of significance) and the following degrees of freedom:

$$df1 : v_1 = 2(nq + 1); \ v_2 = 2np \tag{25}$$

$$df2 : v_1 = 2(np + 1); \ v_2 = 2nq \tag{26}$$

Confidence Limits for the Median. The median is, by definition, the value at which the cumulative binomial probability is equal to 0.5 (i.e., 50% of the values are above and 50% of the values are below the median). Therefore, the 95% CLs for the median are the values above and below which less than 5% of the values are expected to fall, given n trials and a probability of 0.5. These values can be found by calculating the binomial probabilities for each possible rank-value of a given sample population and then finding the upper and lower rank-values that symmetrically include 95% probability (or any other desired probability). The CLs for the median are the actual sample values that correspond to these rank-values. Conveniently, the binomial probabilities have already been calculated in tables found in many statistics books (e.g., Zar, 1984). Thus, to calculate CLs for the median

1. Use a table or calculate the binomial probability for each possible rank-value (including 0) for the given population size, n = C, and $p = q = 0.5$.
2. Pick i, the highest rank-value that has a probability equal to or less than $\alpha/2$.
3. Pick the $j = n - (i + 1)$ rank-value.
4. Order the samples by increasing value.
5. Pick the $(i + 1)^{th}$ and the jth values. These are the CLs for the median value.

Confidence Limits for m *Obtained from the MSS Maximum Likelihood Method.* Using simulated fluctuation tests, Stewart (1994) evaluated the distributions of m's obtained using several of the common methods. He found that the natural logarithms of m's obtained using the MSS maximum likelihood method (Method 5) are approximately normally distributed. From this, Stewart calculated the standard deviation of $\ln(m)$:

$$\sigma_{\ln(m)} \approx \frac{1.225m^{-0.315}}{\sqrt{C}} \qquad (27)$$

where C is the number of cultures. Since $\ln(m)$ is normally distributed, the 95% confidence limits for $\ln(m)$ should be

$$\ln(m) \pm 1.96\sigma_{\ln(m)} \qquad (28)$$

While this is a reasonable approximation, the true confidence limits must be calculated from the actual m and σ of the population, not the experimentally determined m and σ of the sample. Methods to calculate or estimate the correct confidence limits are given in Stewart (1994). A close approximation can be obtained from the following equations (Rosche and Foster, 2000)

$$CL_{+95\%} = \ln(m) + 1.96\sigma(e^{1.96\sigma})^{-0.315} \qquad (29)$$

$$CL_{-95\%} = \ln(m) - 1.96\sigma(e^{1.96\sigma})^{+0.315} \qquad (30)$$

Once the upper and lower limits for $\ln(m)$ are obtained, the upper and lower limits for m are simply the antilogs.

Departures from the Lea-Coulson Model

All the methods for calculating mutation rates discussed previously depend on the Lea-Coulson model of expansion of mutant clones (Lea and Coulson, 1949); therefore, calculated mutation rates will be wrong if the assumptions of the model are violated. However, with some care, several biological meaningful departures can be accommodated, and meaningful mutation rates derived.

Sampling or Low Plating Efficiency. If only a sample of a culture is plated, or if the cells (not just the mutants) have a plating efficiency less than 100%, all clones will be reduced in size by the same relative amount and m will be too small. The observed m can be corrected using Eq. (5) if Method 1 is used, or Eq. (8) if Method 3 is used.

Phenotype Lag. If the expression of a mutant phenotype is delayed for several generations, mutants that arise in the last few generations of growth will result in few mutant progeny, whereas mutants that arise early will contribute a normal number. Thus, the lower end of the distribution will be affected, but, depending on the length of the lag, the upper end will not, resulting in an inflated m. The actual m can be estimated graphically with Method 6 by using only the upper part of the curve and eliminating any obvious jackpots (Rosche and Foster, 2000). If the length of the phenotypic lag is known, Koch (1982) gives a method for estimating m from the quartiles (Method 7).

Selection Against Mutants. If during nonselective growth, mutants grow more slowly than nonmutants, the result is the opposite of what happens if there is a phenotypic lag: mutants that arise in the last few generations of growth will contribute a normal number of mutant progeny, but mutants that arise early will contribute few. This shifts the distribution of mutant numbers from the Luria-Delbrück toward the Poisson (Koch, 1982; Stewart *et al.*, 1990). Koch (1982) gives graphs that can be used to estimate m from the quartiles when the growth rate of mutants ranges from 60 to 90% that of nonmutants. If there is more than one type of mutant and each type has a different growth rate, the distribution can be approximated by assuming there is only one type whose growth rate is the average of the two (Stewart *et al.*, 1990).

Adaptive Mutation. If mutations occur after the cells are plated on selective medium, the distribution of mutant numbers will have a Poisson component and a plot of $\ln(P_r)$ versus $\ln(r)$ [Eq. (15)] will give a curve that is a combination of the Luria-Delbrück and Poisson (Cairns *et al.*, 1988). The two components can be estimated by fitting the experimental values to the combined distributions (Cairns and Foster, 1991).

Other Curve Fitting. Using simulated data, Stewart *et al.* (1990) have determined the effects of several deviations from the Lea-Coulson model on the shape of the $\ln(P_r)$ versus $\ln(r)$ curve. Experimental data can be fit to these curves to test whether a given factor is operative. However, all of the deviant curves are rather similar, so any conclusion that a given factor is distorting the distribution would have to be tested experimentally.

Acknowledgments

Research in the author's laboratory is supported by USPHS grant NIH-NIGMS GM065175.

References

Armitage, P. (1952). The statistical theory of bacterial populations subject to mutation. *J. R. Statist. Soc. B* **14,** 1–40.

Asteris, G., and Sarkar, S. (1996). Bayesian procedures for the estimation of mutation rates from fluctuation experiments. *Genetics* **142,** 313–326.

Bachl, J., Dessing, M., Olsson, C., von Borstel, R. C., and Steinberg, C. (1999). An experimental solution for the Luria Delbrück fluctuation problem in measuring hypermutation rates. *Proc. Natl. Acad. Sci. USA* **96,** 6847–6849.

Cairns, J., and Foster, P. L. (1991). Adaptive reversion of a frameshift mutation in *Escherichia coli. Genetics* **128**, 695–701.

Cairns, J., Overbaugh, J., and Miller, S. (1988). The origin of mutants. *Nature* **335**, 142–145.

Crane, B. J., Thomas, S. M., and Jones, M. E. (1996). A modified Luria-Delbrück fluctuation assay for estimating and comparing mutation rates. *Mutat. Res.* **354**, 171–182.

Drake, J. W. (1970). *The Molecular Basis of Mutation.* Holden-Day, Inc., San Francisco, CA.

Drake, J. W. (1991). A constant rate of spontaneous mutation in DNA-based microbes. *Proc. Natl. Acad. Sci. USA* **88**, 7160–7164.

Glaab, W. E., and Tindall, K. R. (1997). Mutation rate at the hprt locus in human cancer cell lines with specific mismatch repair-gene defects. *Carcinogenesis* **18**, 1–8.

Gurland, J. (1958). A generalized class of contagious distributions. *Biometrics* **14**, 229–249.

Gurland, J. (1963). A method of estimation for some generalized Poisson distributions. *Proc. Int. Symp. on Classical and Contagious Discrete Distributions* 141–158, Pergamon Press.

Jones, M. E. (1993). Accounting for plating efficiency when estimating spontaneous mutation rates. *Mutat. Res.* **292**, 187–189.

Jones, M. E., Thomas, S. M., and Clarke, K. (1999). The application of a linear algebra to the analysis of mutation rates. *J. Theor. Biol.* **199**, 11–23.

Jones, M. E., Thomas, S. M., and Rogers, A. (1994). Luria-Delbrück fluctuation experiments: Design and analysis. *Genetics* **136**, 1209–1216.

Koch, A. L. (1982). Mutation and growth rates from Luria-Delbrück fluctuation tests. *Mutat. Res.* **95**, 129–143.

Koziol, J. A. (1991). A note of efficient estimation of mutation rates using Luria-Delbrück fluctuation analysis. *Mutat. Res.* **249**, 275–280.

Kubitschek, H. E., and Bendigkeit, H. E. (1964). Mutation in continuous cultures. I. Dependence of mutational response upon growth-limiting factors. *Mutat. Res.* **1**, 113–120.

Lea, D. E., and Coulson, C. A. (1949). The distribution of the numbers of mutants in bacterial populations. *J. Genetics* **49**, 264–285.

Luria, S. E. (1951). The frequency distribution of spontaneous bacteriophage mutants as evidence for the exponential rate of phage reproduction. *Cold Spring Harbor Symp. Quant. Biol.* **16**, 463–470.

Luria, S. E., and Delbrück, M. (1943). Mutations of bacteria from virus sensitivity to virus resistance. *Genetics* **28**, 491–511.

Ma, W. T., Sandri, G.v.H., and Sarkar, S. (1992). Analysis of the Luria-Delbrück distribution using discrete convolution powers. *J. Appl. Prob.* **29**, 255–267.

Miller, J. H. (1996). Spontaneous mutators in bacteria: Insights into pathways of mutagenesis and repair. *Annu. Rev. Microbiol.* **50**, 625–643.

Novick, A., and Szilard, L. (1950). Experiments with the chemostat on spontaneous mutations of bacteria. *Proc. Natl. Acad. Sci. USA* **36**, 708–719.

Reddy, M., and Gowrishankar, J. (1997). A genetic strategy to demonstrate the occurrence of spontaneous mutations in non-dividing cells within colonies of *Escherichia coli. Genetics* **147**, 991–1001.

Reyrat, J. M., Pelicic, V., Gicquel, B., and Rappuoli, R. (1998). Counterselectable markers: Untapped tools for bacterial genetics and pathogenesis. *Infect. Immun.* **66**, 4011–4017.

Rice, J. A. (1995). *Mathematical Statistics and Data Analysis.* Wadsworth Publishing Company, Belmont, CA.

Rosche, W. A., and Foster, P. L. (2000). Determining mutation rates in bacterial populations. *Methods* **20**, 4–17.

Sarkar, S., Ma, W. T., and Sandri, G.v.H. (1992). On fluctuation analysis: A new, simple and efficient method for computing the expected number of mutants. *Genetica* **85,** 173–179.

Stewart, F. M. (1994). Fluctuation tests: How reliable are the estimates of mutation rates? *Genetics* **137,** 1139–1146.

Stewart, F. M., Gordon, D. M., and Levin, B. R. (1990). Fluctuation analysis: The probability distribution of the number of mutants under different conditions. *Genetics* **124,** 175–185.

Wierdl, M., Greene, C. N., Datta, A., Jinks-Robertson, S., and Petes, T. D. (1996). Destabilization of simple repetitive DNA sequences by transcription in yeast. *Genetics* **143,** 713–721.

Zar, J. H. (1984). *Biostatistical Analysis*. Prentice Hall, Englewood Cliffs, New Jersey.

[13] Genomic Approaches for Identifying DNA Damage Response Pathways in S. cerevisiae

By MICHAEL CHANG, AINSLIE B. PARSONS, BILAL H. SHEIKH, CHARLES BOONE, and GRANT W. BROWN

Abstract

DNA damage response pathways have been studied extensively in the budding yeast *Saccharomyces cerevisiae*, yet new genes with roles in the DNA damage response are still being identified. In this chapter we describe the use of functional genomic approaches in the identification of DNA damage response genes and pathways. These techniques take advantage of the *S. cerevisiae* gene deletion mutant collection, either as an ordered array or as a pool, and can be automated for high throughput.

Introduction

Genetic studies in the budding yeast *Saccharomyces cerevisiae* have uncovered many components required for proper and efficient execution of the cellular response to DNA damage. Recent advances in genomic, proteomic, and bioinformatic techniques have further increased the utility of *S. cerevisiae* as a model organism for the study of the DNA damage response. In particular, the construction of a complete collection of *S. cerevisiae* gene deletion mutants (Giaever *et al.*, 2002; Winzeler *et al.*, 1999), along with libraries of conditional alleles of essential genes (Kanemaki *et al.*, 2003; Mnaimneh *et al.*, 2004), have allowed for systematic genetic analyses to determine gene function. Although the techniques described in this chapter make use of the haploid gene deletion collection, in principle they can be applied to any

METHODS IN ENZYMOLOGY, VOL. 409
0076-6879/06 $35.00
DOI: 10.1016/S0076-6879(05)09013-0

collection of mutants, with the exception of the barcode-based screening approaches.

Systematic Genome-Wide Screens to Identify Genes Required for Resistance to DNA Damaging Agents

Array-Based Screening

Array-based approaches have been successfully employed to screen for mutants that are sensitive to ionizing radiation (Bennett *et al.*, 2001), methyl methanesulfonate (Begley *et al.*, 2002; Chang *et al.*, 2002), wortmannin (Zewail *et al.*, 2003), bleomycin (Aouida *et al.*, 2004), hydroxyurea, camptothecin (Parsons *et al.*, 2004), oxidative stress (Tucker and Fields, 2004), and etoposide (Baldwin *et al.*, 2005). These screens have typically used the *S. cerevisiae* gene deletion collection, a collection of some 4700 strains, each carrying a deletion of a single nonessential open reading frame (Giaever *et al.*, 2002; Winzeler *et al.*, 1999). This collection has been screened both as haploids (Aouida *et al.*, 2004; Baldwin *et al.*, 2005; Chang *et al.*, 2002; Parsons *et al.*, 2004; Tucker and Fields, 2004; Zewail *et al.*, 2003) and as homozygous diploids (Bennett *et al.*, 2001). The gene deletion collection is available from Open Biosystems (Huntsville, AL; http://www.openbiosystems.com/), EUROSCARF (http://web.uni-frankfurt.de/fb15/mikro/euroscarf/complete. html), and the ATCC (http://www.atcc.org/common/specialCollections/ cydac.cfm), and is supplied as frozen cell suspensions in 96-well plates or as 96 strains stamped on agar plates.

Screens with the gene deletion collection are facilitated by the construction of an ordered array of the mutant strains such that the identity of the mutant at each address on the array is known. This allows for the rapid identification of mutants detected in a given screen. We use an array constructed on 86 mm × 128 mm agar plates formed in single well Omni-Trays (Nunc, Rochester, NY), at a density of 768 colonies per plate. With each strain present in duplicate, this array occupies 16 plates. This array was originally designed for use in synthetic genetic interaction screens (Tong *et al.*, 2001), and the construction of such an array, using the gene deletion collection as supplied in 96-well plates by Open Biosystems, has recently been described in detail (Tong and Boone, 2006). A density of 768 colonies per plate provides a reasonable balance between high density and the ability to reproducibly transfer cells from plate to plate using a replica pinning procedure, but is more suitable for replica pinning using a robotic system. In practice, a lower density, such as 384 colonies per plate, will be easier to work with if the replica pinning is done by hand rather than by a robot. This, of course, necessitates a larger array of some 32 plates. At the other extreme, the use of a contact microarrayer to print and manipulate very high density arrays of up to 9600 colonies per OmniTray has been described, with the advantage that considerably smaller amounts of drug are required to screen the deletion collection in this format (Xie *et al.*, 2005).

Using an ordered array of the gene deletion collection, genome-wide screening to identify genes required for resistance to DNA damaging agents is performed by transferring the yeast strains from the master array onto solid medium containing the given drug and to the same medium lacking the drug. The concentration of drug to be used in a screen can be determined by several approaches. If known drug sensitive mutants are available, a serial-dilution spot assay, described later, can be performed to compare sensitivity to a wild-type strain at several different drug concentrations. If suitable positive controls are not available, a concentration of drug that causes a mild inhibition of growth of the wild-type strain can be used. In either case, pilot studies can be performed with one or several gene deletion collection array plates to test sensitivity at several drug concentrations across several hundred mutant strains. Plates chosen should have several reasonable candidates for sensitive strains, for example, mutants in genes in known DNA-damage response pathways.

The transfer of colonies to the solid medium containing the drug is typically accomplished by a replica pinning procedure, performed manually or with a robot. The sensitivity of each of the strains to a particular drug is assessed by comparing colony size on the plate containing the drug to the size on the no-drug control plate. Since each strain is present twice on the array, the duplicates provide an internal control for variation in pinning. A strain is only scored as positive in a screen if both colonies are small or absent. We perform each screen against the complete array at least three times. Identification of sensitivity in at least two screens serves as a convenient threshold for further confirmation using serial-dilution spot assays, described later. Computational methods for scoring colony size differences have also been described (Tong *et al.*, 2004). One obvious drawback of array-based screening methods is that they consume a lot of drug. In cases where the drug is limiting, competitive growth screening methods, described later, are more suitable. As with any screening procedure, these screens are subject to both false-positives and false-negatives. False-positives (which typically fall in the 11–22% range) are eliminated by confirming the sensitivity of mutants identified in the primary screen by an independent secondary screen (Aouida *et al.*, 2004; Chang *et al.*, 2002). We typically use a serial-dilution assay, again on solid media, as described later. In principle, confirmations could also be performed as growth assays in liquid culture, although without automation the serial-dilution assay has higher throughput. False-negative rates can be estimated by comparing the mutants identified in the primary screen with those mutants known to confer sensitivity to the given drug. Although this would not be possible for a new drug, when this analysis was applied following screens for MMS and bleomycin sensitivity, estimates of 44% and 26% false negatives were obtained (Aouida et al., 2004; Chang et al., 2002).

Manual Replica Pinning

Manual pinning tools can be purchased from V & P Scientific, Inc. (San Diego, CA; http://www.vp-scientific.com/) in both 96- and 384-floating pin formats. Alignment manifolds are also available from V & P Scientific for use in array construction and for precise alignment of array replicas on OmniTray plates. Plastic tip box lids or OmniTrays are used as reservoirs for cleaning and sterilizing the pinning tools.

Pinning Procedure

1. Sterilization of manual pinning tools:
   a. Set up 8 wash reservoirs as follows:
      i. Water reservoirs: 1, 2, 3, 4, 6, 7
      ii. 10% bleach: 5
      iii. 95% ethanol: 8
   b. Move manual pinning tool from reservoirs 1 through 8. Let pinning tool sit in first two water reservoirs for at least ~1 min each, with gentle shaking, to remove cells on the pins. Change first 2 water reservoirs periodically (i.e., every 2 or 3 rounds of washing).
   c. After wash in last (i.e., 95% ethanol) reservoir, allow excess ethanol to drip off the pins, then flame the pins. Allow pinning tool to cool.

2. After sterilizing the pins, replica-pin the mutant strains from the master array onto agar plates containing the selected DNA damaging agent. Make a second replica onto agar plates lacking drug as a control. The two replicas can be made from a single pinning off of the master.

3. Incubate plates at 30° for 1–3 days. Compare growth of colonies on plates containing drug with the growth on plates lacking drug. Score colonies where both of the duplicates show reduced or no growth on the plate that contains the drug. An example of a replica pinning experiment at 768-colony density is shown in Fig. 1.

More sophisticated pinning protocols where multiple replicas are made without returning to the master plate have been described (Aouida et al.,

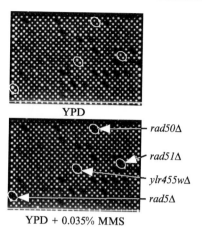

FIG. 1. A screen for methyl methanesulfonate sensitivity using replica pinning. The complete set of haploid yeast deletion mutants was arrayed in duplicate onto 16 plates and pinned onto YPD media or YPD + 0.035% MMS (array plate 11 of 16 is shown). Putative MMS sensitive mutants lead to the formation of smaller colonies when grown on MMS-containing media. Reproduced from Chang *et al.*, 2002, copyright 2002, National Academy of Sciences, USA.

2004). These can produce a dilution effect and reveal sensitivities that are not evident in a single pinning. Although the previous method uses the gene deletion collection of nonessential genes, it could in principle be adapted to screen essential genes. This could be done using heterozygous mutant strains carrying deletions of essential genes to screen for haplo-insufficiency in the presence of drug. Additionally, a collection of strains carrying essential open reading frames under the control of a tetracycline-regulated promoter has been described (Mnaimneh *et al.*, 2004). This collection could be screened for drug sensitivity in the presence of different concentrations of the tetracycline analogue doxycycline in order to reduce expression of essential genes.

Robotic Pinning Systems

There are several robotic systems available for replica pinning. These include the VersArray colony arrayer system (BioRad Laboratories, Hercules, CA), the QBot, QPixXT, and MegaPix (Genetix, Boston, MA), and the Singer Rotor HDA bench top robot (Singer Instruments, Somerset, UK). These systems provide greater precision, reproducibility, and throughput than hand pinning, but at increased expense.

Confirmation of Sensitivity Using Serial-Dilution Spot Assays

1. Grow deletion strains overnight to saturation at 30° in 2 ml yeast extract-peptone-dextrose (YPD).
2. Dilute cultures using sterile water to a concentration of 10^7 cells/ml (or approximately $OD_{600} = 0.5$).
3. Using a multichannel pipettor and a 96-well microtiter plate, further dilute each culture by making 4 serial 10-fold dilutions.
4. Spot 5 μl of each dilution onto medium containing drug and medium lacking drug. Using agar plates formed in single-well OmniTrays, it is possible to analyze 4 dilutions of 12 strains per plate.
5. Incubate plates at 30° for 2–4 days to allow for cell growth before scoring for drug sensitivity. An example of a spot assay is shown in Fig. 2.

Barcode-Based Screening

Because many compounds are in limited supply, a major challenge is to screen the highest number of mutants in the most efficient manner possible while using the least amount of growth medium. Parallel analysis of large numbers of pooled deletion strains in a minimal amount of media is possible due to the two unique molecular barcodes ("uptag" and "downtag") that tag and identify each deletion strain (Shoemaker *et al.*, 1996; Winzeler *et al.*, 1999). This method was originally developed to identify drug targets through

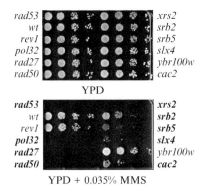

FIG. 2. Confirmation of MMS sensitivity. Putative MMS sensitive strains were grown in YPD overnight at 30°. Serial 10-fold dilutions were spotted onto YPD or YPD + 0.035% MMS and incubated at 30° for 3 days. Strains in **bold** were scored as sensitive. A *rad53* mutant was used as a positive control. Reproduced from Chang *et al.*, 2002, copyright 2002, National Academy of Sciences, USA.

screens of the yeast heterozygote deletion collection for drug-induced haplo-insuffiency (Giaever *et al.*, 1999, 2004; Lum *et al.*, 2004) but it has since been modified for many different applications. In this strategy, presented in schematic form in Fig. 3, strains are pooled and grown in parallel in liquid culture under selective conditions (for example, in the presence of a drug compound). Genomic DNA is extracted from the pool and the barcodes from each strain in the pool are simultaneously amplified using common flanking primers fluorescently labeled with Cy3 and Cy5. This pool of PCR amplimers is then hybridized to high density DNA microarrays

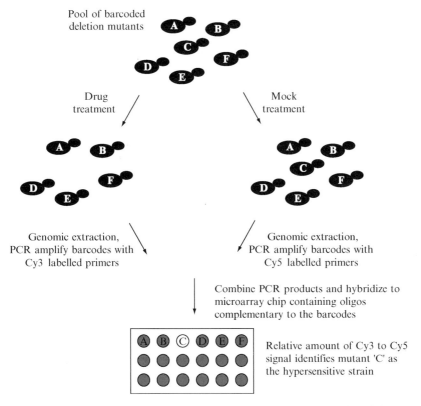

FIG. 3. Parallel analysis of large pools of deletion mutants. Populations of pooled mutant cells each marked with unique molecular barcodes are grown in the presence or absence of a growth inhibitory drug. Genomic DNA is extracted from the pool of mutants and barcodes representing each strain are amplified by PCR using common primers labeled with the fluorescent markers Cy3 or Cy5. Drug sensitive mutants are identified by competitive hybridization of the barcode PCR products to a microarray containing oligos corresponding to each barcode.

containing oligonucleotides corresponding to the barcodes. The relative abundance of each strain in the pool is assessed by the strength of the resulting signals from the microarray readout compared to a mock control. Using this approach, the entire yeast deletion set can be pooled, grown competitively, and quantitatively assessed in small volumes of growth medium (Giaever et al., 2002).

A number of groups have used this strategy to screen the yeast deletion mutant collection for drug hypersensitivity to uncover the pathways and cellular functions affected by drug treatment on a global scale. Giaever and colleagues profiled the nearly complete set of S. cerevisiae deletion strains under a number of different conditions including high salt, sorbitol, galactose, pH 8, minimal media, and nystatin treatment (Giaever et al., 2002). Other groups have used this method to screen DNA damaging agents. For example, Hanway and colleagues analyzed a collection of 2827 homozygous non-essential deletion strains for UV and MMS sensitivity and implicated six genes not previously known to be involved in DNA damage repair pathways in the DNA damage response (Hanway et al., 2002). A separate group also screened the deletion mutant collection for strains hypersensitive to UV radiation (Birrell et al., 2001). This study identified three novel genes that lead to UV sensitivity when deleted, two of which have human orthologs associated with cancer. A screen for sensitivity to ionizing radiation yielded a novel radiation resistance gene, RAD61 (Game et al., 2003). The anticancer agents cisplatin, oxaliplatin, and mitomycin C have also been used in screens (Wu et al., 2004).

While difficult to determine for certain, the overlap between yeast cell array based chemical screens and barcode based screening appears to be around 30–60% (Parsons and Boone, unpublished results). Discrepancies between the array-based and barcode screens are likely a result of intrinsic differences in the assays. The array-based approach is performed on solid media with each strain pinned individually onto medium and allowed to grow without significant competition from other strains. In contrast, chemical-genetic profiling is a competitive fitness assay performed in liquid media. Often, minimum inhibitory drug concentrations vary widely in liquid medium compared to solid medium. In addition, the barcodes are known to carry some mutations; however, they have now been sequenced (Eason et al., 2004) and so new arrays should lead to more accurate profiles.

Protocol for Barcode-Based Chemical-Genetic Screening

1. Pool yeast strains and freeze down in aliquots.

   a. Using a 96-pin hand pinner, pin strains from 96-well stock plates onto YPD agar plates

b. Grow for 2 days at 30°

. c. Scrape cells off plates into YPD + 15% (v/v) glyercol

d. Adjust pool to a cell density between $4 \times 10^8 - 8 \times 10^8$ cells/ml

e. Freeze at −80° in 1 ml aliquots.

2. Thaw pool and dilute to a final concentration of 5×10^5 cells/ml in 5 ml of YPD. If the volume of drug is limiting, it is possible to use smaller volumes of media (less than 1 ml).

3. Expose pool to drug at semi-inhibitory concentration where the growth of the pool is inhibited by approximately 20–30%. Also include a mock-treated solvent control.

4. Incubate at 30°, with shaking, for 14 h

5. Dilute strains to 5×10^5 cells/ml and add fresh compound or solvent to the tubes. Incubate at 30°, with shaking, for 24 h.

6. Prepare genomic DNA from 1.5 ml of culture using yeast DNA extraction kit (MasterPure™ Yeast DNA Purification Kit; Epicentre Cat # MPY80200, Madison, WI) and normalize DNA concentrations to 25 ng/ml.

7. PCR amplify barcodes using 5′ Cy3 and Cy5 labeled common primers. Amplify the drug treated pool with one fluor and the control pool with the other fluor. Primers U1 (GATGTCCACGAGGTCTCT), and U2 (CGTAC-GCTGCAGGTCGAC) are used to amplify the uptags and D1 (CGGTGTC-GGTCTCGTAG) and D2 (ATCGATGAA TTCGAGCT CG) to amplify the downtags. Amplify the uptags and downtags in separate reactions.

PCR reactions:

4 μl gDNA (25 ng/ml)

2 μl Cy3/5-U1 or Cy3/5-D1 (16.5pM/μl)

2 μl U2 or D2 (16.5pM/μl)

67.5 μl Platinum PCR Supermix (Invitrogen; Cat # 11306–016, Grand Island, NY)

94° for 2 min

35 cycles of:

94° for 30 sec

50° for 30 sec

72° for 30 sec

Then 72° for 5 min

4° for ∞.

8. Combine PCR products (uptags and downtags from the drug-treated pool and uptags and downtags for the mock-treated pool) and hybridize to an oligonucleotide chip (we use Agilent custom microarrays) containing oligos complementary to the barcode sequences. All barcode and primer sequences are available at http://www-sequence.stanford.edu/group/yeast_deletion_project/. Barcode oligos can be purchased from Invitrogen (Cat # 40904).

9. Hybridization procedure:
   a. Combine 65 μl of PCR product from each of 4 reactions:
      i. Cy3 labeled uptags from drug-treated pool
      ii. Cy3 labeled downtags from drug-treated pool
      iii. Cy5 labeled uptags from mock-treated pool
      iv. Cy5 labeled downtags from mock-treated pool. Also run a fluor-reversed chip with the drug-treated pool labeled with Cy5 and the mock treated pool labeled with Cy3.
   b. Add 10 μl each blocking mix (up blocking mix: 25 pM/μl each of U1 and U2-RC [GTCGACCTGCAGCGTACG]; down blocking mix: 25 pM/μl each of D1 and D2-RC [CGAGCTCGAATCATCGAT]). These blocking mixes are complementary to the common priming sites of the uptags and downtags, respectively, and are designed to reduce nonspecific hybridization.
   c. Boil at 100° for 4 min, then cool on ice for 5 min.
   d. Add 170 μl hybridization buffer (for 5 ml: 2650 μl 5 M NaCl, 132.5 μl Tris-HCl pH 7.5, 1325 μl 5% Triton X-100, 882.5 μl H_2O, 10 μl 1 M DTT).
   e. Add entire mixture to an Agilent hybridization chamber (Agilent; Cat # G2534A).
   f. Hybridize at 42° for 4 h with slow rotation.
   g. Wash chips by dipping 5 times in 6X SSPE/0.05% TritonX100, then 5 times in 0.06X SSPE at 4°.
   h. Scan chips immediately. We use a GenePix 4000B scanner from Axon instruments (Sunnyvale, CA) to scan the chips and Array-Pro Analyzer from MediaCybernetics (Silver Spring, MD) to analyze the images.

Synthetic Genetic Array Analysis to Identify DNA Damage Response Pathways

Of the ~6000 known or predicted genes in *S. cerevisiae*, about 75% are nonessential (Giaever *et al.*, 2002; Tong *et al.*, 2004; Winzeler *et al.*, 1999). This emphasizes the ability of yeast cells to tolerate individual deletions of most genes, likely reflecting redundant pathways that have evolved to buffer the phenotypic consequences of genetic variation (Hartman *et al.*, 2001). This high degree of genetic redundancy makes it difficult to determine the function of many genes, but studying synthetic genetic interactions can circumvent this problem. A synthetic genetic interaction occurs when a mutation in a gene suppresses, enhances, or modifies the phenotype of a second mutation. In particular, if these two mutations cause cell sickness or cell death, the synthetic genetic interaction is termed synthetic sick or synthetic lethal, respectively. The creation of the *S. cerevisiae* gene deletion mutants has enabled genome-wide, high-throughput synthetic genetic interaction screens using an approach termed synthetic genetic array (SGA) analysis (Tong et al., 2001, 2004). Large-scale SGA analyses have proven to be very useful for predicting gene function because genetic interactions often occur between functionally related genes, and similar genetic interaction profiles tend to identify components of the same pathway (Tong et al., 2004). SGA screens with known DNA-damage response genes successfully identified ELG1 as a gene important for DNA replication and genomic integrity (Bellaoui et al., 2003), and helped reveal sister chromatid cohesion roles for the DNA damage response MRX complex (Warren et al., 2004) and the S-phase checkpoint genes TOF1 and CSM3 (Mayer et al., 2004).

SGA Methodology

In a typical SGA analysis, a query strain containing a mutation in the gene of interest is crossed to each arrayed deletion mutant. The resulting doubly heterozygous diploids are replica-pinned onto sporulation media to yield haploid progeny that have undergone meiotic recombination, which are subsequently replica-pinned onto media that only allow growth of haploid yeast containing both the query mutation and deletion array mutation. Growth rates of the double mutants are assessed to identify synthetic genetic interactions. A detailed protocol to carryout the SGA methodology has recently been described (Tong and Boone, 2006). A schematic diagram of the SGA procedure is presented in Fig. 4.

Because fitness defects of double mutants can also be examined using the barcode approach with a microarray-based readout as described previously for chemical genetic screens, SGA analysis can also be carried out using a pooled population of deletion mutants. In addition, rather than crossing a mutation of interest into the set of ~5000 viable deletion mutants, the query mutation can be introduced by transformation into heterozygous diploids containing the SGA reporter, a method referred to as dSLAM (Pan et al., 2004).

Clustering Analysis

In addition to performing SGA screens with known DNA damage response mutants, the analysis of the genetic interaction data from the greater than 130 SGA screens performed to date has proven to be a useful strategy for the identification of genes involved in the response to DNA damage. Clustering analysis is a useful tool for analyzing quantitative data, and has been used to visualize and analyze the large data sets from multiple SGA screens (Tong et al., 2004). It can be used to find patterns of similar

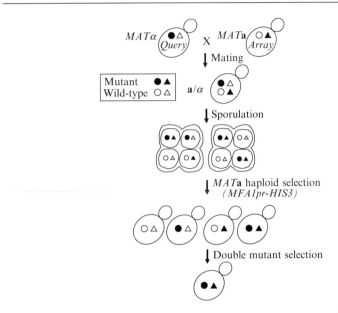

FIG. 4. Synthetic genetic array (SGA) analysis. The query mutant strain is crossed to the ~4600 strains that make up the viable haploid deletion mutant collection. Since the query and the deletion mutants are marked with different dominant selectable marker, diploids can be selected by replica pinning to appropriate media. Diploids are induced to sporulate by pinning to sporulation medium. Spores are germinated on medium that selects for *MATa* haploids, and double mutants are selected from among the haploids. The fitness of the double mutants is scored by comparing their growth to a control cross in which the deletion mutant array is crossed to a wild type strain.

behavior in a set of Items (i.e., genes) in relation to a set of Criteria (i.e., experiments). When clustering, the data is represented in a matrix with Items along one axis, Criteria along the other, and an Item's measurement for a given Criterion at their intersecting coordinate. The aim of clustering is to reorganize the matrix such that Items behaving most similarly along the Criteria axis are next to each other, and Criteria behaving most similarly along the Items' axis are next to each other. In terms of the data discussed here, the "Items" will be deletion strains and the "Criteria" are the different screens, either genetic or chemical. The difference between two Item or Criterion vectors is calculated based on a distance metric. Different metrics may be available in different clustering applications. Users should refer to the documentation of the application they are using to see what metrics are available to them.

In the case of SGA screens, this analysis groups mutants on the basis of the similarity of their genetic interaction profiles. Thus if two deletion mutants display synthetic genetic interactions with the same deletion mutants in a genome-wide analysis, they will likely cluster next to each other. Of particular importance, genes whose products function in the same pathway, biological process, or protein complex frequently cluster together (Tong *et al.*, 2004). Because of this tendency, clustering analysis has predictive value in identifying new genes that function in a given process, and in assigning functions to genes with unknown cellular roles. Clustering analysis helped place *CSM3* in an S-phase checkpoint pathway involving *MRC1* and *TOF1* (Tong *et al.*, 2004), and *RMI1* in a genomic integrity pathway with *SGS1* and *TOP3* (Chang *et al.*, 2005). The relevant portion of a clustering analysis that indicates that *CSM3* has a genetic interaction profile similar to that of *MRC1* and *TOF1*, and that *RMI1* has a genetic interaction profile similar to that of *SGS1* and *TOP3* is presented in Fig. 5.

Cluster 3.0 and Java Treeview

Cluster 3.0 and Java Treeview provide a simple and basic way to perform clustering analysis and view its results. Further, being open source, they are widely accessible at no cost, and there is room for customization and contribution. Cluster 3.0 is an Open Source Clustering Software resource, provided by Michiel de Hoon at the University of Tokyo, with clustering tools available for Windows, Linux, Unix, and Mac OSX and is located at: http://bonsai.ims.u-tokyo.ac.jp/~mdehoon/software/cluster/. Java Treeview (http://genetics.stanford.edu/~alok/TreeView/) is an application developed by Alok Saldanha that is useful for visualizing the hierarchical clustering results.

How to Use Cluster 3.0 and Java Treeview

Cluster 3.0 has a graphical user interface that can be used to load, analyze, and save data. Cluster 3.0 can open tab-delimited data files. The data file must be laid out in a matrix format, where the rows represent Items and columns represent Criteria. (For example, see Table I.)

The first column contains the Item name or label for each gene, and the first row contains the label for each Criterion. The first field of the first row describes what Items are represented along the rows. Each data cell contains the measurement for its respective Item and Criterion. More advanced formatting allows for the assignment of weights to each Item and Criterion. Please refer to the online documentation at http://bonsai.ims. u-tokyo.ac.jp/~mdehoon/software/cluster/manual/index.html for details.

There are several tabs in the Cluster 3.0 interface that allow various tasks to be performed once data is loaded. The "Filter Data" and "Adjust Data" tabs allow filtration and normalization in preparation for clustering,

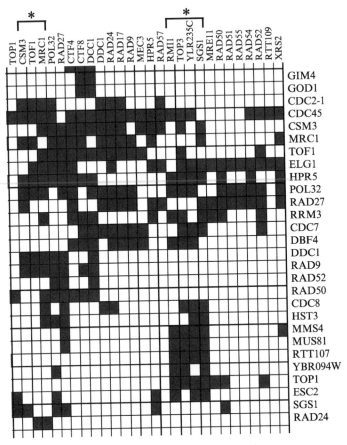

FIG. 5. Clustering analysis can predict gene function. A section of a clustering analysis of synthetic genetic interactions determined by SGA analysis is presented. The rows are the query genes and the columns are the deletion mutant array genes. Synthetic genetic interactions between two genes are represented as dark squares at the intersection of the query gene row and the array gene column. The clusters containing *CSM3, MRC1,* and *TOF1,* and *RMI1, SGS1,* and *TOP3* are indicated by "*".

although this is normally not required for SGA or chemical data. The remaining tabs allow Hierarchical, K-means, and SOM clustering and Principle Component Analysis, respectively. Hierarchical clustering is used most commonly with SGA data. Here, pairs of Items are linked together into clusters based on decreasing similarity. A tree of linkages is built from the bottom up, starting with a cluster containing a pair of very similar Items in a single cluster up to an umbrella cluster containing loosely linked

TABLE I

EXAMPLE OF A DATA FILE FORMATTED FOR CLUSTERING ANALYSIS IN CLUSTER 3.0.
IN THIS TABLE OF DNA DAMAGING AGENT SCREENS A "1" REPRESENTS
SENSITIVITY AND A "0" REPRESENTS NO SENSITIVITY. IN A SIMILAR ANALYSIS OF SGA
DATA, THE VALUES WOULD BE "1" FOR A SYNTHETIC LETHAL OR SYNTHETIC SICK
GENETIC INTERACTION AND "0" FOR NO GENETIC INTERACTION

Gene	HU	CPT	MMS	UV	IR
AAT2	1	0	1	0	0
ADH1	1	1	0	0	0
ADK1	0	0	0	0	1
ADO1	0	0	0	0	1
AKR1	1	1	0	0	1
ANC1	0	0	1	0	1
AOR1	0	0	1	0	0
APN1	0	0	1	0	1
ARD1	0	0	0	0	1
ARO1	0	0	1	0	0
ARO2	1	0	0	0	0
ARO7	0	0	1	0	0
ARV1	0	0	0	0	1
ASF1	1	1	1	1	1

subclusters. This method of clustering gives highly reproducible results and provides trees that describe the hierarchy of clusters based on similarity.

The clustering tabs allow users to select their clustering parameters and metrics. There are several distance metrics available in Cluster 3.0, which include Centered and Uncentered Correlation, Spearman Rank, Kendal Tau, and Euclidean Distance. Distance metrics are used to determine the similarity between two Items or clusters: the less the distance the greater the similarity. The most commonly used similarity metric is based on the Pearson correlation, and this can be applied to the data by choosing Correlation (centered) in the Similarity Metric dialog box. Details on this and the other distance metrics may be found in the Cluster 3.0 online documentation (http://bioinfo.tau.ac.il/man/cluster/html/Distance.html).

Once clustering analysis is executed, Cluster 3.0 automatically generates result files for use with Java Treeview (Warning: result files are generated using the prefix of the input file followed by ".", then the suffix associated with the clustering method. These files are placed in the same directory as the input file. Any pre-existing file with the same name as a result file will automatically be overwritten). After the clustering analysis is completed, the user may save the updated data file.

Once the result files have been generated they may be viewed using Java Treeview. Java Treeview can open result files with a ".cdt" or ".pcl" suffix. The interface consists of several frames. When loading a file, Java Treeview displays the result matrix in the main frame, as well as trees describing relationships in the case of hierarchical clustering. The user may select a region of the result matrix using the mouse or by selecting tree branches and view information on particular spots. The user can set various color schemes, perform analysis, and retrieve basic statistics. The user may also save the matrix data and export the matrix image, trees, and labels to a Postscript or Image file. An example of a clustering analysis using data from screens with methyl methanesulfonate (Chang *et al.*, 2002), hydroxyurea (Parsons *et al.*, 2004), camptothecin (Parsons *et al.*, 2004), UV (Birrell *et al.*, 2001), and ionizing radiation (Bennett *et al.*, 2001), which is excerpted in Table I and is shown in Fig. 6.

As an alternative to Cluster 3.0 and Java Treeview, Spotfire (http://www.spotfire.com) is a web-based software program useful for clustering analysis and visualization. In particular, Spotfire offers a functional genomics platform (Spotfire DecisionSite for Functional Genomics) that contains among many other options, a hierarchical clustering tool. While Spotfire does require a yearly license to be purchased in order to use the software, it is a very user-friendly system and requires virtually no understanding of statistics, bioinformatics, or programming. Users load data from an Excel file and then choose the option "hierarchical clustering" under the Data tab. The program will then offer a prompt to users, asking if they wish the data to be normalized before the clustering analysis. Users are also asked to define which columns in their data set they would like to cluster and which of five different clustering methods and six different similarity measures they would like to use on their data. Once generated, the final visualized cluster can be exported directly as a PowerPoint file.

Genome-Wide Screens for Suppressors of Genomic Instability

As indicated previously, the yeast deletion mutant collection is a useful tool for screening for events that affect growth rate. Mutations that confer a 10-fold decrease in growth are readily detectable either by pinning methods or by competitive growth assays. However, events that occur at low frequency, such as mutations, recombination, or gross chromosomal rearrangements are difficult, if not impossible to detect, by these methods. For example, a mutant displaying a 10-fold increase in mutation rate would produce a positive event only once in 10^5 to 10^6 cells. This number is simply too small to reliably detect in the number of cells that can be transferred by a high-density replica pinning device, or that can be examined in a competitive growth experiment. One means of circumventing this problem is to use strategies that allow the examination of larger numbers of cells. Rather

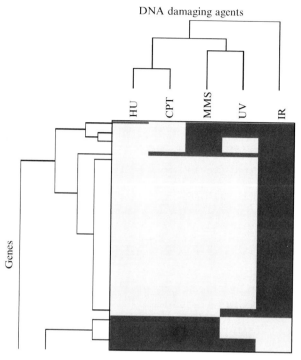

Fig. 6. An example of the output of clustering analysis performed using Cluster 3.0 and visualized using TreeView, as described in the text. Clustering was performed with data from screens with DNA damaging agents, as excerpted in Table I, and a section of the clustergram is presented. The rows are the deletion mutant array genes and the columns are the DNA damaging agents. DNA damaging agent sensitivity for a given gene deletion mutant is indicated by a dark bar at the intersection of the gene row and the DNA damaging agent column. The trees indicate the similarity of the sensitivity profiles of the DNA damaging agents and the genes. In this example, the profile of HU is more similar to that of CPT than it is to that of IR.

than screening for a particular event in the small number of cells that can be replica pinned, it is possible to screen patches of larger numbers of cells. An example of this strategy is the recent screen of the gene deletion collection for mutants that confer increased forward mutation rates (Huang *et al.*, 2003). In this screen, each deletion mutant was patched on solid media and then replica-plated to media containing canavanine. Canavanine resistant colonies result from mutations within the *CAN1* gene, and are readily scored. This strategy was particularly straightforward, as the *CAN1* gene necessary for the screen is already present in the strains that make up the gene deletion collection. However, the strategy can readily be extended to utilize other markers of chromosomal events by using the SGA methodology to introduce the relevant marker into an array of gene deletion mutants. For example, Smith *et al.* introduced markers for a gross chromosomal rearrangement assay into the deletion collection and performed a genome-wide screen using a patching strategy (Smith *et al.*, 2004). In principle, any marker that could be used in a screen can be introduced into the gene deletion collection, including markers to assay different types of recombination, as indicated later. With the exception of minor changes in the selective media, the procedure is as presented schematically in Fig. 4.

Using SGA to Cross Reporters into Deletion Mutant Arrays

1. The strain that is typically used to make crosses with the gene deletion collection is Y5563 (*MATα can1Δ::MFA1pr-HIS3 lyp1Δ ura3Δ0 leu2Δ0 his3Δ1 met15Δ0*).

2. In most instances a PCR strategy can be used to introduce the reporter of interest. In this example, we use PCR to amplify a reporter that has been used to assay recombination between direct repeats (Smith and Rothstein, 1999). This reporter contains two mutant alleles of *leu2* flanking a *URA3* gene. Following amplification by PCR the *leu2ΔEcoRI-URA3-leu3ΔBstEII* reporter is introduced into Y5563 using a standard lithium acetate transformation technique (Gietz and Woods, 2002). Strains carrying the recombination reporter are selected on SD-Ura. Integration into the *leu2* locus is confirmed by PCR.

3. Using the standard SGA procedure (Tong and Boone, 2006; Tong *et al.*, 2001, 2004) the resulting strain, which carries the recombination reporter, is crossed to a *MATa* gene deletion collection array by pinning the recombination strain to YPD medium, pinning the gene deletion array on top of the recombination strain, and incubating for 1 day at 30°. Diploids are selected by pinning to SD/MSG-Ura + G418 and incubating for 2 days at 30°.

4. Diploids are sporulated by pinning to sporulation medium and incubating for 5 days at 22°.

5. Spores are germinated and haploids selected by pinning to SD-His/ Arg/Lys/Ura + canavanine + thialysine plates and incubating for 2 days at 30°. The haploid selection is repeated by pinning again to SD-His/Arg/Lys/ Ura + canavanine + thialysine and incubating for 1 day at 30°.

6. Haploids carrying the gene deletion and the recombination reporter are selected by pinning to SD/MSG-His/Arg/Lys/Ura + canavanine +

thialysine + G418 plates. The resulting strains are then screened for increased recombination frequency.

Screening Using a Patch Assay

1. Use a wild type strain containing the recombination reporter (or other assayable marker) as a negative control and if possible, a known mutant as a positive control.
2. Using a sterile wooden stick, streak the strains to be screened (the deletion mutant array following introduction of the reporter), and the control strains, for single colonies on SD-Ura (or other selection for the reporter to be assayed, as appropriate).
3. Pick a single colony and make an even patch, approximately 1 cm by 1 cm, on YPD. Pick another single colony of the same strain and make a duplicate patch. We find it helpful to include the positive and negative control on each plate.
4. Grow the patches for 24 h at 30°.
5. Using sterile velveteen, replica plate the duplicate patches onto SD-Leu (or the appropriate selection for your reporter system) to select for recombinants. Grow the patches for 36 h at 30°. An example of the patch assay with the recombination reporter is shown in Fig. 7.
6. Identify positives by comparing with the positive and negative controls. Both of the duplicates should score similarly.
7. The strains identified as positives should then be confirmed by a suitable assay. In the example of a recombination screen a fluctuation test would be used (Spell and Jinks-Robertson, 2004).

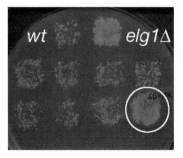

FIG. 7. Patch assay for increased recombination rate. Single colonies were streaked on YPD to form a patch. Following 24 h of growth the patches were replica-plated to the SD-Leu plate shown to detect recombinants. Wild-type and *elg1Δ* strains were included as negative and positive controls, respectively. The patch scored as positive is circled.

Media Recipes

The following recipes are for plates; for liquid media leave out the agar.

YPD: Per liter: 120 mg adenine, 10 g yeast extract, 20 g peptone, 20 g bacto agar (BD Difco), 950 ml water. After autoclaving, add 50 ml of 40% glucose solution.

Drop-out mix (DO): 3 g adenine, 2 g uracil, 2 g inositol, 0.2 g para-aminobenzoic acid, 2 g alanine, 2 g arginine, 2 g asparagine, 2 g aspartic acid, 2 g cysteine, 2 g glutamic acid, 2 g glutamine, 2 g glycine, 2 g histidine, 2 g isoleucine, 10 g leucine, 2 g lysine, 2 g methionine, 2 g phenylalanine, 2 g proline, 2 g serine, 2 g threonine, 2 g tryptophan, 2 g tyrosine, 2 g valine *minus* the indicated supplements.

SD-Ura: Per liter: 6.7 g yeast nitrogen base w/o amino acids (BD Difco), 2 g amino-acids supplement powder mixture (DO – Ura), 100 ml water. Add 20 g bacto agar to 850 ml water. Autoclave separately. Combine and add 50 ml 40% glucose.

SD-Leu: Per liter: 6.7 g yeast nitrogen base w/o amino acids (BD Difco), 2 g amino-acids supplement powder mixture (DO – Leu), 100 ml water. Add 20 g bacto agar to 850 ml water. Autoclave separately. Combine and add 50 ml 40% glucose.

SD/MSG-Ura+G418: Per liter: 1.7 g yeast nitrogen base w/o amino acids or ammonium sulfate (BD Difco), 1 g MSG (L-glutamic acid sodium salt hydrate), 2 g amino-acids supplement powder mixture (DO – Ura), 100 ml water. Add 20 g bacto agar to 850 ml water. Autoclave separately. Combine autoclaved solutions, add 50 ml 40% glucose, cool to ~65°, add 1 ml G418 (Geneticin, Invitrogen; 200 mg/ml in water) stock solution.

Sporulation Medium: Per liter: 10 g potassium acetate, 1 g yeast extract, 0.5 g glucose, 0.1 g amino-acids supplement powder mixture for sporulation (contains 2 g histidine, 10 g leucine, 2 g lysine, 2 g uracil), 20 g bacto agar to 1 liter water. After autoclaving, cool medium to ~65°, add 250 μl of G418 stock solution.

SD-His/Arg/Lys/Ura + canavanine + thialysine: Per liter: 6.7 g yeast nitrogen base w/o amino acids (BD Difco), 2 g amino-acids supplement powder mixture (DO – His/Arg/Lys/Ura), 100 ml water. Add 20 g bacto agar to 850 ml water. Autoclave separately. Combine autoclaved solutions, add 50 ml 40% glucose, cool medium to ~65°, add 0.5 ml canavanine (L-canavanine sulfate salt, Sigma, C-9758; 100 mg/ml in water) and 0.5 ml thialysine [S-(2-aminoethyl)-L-cysteine hydrochloride, Sigma, A-2636; 100 mg/ml in water] stock solutions.

SD/MSG-His/Arg/Lys/Ura + canavanine + thialysine + G418: Per liter: 1.7 g yeast nitrogen base w/o amino acids or ammonium sulfate, 1 g MSG, 2 g amino-acids supplement powder mixture (DO – His/Arg/Lys/

Ura), 100 ml water. Add 20 g bacto agar to 850 ml water. Autoclave separately. Combine autoclaved solutions, add 50 ml 40% glucose, cool medium to ~65°, add 0.5 ml canavanine, 0.5 ml thialysine, and 1 ml G418.

Acknowledgments

We thank Ridhdhi Desai for assistance with the patch assay procedure and Amy Tong for assistance with the SGA diagram. Work in the Brown laboratory is supported by the National Cancer Institute of Canada and the Canadian Institutes of Health Research. Work in the Boone laboratory is supported by the National Cancer Institute of Canada, the Canadian Institutes of Health Research, Genome Canada, and Genome Ontario. G. W. B. is a Research Scientist of the National Cancer Institute of Canada.

References

Aouida, M., Page, N., Leduc, A., Peter, M., and Ramotar, D. (2004). A genome-wide screen in *Saccharomyces cerevisiae* reveals altered transport as a mechanism of resistance to the anticancer drug bleomycin. *Cancer Res.* **64,** 1102–1109.

Baldwin, E. L., Berger, A. C., Corbett, A. H., and Osheroff, N. (2005). Mms22p protects *Saccharomyces cerevisiae* from DNA damage induced by topoisomerase II. *Nucleic Acids Res.* **33,** 1021–1030.

Begley, T. J., Rosenbach, A. S., Ideker, T., and Samson, L. D. (2002). Damage recovery pathways in *Saccharomyces cerevisiae* revealed by genomic phenotyping and interactome mapping. *Mol. Cancer Res.* **1,** 103–112.

Bellaoui, M., Chang, M., Ou, J., Xu, H., Boone, C., and Brown, G. W. (2003). Elg1 forms an alternative RFC complex important for DNA replication and genome integrity. *EMBO J.* **22,** 4304–4313.

Bennett, C. B., Lewis, L. K., Karthikeyan, G., Lobachev, K. S., Jin, Y. H., Sterling, J. F., Snipe, J. R., and Resnick, M. A. (2001). Genes required for ionizing radiation resistance in yeast. *Nat. Genet.* **29,** 426–434.

Birrell, G. W., Giaever, G., Chu, A. M., Davis, R. W., and Brown, J. M. (2001). A genome-wide screen in *Saccharomyces cerevisiae* for genes affecting UV radiation sensitivity. *Proc. Natl. Acad. Sci. USA* **98,** 12608–12613.

Chang, M., Bellaoui, M., Boone, C., and Brown, G. W. (2002). A genome-wide screen for methyl methanesulfonate-sensitive mutants reveals genes required for S phase progression in the presence of DNA damage. *Proc. Natl. Acad. Sci. USA* **99,** 16934–16939.

Chang, M., Bellaoui, M., Zhang, C., Desai, R., Morozov, P., Delgado-Cruzata, L., Rothstein, R., Freyer, G. A., Boone, C., and Brown, G. W. (2005). RMI1/NCE4, a suppressor of genome instability, encodes a member of the RecQ helicase/Topo III complex. *EMBO J.* **24,** 2024–2033.

Eason, R. G., Pourmand, N., Tongprasit, W., Herman, Z. S., Anthony, K., Jejelowo, O., Davis, R. W., and Stolc, V. (2004). Characterization of synthetic DNA bar codes in *Saccharomyces cerevisiae* gene-deletion strains. *Proc. Natl. Acad. Sci. USA* **101,** 11046–11051.

Game, J. C., Birrell, G. W., Brown, J. A., Shibata, T., Baccari, C., Chu, A. M., Williamson, M. S., and Brown, J. M. (2003). Use of a genome-wide approach to identify new genes that control resistance of *Saccharomyces cerevisiae* to ionizing radiation. *Radiat. Res.* **160,** 14–24.

Giaever, G., Chu, A. M., Ni, L., Connelly, C., Riles, L., Veronneau, S., Dow, S., Lucau-Danila, A., Anderson, K., Andre, B., Arkin, A. P., Astromoff, A., El-Bakkoury, M., Bangham, R., Benito, R., Brachat, S., Campanaro, S., Curtiss, M., Davis, K., Deutschbauer, A., Entian, K. D., Flaherty, P., Foury, F., Garfinkel, D. J., Gerstein, M., Gotte, D., Guldener, U., Hegemann, J. H., Hempel, S., Herman, Z., Jaramillo, D. F., Kelly, D. E., Kelly, S. L., Kotter, P., La Bonte, D., Lamb, D. C., Lan, N., Liang, H., Liao, H., Liu, L., Luo, C., Lussier, M., Mao, R., Menard, P., Ooi, S. L., Revuelta, J. L., Roberts, C. J., Rose, M., Ross-Macdonald, P., Scherens, B., Schimmack, G., Shafer, B., Shoemaker, D. D., Sookhai-Mahadeo, S., Storms, R. K., Strathern, J. N., Valle, G., Voet, M., Volckaert, G., Wang, C. Y., Ward, T. R., Wilhelmy, J., Winzeler, E. A., Yang, Y., Yen, G., Youngman, E., Yu, K., Bussey, H., Boeke, J. D., Snyder, M., Philippsen, P., Davis, R. W., and Johnston, M. (2002). Functional profiling of the *Saccharomyces cerevisiae* genome. *Nature* **418**, 387–391.

Giaever, G., Flaherty, P., Kumm, J., Proctor, M., Nislow, C., Jaramillo, D. F., Chu, A. M., Jordan, M. I., Arkin, A. P., and Davis, R. W. (2004). Chemogenomic profiling: Identifying the functional interactions of small molecules in yeast. *Proc. Natl. Acad. Sci. USA* **101**, 793–798.

Giaever, G., Shoemaker, D. D., Jones, T. W., Liang, H., Winzeler, E. A., Astromoff, A., and Davis, R. W. (1999). Genomic profiling of drug sensitivities via induced haploinsufficiency. *Nat. Genet.* **21**, 278–283.

Gietz, R. D., and Woods, R. A. (2002). Transformation of yeast by lithium acetate/single-stranded carrier DNA/polyethylene glycol method. *Methods Enzymol.* **350**, 87–96.

Hanway, D., Chin, J. K., Xia, G., Oshiro, G., Winzeler, E. A., and Romesberg, F. E. (2002). Previously uncharacterized genes in the UV- and MMS-induced DNA damage response in yeast. *Proc. Natl. Acad. Sci. USA* **29**, 29.

Hartman, J. L.t., Garvik, B., and Hartwell, L. (2001). Principles for the buffering of genetic variation. *Science* **291**, 1001–1004.

Huang, M. E., Rio, A. G., Nicolas, A., and Kolodner, R. D. (2003). A genomewide screen in *Saccharomyces cerevisiae* for genes that suppress the accumulation of mutations. *Proc. Natl. Acad. Sci. USA* **100**, 11529–11534.

Kanemaki, M., Sanchez-Diaz, A., Gambus, A., and Labib, K. (2003). Functional proteomic identification of DNA replication proteins by induced proteolysis *in vivo*. *Nature* **423**, 720–724.

Lum, P. Y., Armour, C. D., Stepaniants, S. B., Cavet, G., Wolf, M. K., Butler, J. S., Hinshaw, J. C., Garnier, P., Prestwich, G. D., Leonardson, A., Garrett-Engele, P., Rush, C. M., Bard, M., Schimmack, G., Phillips, J. W., Roberts, C. J., and Shoemaker, D. D. (2004). Discovering modes of action for therapeutic compounds using a genome-wide screen of yeast heterozygotes. *Cell* **116**, 121–137.

Mayer, M. L., Pot, I., Chang, M., Xu, H., Aneliunas, V., Kwok, T., Newitt, R., Aebersold, R., Boone, C., Brown, G. W., and Hieter, P. (2004). Identification of protein complexes required for efficient sister chromatid cohesion. *Mol. Biol. Cell* **15**, 1736–1745.

Mnaimneh, S., Davierwala, A. P., Haynes, J., Moffat, J., Peng, W. T., Zhang, W., Yang, X., Pootoolal, J., Chua, G., Lopez, A., Trochesset, M., Morse, D., Krogan, N. J., Hiley, S. L., Li, Z., Morris, Q., Grigull, J., Mitsakakis, N., Roberts, C. J., Greenblatt, J. F., Boone, C., Kaiser, C. A., Andrews, B. J., and Hughes, T. R. (2004). Exploration of essential gene functions via titratable promoter alleles. *Cell* **118**, 31–44.

Pan, X., Yuan, D. S., Xiang, D., Wang, X., Sookhai-Mahadeo, S., Bader, J. S., Hieter, P., Spencer, F., and Boeke, J. D. (2004). A robust toolkit for functional profiling of the yeast genome. *Mol. Cell.* **16**, 487–496.

Parsons, A. B., Brost, R. L., Ding, H., Li, Z., Zhang, C., Sheikh, B., Brown, G. W., Kane, P. M., Hughes, T. R., and Boone, C. (2004). Integration of chemical-genetic and genetic interaction data links bioactive compounds to cellular target pathways. *Nat. Biotechnol.* **22**, 62–69.

Shoemaker, D. D., Lashkari, D. A., Morris, D., Mittmann, M., and Davis, R. W. (1996). Quantitative phenotypic analysis of yeast deletion mutants using a highly parallel molecular bar-coding strategy. *Nat. Genet.* **14,** 450–456.

Smith, J., and Rothstein, R. (1999). An allele of RFA1 suppresses RAD52-dependent double-strand break repair in *Saccharomyces cerevisiae. Genetics* **151,** 447–458.

Smith, S., Hwang, J. Y., Banerjee, S., Majeed, A., Gupta, A., and Myung, K. (2004). Mutator genes for suppression of gross chromosomal rearrangements identified by a genome-wide screening in *Saccharomyces cerevisiae. Proc. Natl. Acad. Sci. USA* **101,** 9039–9044.

Spell, R. M., and Jinks-Robertson, S. (2004). Determination of mitotic recombination rates by fluctuation analysis in *Saccharomyces cerevisiae. Methods Mol. Biol.* **262,** 3–12.

Tong, A. H., Evangelista, M., Parsons, A. B., Xu, H., Bader, G. D., Page, N., Robinson, M., Raghibizadeh, S., Hogue, C. W., Bussey, H., Andrews, B., Tyers, M., and Boone, C. (2001). Systematic genetic analysis with ordered arrays of yeast deletion mutants. *Science* **294,** 2364–2368.

Tong, A. H., Lesage, G., Bader, G. D., Ding, H., Xu, H., Xin, X., Young, J., Berriz, G. F., Brost, R. L., Chang, M., Chen, Y., Cheng, X., Chua, G., Friesen, H., Goldberg, D. S., Haynes, J., Humphries, C., He, G., Hussein, S., Ke, L., Krogan, N., Li, Z., Levinson, J. N., Lu, H., Menard, P., Munyana, C., Parsons, A. B., Ryan, O., Tonikian, R., Roberts, T., Sdicu, A. M., Shapiro, J., Sheikh, B., Suter, B., Wong, S. L., Zhang, L. V., Zhu, H., Burd, C. G., Munro, S., Sander, C., Rine, J., Greenblatt, J., Peter, M., Bretscher, A., Bell, G., Roth, F. P., Brown, G. W., Andrews, B., Bussey, H., and Boone, C. (2004). Global mapping of the yeast genetic interaction network. *Science* **303,** 808–813.

Tong, A. H. Y., and Boone, C. (2006). Synthetic genetic array (SGA) analysis in *Saccharomyces cerevisiae. Methods Mol. Biol.* **313,** 171–192.

Tucker, C., and Fields, S. (2004). Quantitative genome-wide analysis of yeast deletion strain sensitivities to oxidative and chemical stress. *Comp. Funct. Genomics* **5,** 216–224.

Warren, C. D., Eckley, D. M., Lee, M. S., Hanna, J. S., Hughes, A., Peyser, B., Jie, C., Irizarry, R., and Spencer, F. A. (2004). S-phase checkpoint genes safeguard high-fidelity sister chromatid cohesion. *Mol. Biol. Cell.* **15,** 1724–1735.

Winzeler, E. A., Shoemaker, D. D., Astromoff, A., Liang, H., Anderson, K., Andre, B., Bangham, R., Benito, R., Boeke, J. D., Bussey, H., Chu, A. M., Connelly, C., Davis, K., Dietrich, F., Dow, S. W., El Bakkoury, M., Foury, F., Friend, S. H., Gentalen, E., Giaever, G., Hegemann, J. H., Jones, T., Laub, M., Liao, H., Liebundguth, N., Lockhart, D. J., Lucau-Danila, A., Lussier, M., M'Rabet, N., Menard, P., Mittman, M., Pai, C., Rebischung, C., Revuelta, J. L., Riles, J., Roberts, C. J., Ross-MacDonald, P., Scherens, B., Snyder, M., Sookhai-Mahadeo, S., Storms, R. K., Veronneau, S., Voet, M., Volekaert, G., Ward, T. R., Wysocki, R., Yen, G. S., Yu, K., Zimmerman, K., Philippsen, P., Johnston, M., and Davis, R. W. (1999). Functional characterization of the *S. cerevisiae* genome by gene deletion and parallel analysis. *Science* **285,** 901–906.

Wu, H. I., Brown, J. A., Dorie, M. J., Lazzeroni, L., and Brown, J. M. (2004). Genome-wide identification of genes conferring resistance to the anticancer agents cisplatin, oxaliplatin, and mitomycin C. *Cancer Res.* **64,** 3940–3948.

Xie, M. W., Jin, F., Hwang, H., Hwang, S., Anand, V., Duncan, M. C., and Huang, J. (2005). Insights into TOR function and rapamycin response: Chemical genomic profiling by using a high-density cell array method. *Proc. Natl. Acad. Sci. USA* **102,** 7215–7220.

Zewail, A., Xie, M. W., Xing, Y., Lin, L., Zhang, P. F., Zou, W., Saxe, J. P., and Huang, J. (2003). Novel functions of the phosphatidylinositol metabolic pathway discovered by a chemical genomics screen with wortmannin. *Proc. Natl. Acad. Sci. USA* **100,** 3345–3350.

[14]   Techniques for γ-H2AX Detection

By Asako Nakamura, Olga A. Sedelnikova, Christophe Redon,
Duane R. Pilch, Natasha I. Sinogeeva, Robert Shroff,
Michael Lichten, and William M. Bonner

Abstract

When a double-strand break (DSB) forms in DNA, many molecules of histone H2AX present in the chromatin flanking the break site are rapidly phosphorylated. The phosphorylated derivative of H2AX is named γ-H2AX, and the phosphorylation site is a conserved serine four residues from the C-terminus, 139 in mammals and 129 in budding yeast. An antibody to γ-H2AX reveals that the molecules form a γ-focus at the DSB site. The γ-focus increases in size rapidly for 10–30 min after formation, and remains until the break is repaired. Studies have revealed that small numbers of γ-foci are present in cells even without the purposeful introduction of DNA DSBs. These cryptogenic foci increase in number during senescence in culture and aging in mice. This chapter presents techniques for revealing γ-H2AX foci in cultured cells, in metaphase spreads from cultured cells, in tissues, and in yeast.

Introduction

The finding that histone H2AX becomes phosphorylated at the site of a DSB has enabled researchers to investigate these breaks at levels that are not lethal to the cell. The phosphorylated form of H2AX, named γ-H2AX, forms γ-foci encompassing many megabases of chromatin adjacent to DSB sites, whether they are accidental, incidental, programmed, or intentional. Indeed, it appears that small numbers of γ-foci are present in cells in tissue culture and in mice, indicating that the DSB may be a common aspect of normal metabolism. The γ-foci are easily visualized with antibodies to γ-H2AX with each DSB yielding one focus. The γ-foci are sites of accumulation for many factors involved in DNA repair and chromatin remodeling (Rogakou *et al.*, 1999; Sedelnikova *et al.*, 2004).

Sources of Antibodies to γ-H2AX

There are several types of antibodies useful in studies of DNA DSB formation. The first and still most commonly used is the rabbit polyclonal we developed (Rogakou *et al.*, 1999), which is available commercially from

METHODS IN ENZYMOLOGY, VOL. 409
0076-6879/06 $35.00
DOI: 10.1016/S0076-6879(05)09014-2

several companies and also from us if needed (email to bonnerw@mail.nih. gov). The antibody was raised against the peptide CKATQAS(PO₄)QEY derived from the human sequence. The antibody works well for all mammalian species examined, including the mouse, which contains another serine residue in place of the threonine in the human sequence. When first developed, this antibody also revealed the γ-H2AX orthologues in other divergent species such as the fruit fly and budding yeast. More recent batches of this antibody fail to reveal these orthologues, and researchers may need to prepare one against their epitope of interest. One or more mouse monoclonal antibodies are also available commercially for γ-H2AX. We have developed an antibody against γ-H2A in budding yeast using the peptide AKATKAS(P)QEL, which is available presently from us (email to bonnerw@mail.nih.gov) and in the future commercially. This antibody also recognizes γ-H2A in fission yeast (Nakamura *et al.*, 2004).

Immunocytochemical Detection of γ-H2AX Foci in Mammalian Material

In this section, we present protocols used for γ-H2AX detection in cultured cells (Fig. 1), frozen sections from tissues (Fig. 2), and touchprints from tissues (Fig. 3). Presently, we have not found a good procedure for paraffin-embedded material.

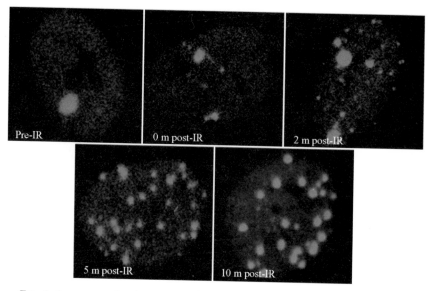

FIG. 1. Immunocytochemical detection of γ-focal growth during the first 10 min after exposure of early passage WI-38 normal human fibroblasts to ionizing radiation. (See color insert.)

Liver   Testes

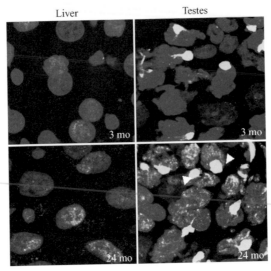

FIG. 2. Immunocytochemical detection of γ-foci in 10-μm frozen sections of liver and testes from young (3 months) and old (24 months) mice, Images shown are collapsed Z-stacks taken with 40× objective, oil immersion. The large areas of γ-H2AX in the testes sections correspond to the X-Y bodies present in late pachyteme spermatocytes. (See color insert.)

γ-H2AX   Merge   53 bp1

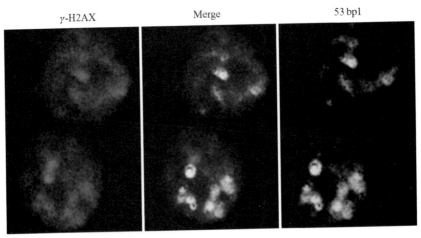

FIG. 3. Immunocytochemical detection of γ-foci and 53 bp1 in touchprints of mouse brain. Images shown are collapsed Z-stacks taken with 40× objective, oil immersion. (See color insert.)

Preparation of Cell Cultures for Immunocytochemistry

Cell growth: The cells of interest are seeded on LabTec 4-well slides at 50,000–100,000 cells/well. If the experiment involves exposure to a chemical agent, a different concentration may be used in each well of a slide. However, for exposure to ionizing radiation, each dose requires a separate slide. Cultures are typically placed in a tissue culture incubator (37°, 5% CO_2) overnight.

Introduction of DNA DSBs: For chemical agents, the culture medium is replaced with medium containing the desired concentration of the agent. For ionizing radiation, the cultures will need to travel to and from the irradiator. Since γ-H2AX does not form at 4°, the cultures may be placed on ice after removing all but a thin layer of the media to keep the cell layer moist. The transport as well as the irradiation is performed with the cells on ice. Upon return, media prewarmed to 37° is added to the wells, and the slides are placed directly on the metal shelves of the incubator for rapid warming. Two types of controls should be included: mock irradiated cultures which are treated exactly like the sample cultures but left outside the irradiator, and irradiated cultures kept on ice and fixed without warming.

Fixation and permeabilization: At the end of the incubation period, the cultures are washed with PBS, and fixed for 20 min with 2% paraformaldehyde in PBS, freshly diluted from a 20% stock (Electron Microscopy Sciences, Hatfield, PA, cat # 15713-S). 70% ethanol chilled to −20° is added for 5 min to permeabilize the cells. At this point, the samples may be stored in 70% ethanol at 4° for up to 1 month.

Preparation of Tissue Touchprints

Preliminaries: PTFE printed slides (Electron Microscopy Sciences, Cat. # 63416-15) are used. The slides contain two 15-mm wells surrounded by a hydrophobic PTFE coating, which restricts the solutions to the well area. In order to work quickly, 2–3 slides per organ are labeled with pencil beforehand. Typically one person removes the organs and a second makes the touchprints.

Touch printing: Upon removal, an organ is sliced into pieces approximately 5 mm × 5 mm. With a strong pair of tweezers and new razor blades, a flat cut is made through a piece of tissue and that surface is pressed onto the slide wells. Touch pressure will need to be adjusted according to the firmness of the tissue. Typically we prepare three slides per tissue with about three touches per well. The rest of the tissue may be quickly frozen on a block of dry ice covered with aluminum foil for use in immunoblotting or in freeze sectioning. After the touch-printed slides have air dried for

15 min to 1 h, they may be stored at −80° after wrapping them securely in aluminum foil to make an air-tight enclosure.

Fixation and permeabilization: Slides stored in the freezer are allowed to warm to 20° in the foil for 20 min. The slides are first immersed for 20 min in 2% paraformaldehyde in PBS to fix the cells and then in 70% ethanol chilled to −20° for 5 min to permeabilize the cells. At this point, the samples may be stored in 70% ethanol at 4° for several days.

Preparation of Frozen Sections

Sections: The same tissue pieces used for touchprints may be saved for frozen sectioning by placing them on dry ice covered with aluminum foil. Frozen sections are prepared. The preferred thickness is 5 μm to ensure that several nuclei do not overlap in the same microscope field. Four sections are placed on each precleaned frosted end slide and annotated as needed with pencil. The preparations are air dried for 15 min to 1 h. These dried but unfixed samples may be stored at −80° after wrapping each slide securely in aluminum foil making an air-tight enclosure. When removed from the freezer, the slides are warmed to 20° in the foil for 20 min.

Fixation and permeabilization: The slides are immersed for 20 min in 2% paraformaldehyde in PBS to fix the cells, twice for 1 h each in PBS, and then in 1% Triton X-100 for 5 min to permeabilize the cells.

Immunocytochemistry, Single Label

Blocking: If the samples have been stored in ethanol, rinse the slides twice with PBS for 5 min each. Blocking solution (8% BSA-PBS, freshly prepared) is added to the samples. With 4-well Lab-Tec slides, 250 μl solution is used per well, the lid is replaced and the slide incubated for 1 h. For touchprints and sections, draw a hydrophobic ring around each section with a PAN PEN (Zymed, Carlsbad, CA), adjust the minimal volume of the solution necessary to fill the ring (usually 50 ul), and incubate for 1 h in a humid chamber (a box with a layer of wet towels). Afterward the samples are rinsed for 5 min with PBS.

Primary antibody: Primary antibody solution is prepared by diluting an aliquot of 100 mg/ml rabbit γ-H2AX antibody stock 800-fold dilution in 1% BSA-PBS. With 4-well Lab-tek slides, 250 μl of primary antibody solution is added per well, the lid replaced, and the slide incubated for 2 h. For touchprints and sections, add the minimal volume necessary to fill the ring and incubate for 2 h in a humid chamber (a box with a layer of wet towels). Afterward the samples are rinsed twice with PBS for 5 min.

Secondary antibody: Secondary antibody solution is prepared by diluting anti-rabbit Alexa fluor 488 (Molecular Probes, Eugene, OR, Cat.# A11034)

500-fold in 1% BSA-PBS. The secondary antibody solution is added to the samples as described previously. After 1 h of incubation, the samples are rinsed three times for 5 min with PBS.

Counterstaining and mounting: A solution of propidium iodide (0.05–0.1 mg/ml) and RNAse A (0.5 mg/ml) in PBS is added to the samples for 30 min at 37° to stain the DNA. Coverslips are placed on the slides using Vectashield mounting medium containing antifading agents (Vector Labs, Burlingame, CA, Cat. H-1000). Slides are stored in a box at 4° for later viewing.

Immunocytochemistry, Double Label

Double labeling of either cultured cells or tissues is performed as single labeling except that the two primary antibodies are applied together for 2 h and the two secondary antibodies, usually Alexa 488 and Alexa 546 (Molecular Probes, Cat.# A11034 and A11030) for 1 h. The nuclei are then counterstained with DAPI mounting medium (1.5 μg/ml in Vectashield).

Combined Immunocytochemistry and FISH on
Metaphase Spreads

In this section, we present protocols used for γ-H2AX detection in mitotic chromosomes combined with the detection of telomere DNA by FISH (Fig. 4).

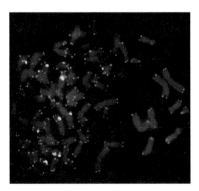

Fig. 4. γ-H2AX foci and telomeres in a metaphase prepared from Hela cell culture 30 min after exposure to 0.6 Gy. (See color insert.)

Preparation of Metaphase Spreads

Cell growth: Proliferating cell cultures grown under standard conditions at 37° in T25 flasks are treated with colcemid to enrich the population for mitotic cells. We found that the conditions for colcemid incubation need to be optimized for each cell type. For primary human strains such as WI-38 (Coriell Cell Repositories, Camden, NJ) and BJ (ATCC, Rockville, MD), we use 100 ng/ml for 3 h. The same conditions are used for mouse primary strains. For immortalized lines such as HeLa or telomerase-infected primary lines, we use 50 ng/ml for 1 h (Stenman *et al.*, 1975). For muntjac cell cultures, we use 100 ng/ml for 18 h (Brown *et al.*, 1996).

Cytospin preparation: At the end of the colcemid incubation, the medium is gently aspirated from the culture leaving mitotic cells undisturbed. Hypotonic buffer (10 mM Tris-HCl [ph 7.4], 40 mM glycerol, 20 mM NaCl, 1.0 mM CaCl$_2$, and 0.5 mM MgCl$_2$) is gently added to the cultures, and is gently aspirated off. Hypotonic buffer is added to the culture for 15 min at 37° to swell the mitotic cells. The amount of buffer added depends on the concentration of cells released into the buffer and is empirically determined. The optimal concentration for cytospinning is 50,000 mitotic cells per ml. We use 1.5 ml hypotonic buffer for primary strains and 3 ml for immortal lines, and adjust the amount if the spreads are too crowded or too sparse. Mitotic cells are loosened by pipeting the buffer over the growth surface several times. The efficiency of this step can be monitored by checking under a microscope for mitotic cells remaining attached in the flask. 200 μl of the mitotic cell suspension is added to the cytospin funnel clamped to a regular glass slide. The samples are cytospun at 1800 rpm for 8 min (Merry *et al.*, 1985).

Fixation: the slides are immersed in 80% EtOH prechilled to −20° for 30 min, briefly rinsed in cold acetone, and air dried to fix the metaphase spreads (Turner, 1982). Samples may be stored in a box at 4° at this stage but it is better to proceed directly with the next steps.

Immunocytochemistry γ-H2AX on Metaphase Spreads
  (FISH Compatible)

The following two procedures were developed for combined immunocytochemistry and FISH, but each can be performed individually.

Blocking: A 1–1.5 cm hydrophobic ring is drawn around the sample area with a Pap Pen (Research Products International Corp., Mt. Prospect, IL) and the slides are hydrated in phosphate buffered saline lacking calcium and magnesium (Ca^{+2}/Mg^{+2} free PBS, 10 mM Na$_2$HPO$_4$, 1.8 mM KHPO$_4$, 137 mM NaCl, 2.7 mM KCl, adjust pH to 7.4 with HCl) for 5 min. 200 μl blocking solution (8% BSA, Ca^{+2}/Mg^{+2} free PBS) is pipeted inside the hydrophobic ring and the samples are incubated in a humid chamber for 1 h. Afterward they are washed with Ca^{+2}/Mg^{+2} free PBS for 5 min.

Primary antibody: 100 μl of primary antibody solution, 5 μg/ml in 10% BSA, (mouse monoclonal anti-γ-H2AX antibody; Upstate 05–636) is pipeted inside the hydrophobic ring and the samples incubated in a humid chamber for 2 h. Afterward they are washed twice with Ca^{+2}/Mg^{+2} free PBS for 5 min each.

Secondary antibody: 100 μl of secondary antibody, 5 μg/ml in 10% BSA, (Alexa-488-conjugated anti-mouse IgG; Molecular Probes) is pipeted inside the hydrophobic ring and the samples incubated in a humid chamber for 1 h (Rogakou et al., 1999).

Telomere FISH on Metaphase Spreads

Preliminaries: the slides are washed three times with Ca^{+2}/Mg^{+2} free TBS for 5 min each. (Note: if only immunocytochemistry of γ-H2AX is desired, the slides are rinsed three times with Ca^{+2}/Mg^{+2} free PBS for 5 min each.) Aqueous mounting medium (Biomedia Corp., Foster City, CA) containing 2–5 ng of DAPI per ml is added inside the hydrophobic ring and a coverslip placed on top. The samples are stored in a box at 4° until ready to view. Obtain a telomere PNA FISH Kit/ Cy3 (Code No. K 5326, DAKO A/S. Denmark). The slides are immersed in 50 mM ethylene glycol-*bis* (succinic acid N-hydroxy-succinimide ester; E-3257, Sigma-Aldrich, St. Louis, MO) (Sedelnikova et al., 2004). We use 2 min for human cell preparations and 3 min for mouse cell preparations. Afterward samples are washed twice with Ca^{+2}/Mg^{+2} free TBS for 5 min each.

Pretreatment: 200 μl of Pre-Treatment Solution (from DAKO FISH kit) is pipeted inside the hydrophobic ring and the samples incubated for 10 min. Afterward they are washed twice with Ca^{+2}/Mg^{+2} free TBS for 5 min each and are dehydrated by sequentially immersing the slides in 70%, 85%, and 100% EtOH for 2 min each. The ethanol is chilled to −20° beforehand, but the immersions are at 20°. Samples may be stored immersed in 100% ethanol at −20°.

Hybridization: The slides are removed from the ethanol and air dried completely (usually about 5 min). 7.5 μl of the kit Telomere Probe Solution (Cy3-conjugated telomere PNA probe) is pipeted onto the cytospin area and a cover slip is placed on top. The slide is placed on a heat block at 85° for 3 min to denature the DNA and then in a humid chamber at room temperature for 1 h for hybridization. Afterward the slides are immersed in kit Rinse Solution to loosen and remove the cover slip, then placed in kit Wash Solution at 65° for 5 min. The samples are dehydrated by sequentially immersing the slides in 70%, 85%, and 100% EtOH for 2 min each as stated previously.

Counterstaining and mounting: The slide is air dried completely. Aqueous mounting medium (Biomedia Corp., Foster City, CA) containing 2–5 ng of DAPI per ml is added to the sample and a coverslip pressed into place. The samples are stored in a box at 4° for storage.

Immunocytochemical Detection of γ-H2AX Foci in Budding Yeast

In this section we present protocols for γ-H2A detection in yeast (Figs. 5 and 6). When first developed, the antibody raised against the human sequence also revealed the γ-H2AX orthologues in other divergent species such as the fruit fly and budding yeast. More recent batches of this antibody fail to reveal these orthologues and researchers may need to

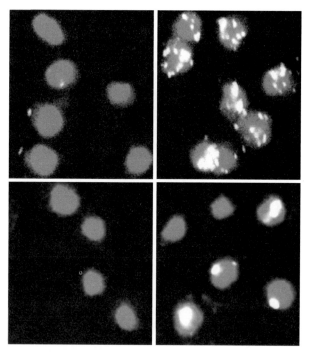

FIG. 5. Immunocytochemical detection of γ-foci in budding yeast formed in response to DNA DSBs. (Upper) Wild-type cells were left untreated (left) or treated with camptothecin (right) for 1 h. (Lower) Yeast cells grown in glucose (left) or galactose (right) for 2 h to induce the production of the HO endonuclease. Note that after HO induction, a single γ-focus is observed in most cells. Detection is with yeast γ-H2A peptide antibody. (See color insert.)

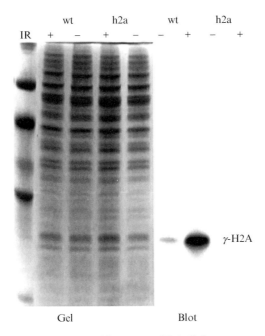

FIG. 6. Immunoblot detection of budding yeast γ-H2A. Cell extracts were prepared by the TCA procedure from irradiated (IR+) and unirradiated (IR−) cells of yeast strain W303 and separated by SDS PAGE. The γ-H2A signal was detected using yeast γ-H2A peptide antibody in yeast strain W303 cells containing (wt) or lacking (h2a) H2A serine 129. The h2a strain is W303 *hta1-s129a hta2-s129a* (Redon *et al.*, 2005).

prepare one against their epitope of interest. We have developed an antibody against γ-H2A in budding yeast using the peptide AKATKAS (P)QEL, which is available presently from us (email to bonnerw@mail.nih.gov) and in the future commercially. This antibody also recognizes γ-H2A in fission yeast (Nakamura *et al.*, 2004).

Cell growth: Yeast media, culture conditions, and manipulations of yeast are as described previously by Egel and Holmberg (1998). The yeast are grown until they reached a density of 2–4 × 10^7 cells/ml.

Formation of DNA DSBs in Budding Yeast

Chemical treatment: Stock solutions of camptothecin (CPT) (Sigma-Aldrich) are prepared at a concentration of 6 mg/ml in dimethylsulfoxyde (DMSO). Stock solutions of hydroxyurea (Sigma-Aldrich) are prepared at 2 M in H$_2$O. Both are diluted into liquid cultures to the desired concentration, using a DMSO-treated control for CPT.

Ionizing radiation: Cells are collected by centrifugation, placed in ice, and exposed to the indicated amount of ionizing radiation from a ^{137}Cs source (Mark I irradiator; J. L. Shepherd and Associates, San Fernando, CA) at a rate of 15.7 Gy/min.

HO endonuclease induction: Cultures are inoculated into 5 ml YPD (2% Bacto Peptone, 1% Bacto Yeast Extract, 2% glucose, pH 5.5) and incubated with aeration overnight (~16 h) at 30°. The cells are sedimented for 2 min at 3000g, rinsed once in 5 ml YP-lactate (2% Bacto Peptone, 1% Bacto Yeast Extract, 3% lactic acid, pH 5.5), resuspended in 5 ml YP-lactate and incubated ~8 h at 30° with aeration. A 2-l flask containing 200 ml of YP-lactate is inoculated with an aliquot of the starter culture sufficient to give 10^7 cells/ml the following morning, as determined empirically. We usually use a 1/100–1/200 dilution. The culture is aerated vigorously (~300 rpm) overnight at 30°. When the culture reaches 10^7 cells/ml, an aliquot, usually 10 or 15 ml, is removed as the uninduced control. Galactose is added to the remaining culture to 2%, 0.1 volume 20% galactose stock, and vigorous aeration continued. Samples are removed for ChIP, DNA analysis, or microscopy as needed. [Note: This protocol was developed in the laboratory of James E. Haber and is described in greater detail at their methods website, (http://www.bio.brandeis.edu/haberlab/jehsite/protocol.html).] It utilizes yeast strains with the HO endonuclease gene regulated with a galactose-inducible promoter (Lee et al., 1998). This protocol uses pregrowth on lactic acid as a carbon source, giving a much more rapid and complete HO endonuclease induction than do protocols using pregrowth with raffinose as the carbon source. Typically, >90% of the *MAT* loci are cleaved within 15–30 min of galactose addition.

Visualization of γ-foci in Budding Yeast

Fixation: Paraformaldehyde is added to 2–5 ml cultures to a final concentration of 4% and incubated at 20° for 1 h with gentle agitation. Cells are washed 5 times with a solution containing 5 mM MgCl$_2$, 40 mM KH$_2$PO$_4$, and resuspended in 0.5 ml of digestion buffer containing 5 mM MgCl$_2$, 40 mM KH$_2$PO$_4$, 1.2 M sorbitol. Yeast lytic enzyme (MP Biomedicals, Costa Mesa, CA, cat no 190123, 100 Unit/ml) is added to 1 mg/ml final concentration. Cells are incubated at 30° for 45–60 min. To prevent cell loss during the protocol, slides precoated with poly-L-lysine are prepared as follows: A hydrophobic ring about 1 cm diameter is drawn on a slide with a Pap Pen. Poly-L-lysine (Sigma-Aldrich, 1 mg/ml) is spotted inside the ring and incubated for 10 min at 20°. The slides are washed twice with distilled water and air-dried for 15–30 min. After the lytic enzyme digestion, cells are washed with digestion buffer and spotted onto the poly-L-lysine coated slides.

Blocking: Without letting the cells dry, the slides are then incubated in blocking buffer (1.5% BSA, 0.5% Tween 20, 0.1%, Triton X-100 in PBS) for 30 min.

Primary antibody: We prepared a primary antibody directed against phosphorylated yeast H2A, which is available from us, and commercially in the near future. The antibody stock is diluted 1000-fold into the blocking buffer and added to the slides for overnight incubation at 20° in a humid chamber. Afterward the slides are washed twice with PBS and once with blocking buffer.

Secondary antibody: Secondary antibody solution, prepared by diluting anti-rabbit Alexa fluor 488 (Molecular Probes) 500-fold in blocking buffer, is added to the samples for 1 h incubation. Afterward the samples are washed three times with PBS.

Counterstaining and mounting: The samples are incubated in a solution containing 0.5 mg/ml RNAse A and 200 μg/ml propidium iodide for 1 h at 37°, and then mounted with antifade medium. The slides are stored in a box at 4°.

Immunoblot Detection of Yeast γ-H2A

Protein Extraction

We have used two methods, one with trichloroacetic acid (TCA) precipitation of total yeast protein, and the other with a sulfuric acid extraction of the histones prior to TCA precipitation. The former is more rapid and needs less starting material but less enriching for the histones.

TCA extraction and precipitation: Aliquots containing 1×10^7 cells are spun in 1.5 ml tubes at 14,000g for 2 s. The pellets are resuspended in 90 μl of 20% TCA and 0.15 g glass beads are added to the tubes. The yeast cells are disrupted by vortexing the tubes for 5 min. After briefly allowing the beads to settle, the liquid supernatant, containing the precipitated histones, is collected and the beads briefly vortexed with 5% TCA. The bead wash is collected and added to the 20% TCA supernatants. The extracts are then centrifuged at 14,000g for 10 min to pellet the histone precipitate. The pellets are washed with cold 100% ethanol, air dried, and stored for immunoblot analysis.

Sulfuric acid extraction: Aliquots containing 4×10^8 cells are harvested by centrifugation at 700g for 5 min and rinsed once with 1 ml YPD in a 1.5-ml tube. To the yeast cell pellet, 0.45 g of glass beads (0.5 mm diameter) and 0.4 ml of 0.2 M H$_2$SO$_4$ are added. The 1.5-ml tubes are placed in a mini-beadbeater-8^{TM} cell disrupter (Biospec Products, Bartlesville, OK) and homogenized for 5 min to disrupt the cells. After 5 min on ice, the extracts are clarified at 14,000g for 15 min at 4°. The supernatants (340 μl each) are collected and 100% w/v TCA added to a final concentration of 20% w/v. After 20 min incubation on ice, the precipitate is collected at 14,000g for 10 min at 4°, washed with ethanol chilled to −20°, and air dried. The proteins extracts obtained are used for 2D gel analysis or immunoblot analysis.

Gel electrophoresis: Electrophoreses is performed in a minislab format apparatus (NOVEX, San Diego, CA) with a premade 4–20% Tris-Glycine gel (Invitrogen, Grand Island, NY, cat. No. EC6028) or custom-made (13.5% polyacrylamide) (Laemmli, 1970) using an SDS buffer. If the air-dried samples are insufficiently washed with ethanol, residual acid will remain and hinder dissolution in the sample buffer. This is indicated by the bromophenol dye turning yellow, and can be remedied by adding concentrated ammonia to restore the blue color. Samples prepared by H_2SO_4 extraction are dissolved in 20 μl of sample buffer, while those prepared by TCA precipitation are dissolved in 90 μl. The sample buffer contains 62.5 mM Tris/HCl pH 6.8, 10% glycerol (v/v), 1% SDS, 5% 2-mercaptoethanol and 5% bromophenol Blue (saturated solution), The samples are vortexed for 2 pulses of 30 s each, heated at 100° for 5 min, kept on ice for 1 min, and then clarified at 14,000g for 10 min at 4°. The supernatants, 15 μl for the H_2SO_4 procedure and 12 μl for the TCA procedure, are loaded into the gel wells, and electrophoresis performed at 180 V until the bromophenol blue reached the bottom of the gel, about 90 min.

Immunoblotting: The gel is blotted onto Polyvinylidene difluoride (PVDF) membranes. The membranes are blocked with 1% BSA, 0.1% Tween-20 in PBS for 1 h, incubated with the γ-H2AX primary antibody at 10,000-fold dilution in the blocking buffer overnight at 4°, washed in PBS, incubated with peroxidase goat anti-rabbit secondary antibody IgG (Calbiochem-Novabiochem Corp., San Diego, CA) at 10,000-fold dilution in 0.1% Tween 20 PBS for 1 h at 20°, and washed in PBS. Anti-γ-binding is visualized by chemiluminescence (ECL RPN 2209, Amersham Pharmacia Biotech, Piscataway, NJ).

Chromatin Immunoprecipitation Using Yeast Anti-γ-H2AX

Crosslinking: Yeast cells (15 ml of a culture at $1–2 \times 10^7$ cells/ml) are sedimented at 3000g for 2 min. The pellet is resuspended in 1 ml 1% formaldehyde and incubated for 15 min at 20°. Glycine is added to a final concentration of 125 mM (0.1 ml of 1.375 M glycine) and incubation continued a further 5 min at 20°. The sample is transferred to a 1.5-ml tube, centrifuged for 20 s, and washed twice in Tris buffered saline (100 mM Tris/HCl, 0.9% w/v NaCl, pH 7.5). The cells are centrifuged and either lysed immediately or stored as a pellet in liquid nitrogen. All subsequent steps are at 4° or on ice.

Lysis: The cell pellet is resuspended in 500 μl of Lysis Buffer-1 M NaCl (50 mM HEPES/KOH pH 7.5, 1 M NaCl, 5 mM EDTA, 1% Triton X-100 and 0.1% Na deoxycholate) plus 5 μl protease inhibitor solution (Protease Inhibitor Cocktail Set III, Calbiochem). An approximately equal volume of glass beads is added. The cell/bead mixture is shaken in a Beadbeater-8 (Biospecs Products, Bartlesville, OK) for 3 min at maximum power, permitted to settle for a minute, and then shaken again for 3 min. The beads are permitted to settle and the lysate transferred to a new microfuge tube. The lysate is sonicated until the DNA has an average fragment size of 1 kb, as determined empirically using trial lysates and monitoring fragment size by agarose gel electrophoresis. We typically use eight, 10 s pulses, with a 30-s cooling period between pulses, using a Microson XL2020 Sonicator (Misonix, Farmingdale, NY) fitted with a microprobe and set to 10% power output. The lysate is cleared by centrifugation for 5 min in a microfuge and the supernatant retained.

Immunoprecipitation: (Conditions described are for the rabbit polyclonal yeast γ-H2AX antibody.) The sample is split, to one-half, 1 μl of anti-γH2AX antibody is added, while the other half of the lysate is used as the no antibody control. The sample is incubated on a mini-rotator (Glas-Col) for 3 h at 20°. Then 15 μl of Protein G Ultra suspension (Pierce) is added and incubation continued on a minirotator overnight at 4°. The Protein G beads are pelleted and the lysate removed.

Bead Washing: The pellet Protein G beads is resuspended in 1 ml of Lysis Buffer-1 M NaCl plus protease inhibitors and incubated for 5 min at 20° on the minirotator. The Protein G beads are pelleted and the buffer removed. This wash is repeated five more times followed by three washes with Lysis Buffer-0.5 M NaCl (50 mM HEPES/KOH pH 7.5, 1 M NaCl, 5 mM EDTA, 1% Triton X-100, and 0.1% Na deoxycholate), one with Wash Solution 3 (10 mM Tris-Cl buffer pH 8.0, 0.25 M LiCl, 0.5% NP-40, 0.5% Na deoxycholate, and 5 mM EDTA) and finally one with 1 × TE pH 8.0.

Crosslink reversal and preparation for PCR: After all traces of wash solution are removed from the Protein G beads, they are resuspended in 100 μl of 1 × TE pH 8.0, incubated at 65° for 8 h to reverse the cross-links, and then heated to 95° for 20 min. The sample is now ready for analysis. Typically for this size sample, use 1 μl in a standard PCR reaction (Shroff *et al.*, 2004.

References

Brown, K. D., Wood, K. W., and Cleveland, D. W. (1996). The kinesin-like protein CENP-E is kinetochore-associated throughout poleward chromosome segregation during anaphase-A. *J. Cell Sci.* **109**, 961–969.

Egel, R., and Holmberg, S. (1998). Cultivation of yeast cells: Saccharomyces cerevisiae and Schizosaccharomyces pombe. *In* "Cell Biology" (J. E. Celis, ed.), 2nd Ed., Vol. 1, pp. 421–430. Academic Press, San Diego.

Laemmli, U. K. (1970). Cleavage of structural proteins during assembly of the head of bacteriophage T4. *Nature* **227**, 680–685.

Lee, S. E., Moore, J. K., Holmes, A., Umezu, K., Kolodner, R. D., and Haber, J. E. (1998). Saccharomyces Ku70, Mre11/Rad50 and RPA proteins regulate adaptation to G2/M arrest after DNA damage. *Cell* **94**, 399–409.

Merry, D. E., Pathak, S., Hsu, T. C., and Brinkley, B. R. (1985). Anti-kinetochore antibodies: Use as probes for inactive centromeres. *Am. J. Hum. Genet.* **37**, 425–430.

Nakamura, T. M., Du, L. L., Redon, C., and Russell, P. (2004). Histone H2A phosphorylation controls Crb2 recruitment at DNA breaks, maintains checkpoint arrest, and influences DNA repair in fission yeast. *Mol. Cell. Biol.* **24**, 6215–6230.

Redon, C., Pilch, D., and Bonner, W. M. (2005). Genetic analysis of *Saccharomyces cerevisiae* H2A serine 129 mutant suggests a functional relationship between H2A and the sister chromatid cohesion partners Csm3-Tof1 for the repair of Topoisomerase I-induced DNA damage. *Genetics*, 2005 (E-pub ahead of print).

Rogakou, E. P., Boon, C., Redon, C., and Bonner, W. M. (1999). Megabase chromatin domains involved in DNA double-strand breaks *in vivo*. *J. Cell Biol.* **146**, 905–915.

Sedelnikova, O. A., Horikawa, I., Zimonjic, D. B., Popescu, N. C., Bonner, W. M., and Barrett, J. C. (2004). Senescing human cells and ageing mice accumulate DNA lesions with unrepairable double-strand breaks. *Nature Cell Biol.* **6**, 168–170.

Shroff, R., Arbel-Eden, A., Pilch, D. R., Ira, G., Bonner, W. M., Haber, J., and Lichten, M. (2004). Distribution and dynamics of chromatin modification induced by a defined DNA double-strand break. *Curr. Bio.* **14**, 1703–1711.

Stenman, S., Rosenqvist, M., and Ringertz, R. (1975). Preparation and spread of unfixed metaphase chromosomes for immunofluorescence staining of nuclear antigens. *Exptl. Cell Res.* **90**, 87–94.

Turner, B. R. (1982). Immunofluorescent staining of human metaphase chromosomes with monoclonal antibody to histone H2B. *Chromosoma* **87**, 345–357.

Further Reading

Rogakou, E. P., Pilch, D. R., Orr, A. H., Ivanova, V. S., and Bonner, W. M. (1998). DNA double-stranded breaks induce histone H2AX phosphorylation on serine 139. *J. Biol. Chem.* **273**, 5858–5868.

[15]   Methods for Studying the Cellular Response to DNA Damage: Influence of the Mre11 Complex on Chromosome Metabolism

By Jan-Willem F. Theunissen and John H. J. Petrini

Abstract

Dramatic progress in understanding the mediators and mechanisms of chromosome break metabolism has been made in recent years. As a result, the links between disease and defects in chromosome dynamics have become clearer. In this chapter, we discuss techniques employed in our laboratory to study chromosome break metabolism, which include assessments at the molecular and cellular level. In our laboratory, we use the Mre11 complex as a tool to study this process, but the techniques discussed are of general relevance.

Introduction

DNA damage, whether caused by extrinsic or intrinsic factors, evokes a cellular response that can lead to a number of outcomes including cell cycle checkpoint activation, DNA repair, and apoptosis. In this chapter we discuss a repertoire of techniques to study the cellular response to DNA double-strand breaks (DSBs).

In our laboratory we study the highly conserved Mre11 complex, consisting of Mre11, Rad50, and Nbs1. For the most part, the techniques in this chapter were established or refined from pre-existing methodologies to study phenotypic outcomes of Mre11 complex mutations in mammals. As this chapter is focused on the methods applied, the reader is directed to a recent review for a brief synopsis of the data obtained (Stracker *et al.*, 2004). In general, the described procedures can be used to study any protein involved in the cellular response to DSBs (Bakkenist and Kastan, 2004; Petrini and Stracker, 2003). Where appropriate, we included sample data sets that might be obtained using the protocols described.

After outlining preparation of murine cells for tissue culture, we will present protocols to study checkpoint deficiency and chromosomal instability. We will end with immunofluorescence protocols in human and mouse cells to study recruitment of DNA repair proteins to DSBs. For each protocol we will provide investigators with tools to analyze their results quantitatively.

METHODS IN ENZYMOLOGY, VOL. 409
0076-6879/06 $35.00
DOI: 10.1016/S0076-6879(05)09015-4

This chapter focuses predominantly on cell-based assays. In general, the clastogenic treatments employed create supraphysiologic levels of chromosome breakage. This may limit the scope of interpretations that can reasonably be drawn. Nevertheless, the data are generally predictive of, and consistent with phenotypic outcomes at the organismal level.

Derivation of Primary and Transformed Cells

Preparation of Mouse Embryonic Fibroblasts (MEFs)

MEFs are harvested from 13.5–15.5 day old mouse embryos by timed pregnancy. For timed pregnancies, mate one female mouse, between 2 and 5 months of age, with one male mouse, at least 3 months of age. Since the estrus cycle is 4 days long and insemination usually takes place around midnight, check for post coitus semen plugs before 11:00 A.M. for 4 days. More details on breeding techniques can be found in *Mouse Genetics* (Silver, 1995) (or on the web at http://www.informatics.jax.org/silver).

Isolate embryos 13 to 15 days after plug detection according to the method outlined later. A similar protocol is available in *Manipulating the Mouse Embryo* (Nagy, 2003).

Materials

70% ethanol and PBS.

Culture medium: DMEM (high glucose; Invitrogen, Carlsbad, CA), 10% fetal calf serum (FCS), L-glutamine (Invitrogen), and streptomycin/penicillin.

Trypsin solution: 100 mM NaCl, 5 mM dextrose, 0.75 mM Na$_2$HPO$_4$ •7H$_2$O, 5 mM KCl, 3.5 mM KH$_2$PO$_4$, 1.5 mM EDTA, 2.5 g/l trypsin (trypsin [1:250], powder, Invitrogen), 25 mM Trizma base (Sigma, St. Louis, MO), bring up to pH 7.6 with HCl. An equivalent trypsin solution containing EDTA can also be used.

Freezing solution: 10% DMSO (Hybri-Max Sigma D2650), 10% FCS, and 80% culture medium.

Sterile dissection instruments (forceps and scissors), 15-ml screw cap plastic conical tubes, Petri dishes, 1.5-ml snap cap tubes, and 10-cm or 15-cm tissue culture dishes.

Method

1. Euthanize a pregnant female using the method recommended at your institution. Wet the abdomen with 70% ethanol. Make a low horizontal abdominal incision with scissors into the skin. Grasp the skin

above and below this incision separately with your gloved hands, and pull both skin parts in opposite directions to expose the abdominal wall.

2. Make a low horizontal incision with clean scissors into the abdominal wall, dissect out the uterine horns containing the embryos and put these into a Petri dish containing ~7 ml of PBS.

3. Cut through the uterus between each embryo, and open the amnion of each embryo. Dissect out the embryos and transfer each one to a separate Petri dish containing ~7 ml of PBS.

4. Remove the red tissues (heart, liver, and spleen) from each embryo using two fine forceps. Cut a little piece of the tail or head for genotyping purposes, and transfer this piece of tail to a 1.5-ml tube for genomic DNA preparation.

5. Rinse as much blood as possible from the embryo by moving it back and forth in the Petri dish with forceps, and transfer each embryo to a 15-ml screw cap conical tube containing 3 ml of trypsin solution. Incubate these embryos overnight at 4° to allow diffusion of the trypsin into the embryos. At this temperature the trypsin is nearly inactive.

6. The next day, aspirate excess trypsin solution off carefully, without disturbing the embryo.

7. Put conical tubes at 37° to activate the trypsin and incubate for 45 min. Alternatively you can place the embryos with the trypsin solution from Step 5 directly after harvest at 37° for 75 min.

8. Add 7 ml of culture medium and pipette vigorously (about 10 times) to break up the embryo until you obtain a suspension. For each embryo, plate this suspension onto one 15-cm or three 10-cm tissue culture plates.

9. Freeze cells down when these plates are subconfluent. If the plates are too confluent at the time of freeze down, plating efficiency upon thawing will be low. Alternatively you can passage these MEFs 1:3 when the plates reach confluency. After replating, these cells are at passage 1 (p1). Each time the cells are passaged 1:3, the passage number increases by 1.

Preparation of Ear Fibroblasts (Shao et al., 1999)

Materials

Anesthetic: avertin (dose [ml] = [weight (g) ×2] / 100) or alternative.
70% ethanol and PBSK: PBS with 100 μg/ml of kanamycin.
Collagenase/dispase mix: collagenase D (Roche, Indianapolis, IN) and dispase neural protease from Bacillus polymyxa grade II (Roche) both at 4 mg/ml in DMEM.
Culture medium: DMEM (high glucose) with 10% FCS, L-glutamine, nonessential amino acids and antibiotic-antimycotic (Invitrogen).

Sterile dissection instruments (forceps and scissors), #11 scalpel blades, 6- and 10-cm tissue culture dishes, 70-μm cell strainers (Falcon, BD Biosciences, San Jose, CA), 50-ml screw cap plastic conical tubes.

Method

1. Anesthetize mouse, and wet the ear with 70% ethanol.
2. Cut off the dorsal portion (0.6 cm^2) of 1 or 2 ears with sterile scissors. Rinse ears with ethanol, and place them with forceps in a 6-cm tissue culture dish containing sterile PBSK.
3. Once in the tissue culture hood, rinse the ears twice with PBSK.
4. Aspirate off all PBSK, and mince the ears with two sterile #11 scalpel blades into small squares (2 mm^2). Mincing is not required, but increases cell number yield.
5. Add 2 ml of the collagenase/dispase mix, move the dish back and forth to submerge the ear pieces, and incubate in 37° incubator for 45 min.
6. Add 3 ml of culture medium (with 5× of antibiotic-antimycotic) and incubate overnight in 37° incubator.
7. Next day, dissociate the ears by pipetting up and down at least five times.
8. Place a cell strainer on a 50-ml conical tube, pass the ear suspension through this strainer with a pipette, remove the cell strainer, and spin the cells down in a swinging bucket centrifuge at ∼400g for 5 min at RT.
9. Remove the collagenase/dispase mix by aspiration, and plate the cells on one 10-cm dish containing ∼10 ml of culture medium.
10. The cells should reach confluency after ∼3 days.

NOTE
Essentially the same method can be used to prepare tail fibroblasts. Use 1 cm of tail in the aforementioned protocol.

Preparation of Splenocytes and Thymocytes

Materials

PBS, and culture medium: RPMI, 10% FCS, L-glutamine, and streptomycin/penicillin.
Dissection instruments (forceps and scissors), 70-μm cell strainer (Falcon), plungers of 10-ml syringes, 50-ml screw cap plastic conical tubes, and tissue culture flasks.

Method

1. Harvest the spleen or thymus. Place tissue in a cell strainer positioned on top of a 50-ml conical tube.
2. Smash each tissue with a plunger.
3. Wash the cells through the cell strainer by adding 10 ml of PBS with a pipette.
4. Remove the cell strainer, and spin the cells down in a swinging bucket centrifuge at $\sim$400g for 5 min at RT.
5. Remove the supernatant, add 10 ml of culture medium, resuspend the cell pellet with a pipette, and spin the cells down in a swinging bucket centrifuge at $\sim$400g for 5 min at RT.
6. Repeat Step 5 once.
7. Plate your cells at 1×10^6 cells/ml in a tissue culture flask containing culture medium.

Derivation of SV40 Transformed Cells

The method we outline to transform primary ear fibroblasts or MEFs is also a good general procedure for recombinant DNA transfection of these cells.

Materials

pX-8 (SV40 plasmid with inactivated origin) (Fromm and Berg, 1982).
Fugene transfection reagent (Roche) or alternative transfection reagent such as calcium phosphate.
Tissue culture plates, 5-ml polystyrene round bottom tubes, DMEM, and culture medium (as in Preparation of MEFs, shown previously).

Method

1. Plate 1×10^6 primary MEFs on a 10-cm tissue culture plate the day before transfection.
2. The next day, replace the culture media with fresh media.
3. For each 10-cm plate, prepare the transfection mixture as follows:

- Add 1 ml of DMEM (serum free) to a 5-ml round bottom tube.
- Add 20 μl of Fugene transfection reagent directly to the medium without touching the wall of the tube. Gently mix with 1-ml pipette.
- Incubate at RT for 5 min.
- Add 10 μg of pX-8 DNA to a second 5-ml tube.
- Add the Fugene/DMEM mix drop wise to the DNA.
- Incubate at RT for 15 min.

4. Add the Fugene DMEM DNA mixtures drop wise to each 10-cm dish, shake the dish gently back and forth, and place it in an incubator.
5. The next day, passage each 10-cm dish onto one 15-cm dish.
6. Passage these dishes 1:5 when they reach confluency, and wait for rapidly growing (i.e., transformed) colonies to form. Colony formation will take between 2 weeks and one month. Since the transformed cells grow more rapidly than the primary cells, no selection is required.

IR-Induced Cell Cycle Checkpoints

Reviews and protocols for IR-induced cell cycle checkpoints (Fig. 1) have recently been published in *Methods in Molecular Biology,* Volumes 280 and 281 (Schonthal, 2004a,b). Rather than omitting these checkpoint protocols in this chapter we decided to add data analysis tools.

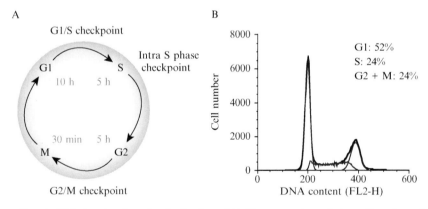

FIG. 1. IR-induced cell cycle checkpoints. (A) Representation of the cell cycle and the checkpoints discussed in this section. The approximate duration of each cell cycle phase (in h or min) in asynchronously proliferating early phase MEFs is depicted. (B) DNA content histogram. In this graph containing 75,000 asynchronous p3 MEFs cell number is plotted against DNA content (FL2-H). The Watson Pragmatic model in the FlowJo cell cycle platform was used to calculate the percentage of cells in G1, S, and G2 + M. Assuming that the cells in this asynchronous population are growing at about the same rate, the fraction of cells in a given phase (the ~ percentage of mitotic cells can be derived from Fig. 4C) multiplied by the total doubling time gives you the length of that phase. (See color insert.)

G1/S Checkpoint Assay

Introduction

One of the first cell cycle checkpoints characterized in mammalian cells was the p53-dependent G1/S checkpoint (Kuerbitz *et al.*, 1992). In order to assay for this checkpoint, the percentage of S-phase cells is determined 10–20 h after IR treatment in a population of asynchronous cells. Whereas p53-proficient cells arrest at the G1/S border upon IR treatment, p53-deficient cells enter S phase prematurely.

In the initial study (Kuerbitz *et al.*, 1992), the percentage of S-phase cells was determined by flow cytometry on cells labeled for 4 h with the thymidine analog BrdU 17 h after IR treatment. The initiation (8 to 24 h after IR) and length (3 to 4 h) of the BrdU pulse need to be established empirically (e.g., Kuerbitz *et al.*, 1992; Xu and Baltimore, 1996). Both parameters depend upon the cell cycle kinetics of the cells under study. G1/S checkpoint activation can be observed when cells are exposed to IR doses as low as 2 Gy.

Materials

p2-p5 MEFs or ear fibroblasts. SV40-transformed cells should not be used in this assay, since p53 proficiency is required for a normal checkpoint response.

PBS, culture medium, and trypsin solution (as described earlier).

10 mM BrdU (Sigma B9285) stock in PBS. Aliquots can be stored at $-20°$ for 6 months. BrdU is light sensitive, so restrict light exposure as much as possible when executing the experiment by covering the samples with aluminum foil during long incubation steps (Steps 9, 19, and 22).

70% ethanol cooled at $-20°$.

α-BrdU-FITC antibody (BD Biosciences, San Jose, CA, catalog # 556028). FITC is light sensitive, so restrict light exposure as much as possible.

Propidium iodide solution: 25 μg/ml propidium iodide and 100 μg/ml RNase A in PBS. Propidium iodide is also light sensitive, so restrict light exposure as much as possible.

HCLT: 0.1 M HCl with 0.5% Tween 20.

PBSTB: PBS with 0.5% Tween and 1% BSA.

10-cm tissue culture plates, 15-ml screw cap plastic conical tubes, 1.5-ml snap cap tubes, and 5-ml round bottom tubes for flow cytometry.

Heating block at $-100°$.

A^{137}Cs irradiator at $\sim$2 Gy/min is used as a source for IR. An IR dose of 0.1 Gy generates about 4 DSBs in a human diploid genome (Cedervall *et al.*, 1995).

Controls

A $p53^{-/-}$ MEF culture is recommended as a positive control for a defective G1/S checkpoint.

Include controls for BrdU labeling (Data analysis section contains more details):

- Dish of cells labeled with BrdU and stained with α-isotype-FITC antibody.
- And/or dish of cells not labeled with BrdU and stained with α-BrdU-FITC antibody.

Mock treatment: dishes of cells prepared and handled identically to the irradiated samples, except for lack of IR treatment.

Method

1. For each treatment seed three 10-cm tissue culture dishes each with 0.8–1×10^6 MEFs. After 24 h the cells should be growing logarithmically and reaching subconfluency.

2. 24 h after plating, irradiate the cells with 10 Gy of IR. Mock treat a matched set of cells at the same time.

3. 14 h after treatment, pulse-label cells with 10 μM BrdU for 4 h.

4. Harvest cells from each dish using 1 ml of trypsin solution and transfer to individual 15-ml screw cap plastic conical tubes. Inactivate the trypsin by adding 4 ml of culture medium.

5. Pellet the cells in a swinging bucket centrifuge at $\sim400g$ for 5 min at RT.

6. Decant the supernatant, resuspend the cells in 5 ml of PBS, and re-spin at $\sim400g$ for 5 min at RT.

7. Aspirate PBS down to 250 μl and resuspend the cell pellets. If cells are not well resuspended, the cells will clump when treated with 70% ethanol during fixation.

8. Fix the cells by adding 3 ml of $-20°$ 70% ethanol while gently vortexing at 30% strength.

9. Place at $-20°$ for at least 2 hr in aluminum foil to restrict light exposure. The cells can be stored at $-20°$ for up to one week.

10. Pellet fixed cells by centrifugation at $\sim400g$ for 10 min at RT, and remove the ethanol by aspiration.

11. Resuspend the cells in 1 ml of PBS, and transfer the cells to 1.5-ml microcentrifuge tubes for ease of handling.

12. Spin the cells down in a tabletop centrifuge at $\sim700g$ for 5 min at $4°$, and remove all supernatant with a 1-ml pipette.

13. Add 0.5 ml HCLT while gently vortexing at 30% strength, then incubate on ice for 10 min.

14. Add 0.5 ml of water. Spin at ∼2500g for 5 min at 4°, and remove all supernatant with 1-ml pipette.

15. Add 1 ml of water while gently vortexing at 30% strength, and incubate at 100° for 10 min by placing the 1.5-ml tubes in a heating block.

16. Transfer to ice after boiling. Add 500 μl of PBSTB. Spin at ∼2500g for 5 min at RT, and remove all supernatant with 1-ml pipette.

17. Add 750 μl of PBSTB while gently vortexing at 30% strength. Spin at ∼2500g for 5 min at RT, and remove all supernatant with 1-ml pipette.

18. Resuspend in 90 μl of PBSTB + 10 μl of α-BrdU-FITC. Stain controls as outlined in the Controls section.

19. Cover the samples with aluminum foil to limit light exposure, and incubate at RT for at least 30 min.

20. Add 750 μl of PBSTB. Spin at ∼2500g for 5 min at RT, and remove all supernatant with 1-ml pipette.

21. Repeat Step 20 once.

22. Resuspend in 100–300 μl of propidium iodide solution and transfer cell suspension to 5-ml round bottom tube. The volume depends on the cell number: 0.5 $\times 10^6$ cells/250 μl runs easily on the flow cytometer. Cover the samples with aluminum foil to limit light exposure and incubate at RT for at least 15 min.

23. Run 1.0 $\times 10^5$ events for each sample on a FACS Calibur (BD Biosciences). Acquire α-BrdU-FITC in a logarithmic FL1-H channel and DNA content (propidium iodide) in a linear FL2-H or FL3-H channel. Also include FL2-Area/FL2-Width or FL3-Area/FL3-Width in your acquisition (Data analysis section contains more details).

Data Analysis

In this section, we will describe a procedure to process acquired data. First, the data is imported into FlowJo (Tree Star, Ashland, CO) or comparable flow cytometry software. As outlined in the Method section the α-BrdU-FITC signal is acquired in the FL1-H channel, whereas DNA content is acquired in the FL3-H channel. For more details on flow cytometry we recommend *Practical Flow Cytometry* (Shapiro, 2003).

Within the FlowJo platform, the dual FL2-Area/FL2-Width or FL3-Area/FL3-Width graph is used to eliminate aneuploid cells and nuclei doublets from further analysis (Fig. 2A). This elimination procedure can reduce the number of cells as much as 2.5 fold in $p53^{-/-}$ deficient cells, primarily due to aneuploidy (number of cells after gating, Table I). Subsequently the α-isotype-FITC control sample is used to set the gate for BrdU-positive cells (Fig. 2B). BrdU-positive cells are defined as those cells with FL1-H signals above the gate cut-off.

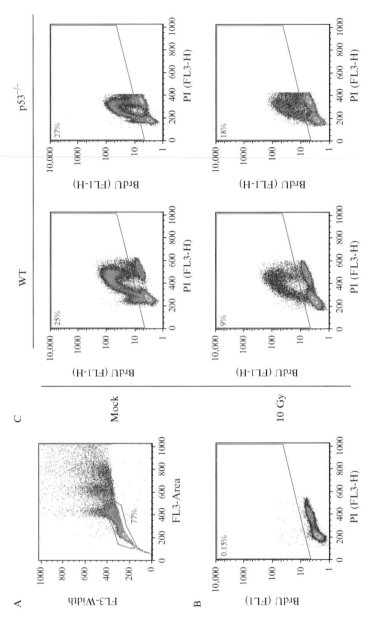

FIG. 2. (A). FL3-Width versus FL3-Area. The red gate is devoid of aneuploid cells and nuclei doublets and the percentage of cells in this gate is depicted in the graph. (B, C) α-BrdU-FITC (FL1-H) versus propidium iodide (PI) (FL3-H). The red gate contains α-BrdU-FITC positive cells and the percentage in this gate is depicted in the upper left hand corner of the graph. (B) Isotype control sample. (C) Mock and 10 Gy irradiated wild type (WT) and *p53*[−/−] cells. (See color insert.)

TABLE I
G1/S CHECKPOINT

Genotype	Treatment	Sample #	# Cells analyzed	# Cells after gating	BrdU + %	Normalized BrdU + %
WT	Mock	1	100,000	78,300	25	25/**26** = 0.96
WT	Mock	2	100,000	80,100	27	27/**26** = 1.04
WT	Mock	3	100,000	77,700	26	26/**26** = 1.00
				Average:	**26**	**1.00**
WT	IR	1	100,000	75,700	9	9/**26** = 0.34
WT	IR	2	100,000	74,800	7	7/**26** = 0.29
WT	IR	3	100,000	75,100	8	8/**26** = 0.29
				Average:	**8**	**0.31**
p53$^{-/-}$	Mock	1	100,000	54,700	21	21/**25** = 0.84
p53$^{-/-}$	Mock	2	100,000	55,700	28	28/**25** = 1.10
p53$^{-/-}$	Mock	3	100,000	54,800	27	27/25 = 1.07
				Average:	**25**	**1.00**
p53$^{-/-}$	IR	1	100,000	42,600	17	17/25 = 0.68
p53$^{-/-}$	IR	2	100,000	42,800	18	18/**25** = 0.72
p53$^{-/-}$	IR	3	100,000	43,700	18	18/**25** = 0.72
				Average:	**18**	**0.71**

Alternatively one can use the sample that was not labeled with BrdU and stained with the α-BrdU-FITC antibody to define the gate cut-off. Finally we apply the gate to our samples (Fig. 2C), and the percentage of BrdU positive cells in each sample is tabulated (Table I).

In order to compare the IR-induced G1/S checkpoint between different genotypes the data from FlowJo is normalized for each genotype by dividing all BrdU positive percentages by the average untreated BrdU-positive percentage (Table I).

To determine whether the differences between two genotypes are statistically significant, the Wilcoxon rank sum test is applied on the normalized IR treated triplicate samples from at least 3 experiments (at least 9 data points per genotype) (Table II).

Mstat software (written by Dr. Norman Drinkwater; http://mcardle. oncology.wisc.edu/mstat) can be used for this statistical test. In Mstat define a new variable for each genotype by choosing New from the Data menu. In the dialog box that appears, enter a variable name, select the Variable button, and type the data in the Values box as described in the Mstat tutorial (choose Help on Mstat in Help menu). When the data is entered, click on the Add button if you wish to define additional variables or the Done button. After entering the data for all genotypes, select Wilcoxon in the Test menu. When the dialog box appears select the two

TABLE II
WILCOXON RANK SUM TEST FOR G1/S CHECKPOINT

Experiment #	Sample #	WT genotype Normalized BrdU + % after IR	p53$^{-/-}$ genotype Normalized BrdU + % after IR
1	1	0.34	0.68
1	2	0.29	0.72
1	3	0.29	0.72
2	4	0.18	0.53
2	5	0.18	0.77
2	6	0.20	0.61
3	7	0.19	0.75
3	8	0.18	0.71
3	9	0.18	0.91

P (two-sided) = 2×10^{-5}.

genotypes that you want to compare in a two-sided test (i.e., test in which the direction for the difference between both populations is not specified).

Radioresistant DNA Synthesis

Introduction

The Mre11 complex and ATM are both required to transiently inhibit DNA replication when cells in S phase are irradiated. Radioresistant DNA synthesis (RDS) is an intrinsic feature of human and murine ATM and Mre11 complex deficient cells, and is indicative of a defect in the intra-S phase checkpoint (e.g., Painter, 1981; Theunissen et al., 2003).

In order to assay for RDS, asynchronous cells are prelabeled with [Methyl-^{14}C]-thymidine. After prelabeling, these cells are irradiated, and shortly after irradiation labeled with [Methyl-^{3}H]-thymidine. The resulting ^{3}H/^{14}C ratio is a measure of the rate of DNA synthesis. A higher rate of DNA synthesis after IR in cells of your genotype of interest compared to wild-type indicates RDS.

An accurate estimate of the cell number can only be achieved by incorporating ^{14}C prelabeling due to the variability in cell number generated by execution of the outlined method. Furthermore, omission of ^{14}C prelabeling precludes calculation of the rate of DNA synthesis after IR treatment as the ^{14}C count is the denominator in the rate of DNA synthesis.

Materials

Primary or SV40 transformed MEFs or ear fibroblasts. SV40 transformed cells provide more reproducible data than primary cells, most likely due to differences in proliferation rates between both cell types.

PBS, culture medium and trypsin solution (as in Preparation of MEFs).

[Methyl-^{14}C]-thymidine (50 μCi from GE Healthcare, Piscataway, NJ, catalog # CFA532) and [Methyl-^{3}H]-thymidine (1 mCi from GE Healthcare, catalog # TRK758).

Ice-cold 100% ethanol and 10% (w/v) trichloroacetic acid (TCA).

FH 225V 10-place filter manifold, 25 mm (GE Healthcare, product code 80629508).

25-mm glass microfibre filters (Whatman, Florham Park, NJ, catalog # 1822 025).

Scintillation vials each with 5 ml of liquid scintillation (LSC) cocktail (Aquasol from PerkinElmer, Wellesley, MA, catalog # 6NE9349).

A ^{137}Cs irradiator at ~2 Gy/min is used as a source for IR.

Controls

One plate with cells should be labeled with ^{14}C only. This enables determination of the fraction of ^{14}C energy that falls into the high-energy window (Data analysis section contains more details).

When performing this assay, use of *Atm*$^{-/-}$ SV40 transformed MEFs is recommended as a positive control for RDS.

Mock treatment: dishes of cells prepared and handled identically to the irradiated samples, except for lack of IR treatment.

Method

1. For each treatment, plate 2.5 × 10^5 cells onto one 10-cm dish and incubate in 37° incubator for 24 h.

2. 24 h after plating, label cells with 0.0125 μCi/ml [methyl-^{14}C]-thymidine for 24 h. The required length of labeling should at least be equal to the doubling time.

3. After the 24 h labeling period, remove ^{14}C-containing culture medium, and replate the cells from each 10-cm dish onto five 6-cm plates.

4. 24 h after plating, mock treat or irradiate the plates. Use 10 Gy of IR for a single treatment, or a range of 5–20 Gy of IR for multiple doses.

5. 1 h after IR treatment, remove the culture medium, and label cells by adding [methyl-^{3}H]-thymidine to the culture medium at 4 mCi/ml for 20 min. For an RDS assay with multiple time points, remove medium after 15 min–2 h.

6. Remove ^{3}H-containing medium from cells. Rinse once with 2 ml of PBS. Add 1 ml of trypsin solution.

7. Perform TCA precipitation on trypsinized cells as follows. The valves of the 10-place manifold should be closed and the manifold should be under vacuum. Wet glass microfibre filters with ice-cold 10% TCA and place 10 filters on the filter support grids of the 10-place filter manifold. Place the chimney weights on the filters.

8. Wash each filter with ice-cold 10% TCA by adding 2 ml into the chimney weight and opening each valve for about 10 s.

9. For each treatment, add the trypsinized cells from the five 6-cm dishes into separate chimney weights. Open each valve slowly for about 10 s.

10. Wash each filter with ice-cold 10% TCA as in Step 8.

11. Finally wash each filter with ice-cold 100% ethanol by adding 3 ml into the chimney weight and opening each valve for about 10 s.

12. Place each dry filter into a scintillation vial containing LSC-cocktail.

13. Repeat Steps 7–12 in order to process all samples.

14. For counting your samples use a dual program on a scintillation counter (Beckman Coulter, Fullerton, CA):

- The first program should contain a low-energy range (0–18.6 keV) that will include ^{3}H and some ^{14}C energy.

- The second program should contain a high-energy range (18.6–156 keV) that will only include ^{14}C energy.

NOTE

To increase reproducibility after ^{3}H labeling one can add cold 2.5 mM thymidine (Sigma T1895) to the rinse medium and the trypsin solution.

Data Analysis

Since the energy emission spectra of ^{3}H and ^{14}C overlap, adjusted ^{3}H and ^{14}C counts need to be calculated to transfer some of the ^{14}C energy from the low-energy window to the high-energy window. To adjust the raw low energy and high energy counts from Table III, the counts from the five control plates labeled with ^{14}C only are used to calculate the average fraction of ^{14}C energy that fell into the high-energy window: divide the average count from the high-energy window by the average count from the low- and high-energy windows combined (Table IV). This fraction (called ^{14}C fraction) should be between 60 and 80%.

The adjusted ^{3}H counts = (low energy counts) – [(high energy counts) × (1 – ^{14}C fraction)/(^{14}C fraction)]. And the adjusted ^{14}C counts = (high energy counts)/^{14}C fraction (Table IV).

In order to look at the rate of DNA synthesis, calculate the ratio of adjusted ^{3}H to ^{14}C counts. In order to compare RDS between different genotypes, these ratios need to be normalized for each genotype by dividing the mock and IR-treated ratios by the average mock ratio (Table III). Whereas in irradiated wild-type cells the normalized ratio is equal to 0.38,

TABLE III
INTRA S PHASE CHECKPOINT

Genotype	Treatment	Low energy counts (CPM)	High energy counts (CPM)	Adjusted ^{3}H counts (CPM)	Adjusted ^{14}C counts (CPM)	^{3}H/^{14}C ratio	Normalized ^{3}H/^{14}C ratio
WT	0 Gy	37,893	6587	34,933	9547	3.66	3.66/**4.46** = 0.82
WT	0 Gy	47,982	6899	44,882	9999	4.49	4.49/**4.46** = 1.01
WT	0 Gy	51,379	6863	48,296	9946	4.86	4.86/**4.46** = 1.09
WT	0 Gy	52,531	6844	49,456	9918	4.99	4.99/**4.46** = 1.12
WT	0 Gy	48,580	7258	45,320	10,519	4.31	4.31/**4.46** = 0.97
					Average:	**4.46**	**1.00**
WT	10 Gy	19,003	6665	16,008	9660	1.66	1.66/**4.46** = 0.37
WT	10 Gy	22,090	6699	19,081	9708	1.97	1.97/**4.46** = 0.44
WT	10 Gy	15,615	6906	12,512	10,008	1.25	1.25/**4.46** = 0.28
WT	10 Gy	20,023	7092	16,837	10,278	1.64	1.64/**4.46** = 0.37
WT	10 Gy	21,895	6937	18,779	10,054	1.87	1.87/**4.46** = 0.42
					Average:	**1.68**	**0.38**
Atm-/-	0 Gy	67,902	6337	65,055	9183	7.08	7.08/**6.48** = 1.09
Atm-/-	0 Gy	69,506	7075	66,327	10,253	6.47	6.47/**6.48** = 1.00
Atm-/-	0 Gy	80,418	7487	77,054	10,850	7.10	7.10/**6.48** = 1.10
Atm-/-	0 Gy	55,468	6662	52,475	9654	5.44	5.44/**6.48** = 0.84
Atm-/-	0 Gy	68,583	7126	65,381	10,328	6.33	6.33/**6.48** = 0.98
					Average:	**6.48**	**1.00**
Atm-/-	10 Gy	42,593	6302	39,761	9133	4.35	4.35/**6.48** = 0.67
Atm-/-	10 Gy	34,830	6680	31,828	9682	3.29	3.29/**6.48** = 0.51
Atm-/-	10 Gy	41,117	6227	38,319	9024	4.25	4.25/**6.48** = 0.65
Atm-/-	10 Gy	38,482	6637	35,500	9619	3.69	3.69/**6.48** = 0.57
Atm-/-	10 Gy	38,870	6919	35,761	10,028	3.57	3.57/**6.48** = 0.55
					Average:	**3.83**	**0.59**

TABLE IV
^{14}C FRACTION

Sample #	Low energy counts (CPM)	High energy counts (CPM)
1	3087	6988
2	2985	6768
3	3202	7243
4	3140	6623
5	643	1172
Average	2611	5759

^{14}C Fraction: $2611/(2611 + 5759) = 0.69$.

this ratio is increased up to 0.59 in $Atm^{-/-}$ cells, indicating RDS in the latter genotype.

To determine whether the differences between two genotypes are statistically significant, the Wilcoxon rank sum test is applied on the five normalized IR treated samples from at least 2 experiments (at least 10 data points per genotype) by using Mstat software as described in the Data analysis section of the G1/S Checkpoint Assay.

G2/M Checkpoint Assay

Introduction

ATM and the Mre11 complex are also required for the G2/M check-point. Initially the assay used for this checkpoint consisted of preparing and counting metaphases at different time points after IR treatment (Zampetti-Bosseler and Scott, 1981). A failure to decrease the number of mitotic cells after IR in your genotype of interest compared to wild type indicated a defect in this checkpoint.

Recently Xu et al. developed a rapid and sensitive flow cytometric assay in which propidium iodide is used as a DNA content marker to delineate G2 and mitotic cells from G1 and S-phase cells, and phosphorylated histone H3 is used as a mitotic marker to distinguish mitotic from G2 cells (Xu et al., 2001). At early time points following IR treatment (30 to 60 min after IR), one can use this assay to determine whether cells, which were in G2 at the time of IR, arrest in G2 or progress into mitosis. Wild-type G2 cells normally arrest in G2 for several hours, whereas checkpoint defective cells (e.g., $Atm^{-/-}$ cells) fail to arrest properly and enter mitosis despite the presence of DNA damage. Wild-type cells that arrest in G2 eventually recover and enter mitosis approximately 12 h after IR.

Below we will describe in detail an assay to monitor the early G2/M checkpoint. However one can also monitor the so-called late G2/M checkpoint. This checkpoint occurs approximately 24 h after IR, is molecularly distinct, reflects damage incurred during S phase, and is enhanced by RDS. This late G2/M checkpoint is assayed by staining the cells with propidium iodide and determining the percentage of total cells with a 4N DNA content (G2 and mitotic cells combined) (Xu *et al.*, 2002).

Unfortunately some investigators fail to make a distinction between the early and late G2/M checkpoints. In addition, we do not recommend a 5–12 h nocodazole treatment after IR when assaying for the early G2/M checkpoint, as cells that were in S phase at the time of IR treatment (Fig. 1) are also included in this procedure.

Material

Primary or SV40 transformed MEFs or ear fibroblasts. Both primary and SV40 transformed cells provide reproducible data.

PBS, culture medium and trypsin solution (as in Preparation of MEFs).

10-cm tissue culture tubes, 15-ml screw cap plastic conical tubes, 1.5-ml snap cap tubes, and 5-ml polystyrene round bottom tubes.

70% ethanol cooled at $-20°$.

Dilution buffer: 1% FCS and 0.1% Triton X-100 in PBS.

α-Histone H3-Ser10-P antibody (Upstate Biotechnology, Charlottesville, VA, catalog # 06570).

FITC-conjugated goat α-rabbit IgG antibody (Jackson Immunologicals, West Grove, PA). FITC is light sensitive, so restrict light exposure as much as possible.

Propidium iodide solution: 25 μg/ml propidium iodide and 100 μg/ml RNase A in PBS. Propidium iodide is also light sensitive, so restrict light exposure as much as possible.

$A^{137}Cs$ irradiator at $\sim$2 Gy/min is used as a source for IR.

Controls

- Dish of cells not labeled with α-histone H3-Ser10-P antibody, but labeled with goat α-rabbit IgG-FITC.
- And/or dish of cells labeled with α-histone H3-Ser10-P antibody, but not labeled with goat α-rabbit IgG-FITC (Data analysis section contains more details).
- Mock treatment: dishes of cells prepared and handled identically to the irradiated samples, except for lack of IR treatment.

Method

1. Plate in triplicate 7.5×10^5 primary MEFs or 5×10^5 SV40 transformed MEFs on 10 cm dishes. After 24 h the cells should be growing logarithmically and reaching subconfluency.

2. Mock treat or irradiate 24 h after plating with 1–20 Gy of IR.

3. Let the cells recover for 1–2 h in the $37°$ incubator.

4. Harvest the cells by trypsinization. Inactivate the trypsin with culture medium, and transfer the cells to 15-ml screw cap tubes. For cells that do not adhere well, the culture medium should also be collected.

5. Spin cells down in a swinging bucket centrifuge at $\sim400g$ for 5 min at RT, and remove medium by aspiration. Resuspend cells in 5 ml of PBS.

6. Spin cells down as in Step 5, and remove PBS by aspiration. Resuspend the cells in 250 μl of PBS.

7. Fix the cells by adding 3 ml of ice-cold 70% ethanol while gently vortexing at 30% strength.

8. Incubate at $-20°$ for at least 2 h.

9. Spin cells down as in Step 5, and remove ethanol by aspiration. Add 1 ml of dilution buffer while gently vortexing at 30% strength and transfer cell suspensions to 1.5-ml tubes for ease of handling.

10. Spin cells down in a tabletop centrifuge at $\sim400g$ for 5 min at RT, and remove dilution buffer with a 1-ml pipette.

11. Resuspend the cells in 200 μl of dilution buffer containing 1:200 α-Histone H3-Ser10-P. Incubate at RT for 1 h.

12. Spin cells down as in Step 10, and remove dilution buffer with a 1-ml pipette.

13. Resuspend the cells in 1 ml of dilution buffer, and repeat Step 12.

14. Resuspend the cells in 200 μl of dilution buffer containing 1:200 goat α-rabbit IgG-FITC. Cover the samples with aluminum foil to limit light exposure and incubate at RT for 1 h.

15. Spin cells down as in Step 10, and remove dilution buffer with a 1-ml pipette.

16. Resuspend the cells in 1 ml of dilution buffer, and repeat Step 15.

17. Resuspend the cells in 300 μl of propidium iodide solution. The volume depends on the cell number: 0.5×10^6 cells/250 μl runs easily on the flow cytometer. Cover the samples with aluminum foil to limit light exposure and incubate at RT for at least 15 min. You can store the cells covered in aluminum foil at $4°$ for up to 24 h before analysis.

18. Run the samples on a FACS Calibur (BD Biosciences). Acquire FITC in logarithmic FL1-H, and propidium iodide in linear FL3-H. Also include FL3-Area/FL3-Width (data analysis section contains more details).

Data Analysis

In this section, we will discuss a procedure to process acquired data. First, the data is imported in FlowJo (Tree Star) or comparable flow cytometry software. As outlined in the Method section, the Histone H3 signal is acquired in the FL1-H channel by virtue of goat α-rabbit IgG-FITC staining, whereas DNA content is acquired in the FL3-H channel.

Within the FlowJo platform the dual FL3-Area/FL3-Width graph is used to eliminate aneuploid cells and nuclei doublets from further analysis (Fig. 3A). This elimination procedure can reduce the number of cells as much as 1.5-fold (number of cells after gating, Table IV). Subsequently the no α-histone H3-Ser10-P antibody control sample is used to set the gate for histone H3 positive cells (Fig. 3B). Alternatively one can use the sample that was not labeled with α-rabbit IgG-FITC. Finally we apply this gate to our samples (Fig. 3C), and the percentage of Histone H3-Ser10-P positive cells is tabulated (Table V).

In order to compare the IR-induced G2/M checkpoint between different genotypes, the data are normalized for each genotype by dividing all acquired percentages by the average untreated percentage (Table V).

Metaphase Spreads

Introduction

Defective cell cycle checkpoints, especially in S phase are highly correlated with chromosomal instability (e.g., Myung *et al.*, 2001). Chromosomal instability in murine hypomorphs of the Mre11 complex (*Mre11*$^{ATLD1/}$ ^{ATLD1} and *Nbs1*$^{\Delta B/\Delta B}$) consists primarily of chromatid and chromosomal breakage (Theunissen *et al.*, 2003).

The majority of cytogenetic techniques to study chromosomal instability involve staining and microscopic observation of metaphase chromosomes (Rooney, 2001a,b). To increase the yield of mitotic cells containing metaphase chromosomes, cells are treated with an agent that inhibits tubulin polymerization (e.g., colcemid/demecolcine) or causes microtubule depolymerization (e.g., nocodazole).

Material

Colcemid. Make a 10,000× stock (0.74 mg/ml) in water. The final 1× concentration is 74×10^{-6} mg/ml $= 2 \times 10^{-7}$ M. The duration of treatment should be optimized for each cell type and is dependent upon doubling time. Whereas a short treatment period on slowly dividing cells does not effectively enrich for mitotic cells, a long treatment period on rapidly dividing cells causes significant chromosome hypercondensation.

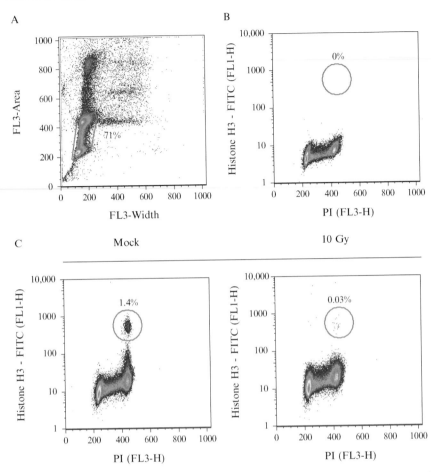

Fig. 3. G2/M checkpoint. (A) FL3-Area versus FL3-Width. The red gate is devoid of aneuploid cells and nuclei doublets and the percentage in this gate is depicted in the graph. (B, C) Histone H3 – FITC (FL1-H) versus propidium iodide (PI) (FL3-H). The red gate encircles Histone H3 – FITC positive cells and the percentage in this gate is depicted in the graph. (B) Control sample not stained with α-Histone H3 antibody (C) Mock and 10 Gy irradiated wild type (WT) cells taken 1 h after treatment. (See color insert.)

75 mM KCl. Add 1.1 g of KCl to 200 ml of ultrapure water (milli Q from Millipore, Billerica, MA). Make fresh and prewarm at 37° for 15 min. This hypotonic solution will increase cell volume, and ultimately facilitate spreading of chromosomes onto the glass slides.

Fixative: 75% methanol, 25% acetic acid. Make fresh and pre-cool on ice for 30 min.

TABLE V
G2/M CHECKPOINT

Genotype	Treatment	Sample #	# Cells analyzed	# Cells after gating	FITC+ (%)	Normalized FITC+
WT	Mock	1	10,0000	76,021	1.4	1.4/**1.4** = 1.04
WT	Mock	2	10,0000	78,030	1.2	1.2/**1.4** = 0.91
WT	Mock	3	10,0000	75,732	1.4	1.4/**1.4** = 1.05
				Average:	**1.4**	**1.00**
WT	IR	1	10,0000	76,922	0.03	0.03/**1.4** = 0.02
WT	IR	2	10,0000	76,055	0.01	0.03/**1.4** = 0.01
WT	IR	3	10,0000	73,577	0.01	0.01/**1.4** = 0.01
				Average:	**0.01**	**0.01**

A 0.4% stock solution of Giemsa (Sigma GS500), from which a 20-fold dilution (0.02% final concentration) is freshly prepared and filtered through 3-mm chromatography paper (Whatman).

Microscope slides placed at $-20°$ ($25 \times 75 \times 1$ mm; Superfrost/Plus Microscope slides from Fisher, Hampton, NH), 22×50 mm cover glass (No.1 from Corning, NY), Permount (Fisher), and nail polish. A ^{137}Cs irradiator at ~2 Gy/min is used as a source for IR. An IR dose of 0.1 Gy generates about 4 DSBs in a human diploid genome (Cedervall et al., 1995).

If using primary splenocytes; lipopolysaccharide (LPS): make a 25 mg/ml stock in water. Store frozen aliquots of LPS at $-80°$.

Method for Preparing Cells
I. Adherent cells (e.g., MEFs, ear fibroblasts)

1. Plate 1×10^6 cells on a 10-cm dish. After 24 h the cells should be growing logarithmically and reaching subconfluency.
2. At least 24 h after plating, mock treat or irradiate the cells with 0.5–2 Gy.
3. 1 to 20 h after treatment, add colcemid to a final concentration of $2 \times 10^{-7} M$. For MEFs or mouse ear fibroblasts, treatment time is typically 4 h.
4. After colcemid treatment, transfer the culture medium, which may contain detached mitotic cells, to a 15-ml conical tube. Trypsinize and transfer the adherent cells to the same 15-ml conical tubes that contain the detached cells. Spin cells down at ~400g for 5 min at RT.
5. Remove the culture medium by aspiration, and resuspend the cells in 150 μl of PBS.
6. Add 10 ml of $37°$ 75 mM KCl. The first 2 ml should be added drop wise to the cells. Incubate this solution at $37°$ for 15 min.

II. Primary splenocytes

1. Seed 5–7.5 × 10⁶ cells at 500,000 cells/ml in a tissue culture flask, and stimulate primary splenocytes (as harvested in Preparation of Splenocytes and Thymocytes) with LPS at a final concentration of 25 μg/ml.

2. 48 h after stimulation, mock treat or irradiate the cells with 0.5–2 Gy.

3. 1 to 20 h after treatment, add colcemid to a final concentration of 2×10^{-7} M.

4. 3 h later, pipette all cells into a 15-ml conical tube and spin at ∼400g for 5 min at RT.

5. Remove the culture medium by aspiration, and resuspend the cells in 150 μl of PBS.

6. Add 10 ml of 37° 75 mM KCl. The first 2 ml should be added dropwise to the cells. Incubate this solution at 37° for 7 min.

Method for Fixation, Slide Preparation, and Staining

1. After the incubation in KCl, add without mixing a layer of 5 ml ice-cold fixative to the top of the conical tube with a pipette. Pellet the cells by spinning at ∼275g for 5 min at RT. If fixative is not added at this step, the cells will clump during subsequent fixation.

2. Remove the supernatant by gentle aspiration down to 250 μl, and flick the tube to resuspend the cell pellet. Add 10 ml of fixative with a pipette. The first 2 ml should be added dropwise to the cells while flicking the tube. Incubate this solution on ice for 5 min.

3. Spin the cells down at ∼275g for 5 min at RT, and remove fixative by aspiration. Add 10 ml of fresh fixative and flick the tube to resuspend the cell pellet.

4. Repeat Step 3 once.

5. After the last wash, incubate the cells for about 15 min. Alternatively the cells can be stored at −20° for up to one month.

6. Spin the cells down at ∼275g for 5 min at RT, and remove fixative by aspiration. Resuspend the fixed cells in 500 μl of fixative with a pipette.

7. Remove the microscope slides from −20°. With a 200-μl pipette, drop 2 to 3 drops of cells from a height of 1–2 in onto a clean, prechilled glass slide. Wait until the slide is visually dry. Look under a phase microscope to determine what the cell density on the slide is. If needed, adjust the cell concentration with fixative.

8. Leave the slides overnight at RT to let them dry.

9. The next day, stain the slides in a solution of fresh 0.02% Giemsa for 10 min.

10. Rinse the slides under a gentle stream of water or dunk the slides into a Coplin jar filled with water and incubate for 1 min. Let the slides dry

for at least 1 h and mount the slides with a few drops of Permount and cover glass. Seal the slides at least 15 min later with nail polish to prevent the Permount from drying out. Alternatively stain the slides with DAPI as outlined in the Immunofluorescence section.

Statistical Concepts

Before discussing how to score chromosomal aberrations, we will briefly describe some statistical concepts. In the current literature a P value below 0.05 is generally accepted as an indicator for statistical significance. The P value or significance level is defined as the probability of obtaining the observed result or a more extreme result in the scenario in which both genotypes are equal. If the P value is higher than 0.05 between genotypes A and B, one could argue that the sample sizes were too small to detect significant differences. Before performing an experiment, one should consider the power of the statistical test that will be used. Power is defined as the probability that an experiment of a given sample size will detect statistical significance between two groups. Usually the power of a statistical test should be higher than 90%.

In the example in Fig. 4A the sample sizes for genotypes A and B are 20 metaphases. Six abnormal metaphases are detected for genotype A (0.3 or 30% chromosomal instability) and 11 for genotype B (0.55 or 55% chromosomal instability). The P-value for this experiment is calculated in Mstat (http://mcardle.oncology.wisc.edu/mstat). In Mstat, enter the 2 by 2 table from Fig. 4A by selecting New from the Data menu. In the dialog box that appears, enter a variable name, select the Table button under Variable type, and type the table in the Values box as outlined in the Mstat tutorial (choose Help on Mstat in Help menu). When the data are entered, click on the Done button. Select Fisher's exact in the Test menu. When the dialog box appears choose the entered table and select the two-sided button. When performing the Fisher's exact test for this experiment in Mstat the P value is 0.20, and thus there is no statistical difference between both genotypes.

As mentioned in the first paragraph, one should consider the power of the statistical test. Using a power value of 90% and a significance level of 0.05, one can determine the required sample sizes to detect statistical significance for the experiment in Fig. 4A. In Mstat select Sample Size Fisher in the Design menu. In the dialog box that appears, type 0.05 for Significance Level, select two-sided, 0.9 for Power, 0.3 for the Lower Proportion (genotype A with 30% chromosomal instability), and 0.55 for the Higher Proportion (genotype B with 55% chromosomal instability). After clicking OK, you can see that the sample sizes for genotypes A and B should be equal to 88 to detect statistical significance.

A

Genotype	Sample size	# of normal metaphases	# of aberrant metaphases	% of chrom. instability
A	20	14	6	30
B	20	9	11	55

P (two-sided test) = 0.20
Sample sizes for significance level of 0.05 and
power of 90% = 88

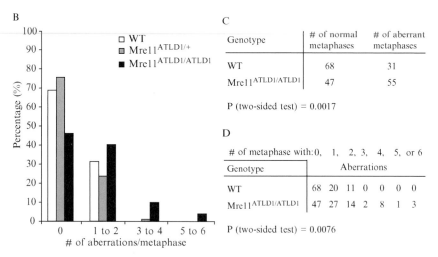

C

Genotype	# of normal metaphases	# of aberrant metaphases
WT	68	31
Mrel1$^{ATLD1/ATLD1}$	47	55

P (two-sided test) = 0.0017

D

Genotype	# of metaphase with: 0, 1, 2, 3, 4, 5, or 6 Aberrations						
WT	68	20	11	0	0	0	0
Mrel1$^{ATLD1/ATLD1}$	47	27	14	2	8	1	3

P (two-sided test) = 0.0076

FIG. 4. Normal and aberrant metaphases in p3 ear fibroblasts harvested 5 h after 1 Gy of IR (A) Fisher's exact test on the number of normal and aberrant metaphases (1 or more aberrations/metaphase) between A and B. Includes sample size calculation for groups A and B. (B) Percentage of metaphases versus the number of aberrations/metaphase. The metaphases are subdivided in four categories of 0, 1 to 2, 3 to 4, and 5 to 6 aberrations/metaphase. (C) Fisher's exact test on the number of normal and aberrant metaphases (1 or more aberrations/metaphase) between wild type (WT) and *Mrel1$^{ATLD1/ATLD1}$* . The P-value is indicated for a two-sided test. (D) Chi-square test on number of metaphases with 0, 1, 2, 3, 4, 5, or 6 aberrations between WT and *Mrel1$^{ATLD1/ATLD1}$* ear fibroblasts. The P-value is indicated for a two-sided test.

Data Analysis

When scoring for aberrations we advise you to blind your samples. Secondly, score at least 75 metaphases from each genotype or perform a power calculation as described in the previous paragraph on a small sample set in a pilot experiment. Score metaphases as normal or aberrant on the microscope. If your microscope camera is capable of acquiring high-resolution pictures, scoring in Photoshop (Adobe) is possible. For chromosome counts, tick each chromosome off in Photoshop with the paintbrush tool while using a counting device.

In this paragraph we will discuss chromosomal instability in mouse cells. Each mouse metaphase contains 40 chromosomes, all of which are acrocentric (i.e., centromere situated close to one chromosome end). Figure 5 depicts metaphases that illustrate the types of aberrations one might see. In $Atm^{-/-}$ and $Mre11^{ATLD1/ATLD1}$ cells, the increase in chromosomal instability compared to wild type is primarily due to the presence of chromatid and chromosomal breaks that are detected as such (black arrowheads in Fig. 5B and D), or result in the formation of fragments (white arrowheads in Fig. 5B and E). Aside from breaks and fragments, metaphases can also carry end-to-end fusions or exchanges (black and white arrows in Fig. 5, respectively).

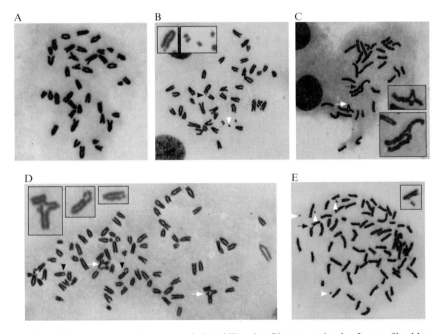

FIG. 5. Spontaneous chromosomal instability in Giemsa stained p3 ear fibroblast metaphases. Black arrowheads: chromatid or chromosomal breaks; white arrowheads: fragments; black arrows: end-to-end fusions; white arrows: exchanges (exchanges counted as two chromosomes). Some aberrations are magnified and represented separately for clarity (A) Wild type (WT) metaphase; 39 chromosomes. (B) $Mre11^{ATLD1/ATLD1}$; 40 chromosomes (not counting four fragments). (C) $Rad50^{S/S}$ $Mre11^{+/ATLD1}$; 34 chromosomes. (D) $Rad50^{S/S}$; 80 chromosomes. (E) $Rad50^{S/S}$ $Mre11^{ATLD1/ATLD1}$; 78–80 chromosomes.

In addition cultured mouse ear fibroblasts and MEFs undergo spontaneous aneuploidy. The metaphases in Fig. 5D and E each contain 80 chromosomes. A large portion of metaphases will also contain fewer than 40 chromosomes (Fig. 5A and C). Acquisition of metaphase pictures in which chromosomes overlap (Fig. 5E) or are over-condensed is not advised, since it interferes with proper scoring.

Once the data are recorded in a spreadsheet, the data can be graphed as depicted in Fig. 4B or tabulated (Theunissen *et al.*, 2003). Categorizing the metaphases according to the number of aberrations/metaphase is recommended over dividing the total number of aberrations by the total number of metaphases analyzed. In the former calculation less information is lost compared to the latter one.

To determine whether there are statistically significant differences between genotypes, the Fisher's exact test is applied on the 2 by 2 table as in Fig. 4C. Alternatively one can use the chi-square test on a 2 by $\times$ table, in which $\times$ is equal to the number of aberrations/metaphase (Fig. 4D). In Mstat, select Chi square in the Test menu.

Immunofluorescence

The cellular response to DSBs can be visualized by performing immunofluorescence experiments on proteins involved in this process, since they migrate to the sites of damage and accumulate into discrete foci (Maser *et al.*, 1997; Nelms *et al.*, 1998).

At early time points proceeding DSB formation (20 min to 2 h), the coalescence of certain DSB-associating proteins into discrete foci is obscured by their abundant nucleoplasmic pools (e.g., members of the Mre11 complex). In order to visualize localization of such abundant proteins into early foci, gentle pre-extraction techniques have been developed (Fig. 6A; Section on indirect immunofluorescence with pre-extraction) (Mirzoeva and Petrini, 2001, 2003). Appearance of these foci can be used to study the order and dependence of protein recruitment to DSBs.

At later time points after DSB formation (4 to 12 h post IR treatment), large discrete IR-induced foci (abbreviated as IRIF) can be detected without pre-extraction even if the protein is abundant (Figs. 6B and 7; Section on indirect immunofluorescence without pre-extraction) (Maser *et al.*, 1997; Williams *et al.*, 2002). The occurrence and number are affected by both the DSB repair proficiency of the cells under study and the IR dose. Although these late foci participate in DSB metabolism, they do not provide insight into recruitment of proteins to DSBs, as most DSBs are repaired within 90 min after induction. Most likely these late foci

A   Early foci   B   Late foci

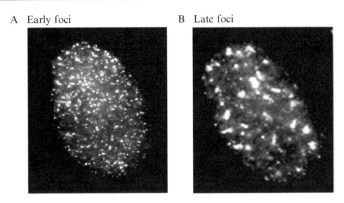

FIG. 6. Early and late foci of Mre11 in primary human fibroblasts. (A) Pre-extraction 20 min after IR. (B) Pre-extraction 8 hr after IR. Indirect immunofluorescence without pre-extraction gives the same localization pattern 8 hr after IR.

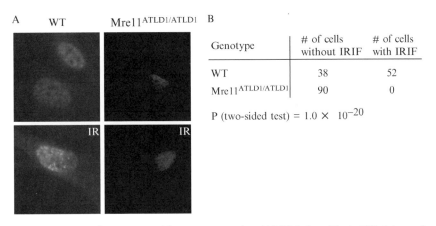

Genotype	# of cells without IRIF	# of cells with IRIF
WT	38	52
Mre11$^{ATLD1/ATLD1}$	90	0

P (two-sided test) = 1.0×10^{-20}

FIG. 7. Immunofluorescence without pre-extraction. (A) IR-induced foci of Nbs1 in mock and IR-treated wild type (WT) and $Mre11^{ATLD1/ATLD1}$ MEFs 8 h after 12 Gy of IR. (B) Fisher's exact test on the number of IRIF negative and positive cells 8 h after 12 Gy. The P-value is indicated for a two-sided test.

represent irreparable or slowly repaired lesions (Petrini and Stracker, 2003).

Controls

In single labeling experiments, prepare controls by omitting the primary antibody or replacing it with preimmune serum.

In double labeling experiments, prepare two controls by omitting one or the other primary antibody and including both secondary antibodies.

In double labeling experiments, also prepare controls for cross-reactivity of each secondary antibody against primary antibodies of different species origin.

Indirect Immunofluorescence Without Pre-Extraction

This protocol is for adherent cells that can attach to and grow on glass slides. Cells in suspension will require a cytospin or some other method of attachment as described in *Using Antibodies: A Laboratory Manual* (Harlow and Lane, 1999).

Materials

Human or murine (MEFs or ear fibroblasts) adherent cells.
Slides:
- 4- or 6-well slides from Erie Scientific Co. (Portsmouth, NH, catalog # 10-749S). Multiple antibodies can be processed on the same slide. Sterilize by autoclaving or UV treatment.
- Alternatively 25 × 75 × 1 mm Superfrost/Plus Microscope slides (Fisher Scientific), or cover glass 18 mm sq or 22 × 50 mm (No. 1 thickness from Corning). When plating cells on cover glass, make a circle around the border of the glass using a liquid-repellent slide marker pen (Electron Microscopy Sciences, Hatfield, PA, catalog # 713010) or silicone grease and UV treat for 5 min to sterilize.

For methanol fixation: methanol and acetone, both at $-20°$.

For paraformaldehyde (PFA) fixation; 4% PFA solution.
- Add 4 g PFA in 90 ml of sterile water in a fume hood.
- Dissolve PFA by adding 0.5 ml of 10 M NaOH, stirring, and heating to about $60°$.
- Filter the solution through 3-mm chromatography paper (Whatman) paper and cool the solution down on ice.
- Add 10 ml of 10× PBS.
- When cool, pH the solution to pH 7.5 with a small amount of sulfuric acid. Add sucrose to 2%.
- Important note: PFA must be used the day you make it. Alternatively, prepare a large batch, aliquot, and store at $-80°$ for a few months.

For PFA fixation; Permeabilization Solution
- 50 mM NaCl
- 3 mM MgCl$_2$

- 200 mM sucrose
- 10 mM HEPES pH 7.9
- 0.5% TX-100.

Blocking solution: 10% FCS in PBS

Antibody dilution buffer (ADB): 2–5% FCS in PBS or 1% BSA in PBS.

For 10 ml of fluorescent mounting medium:

- Dissolve 10 mg of p-phenylene diamine (Sigma D9542) in 1 ml of water.
- Vortex vigorously for 2–3 min to dissolve completely.
- Add 1 ml of 10× PBS and 8 ml of Glycerol. Mix well.
- Aliquot in amber (opaque) tubes. Store at −80° and use for up to 6 months.

Coplin jars, humidified chamber, plastic coverslips.

DNA counterstaining agent; a stock of 1 mg/ml of 4′,6′diamino-2-phenylindole (DAPI) in water.

Method

I. Growth of cells on slides or cover glasses:

This will require sterile slides or cover glass in some type of sterile dish. One can fit 5 slides in a 15-cm dish or six 18 mm sq cover glasses in the wells of a 6-well dish.

1. Trypsinize a 10-cm plate of subconfluent cells.

2. Make a cell dilution of 2.5×10^4 cells/ml and spot 100 μl on each well of a 4- or 6-well slide or 200 μl on 18 mm sq cover glass. If the cells do not adhere, gelatinize the glass or treat it with alcian blue (Section Alcian Blue Treatment to Increase Adherence of Cells to Glass). After 24 h the cells should be growing logarithmically and reaching subconfluency. The number of cells should be optimized for each cell type.

3. After at least 24 h, mock treat or irradiate the cells with 12 Gy or lower doses of IR.

II. Fixation

For each antibody, determine empirically what fixation method works best.

Methanol Fixation:

1. 6–10 h after IR treatment remove media with a brief wash at RT in a Coplin jar containing PBS.

2. Immerse the slides in a Coplin jar containing methanol (−20°) for 20 min.

3. Immerse the slides in a Coplin jar containing acetone (–20°) for 10–20 s, and allow the slides to dry briefly.

4. Wash slides 3 times for 5 min each at RT in a Coplin jar containing PBS.

Paraformaldehyde Fixation:

1. 6–10 h after IR treatment remove media with a brief wash at RT in a Coplin jar containing PBS.

2. Fix cells for 10 min at RT in a Coplin jar containing a fresh PFA solution.

3. Wash slides 2 times for 5 min each at RT in a Coplin jar containing PBS.

4. Permeabilize the cells in the permeabilization solution for 5 min at RT.

5. Wash slides 3 times for 5 min each at RT in a Coplin jar containing PBS. Slides can be stored overnight in PBS at 4°.

III. Immunodetection

1. Incubate the cover glasses or slides in blocking solution for 1 h at RT or overnight at 4° in a humidified chamber. Use a solution of 10% FCS in PBS. Other blocking solutions may work as well.

2. Wash slides 3 times for 5 min each at RT in a Coplin jar containing PBS.

3. Dilute the primary antibody of interest in ADB and add 75 μl to each cover glass. For multi-well slides it typically takes 25 μl of antibody solution to cover the cells from one well. For regular microscope slides add 200 μl of antibody solution and cover the slide with a plastic cover slip to ensure even coverage.

4. Incubate for 1 h at RT or overnight at 4° in a humidified chamber.

5. Wash 3 times in PBS for 5 min each at RT.

6. Dilute the secondary antibody as outlined in point 3.

7. Incubate for 45 min at RT.

8. Wash 3 times in PBS for 5 min each at RT.

9. After the last wash, counter stain with 0.1 μg/ml DAPI for 1 min at RT.

10. Wash once in PBS for 1 min at RT to remove excess DNA stain.

11. Mount cover glasses onto slides: add a few drops of fluorescent mounting medium on a 25 × 75 mm slide. When placing the cover glass on the slide, the cells should end up between the glass surfaces.

12. Seal the cover glass onto the slide with nail polish.

13. View under epifluorescence.

Cover the slides with aluminum foil to limit light exposure. Stored at 4°, the slides keep reasonable signal for 2 to 4 weeks.

*Indirect Immunofluorescence with Pre-Extraction (*In Situ *Cell Fractionation)*

This protocol is optimized for adherent human cells.

1. Growth of cells (preferentially on cover glass) and Immunodetection performed as in Section Indirect Immunofluorescence Without Pre-extraction.
2. Pre-extraction and Fixation.

Materials

Adherent human cells: 37Lu or IMR90 normal human diploid fibroblasts. HeLa cells are more problematic since they are less flat than primary fibroblasts.
Cytoskeleton buffer (made fresh):

- 10 mM PIPES, pH 6.8
- 100 mM NaCl
- 300 mM sucrose
- 3 mM MgCl$_2$
- 1 mM EGTA
- 0.5% Triton X-100 (v/v).

Cytoskeleton stripping buffer (made fresh):

- 10 mM Tris-HCl, pH 7.4
- 10 mM NaCl
- 3 mM MgCl$_2$
- 1% Tween 40 (v/v)
- 0.5% sodium deoxycholate (w/v).

Streck tissue fixative:

- 150 mM 2-bromo-2-nitro-1, 3-propanediol (Sigma B0257)
- 108 mM diazolidinyl urea (Sigma D5146)
- 10 mM Na Citrate
- 50 mM EDTA
- pH to 5.7
- Aliquot and freeze at $-20°$.

Permeabilization solution:

- 100 mM Tris-HCl, pH 7.4
- 50 mM EDTA
- 0.5% Triton X-100 (v/v).

Note: composition of extraction buffers as in Nickerson *et al.*, 1990.

Method for Pre-Extraction and Fixation

1. 10 min–2 h after treatment, wash the cover glasses twice with ice-cold PBS.
2. Treat with ice-cold cytoskeleton buffer for 5 min on ice.
3. Aspirate and add ice-cold cytoskeleton stripping buffer for 5 min on ice.
4. Wash three times for 5 min each with ice-cold PBS.
5. Fix cells in Streck tissue fixative for 30 min at RT.
6. Wash in PBS for 5 min at RT.
7. Permeabilize in permeabilization solution for 15 min at RT.
8. Wash 3 times in PBS for 5 min each at RT.

Alcian Blue Treatment to Increase Adherence of Cells to Glass

Treat slides/cover glass for at least 1 h with a 1 mg/ml solution of Alcian Blue 8 GX (Sigma # A5268).

Wash slides thoroughly with water until all traces of dye are gone; some microscopic specks of dye might remain, but these do not interfere with subsequent steps.

Sterilize by dipping in 95% ethanol and incubate the slides at RT until they are visually dry before plating the cells.

Data Analysis

In essence data analysis is similar to the procedure outlined for metaphase spreads in the Metaphase Spreads section. Briefly, we recommend blinding the samples, scoring at least 75 randomly acquired cells for each genotype on the microscope or in Photoshop, and executing statistical analyses as described for metaphase spreads. For comparisons between genotypes, images should be acquired on the microscope with the same camera exposure time, and then analyzed in Photoshop without any or uniformly applied manipulations to each image. In addition, scoring more than one well/slide per treatment is recommended, due to possible well/slide variability.

In the experiment in Fig. 7 we scored for the presence of IR-induced foci of Nbs1 in untreated and irradiated wild type and *Mre11*$^{ATLD1/ATLD1}$ MEFs. We defined cells with IRIFs as those that contain at least 10 bright foci, as seen in the IR-treated wild type cell in Fig. 7A. The data for 90 cells for each genotype was categorized in a 2 by 2 table and a Fisher's exact test was executed in Mstat as described in the Metaphase Spreads section (Fig. 7B).

Acknowledgments

This work was supported by GM56888 and the Joel and Joan Smilow Initiative. J. W. F. T. is supported by a Frank Lappin Horsfall, Jr. Fellowship. The authors are grateful for critical reading of the manuscript by members of the Petrini laboratory. We apologize for not being able to cite comprehensibly due to space constraints.

References

Bakkenist, C. J., and Kastan, M. B. (2004). Initiating cellular stress responses. *Cell* **118**, 9–17.

Cedervall, B., Wong, R., Albright, N., Dynlacht, J., Lambin, P., and Dewey, W. C. (1995). Methods for the quantification of DNA double-strand breaks determined from the distribution of DNA fragment sizes measured by pulsed-field gel electrophoresis. *Radiat. Res.* **143**, 8–16.

Fromm, M., and Berg, P. (1982). Deletion mapping of DNA regions required for SV40 early region promoter function *in vivo*. *J. Mol. Appl. Genet.* **1**, 457–481.

Harlow, E., and Lane, D. (1999). "Using Antibodies: A Laboratory Manual" Cold Spring Harbor Laboratory Press, Cold Spring Harbor, N.Y.

Kuerbitz, S. J., Plunkett, B. S., Walsh, W. V., and Kastan, M. B. (1992). Wild-type p53 is a cell cycle checkpoint determinant following irradiation. *Proc. Natl. Acad. Sci. USA* **89**, 7491–7495.

Maser, R. S., Monsen, K. J., Nelms, B. E., and Petrini, J. H. (1997). hMre11 and hRad50 nuclear fociare induced during the normal cellular response to DNA double-strand breaks. *Mol. Cell. Biol.* **17**, 6087–6096.

Mirzoeva, O. K., and Petrini, J. H. (2001). DNA damage-dependent nuclear dynamics of the mre11 complex. *Mol. Cell. Biol.* **21**, 281–288.

Mirzoeva, O. K., and Petrini, J. H. (2003). DNA replication-dependent nuclear dynamics of the Mre11 complex. *Mol. Cancer Res.* **1**, 207–218.

Myung, K., Datta, A., and Kolodner, R. D. (2001). Suppression of spontaneous chromosomal rearrangements by S phase checkpoint functions in *Saccharomyces cerevisiae*. *Cell* **104**, 397–408.

Nagy, A. (2003). "Manipulating the Mouse Embryo: A Laboratory Manual," 3rd Ed. Cold Spring Harbor Laboratory Press, Cold Spring Harbor, N.Y.

Nelms, B. E., Maser, R. S., MacKay, J. F., Lagally, M. G., and Petrini, J. H. J. (1998). *In situ* visualization of DNA double-strand break repair in human fibroblasts. *Science* **280**, 590–592.

Nickerson, J. A., Krockmalnic, G., He, D. C., and Penman, S. (1990). Immunolocalization in three dimensions: Immunogold staining of cytoskeletal and nuclear matrix proteins in resinless electron microscopy sections. *Proc. Natl. Acad. Sci. USA* **87**, 2259–2263.

Painter, R. B. (1981). Radioresistant DNA synthesis: An intrinsic feature of ataxia telangiectasia. *Mutat. Res.* **84**, 183–190.

Petrini, J. H., and Stracker, T. H. (2003). The cellular response to DNA double-strand breaks: Defining the sensors and mediators. *Trends Cell. Biol.* **13**, 458–462.

Rooney, D. E. (2001a). "Human cytogenetics: Constitutional analysis: A practical approach," 3rd ed. Oxford University Press, Oxford.

Rooney, D. E. (2001b). "Human cytogenetics: Malignancy and acquired abnormalities: A practical approach," 3rd ed. Oxford University Press, Oxford; New York.

Schonthal, A. (2004a). "Checkpoint Controls and Cancer -Volume 1: Reviews and Model Systems," Humana Press, Totowa, NJ.

Schonthal, A. (2004b). "Checkpoint Controls and Cancer -Volume 2: Activation and Regulation Protocols," Humana Press, Totowa, NJ.

Shao, C., Deng, L., Henegariu, O., Liang, L., Raikwar, N., Sahota, A., Stambrook, P. J., and Tischfield, J. A. (1999). Mitotic recombination produces the majority of recessive fibroblast variants in heterozygous mice. *Proc. Natl. Acad. Sci. USA* **96,** 9230–9235.

Shapiro, H. M. (2003). "Practical Flow Cytometry," 4th ed. Wiley-Liss, Hoboken, N.J.

Silver, L. M. (1995). "Mouse Genetics: Concepts and Applications." Oxford University Press, New York.

Stracker, T. H., Theunissen, J. W., Morales, M., and Petrini, J. H. (2004). The Mre11 complex and the metabolism of chromosome breaks: The importance of communicating and holding things together. *DNA Repair (Amst)* **3,** 845–854.

Theunissen, J. W., Kaplan, M. I., Hunt, P. A., Williams, B. R., Ferguson, D. O., Alt, F. W., and Petrini, J. H. (2003). Checkpoint failure and chromosomal instability without lymphomagenesis in Mre11(ATLD1/ATLD1) mice. *Mol. Cell* **12,** 1511–1523.

Williams, B. R., Mirzoeva, O. K., Morgan, W. F., Lin, J., Dunnick, W., and Petrini, J. H. (2002). A murine model of nijmegen breakage syndrome. *Curr. Biol.* **12,** 648–653.

Xu, B., Kim, S., and Kastan, M. B. (2001). Involvement of Brca1 in S-phase and G(2)-phase checkpoints after ionizing irradiation. *Mol. Cell. Biol.* **21,** 3445–3450.

Xu, B., Kim, S.-T., Lim, D.-S., and Kastan, M. B. (2002). Two molecularly distinct G2/M checkpoints are induced by ionizing irriadation. *Mol. Cell. Biol.* **22,** 1049–1059.

Xu, Y., and Baltimore, D. (1996). Dual roles of ATM in the cellular response to radiation and in cell growth control. *Genes Dev.* **10,** 2401–2410.

Zampetti-Bosseler, F., and Scott, D. (1981). Cell death, chromosome damage and mitotic delay in normal human, ataxia telangiectasia and retinoblastoma fibroblasts after x-irradiation. *Int. J. Radiat. Biol. Relat. Stud. Phys. Chem. Med.* **39,** 547–558.

[16]   Detecting Repair Intermediates *In Vivo*: Effects of DNA Damage Response Genes on Single-Stranded DNA Accumulation at Uncapped Telomeres in Budding Yeast

By MIKHAJLO K. ZUBKO, LAURA MARINGELE,
STEVEN S. FOSTER, and DAVID LYDALL

Abstract

Single-stranded DNA (ssDNA) is an important intermediate in many DNA repair pathways. Here we describe protocols that permit the measurement of ssDNA that has arisen in the yeast genome *in vivo*, in response to telomere uncapping. Yeast strains defective in DNA damage response (DDR) genes can be used to infer the roles of the corresponding proteins in regulating ssDNA production and in responding to ssDNA. Using column based methods to purify yeast genomic DNA and quantitative amplification of single-stranded DNA (QAOS) it is possible to measure ssDNA at numerous single copy loci in the yeast genome. We describe how to measure ssDNA in synchronous cultures of *cdc13-1* mutants, containing a temperature sensitive mutation in an essential telomere capping protein, and in asynchronous cultures of *yku70Δ* mutants also defective in telomere capping.

Introduction

Damage in genomic DNA activates the DDR, inducing DNA repair and signaling cell cycle arrest. ssDNA is an important intermediate in many DNA repair pathways and an important stimulant for checkpoint-pathway-dependent cell cycle arrest (Vaze *et al.*, 2002; Zou and Elledge, 2003; Zubko *et al.*, 2004). Telomeres, at the ends of chromosomes, are usually capped and therefore do not induce DNA repair or activate checkpoint pathways. However, uncapped telomeres are potent inducers of DNA repair and cell cycle arrest and in many cases are associated with the accumulation of excessive ssDNA (Lydall, 2003).

In budding yeast *CDC13* encodes an essential telomere capping protein, and *cdc13-1*, a temperature sensitive allele, is a useful tool to examine interactions between the DDR and uncapped telomeres (Garvik *et al.*, 1995; Lydall and Weinert, 1997). The budding yeast Yku70/Yku80 heterodimer is a nonessential telomere capping protein complex, that also plays a

METHODS IN ENZYMOLOGY, VOL. 409
Copyright 2006, Elsevier Inc. All rights reserved.

critical role in the non-homologous end joining pathway of DNA repair (Fisher and Zakian, 2005).

A failure in telomere capping in *cdc13-1* and *yku70Δ* mutants leads to extensive accumulation of ssDNA near telomeric ends (Garvik *et al.*, 1995; Gravel *et al.*, 1998; Maringele and Lydall, 2002). By combining *cdc13-1* or *yku70Δ* mutations, with mutations in DNA repair genes (e.g., *exo1Δ*) of checkpoint genes (e.g., *chk1Δ, dun1Δ, mec1Δ, rad9Δ, rad17Δ, rad24Δ, rad53Δ*) it is possible to determine the roles of the corresponding DDR gene products in activating or inhibiting ssDNA production at uncapped telomeres and in signaling cell cycle arrest (Booth *et al.*, 2001; Jia *et al.*, 2004; Lydall and Weinert, 1995; Maringele and Lydall, 2002; Zubko *et al.*, 2004). QAOS is a real-time PCR based method that we use to measure ssDNA accumulation in telomere capping mutants (Booth *et al.*, 2001). QAOS can also be used to measure ssDNA accumulating at DSBs *in vivo* (van Attikum *et al.*, 2004).

Experimental Outline

We describe methods to measure ssDNA accumulation during a single cell cycle of synchronized *cdc13-1* mutants. We also describe the use of a conditional degron-based method to inactivate proteins that are important to the vitality of *cdc13-1* mutants in synchronous cultures. Finally we describe how to measure ssDNA accumulation over many generations in *yku70Δ* mutants.

For synchronous cultures, *MAT**a** bar1 cdc13-1 cdc15-2* cells are arrested in G1 by α factor at the permissive temperature 23° (or 20°). After removal of α factor, the cells are released from the G1-arrest to 36° (the restrictive temperature for *cdc13-1* and *cdc15-2* mutations). Cells proceed synchronously through S phase and undergo telomere uncapping because of the *cdc13-1* defect. Flow cytometric analysis shows that DNA replication is generally complete by 60 min (Lydall and Weinert, 1995). Checkpoint-proficient cells arrest in late S/G2 (the *cdc13-1* stage) but cells deficient on checkpoint genes (e.g., *rad9Δ*) continue cell cycle progression and arrest only at late nuclear division due to the *cdc15-2* mutation (by 80 min about half of the checkpoint defective cells reach late nuclear division) (Lydall and Weinert, 1995). Aliquots of cells are harvested at intervals to examine cell cycle position, and viability and to extract DNA.

To investigate the role of essential genes, or genes that are important for the vitality of *cdc13-1* mutants, conditional inactivation, rather than deletion, of gene products is required. We have recently used a temperature sensitive degron cassette to conditionally inactivate the DNA repair protein Rad50, in *MAT**a** bar1 cdc13-1 cdc15-2* cells (Foster *et al.*, submitted). In such strains temperature-dependent ubiquitylation of exposed lysine residues on the surface of the degron cassette causes degradation of Rad50 to be degraded at 37° (Sanchez-Diaz *et al.*, 2004).

To examine ssDNA accumulation in *yku70Δ* mutants at 37° it is necessary to grow cells in liquid culture exponentially for up to 12 h (Maringele and Lydall, 2002). Prior to performing experiments it is useful to know the length of a cell-cycle at 37° (usually 1.5 h for strains in our background) and also which strains stop dividing at restrictive temperature, and how rapidly.

Methods

Synchronization of Cultures with Alpha factor

The following protocol is reproducible for *bar1* W303 strains. Routinely one person is able to handle four cultures in parallel.

Isolation of Single Colonies. Streak out yeast strains for single colonies from frozen stocks or fresh plates on YEPD: yeast extract [10 g/l], peptone [20 g/l], dextrose [2%] adenine [50 mg/l]; autoclave the yeast extract and peptone (and 20 g bacto agar, if plates, rather than liquid are required) in 950 ml water, after autoclaving add 5 ml 1% adenine and 50 ml 40% glucose. Adenine is present because most of our strains possess the revertible *ade2-1* mutation. Incubate at 23° for 2–4 days, or 20° for 3–5 days.

BAR1/bar1 Assays. For efficient synchronization, *MAT**a** bar1* mutants are used since they are more responsive to α factor. The expression of the *bar1* phenotype is unstable, for unknown reasons, so each culture for a new experiment is first tested by a *bar1* assay. It is useful to have positive (*bar1*) and negative (*BAR1*) controls. For most strains, checking 3–4 individual colonies from each strain is sufficient to find a proper one. Rarely, for some "recalcitrant" strains, checking up to 20 individual colonies is necessary.

Culture Inoculation. Inoculate single colonies in 2 ml YEPD in a 15 ml glass tube and grow until saturated on New Brunswick TC7 roller at 23° (2 days). Saturated cultures are stored at 4° for several days.

Plaque Preparation and Testing. For as many cultures as are to be tested by bar assays place about 300 μl of melted 1% agar (in sterile water or YEPD) in 1.5 ml Eppendorf tubes and keep tubes in heating block at 45° to prevent solidification of agar. One culture at a time, add 2 μl of saturated culture to the 45° agar, vortex briefly to mix, and pour the mix onto one-quarter or one-half of prelabeled solid YEPD agar plate. Use other segments of the plate for other cultures. When finished spreading all cultures, let them solidify. At this time, prepare a thick suspension of a freshly grown *MATα* tester strain by resuspending cells in a small volume of sterile water (50–200 μl, depending on the number of *MAT**a*** strains analyzed). Add a 3 μl drop of the *MATα* cell suspension onto the center of each solidified *MATa* plaque and let drops dry. Incubate plates at 23° for 2 days. Transparent halos are formed by *bar1 MATa* cells close to the *MATα* cells, whereas *BAR1 MATa* cells do not form such halos.

Culture Inoculation. In the afternoon (~4:00 P.M.), inoculate 150 μl of the saturated *bar1* cultures into conical flasks containing 120 ml of YEPD (use 300 μl inoculums for checkpoint proficient strains that grow more slowly). Incubate cultures overnight at 23° with good aeration. If growth temperature of 20° is necessary, the inoculation should be done earlier (~12:00 P.M.).

Adjustment of Cell Density. In the morning (~7:00 A.M.), count total bud numbers in cultures using a hemocytometer. Dilute cultures with fresh YEPD medium to get 250 ml of culture at 8×10^6 buds/ml. Return to shaking water bath at 23° for a further 2.5 h (growth at 20° requires 3–4 h). At this stage cultures have densities about 1.5×10^7 buds/ml. Take samples from these cultures as asynchronous controls for DAPI staining (see Determination of Cell Cycle Position later, 1 ml each).

G1 Arrest. Add α factor to 20 n*M* and culture for a further 2.5 h at 23° (α factor is from Sigma, St. Louis, MO [T6901], 1:25,000 dilution of 5×10^{-4} *M*; stock is 0.842 mg/ml in H₂O, stored in aliquots at −20°). At 20°, α factor arrest requires about 4–5 h. Examine the cultures using microscope to ensure that the majority of cells (>90%) have arrested cell division and formed schmoos (pear-like cell shapes). Note: Even though a large proportion of the culture look like schmoos, many will still be in pairs, due to a failure of cell separation after the previous cytokinesis. These cells are still arrested in G1 and are perfectly acceptable to use for synchronous cultures. At this stage cultures usually contain about 3×10^7 buds/ml.

G1 Release

α *Factor Removal.* To remove α factor after G1 arrest, spin cultures down in a centrifuge (Beckman, Fullerton, CA, GS-6R Centrifuge, 1000 rpm, 4 min in 250-ml centrifuge tubes) and discard supernatant. Note this time point as − 40 min, for better reproducibility between experiments. Wash the cell pellet twice, each time resuspending in 50 ml YEPD and respinning in 50 ml Falcon tubes for 3 min at 2000 rpm. Resuspend final pellet in 150 ml of YEPD at 23° in 500-ml flask. Leave the culture on the bench until time 0 is reached (40 min after the culture was first spun down). At this point remove first samples for DNA and cell-cycle analyses (as indicated in Determination of Cell Viability and Determination of Cell Cycle Position, later).

Incubation at 36°. Add 125 ml of prewarmed YEPD (51°) to bring the temperature to 36°. Place cultures at 36° water baths and culture with intensive shaking for 4 h.

Sample Collection During Incubation at 36°. Samples are taken for analysis of cell cycle position (20 min intervals, 1 ml from each culture to Eppendorf tubes), cell viability (40 min intervals, 20 μl from each culture to 2 ml of sterile water in 15 ml glass tubes) and for DNA extraction (40 min intervals, 25 ml from each culture to 50 ml Falcon tubes containing 2.5 ml of 0.5 M EDTA (pH = 8.0) and 250 μl of 10% sodium azide. This is most readily achieved using a 25 ml pipette to remove 27 ml from the culture at each 40 min time point, and a 1 ml pipette at the other time points. Cells collected (1 ml) for analysis of cell cycle position are spun for 7–10 s in microcentrifuge at high speed, resuspended in 0.5 ml of 70% ethanol, and kept at 4° until analysis. Cells collected (25 ml) for DNA are centrifuged for 5 min at 2,000 rpm at 4°, the pellet resuspended in 1 ml of ice-cold sterile water, transferred to 1.5-ml Eppendorf tube, and centrifuged 7–10 s at high speed. After discarding supernatant, the pellet is frozen at −70° or below, until DNA extraction.

Conditional Degradation of Essential Proteins in Synchronous Cultures

Degron strains, in which My Favorite Protein (Mfp) is selectively degradeable at 37°, are created as described by K. Labib and colleagues (Sanchez-Diaz *et al.*, 2004). All procedures for synchronous cultures are as described previously, except that since protein degradation depends on *GAL::UBR1* overexpression, all strains are grown initially in or on YE-PRaff medium, which contains 2% raffinose, rather than 2% glucose. At the stage of G1 arrest the YEPRaff medium is supplemented with additional 2% galactose, to induce efficient *UBR1* expression and efficient Mfp degradation after the G1 release. All subsequent steps are in YEPRaff/Gal medium. Because cells grow slowly in YEPRaff/Gal the time course is slightly longer for these experiments.

Culture Inoculation. Inoculate 200 μl of *bar1* culture into 120 ml YEPRaff at about 2:00 P.M. in 500-ml conical flask and incubate overnight in a shaking water bath at 23° with good aeration.

G1 Arrest. Add α factor to 20 nM and culture for a further 2 h at 23°. Check whether cells are arrested using a microsope and add galactose to a final concentration of 2%. (Thus the medium is now YEPRaff/Gal.) Leave for a further hour at 23°.

G1 Release and Shift to High Temperature. As in the G1 Release section, except the media used is yeast YEPRaff/Gal and the final temperature is 37°, and the time course is continued until 320 min.

Note: Cells are plated on YEPD for cell viability determination.

Growth of Asynchronous Populations at 37°

Since it is necessary to take numerous, sequential, 27-ml samples from each culture, ensure that a large enough volume for each culture is inoculated.

Culture Preparation and Inoculation. Streak out cells for each strain onto fresh YEPD plates and grow at 20° for 1–2 days, until you have enough cells for experiment. Early in the morning, inoculate patches of each strain into the appropriate volume of liquid YEPD (see later) to a final concentration of about 1×10^7 cells/ml. The amount of cells inoculated and the volume of media are different from strain to strain, depending on which strains will stop dividing and which will not. For wild-type and strains that divide at 37° (for example *yku70Δ exo1Δ*, and *yku70Δ chk1Δ*) inoculate 55 ml YEPD. For strains that stop dividing after about 5 h (for example, *yku70Δ* or *yku70Δ rad24Δ*), inoculate 80 ml YEPD. For *cdc13-1* or other mutants that stop dividing rapidly at 37°, inoculate 150 ml. If other mutants are used, we recommend testing the growth potential for each strain at restrictive temperature, prior to the ssDNA detection experiment.

Shift to High Temperature. Keep one 1 bottle of YEPD at 37°. After approximately 2 h growth at 20° in liquid culture, take the first sample (T_0) of 25 ml for ssDNA measurements and another 2 ml for additional parameters (viability and cell-cycle position, see sections Determination of Cell Viability and Determination of Cell Cycle Position, later). Increase the temperature in the shaking water bath to 37°. Add 37° prewarmed YEPD as follows: 30 ml for strains that grow at 37° (1:2 dilution), 50 ml for strains that stop dividing after 5 h (1:2 dilution), and 80 ml for *cdc13-1* strains (1:3 dilution).

Sample Collection and Dilution of Cultures. Every 1.5 h, remove 27 ml from cultures for measurements and dilute immediately with 37° YEPD as follows: for strains that divide at 37°, add 30 ml warm YEPD, for strains that stop division after about 5 h at 37°, add 60 ml at 1.5 h, and 50 ml at 3 h. Do not add any YEPD at 4.5 h or later time points. For *cdc13-1* strains do not dilute the culture.

Determination of Cell Viability

10,000-fold dilutions are convenient for counting cell viability. Plating is performed between sample collection time points. Vigorously vortex tubes containing 20 μl of culture in 2 ml water for a few seconds. Remove 20 μl from each tube to a new tube containing 2 ml water (creating 10,000-fold dilution). Vortex and spread 100 μl from this final dilution onto half of a YEPD plate using glass 1 ml pipette. Spread another 100 μl onto the other half of the plate (the split-plate approach helps to overcome occasional contamination). At late times (from 160 min) we also plate the 100 fold higher concentration of cells (from the first dilution tube), since some cells (like *cdc13-1 rad9Δ* or other checkpoint mutants) show rapid decrease in viability. Incubate plates for 3 days at 23° and count all visible colonies. The number of colonies formed at time 0 defines 100% viability.

Determination of Cell Cycle Position

Staining with DAPI. Spin cells fixed in 70% ethanol for about 10 s at full speed in a bench microcentrifuge. Wash twice by resuspending pellet in 0.5 ml of sterile water and respinning. Resuspend cells in 200–500 μl of 0.2 μg/ml DAPI (4,6-diamidino-2-phenylindole, Sigma D9542; water solution made by dilution [1 μl:50 ml] of 10 mg/ml stock, stored in the dark at 4°). Cells can be stored at 4° for several months.

Sonication. Sonicate at an amplitude of 5 microns to separate cells (Soniprep 150 sonicator, SANYO Gallenkamp PLC, Loughborough, UK). At 0–40 min, cells are difficult to separate, so sonicate for 10 s. At 40–100 min, sonicate for 5 s, at 120–160 min sonicate for 3s, and at later time points, sonicate for 1 s.

Scoring Cell Cycle Position. Place 8–10 μl of culture on a slide, cover with cover slip, and examine on a fluorescence microsope (100× objective, oil-immersion). Count at least 100 cells, using a multicounter, and determine number of cells at different stages of cell cycle (see Fig. 1).

Measurement of ssDNA

Isolation of DNA. The DNA isolation method we use is based on a previously published method (Wu and Gilbert, 1995) and the Qiagen Genomic DNA handbook (Qiagen, Valencia, CA). We are able to perform 16 DNA preparations, in two batches of 8, over an 8 h working day. The number of preparations is primarily limited by the number of tubes that fit in the fixed angle centrifuge rotor. Perform all stages on ice, unless otherwise stated, and precool centrifuges.

BUFFERS/SOLUTIONS NEEDED
   Isopropanol
   70% V/V ethanol in water
   TE: 10 m*M* TrisHCl (pH 8.0), 1 m*M* EDTA

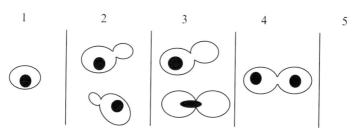

FIG. 1. Cell cycle analysis. Classifications are 1: No bud, single cell, single nucleus. 2: Bud <50% diameter of mother cell. Single nucleus. 3: Bud >50% diameter of mother cell. Single nucleus. 4: Two buds, two nuclei. 5: None of the other types.

NIB, Nuclei isolation buffer: 17% (V/V) glycerol, 50 mM MOPS, 150 mM potassium acetate, 2 mM MgCl$_2$, 500 μM spermidine, 150 μM spermine. For 1 litre: 170 ml glycerol, 10.463 g MOPS, 14.72 g potassium acetate, 2 ml 1 M MgCl$_2$, 0.55 ml 0.9 M spermidine (stored at −70°), 52 mg spermine (added as dry powder); adjust to 1 liter by water, dissolve all components, adjust pH to 7.2, filter sterilize, and store in 50-ml aliquots, in dark at 4°.

Sodium azide: 10% (w/v) in water, store at RT.

G2 with RNAse: 800 mM guanidine hydrochloride, 30 mM EDTA, 30 mM Tris-HCl, 5% Tween 20, 0.5% Triton X-100, pH 8.0, 200 μg/ml RNAse A (1 ml of 10 mg/ml RNAse per 50-ml tube of G2, can be stored at 4° for several weeks).

10 mg/ml RNAse A (dissolve pancreatic RNAse at 11 mg/ml in 0.01 M sodium acetate, pH 5.2). Place in boiling water bath for 15 min and then allow to cool to room temperature. Adjust pH by adding 0.1 volume of 1 M Tris-HCl (pH 7.4). Aliquot and store at −20°. Do not freeze/thaw.

20 mg/ml Proteinase K (dissolve in sterile water. Store in aliquots at −20° or −80°).

QBT: 750 mM NaCl, 50 mM MOPS, 15% isopropanol, 0.15% Triton X-100, pH 7.0.

QC: 1 M NaCl, 50 mM MOPS, 15% isopropanol, pH 7.0.

QF: 1.25 M NaCl, 50 mM Tris-HCl, 15% isopropanol, pH 8.5.

ICE-COLD ACID-WASHED GLASS BEADS. Stratech (Cat No 11079 105) 0.425- to 0.6-mm in diameter, washed for 1 h in concentrated nitric acid, rinsed with milliQ water until pH is neutral, baked to dryness, and stored at 4°.

PELLET RESUSPENSION. Thaw cell pellet (kept at −80°) on ice and resuspend the pellet in 1.0 ml NIB and transfer to 2-ml screw cap microtube with skirt (Sarstedt Ltd., Leicester, UK, Cat. No 72.694.006).

CELL WASHING. Harvest cells in screw cap tube. Spin 7-10 s in a microcentrifuge at top speed, pour off supernatant, and resuspend in 600 μl of NIB by vortexing.

GLASS BEAD ADDITION. Add ice cold, acid washed glass to about 1.5-ml mark on the tube (use 1.5 ml Eppendorf tube to scoop and transfer beads).

CELL BREAKAGE. To break cells, we use a Fast Prep^{TM} FP120 machine (BIO 101, SAVANT, Holbrook, NY). More than 90% cell breakage is necessary and this can be monitored by scoring the fraction of ghost cells by phase contrast microscopy. Cells arrested for a long time at 36° are more susceptible to breakage. For time points 0 and 40 min, 8 or more breakage cycles are necessary (each session lasts 5 s at 5.5 power setting and is followed by incubation for 30–60 s on ice water for cooling). For time points 80 and 120 min 6 bursts, and for 160–240 min 5 bursts of breakage are recommended.

HARVESTING NUCLEI. First prepare Eppendorf tubes by removing the caps and cutting the bottoms from tubes with a dog nail clipper (purchased from pet store) or razor blade/scalpel. Puncture the bottom of Sarstedt tubes containing glass beads with a 21-gauge syringe needle, place tightly in the top of pre-prepared Eppendorf tubes. Place pairs of tubes in a Falcon 352059 round bottomed 15-ml tube and spin 2 min in precooled (about 2000 rpm). Nuclei and broken cells collect at the bottom of the 15-ml tube. Wash the remaining broken cells from beads by adding 1 ml NIB to the top the glass beads in the Sarstedt tube, and spin (about 2000 rpm, 2 min). Repeat washing step and let liquid and nuclei accumulate in bottom of Falcon tube. Finally, discard the Eppendorf/Sarstedt tubes using tweezers and vortex the 15-ml Falcon tube briefly (its content is about 2.5 to 3 ml of cell debris at this stage).

HARVESTING NUCLEI AND CELL DEBRIS. Spin Falcon tubes in a fixed angle F0850 rotor (containing appropriate sleeves) in Beckman Allegra 64R centrifuge ($6500g$, 20 min, $4°$). The lids do not fit on the tubes in the rotor so save them in a clean place. Discard the supernatant; remove the residual liquid by aspiration.

TREATMENT WITH RNASE. Fully resuspend cell/nuclear pellet in 2 ml G2/RNAse. Vortexing or careful pipetting are helpful at this stage. Put the caps back on the tubes. Incubate at $37°$ for 30 min with gentle mixing every 10 min, or so.

TREATMENT WITH PROTEINASE K. Add 50 μl of 20 mg/ml proteinase K. Incubate for one hour at $37°$, mixing tubes periodically.

CLEARING THE LYSATE. Remove caps from the tubes and spin in F0850 rotor ($6500g$, 10 min, $4°$).

QIAGEN COLUMN EQUILIBRATION AND QBT PREPARATION. While tubes are spinning place Qiagen 20-g columns in rack and pre-equilibrate by adding 1 ml of QBT solution to top of column, and let flow through by gravity. In addition place 2 ml QBT in each of 8 clean Falcon 352059 tubes.

BINDING DNA TO QIAGEN COLUMN. Pour supernatant, containing DNA, into tubes containing the 2 ml of QBT solution. Vortex briefly. Pour part of DNA/G2/QBT mix onto the top of pre-equilibrated Qiagen 20-g column. Allow solution to flow through the column by gravity, discarding the flow through, until all DNA/G2/QBT mix has been passed onto column.

WASHING COLUMN. Wash each column 3 times with 1 ml of QC solution.

DNA ELUTION. Add 1 ml of prewarmed ($50°$) QF solution to top of column. At this point, start collecting flow through in 15-ml Falcon 352059 tubes (keep caps). When the first 1 ml has entered column complete elution by adding another 1 ml of prewarmed QF solution.

PRECIPITATION OF DNA. Add 0.7 vol (1.4 ml) of isopropanol (room temperature) to DNA solution and vortex the mixture. Spin in F0850 rotor (7700g, 20 min, 4°). In most, but not all, cases a small tight pellet is visible at this stage. It is useful to mark the top of the tube for orientation in the centrifuge, to help detect the position of the pellet. Discard supernatant carefully; invert the tube onto paper towel to remove as much liquid as possible.

WASHING PELLET. Add 1 ml of 70% ethanol to DNA, vortex the tube gently, and spin in F0850 rotor (7700g, 20 min, 4°). Discard supernatant as above, spin down any ethanol remaining by respinning for 1 min, and remove the rest of the supernatant by aspiration with pipette tip.

DISSOLVING DNA. Add 400–600 μl TE to DNA pellet, tightly cap the tube, and incubate on New Brunswick TC7 roller at 23° for overnight or longer.

MEASURING YIELD OF DNA. Dilute 10 μl of DNA to 90 μl of TE, vortex, and leave overnight at 4° to equilibrate.

Measure yield in comparison to yeast genomic DNA standards by real time PCR.

Alternatively, a spectrophotometer can be used, especially for preparing DNA standards for the first time. For all real-time PCR measurements, we set up samples in triplicate, in 96-well plates according to Table I.

REAL TIME PCR. Preparation of Oligonucleotides. Dissolve lyophilised oligonucleotides (Sigma Genosys) in TE at 200 μM, by adding 5 μl TE per nmol of oligonucleotide supplied. Once resuspended this stock is stored at $-20°$. To make working (100×) stocks, also stored at $-20°$, dilute primers further with TE according to Table II.

DNA ADDITION. Add 10 μl of each DNA standard (STD), in triplicate, to the bottom of 96-well PCR plate (we use standards with concentrations 0.4 ng/μl, 2 ng/μl and 4 ng/μl in TE buffer). Add 10 μl TE for the no template control (NTC) and all 10-fold diluted samples of DNA. Use a separate filter-protected tip for each triplicate.

PREPARING PCR MASTER MIX. The PCR master mix is made at 1.66× final concentration, and 15 μl of this is added to each 10 μl sample of DNA. In 2-ml screw cap Eppendorf tube containing the appropriate amount of sterile water, add all primers and probes, tightly cap the tube, incubate at 95° for 5 min followed by incubation on ice for 3 min, and then add the remaining components described in Table III. Put the cap onto the tube and invert tube about 20 times to mix the compounds, spin briefly in microfuge to collect the contents and keep the mix on ice.

ADDING PCR MASTER MIX TO DNA. Add 15 μl of PCR master mix to each DNA sample using an automatic single-channel pipette with filter-protected tip (Finnpipette, Labsystems, Thermo Life Sciences,

TABLE I
96 WELL PLATE SET UP

	1	2	3	4	5	6	7	8	9	10	11	12
A	A/0	A/0	A/0	A/40	A/40	A/40	A/80	A/80	A/80	A/120	A/120	A/120
B	A/160	A/160	A/160	A/200	A/200	A/200	A/240	A/240	A/240	B/0	B/0	B/0
C	B/40	B/40	B/40	B/80	B/80	B/80	B/120	B/120	B/120	B/160	B/160	B/160
D	B/200	B/200	B/200	B/240	B/240	B/240	C/0	C/0	C/0	C/40	C/40	C/40
E	C/80	C/80	C/80	C/120	C/120	C/120	C/160	C/160	C/160	C/200	C/200	C/200
F	C/240	C/240	C/240	D/0	D/0	D/0	D/40	D/40	D/40	D/80	D/80	D/80
G	D/120	D/120	D/120	D/160	D/160	D/160	D/200	D/200	D/200	D/240	D/240	D/240
H	STD1	STD1	STD1	STD2	STD2	STD2	STD3	STD3	STD3	NTC	NTC	NTC

96-well plate set up for PCR to measure total DNA or ssDNA from 4 yeast strains (A, B, C and D) sampled every 40 minutes. Each PCR is performed in triplicate. STD wells contain standards. NTC contain no template control.

TABLE II
REAL-TIME PCR PRIMER PREPARATION

	100× stock (μM)	200 μM stock (μl)	TE (μl)
Tagging primers	3	6	394
Tags	30	60	340
Forward/Reverse primers	30	60	340
Probes	20	40	360

TABLE III
MASTER MIX PREPARATION

	Final conc.	Stock conc.	Volume (μl)
Water			960
Tagging primer	30 nM	3 μM (100×)	25
Tag	300 nM	30 μM (100×)	25
Reverse primer	300 nM	30 μM (100×)	25
Probe	200 nM	20 μM (100×)	25
Ex Taq buffer	0.97×	10×	245
dNTPs	194 nM	2.5 mM	195
Ex Taq Polymerase	0.024 u/μl	5 units/μl	12
Total			1512

The above mixture is sufficient to add 15 μl master mix to 96 individual 10 μl DNA samples (and contains 5% extra, to cover pipetting errors). This recipe permits each 250 unit batch of Takara Ex Taq polymerase to be used for 4 individual 96-well plates.

Basingstoke, UK). Using a single tip to add the Master Mix to all 96 samples speeds addition. Try to hold the pipette at the same position for loading every well and direct the master mix onto internal wall of wells, to minimize cross-contamination. When finished, cover slots of the plate with caps or adhesive film prior to PCR.

SETTLING PLATE. Centrifuge the 96-well plate in a centrifuge (2000 rpm, 1 min) to settle the reaction mixtures.

RUNNING REAL TIME PCR. Place the 96-well plate in real time PCR machine (ABI 7700 Sequence Detection System, PE Biosystems, Warrington, UK). We amplify the *PAC2* locus to measure DNA concentrations. Primers are shown in Table IV. *PAC2* amplification requires no tagging primer or Tag: instead forward, reverse primers and a fluorescent probe are used.

PCR conditions are Step 1: 94° 5 min (1 cycle).

Step 2: 95°, 15 s; 63°, 1 min (40 cycles).

ANALYSIS OF REAL-TIME PCR DATA AND CONCENTRATION ADJUSTMENT. After PCR, use the standards to determine the concentration of DNA in

TABLE IV
OLIGONUCLEOTIDES FOR MEASURING TOTAL DNA AND ssDNA ON THE TG-STRAND NEAR TELOMERES

Primer	Loci	Sequence	Type of primer
M 442	PAC2	AATAACGAATTGAGCTATGACACCAA	Forward
M 573	PAC2	AGCTTACTCATATGGATTTCATACGACTT	Reverse
M 326	PAC2	CTGCCGCGTTGGTCAAGCCTCAT	Probe
M 239	–	AACCAGCGCAGCGGCAT**GTGT**	Tag
M 272	PDA1	AACCAGCGCAGCGGCAT**gtg**atgatggaa	Tagging
M 320	PDA1	TGGAGATGGCTTGTGACGCCTTG	Reverse
M 698	PDA1	TGGCATTCTCGATACCGACAGCAATGGCCTC	Probe
M 313	–	GATCTCGAGCTCGATATCGGATCC**ATT**	Tag
M 314	YER186C	GATCTCGAGCTCGATATCGGATCC**att**tacgggcgga	Tagging
M 96	YER186C	AATCTCGCCTAACAAAAAAGGCTTCTTAGTG	Reverse
M 325	YER186C	AGGCAAATAACGGCAAGCCCTCTCC	Probe
M 315	–	AAGGAGCGCAGCGCCTGT**ACCA**	Tag
M 444	YER188W	AAGGAGCGCAGCGCCTGTA**cca**tagcgtgat	Tagging
M 319	YER188W	AACGTACAGGTTACGATCGCGTCATTTTA	Reverse
M 427	YER188W	TAGCCGTTATCATCGGGCCCAAAACCGTATTCATTG	Probe
M 315	–	AAGGAGCGCAGCGCCTGT**ACCA**	Tag
M 513	X-repeat	AAGGAGCGCAGCGCCTGTA**cca**catttaatatct	Tagging
M 512	X-repeat	ATTGAGTGGATAGTAGGATGGTGAAAAAGTGGTATAACG	Reverse
M 510	X-repeat	TCATTCGGCGGCCCCAAATATTGTATAACTGCCC	Probe
M 520	–	TGCCCTCGCATCGCTCTC**GAA**	Tag
M 521	Y'-5000	TGCCCTCGCATCGCTCTC**gaa**acaaagtcagtga	Tagging
M 517	Y'-5000	GTCCTGGAACGTTGTCACGAAAAAGC	Reverse
M 516	Y'-5000	TGCTAGGCCGAACGACAGCTCTACGATGCGTACTT	Probe
M 316	–	TGCCCTCGCATCGCTC**ACA**	Tag
M 243	Y'-600	TGCCCTCGCATCGCTCTC**aca**gccctatcag	Tagging
M 237	Y'-600	GAGATCAGCTTGCGCTGGGAGTTACC	Reverse
M 526	Y'-600	ACAGGAATGCCGTCCAATGCGGCACTTTAGA	Probe

Oligonucleotides are classified at tags, tagging primers, probes, forward and reverse primers. For tagging primers the lower case sequence corresponds to the region that hybridizes to the genome. The bold regions overlap between the tag and the tagging primer.

the experimental samples. Export the data into an excel spreadsheet. Ignoring any individual samples that differ by more than two-fold (1 C_t) from other duplicates, use the mean value of the triplicates to estimate the DNA concentration. Add sufficient TE to all non-diluted samples (see section, Dissolving DNA) to adjust concentration to 2 ng/μl, and vortex. Leave samples overnight at 4° to equilibrate, then check the concentration by repeating real time PCR measurement of *PAC2* locus. At this point all samples should contain similar amounts of DNA (ideally 2 ng/μl and certainly between 1 and 3 ng/μl). The precise amount of DNA in each sample is used to generate a "correction factor" that corrects the single stranded DNA measurements made later.

Once DNA concentrations have been equalized, it is convenient to use a multichannel pipetter to create a set of 96-well plates that contain 10 μl samples in rows A to G, arranged as in Table I. The plates are covered with adhesive film and stored at −20° or −80° until needed. When needed plates are thawed and ssDNA standards (see section, Preparing ssDNA Standards) added to row H. After this the appropriate PCR master mix is added to each well.

MEASURING ssDNA BY QAOS. QAOS has been described in detail (Booth *et al.*, 2001). In brief, a tagging primer, consisting of a genomic locus-specific sequence and an artificial sequence (tag), binds to ssDNA, but not dsDNA, at low temperature (T_m between 16° and 45°). A novel hybrid sequence, containing the tag and part of the yeast genome, is formed during a round of the primer extension. The amount of hybrid sequence produced is proportional to the initial amount of ssDNA and is measured by real time PCR.

We use Taqman probes (Sigma Genosys, Pampisford, UK) and the ABI 7700 machine for real-time PCR, but there is no reason why other real time PCR machines or primer/probe designs should not work.

Primers to detect ssDNA on the TG (3′) strand near telomeres are shown in Table IV and to detect ssDNA on the AC (5′) in Table V. With the exception of *PAC2* probes, which are labeled with VIC, all probes are labeled with FAM. All probes are quenched with TAMRA. ssDNA measurements are similar to the determination of DNA concentrations in the above paragraph, except that the standards are all at 2 ng/μl but contain different proportions of ssDNA.

PCR conditions to measure ssDNA are

Step 1: 40° 5 min (1 cycle).
Step 2: ramp to 72° at 2°/min (1 cycle).
Step 3: 94° 4 min (1 cycle).
Step 4: 95°, 15 s; 67°, 1 min (40 cycles).

TABLE V
OLIGONUCLEOTIDES FOR ssDNA ON THE AC-STRAND NEAR TELOMERES

Primer	Loci	Sequence	Type of primer
M 315	–	AAGGAGCGCAGCGCCTGTA**CCA**	Tag
M 271	*PDA1*	AAGGAGCGCAGCGCCTGTA**cca**tctatctgttg	Tagging
M 288	*PDA1*	ACGGCTTTCACTGAGGCACCTCTC	Reverse
M 698	*PDA1*	TGGCATTCTCGATACCGACAGCAATGGCCTC	Probe
M 313	–	GATCTCGAGCTCGATATCGGATCC**ATT**	Tag
M 311	*YER186C*	GATCTCGAGCTCGATATCGGATCC**att**ctctcctta	Tagging
M 321	*YER186C*	GCAAGTAGGAAGCATCCCTTCAAGTCATT	Reverse
M 325	*YER186C*	AGGCAAATAACGGCAAGCCCTCTCC	Probe
M 233	–	ATGCCCGCACCGCCTCA**TTG**	Tag
M 242	*Y'-600*	ATGCCCGCACCGCCTCA**ttg**cgctggga	Tagging
M 236	*Y'-600*	CCGAAATGTTTTATTGCAGAACAGCCCTAT	Reverse
M 526	*Y'-600*	ACAGGAATGCCGTCCAATGCGGCACTTTAGA	Probe

This table is similar to Table IV except that probes are designed to hybridize to the AC (5') strand near telomeres.

TABLE VI
PREPARATION OF ssDNA STANDARDS

ssDNA (%)	Total (μl)	Boiled DNA (μl)	Non boiled DNA (μl)
51.2	500	256	244
3.2	500	16	484
0.8	500	4	496

After PCR, the mean value of ssDNA in each sample is plotted (ignoring individual samples that differed from the other duplicates more than two-fold, and using the correction factor calculated previously).

PREPARING ssDNA STANDARDS. To ensure standard DNA does not contain ssDNA generated during DNA replication, DNA is prepared from G1 arrested wild-type cells (DLY62: *bar1 CDC13* cells) and adjusted to 2 ng/μl. At least 1600 μl is necessary.

DENATURING DNA. Boil 300 μl DNA in a screw cap Eppendorf tube for 7 min at 98°, then place on ice water. Every 15 s, vortex for 2 s and return to ice. Do this 3–4 times and then leave on ice for at least 5 min.

MIXING NATIVE AND BOILED DNA. To make ssDNA standards mix different amounts of boiled and non-boiled DNA as shown in Table VI. Vortex standards for at least 1 min to distribute ssDNA, centrifuge briefly, vortex briefly, and centrifuge again. Keep standards on ice for 15–30 min, vortex them again for 1 min, centrifuge briefly, and use for measurements. Standards are kept at $-20°$, defrosted on ice, and vortexed for 15–20 s before use.

Acknowledgments

Our lab is supported by the Wellcome Trust (054371 and 075294). Steven Foster is supported by the award of a BBSRC committee studentship.

References

Booth, C., Griffith, E., Brady, G., and Lydall, D. (2001). Quantitative amplification of single-stranded DNA (QAOS) demonstrates that cdc13-1mutants generate ssDNA in a telomere to centromere direction. *Nucleic Acids Res.* **29**, 4414–4422.

Fisher, T. S., and Zakian, V. A. (2005). Ku: A multifunctional protein involved in telomere maintenance. *DNA Repair (Amst.)* **4**, 1215–1226.

Garvik, B., Carson, M., and Hartwell, L. (1995). Single-stranded DNA arising at telomeres in cdc13 mutants may constitute a specific signal for the RAD9 checkpoint. *Mol. Cell. Biol.* **15**, 6128–6138.

Gravel, S., Larrivee, M., Labrecque, P., and Wellinger, R. J. (1998). Yeast Ku as a regulator of chromosomal DNA end structure. *Science* **280**, 741–744.

Jia, X., Weinert, T., and Lydall, D. (2004). Mec1 and Rad53 inhibit formation of single-stranded DNA at telomeres of *Saccharomyces cerevisiae cdc13-1* mutants. *Genetics* **166**, 753–764.

Lydall, D. (2003). Hiding at the ends of yeast chromosomes: Telomeres, nucleases and checkpoint pathways. *J. Cell. Sci.* **116**, 4057–4065.

Lydall, D., and Weinert, T. (1995). Yeast checkpoint genes in DNA damage processing: Implications for repair and arrest. *Science* **270**, 1488–1491.

Lydall, D., and Weinert, T. (1997). Use of cdc13-1-induced DNA damage to study effects of checkpoint genes on DNA damage processing. *Methods Enzymol.* **283**, 410–424.

Maringele, L., and Lydall, D. (2002). EXO1-dependent single-stranded DNA at telomeres activates subsets of DNA damage and spindle checkpoint pathways in budding yeast yku70Δ mutants. *Genes Dev.* **16**, 1919–1933.

Sanchez-Diaz, A., Kanemaki, M., Marchesi, V., and Labib, K. (2004). Rapid depletion of budding yeast proteins by fusion to a heat-inducible degron. *Sci. STKE* **2004**, PL8.

van Attikum, H., Fritsch, O., Hohn, B., and Gasser, S. M. (2004). Recruitment of the INO80 complex by H2A phosphorylation links ATP-dependent chromatin remodeling with DNA double-strand break repair. *Cell* **119**, 777–788.

Vaze, M. B., Pellicioli, A., Lee, S. E., Ira, G., Liberi, G., Arbel-Eden, A., Foiani, M., and Haber, J. E. (2002). Recovery from checkpoint-mediated arrest after repair of a double-strand break requires Srs2 helicase. *Mol. Cell.* **10**, 373–385.

Wu, J. R., and Gilbert, D. M. (1995). Rapid DNA preparation for 2D gel analysis of replication intermediates. *Nucleic Acids Res.* **23**, 3997–3998.

Zou, L., and Elledge, S. J. (2003). Sensing DNA damage through ATRIP recognition of RPA-ssDNA complexes. *Science* **300**, 1542–1548.

Zubko, M. K., Guillard, S., and Lydall, D. (2004). Exo1 and Rad24 differentially regulate generation of ssDNA at telomeres of *Saccharomyces cerevisiae cdc13-1* mutants. *Genetics* **168**, 103–115.

[17]   Analysis of Non-B DNA Structure at Chromosomal Sites in the Mammalian Genome

By Sathees C. Raghavan, Albert Tsai,
Chih-Lin Hsieh, and Michael R. Lieber

Abstract

Changes at sites of genetic instability ultimately involve DNA repair pathways. Some sites of genetic instability in the mammalian genome appear to be unstable because they adopt a non-B DNA conformation. We describe two structural approaches for determination of whether a genomic region is configured in a non-B DNA conformation. Our studies indicate that at least some chromosomal fragile sites can be explained by such altered DNA conformations. One of the methods that we describe is called the bisulfite modification assay. This is a powerful assay because it provides information on individual DNA molecules. The second approach uses preexisting DNA structural reagents, but describes our specific application of them to analysis of DNA *in vivo*.

Introduction

The mechanisms of chromosomal translocations in mammalian cells have been largely undefined. Recent progress on the most common translocation in human cancer, t(14;18), highlights interesting issues in DNA structure at chromosomal fragile sites (Raghavan and Lieber, 2004; Raghavan *et al.*, 2004a, 2005a,b).

There are a variety of methods to detect non-B DNA structures in the DNA (Lilley and Dahlberg, 1992a,b; Sinden, 1994; Soyfer and Potaman, 1996). Many of these probing agents are used to detect single-stranded regions within non-B DNA structures. The methods generally used in the literature to detect such single-stranded regions within non-B DNA structures include potassium permanganate or osmium tetroxide assays (reacts with thymine and to a lesser extent to cytosine, when DNA is single stranded) (Boublikova and Palecek, 1990; Duncan *et al.*, 1994; Rahmouni and Wells, 1989; Sasse-Dwight and Gralla, 1989), P1 nuclease (Romier *et al.*, 1998) or S1 nuclease (Hentschel, 1982; Johnston, 1988), bromoacetaldehyde (Kohwi and Kohwi-Shigematsu, 1988; Sinden, 1994), haloacetaldehyde (reacts with adenine) (Kohwi and Kohwi-Shigematsu, 1988; Lilley, 1983; Secrist *et al.*, 1972), carbodiimide (single-strand specific T and G modification) (Hale

METHODS IN ENZYMOLOGY, VOL. 409
0076-6879/06 $35.00
DOI: 10.1016/S0076-6879(05)09017-8

et al., 1980), formaldehyde crosslinking and diethyl pyrocarbonate (reacts with adenine and to a lesser extent with guanine) (Johnston, 1988; Leonard *et al.*, 1971). Other probes for DNA structure include psoralen crosslinking (Ussery and Sinden, 1993), the DMS assay (specific to G base) (Hanvey *et al.*, 1988; Johnston, 1988), antibodies (Agazie *et al.*, 1994; Lee *et al.*, 1987; Nordheim *et al.*, 1981), gel mobility shift assays (Raghavan *et al.*, 2004b; Sakamoto *et al.*, 1999), two-dimensional gel electrophoresis (Sinden, 1994) or electron microscopy (Griffith *et al.*, 1999; Sakamoto *et al.*, 1999) for large DNA sequences. (Smaller DNA sequences can be analyzed by NMR and circular dichroism.) Bisulfite is another chemical probe that has been used for studying the DNA structure (cruciform structures) (Gough *et al.*, 1986), but to only a limited extent.

In this chapter, we describe some of the chemical probes used in our studies (Raghavan *et al.*, 2004a,b, 2005c). Though these methods have been described previously, we have modified them in order to get individual DNA molecule information (especially in the case of the bisulfite modification assay).

Specific Methods

Bisulfite Modification Assay

Chemistry. Any cytosine on a region of single-stranded DNA (ssDNA) or an unpaired region of DNA can form an adduct across the C5-C6 bond, which reacts with sodium bisulfite. However, bisulfite will not react with cytosines when the DNA is double-stranded (Gough *et al.*, 1986; Hayatsu, 1976; Hayatsu *et al.*, 1970; Shapiro *et al.*, 1970, 1973, 1974; Raghavan *et al.*, 2004a; Yu *et al.*, 2003). The chemistry of the bisulfite reaction with cytosine is described as follows (Fig. 1). In the first step, the bisulfite moiety is added to the C5-C6 double bond of cytosine through a reversible reaction resulting in the formation of a cytosine sulfonate. This step is strongly favored by acidic pH, such as pH 5.2 (but can be done at pH as high as 6.0 with little appreciable difference in our studies). In the second step, cytosine sulfonate is hydrolytically deaminated to generate a uracil sulfonate. This step is irreversible. In the final step, desulfonation leads to conversion of uracil sulfonate into uracil. This step requires a basic pH, such as pH 8 to 9 in our assays. Upon PCR amplification and DNA sequencing this C to U conversion can be read as C to T conversion. Then the template strand can be individually identified based on the position of the cytosines when any of the cytosines are converted on a given DNA molecule (Fig. 2).

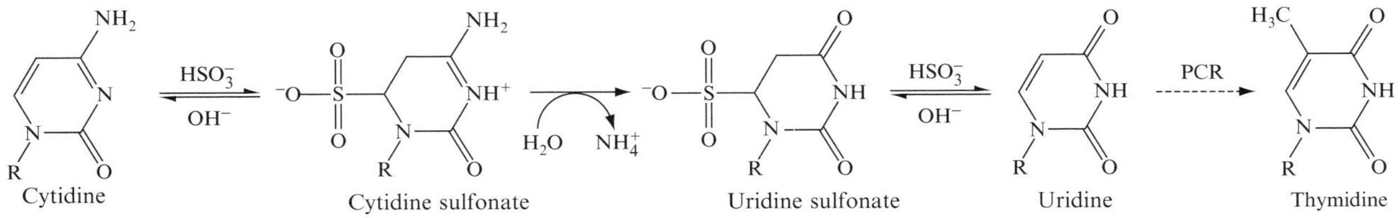

FIG. 1. Chemistry of bisulfite-catalyzed modification of cytosines into uracil. The details of each step of the reaction are described in the text.

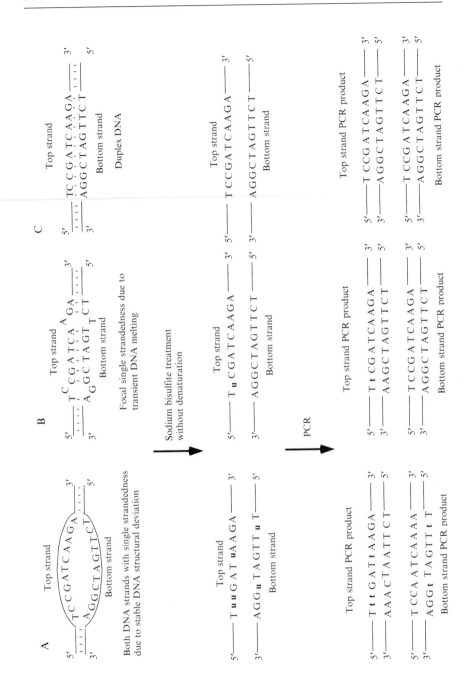

Procedure

1. Extraction of DNA: In our studies, we have used chromosomal DNA extracted by nondenaturing methods from various cell lines. In brief, the cells are collected by centrifugation and washed with 1× PBS. Chromosomal DNA is extracted by lysing the cells in TE (10 mM Tris [pH 8.0] and 1 mM EDTA) and 0.5% SDS, followed by proteinase K digestion (20 µg/ml) at 37° overnight. The genomic DNA is phenol, phenol:chloroform (50:50), and chloroform extracted, followed by precipitation in chilled 100% ethanol. DNA is rinsed in 70% ethanol and air dried. The pellet is resuspended in TE to a final concentration of ~1 µg/µl. Prior to the bisulfite treatment, the chromosomal DNA is digested with EcoRI to reduce the size of the DNA. Then the digested DNA is purified by phenol:chloroform extraction and ethanol precipitation.

In experiments where plasmid DNA is used, the DNA is isolated by a nondenaturing method, and purified by cesium chloride density gradient centrifugation (Sambrook *et al.*, 1989). Following removal of ethidium bromide by butanol extraction, the plasmid DNA is precipitated with chilled ethanol and is washed three times with 70% ethanol. The pellet is dissolved in TE (pH 8.0). Minichromosomal DNA from mammalian cells is harvested after a 42-h incubation at 37° by the Hirt harvest method (nondenaturing). Briefly, the cells are sedimented by centrifugation and washed once in 1× PBS at room temperature. Cells are then resuspended in Hirt buffer (0.01 M Tris [pH 7.9]; 0.01 M EDTA), lysed by SDS (0.6%), and incubated at 4° for 24 h after the addition of NaCl (1.25 M). Proteinase K is added (100 µg/ml) to the supernatant, which is then incubated at 37° overnight. Then the minichromosomes are extracted with phenol/chloroform and precipitated with 1/10 volume of 3 M sodium acetate (pH 7.0) and 2.5 volume of ethanol. The DNA is recovered and resuspended in 25 µl of TE (pH 8.0).

2. Sodium bisulfite treatment and purification of the DNA (see Fig. 3): Approximately 5 µg (2–20 µg work fine) of EcoRI-digested chromosomal DNA is resuspended in a 30-µl volume for the bisulfite modification assay. In cases where minichromosomal DNA is being analyzed, Hirt harvest from mammalian cells is resuspended in 30 µl and used for the bisulfite reaction. In cases where plasmid DNA is being analyzed, 1 µg of DNA is

---

FIG. 2. Schematic representation of outcomes of bisulfite treatment based on the initial DNA conformation. In panel (A), an example where both strands of DNA are single-stranded are shown. In panel (B), an example of single-strandedness due to random breathing is shown. In panel (C), the outcomes of bisulfite treated DNA on a duplex DNA are shown. The cytosine to uracil (or thymine) are shown in bold, lower case letters.

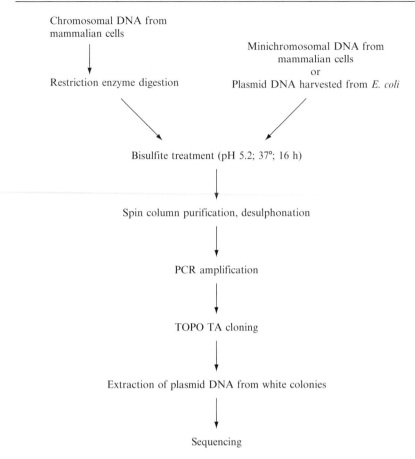

Chromosomal DNA from
mammalian cells

Minichromosomal DNA from
mammalian cells
or
Plasmid DNA harvested from *E. coli*

Restriction enzyme digestion

Bisulfite treatment (pH 5.2; 37°; 16 h)

Spin column purification, desulphonation

PCR amplification

TOPO TA cloning

Extraction of plasmid DNA from white colonies

Sequencing

FIG. 3. Schematic representation of the bisulfite modification assay procedure to detect single-strandedness on chromosomal DNA, minichromosomal DNA, or plasmid DNA. Flow chart of the bisulfite modification assay procedure.

harvested from *E. coli* and resuspended in 30 μl and used for the bisulfite treatment. In a fresh 500-μl tube, 12.5 μl of 20 mM hydroquinone and 457.5 μl of 2.5 M sodium bisulfite (pH 5.2) is mixed. The pH of the sodium bisulfite is adjusted to 5.2 by adding the required amount of sodium hydroxide. Then the DNA is added to the sodium bisulfite-hydroquinone mix, and the tube is covered with an aluminum foil. The reaction is incubated for 14–16 h at 37° (there is no significant difference in the bisulfite conversion of cytosines at 37° or 55°; however, we routinely used 37°).

The DNA is then purified using a Wizard DNA Clean-Up Kit (Promega, Madison, WI), according to the manufacturer's protocol. The bisulfite modified DNA is then desulfonated with 0.3 M NaOH for 15 min at 37°, and the DNA is recovered by ethanol precipitation. The purified DNA is resuspended in 30 μl of 10 mM Tris, 1 mM EDTA (pH 8.0) and stored at −20°.

3. PCR amplification of fragment of interest. The fragments of interest are PCR amplified from the bisulfite treated DNA with appropriate primers, using standard PCR conditions. The PCR reactions are performed using a thermocycler (Stratagene, La Jolla, CA) under the following conditions: 1× reaction mix (10 mM Tris [pH 8.3], 50 mM KCl, 0.01% gelatin) with 1.5 mM MgCl$_2$, 400 μM dNTPs, and 1 unit of Taq polymerase in a volume of 10 μl. Amplification is carried out for 35 cycles. The annealing temperature used is standardized according to the primer used in each specific PCR reaction. The PCR products are resolved on an agarose gel, and the correct-sized fragments are recovered using a Gene Clean Kit (BIO101, Inc., Carlsbad, CA).

4. Cloning and sequencing of the gene of interest. Purified PCR products are cloned using the TOPO-TA cloning Kit (Invitrogen, Grand Island, NY). Plasmid DNA from each clone of interest is purified using standard protocol or using the Gene Elute Plasmid Mini Prep Kit (Sigma, St. Louis, MO). Sequencing reactions are carried out using the SequiTherm Excel II sequencing kit (Epicentre Technologies, Madison, WI) and MWG thermal cycler model Primus 96 Plus (MWG Biotech, High Point, NC). Automated sequencing is carried out using the Li-Cor DNA Analyzer model 4200 (Li-Cor, Lincoln, NE). In each experiment, a minimum of 12 clones and maximum of 100 clones are sequenced to reach the final conclusion.

Precautions

1. The primer sites are also subject to bisulfite conversion, and, as a result, some molecules lose the ability to serve as PCR templates; however, this source of background conversion is rare and random in location. Therefore, it is important to select DNA strand sequences with fewer numbers of G's for the region of priming.

2. For DNA structure studies, the DNA must be extracted using nondenaturing methods to preserve the non-B DNA structure.

3. Desulphonation of uracil sulfonate into uracil is critical.

4. In some rare cases, the same parental molecule may get amplified many times and may result in incorrect interpretations. Duplicate molecules are excluded based on their identical sites of background conversion.

Discussion. We have chosen consecutiveness of C's as a criterion to define significant lengths of single-strandedness throughout our studies (Raghavan *et al.*, 2004a,b, 2005c). Such consecutive C's can have intervening nucleotides, of course, such as CNNCCNNCCCNNC; yet if all of the C's are converted to U (and hence to T at the PCR step), then this suggests a region of significant single-strandedness. We use the criterion of ≥7 C's as our threshold for significant lengths to indicate single-strandedness. Thresholds of 5 and 6 give similar results, but thresholds of less than 5 are complicated by the conversion frequency of duplex DNA. Such conversion may be due to the natural degree of breathing of the DNA over the course of the bisulfite incubation period.

Different conditions are used to explore whether the bisulfite reaction conditions contribute towards the single-strandedness observed at the bcl-2 Mbr. When the bisulfite reaction is carried out at 55° or 37°, it does not change the conversion frequency significantly. However, we did not find much reactivity when the reaction was carried out at room temperature. Similarly, when the reaction pH is increased from 5.2 to 6.0, there is no significant difference in the pattern of single-strandedness for the regions that we have studied.

Though the bisulfite modification is clearly lower in the neighboring regions around the Mbr, we sequenced nine additional regions selected at random from the human genome to evaluate the specificity of extensive bisulfite conversion to the bcl-2 Mbr. Results show that seven out of the nine regions have 4-fold less conversion compared to the bcl-2 Mbr (see Suppl. Fig. 5A and B in Raghavan *et al.*, 2004a). In the remaining two regions, FEN-1 and pseudo-GAPDH, the C conversion frequency is less than for the bcl-2 Mbr. This suggests that, although there is some background conversion present at random locations in the genome (probably due to random or preferred sites of DNA breathing), the long regions of continuous bisulfite reactivity at the bcl-2 Mbr are due to a consistent and substantial structural deviation in about one-fourth of the Mbr alleles. We also find that there is no correlation between GC-content and bisulfite conversion frequency, as assessed on a wide variety of regions (data not shown), including the highly GC-rich immunoglobulin class switch regions (Yu *et al.*, 2003).

Full Molecule Bisulfite Modification Assay

The bisulfite modification assay described thus far provides information about only one strand of a duplex DNA molecule at a time. However, it is useful to get information about both strands of the same molecule, as that dramatically helps to predict the nature of the non-B DNA structure. One may think of this as "full-molecule" information, rather than the "half-molecule" information obtained when the two strands must be separated. Full-molecule analysis is not possible for chromosomal DNA, as the two DNA strands must be amplified for the bisulfite modification assay. Full-molecule analysis is possible in the case of *E. coli* plasmids bearing regions such as the bcl-2 Mbr, but on a very limited scale because most of the bacterial transformants contain plasmids that reflect information only from the top or the bottom strand of the original bisulfite-treated DNA (Raghavan *et al.*, 2005c).

Procedure. For the full-molecule bisulfite modification assay, sodium bisulfite treatment is carried out as described previously. In our study, we used a plasmid containing the DNA fragment of interest for bisulfite treatment (Raghavan *et al.*, 2005c). Following bisulfite treatment, the plasmid DNA is digested with suitable restriction enzymes. The fragment of interest is then gel purified and cloned into a new plasmid backbone (no bisulfite treatment on this vector fragment). The ligated products are transformed into a uracil glycosylase null strain of *E. coli*, BD1528 (CGSC strain 7221, Genetic Stock Center, Yale U., New Haven, CT), to obtain colonies that contain plasmid DNA with both top and bottom strand modification. The plasmid DNA is then extracted from such colonies and retransformed into *E. coli* to separate top and bottom strands derived from each single DNA molecule. Plasmid DNA is extracted from at least eight randomly selected colonies of these secondary transformations and sequenced using T7 and T3 primers.

Discussion. We have fully characterized a small number of molecules on both strands (see Fig. 3A and B in Raghavan *et al.*, 2005c). We find that the single-stranded regions for full-molecules correspond extremely well to the collective information obtained for populations of half-molecules. In particular, the regions of single-strandedness on the top and bottom portions match very well with the population information for the chromosomal DNA as well (see Fig. 1 in Raghavan *et al.*, 2005c). This information confirms that the collective half-molecule information is very similar, if not indistinguishable, to full-molecule information.

Potassium Permanganate and Osmium Tetroxide Chemical Probing

Osmium tetroxide (OsO_4) and potassium permanganate ($KMnO_4$) have been used previously to detect single-stranded regions of DNA *in vitro* (Lilley and Dahlberg, 1992a,b; Palecek, 1992). $KMnO_4$ has also been used to detect short single-stranded regions by permeation of living mammalian cells (Duncan *et al.*, 1994; Michelotti *et al.*, 1996). We use the $KMnO_4$ or OsO_4 treatment directly on cells for characterizing the bcl-2 Mbr because it can capture single-stranded regions that might exist even in the context of chromatin (Raghavan *et al.*, 2004a).

Chemistry

$KMnO_4$ and OsO_4 react with thymine in single-stranded oligonucleotides and DNA to yield cis-thymine glycol (5,6-dihydroxy-5,6-dihydrothymine) (Fig. 4). In addition to the thymine glycol, formation of cytosine glycol and 5,6-dihydroxycytosine also has been reported (Lilley and Dahlberg, 1992a). In the presence of tertiary amines, such as pyridine or TEMED, OsO_4 rapidly forms stable complexes with pyrimidine bases in single-stranded DNA (Boublikova and Palecek, 1990). Both OsO_4 and $KMnO_4$ react with the C5-C6 double bond of thymine (Fig. 4) when it is in ssDNA, but in the case of duplex DNA (B-DNA), the target C5-C6 double bond, which is in the major groove, is not accessible to these chemicals (Lilley and Dahlberg, 1992a).

Procedure

1. Treatment of cells with $KMnO_4$ or OsO_4 and purification of DNA (see Fig. 5). Cells are pelleted and washed twice in 1/10 culture volume of phosphate buffered saline at $4°$. Cells are resuspended in buffer ($\sim 1.5 \times 10^7$ cells/ml) containing 15 mM Tris-HCl (pH 7.5), 60 mM KCl, 15 mM NaCl, 5 mM $MgCl_2$, 0.5 mM EGTA, 300 mM sucrose, 0.5 mM DTT, and 0.1 mM PMSF. $KMnO_4$ from a freshly prepared stock (0.2 M) was added to a final concentration of 40 mM and incubated for 30 s on ice, whereas the control received an equal volume of TE. For OsO_4, different concentrations (0, 2, 5 mM) are added (along with 2 mM TEMED) to the cells and incubated for 15 min at room temperature. Reactions are stopped using equal volumes of stop solution (50 mM EDTA, 1% SDS, 400 mM 2-mercaptoethanol), which are also added to the untreated sample. DNA is recovered after two phenol:chloroform and one chloroform extraction. The DNA is then precipitated by ethanol precipitation with 400 mM NaCl. Piperidine cleavage is performed as described (Ausubel *et al.*, 1996). A positive control is prepared by digesting the chromosomal DNA with restriction enzyme instead of chemical treating and use for primer extension and LM-PCR.

2. Primer extension reaction on chemically modified DNA. Primer extension reactions are carried out by mixing the following on ice: 3 μg of previous piperidine cleaved DNA sample, $1\times$ thermo polymerase buffer [10 mM KCl, 10 mM $(NH_4)_2SO_4$, 20 mM Tris HCl (pH 8.8), and 0.1% Triton X-100], 4 mM $MgSO_4$, 200 $\mu$$M$ dNTPs, 0.25 $\mu$$M$ end-labeled oligomers, and 0.4 U vent (exo-) polymerase (NEB) after layering with

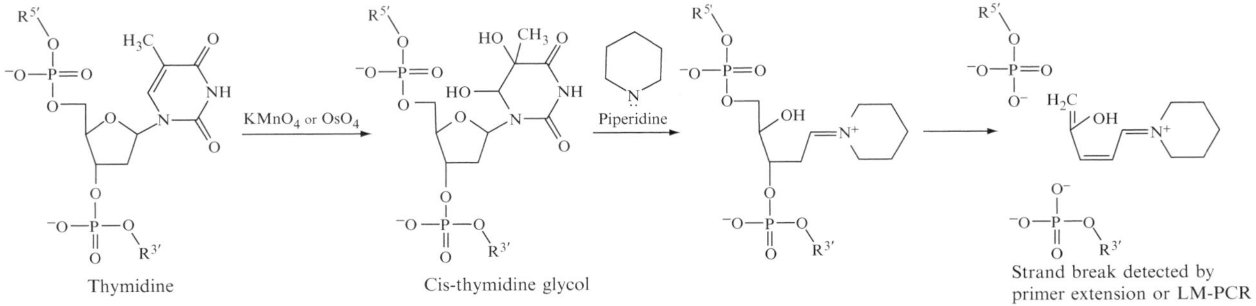

FIG. 4. Chemistry of the induction of nicks on DNA by $OsO_4/KMnO_4$ probing. Upon incubation with $OsO_4/KMnO_4$ the thymidine is converted to a cis-thymidine glycol (step 1). Incubation of this complex in the presence of piperidine at 90° will lead to the replacement of thymine by piperidine (step 2). In the final step (step 3), two nicks are generated, as shown, to release the piperidine moiety.

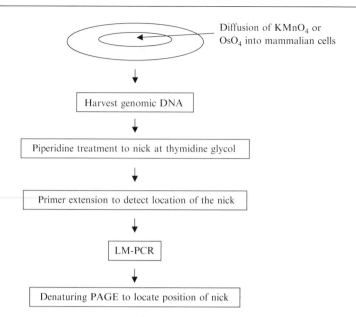

FIG. 5. Flow chart of the steps involved in the detection of OsO₄/KMnO₄ sensitivity within the mammalian cells.

mineral oil. Single-primer extension (linear amplification) of the primer is carried out on a PCR machine with the following conditions: 95° for 6 min, 55° for 3 min and 75° for 5 min (1 cycle). The products are purified by phenol:chloroform extraction, and the DNA is recovered by ethanol precipitation. The air dried pellet is resuspended in 15 μl of TE. By designing a proper set of primers for the top or bottom strands, one can also obtain strand-specific information.

   3. Ligation-mediated polymerase chain reaction.

   LM-PCR is performed as follows. In a 10-μl final volume, 50 μM linker is ligated with the primer-extended DNA (5 μl) in T4 DNA ligase buffer (1×, 50 mM Tris-HCl [pH 7.5], 10 mM MgCl₂, 10 mM DTT, 1 mM ATP, 25 μg/ml BSA). For each ligation, 1 U (1 U/μl stock) T4DNA ligase (Roche, Indianapolis, IN) is used and incubated for 20 h at 16°. The products are purified by phenol chloroform extraction, and the ligated DNA is recovered by ethanol precipitation. The air dried pellet is resuspended in 15 μl of TE. For amplification of the ligated products, one of the primers used is end-labeled with [γ-³²P]ATP (3000 Ci/mmol, PerkinElmer Life Sciences, Boston, MA) using T4 polynucleotide kinase (New England BioLabs, Beverly, MA), according to the manufacturer's instruction. Unincorporated radioisotope is removed by G-25 Sepharose (Amersham

Biosciences, Piscataway, NJ) spin-column chromatography. The PCR reactions are performed using a thermocycler (Stratagene, La Jolla, CA) under the following conditions: 1× reaction mix (10 mM Tris-HCl [pH 8.3], 50 mM KCl, 0.01% gelatin) with 1.5 mM MgCl$_2$, 400 μM dNTPs, and 2 units of Taq per 10 μl reaction.

4. Denaturing polyacrylamide gel electrophoresis to resolve the hot LM-PCR products. The previous PCR reactions are stopped by addition of the stop dye (47.5% formamide, 10 mM EDTA, 0.025% xylene cyanol, 0.025% bromophenol blue). The reaction products are resolved on an 8% (or 6%) denaturing polyacrylamide gel. The gel is dried and exposed to a phosphorimager screen, and the signal is detected using a Molecular Dynamics PhosphorImager 445SI (Molecular Dynamics, Sunnyvale, CA) and analyzed with ImageQuant software (v5.0).

Precautions

1. Duration and concentration of chemical treatment. This is very critical, especially for KMnO$_4$. Longer incubation and use of higher concentrations will result in DNA degradation.
2. Selection of appropriate primers is very critical. A/T rich primers may result in no primer extension products. Also, it is important to design overlapping primers with higher TM (60° or above) for primer extension and PCR.
3. It is important to standardize primer extension conditions for each primer separately.
4. At least 25 μM or higher linker concentration should be used for ligation.
5. The highest possible annealing temperature during the final PCR should be used. This will help to minimize nonspecific products.
6. Use restriction enzyme induced double-strand break as a positive control for the primer extension and LM-PCR.

Discussion. The results obtained after incubating the mammalian cells with increasing concentrations of OsO$_4$ or KMnO$_4$ can show that a region is highly sensitive compared to the neighboring DNA regions and to other genomic locations.

Conclusion

Both bisulfite assay and KMnO$_4$/OsO$_4$ assay are very useful techniques to study DNA structures within the human genome. However, both of these techniques have their own advantages and disadvantages. KMnO$_4$/ OsO$_4$ assays are very useful for giving information about the DNA structure within cells without interfering with the chromatin organization.

However, this is a bulk method. In contrast, bisulfite modification assays are extremely useful in giving information on an individual DNA molecule level. This assay, however, requires extraction of DNA from the cells.

Acknowledgments

We thank Dr. Ian Haworth, Dr. David M. Lilley, Dr. Hikoya Hayatsu, Dr. Kefei Yu, Ms. Sabrina Houston, and Dr. Geoffrey Cassell for helpful comments or discussions during the development of this approach.

References

Agazie, Y. M., Lee, J. S., and Burkholder, G. D. (1994). Characterization of a new monoclonal antibody to triplex DNA and immunofluorescent staining of mammalian chromosomes. *J. Biol. Chem.* **269,** 7019–7023.

Ausubel, F. M., Brent, R., Kingston, R. E., Moore, D. D., Seidman, J. G., Smith, J. A., and Struhl, K. (1996). "Current Protocols in Molecular Biology." John Wiley & Sons, New York.

Boublikova, P., and Palecek, E. (1990). Osmium tetroxide, N,N,N',N'-tetramethylethylenediamine, a new probe of DNA structure in the cell. *FEBS Let.* **263,** 281–284.

Duncan, R., Bazar, L., Michelotti, G., Tomonaga, T., Krutzsch, H., Avigan, M., and Levens, D. (1994). A sequence-specific, single-stranded binding protein activates the far upstream element of c-myc and defines a new DNA-binding motif. *Genes Dev.* **8,** 465–480.

Gough, G. W., Sullivan, K. M., and Lilley, D. M. (1986). The structure of cruciforms in supercoiled DNA: Probing the single-stranded character of nucleotide bases with bisulphite. *EMBO J.* **5,** 191–196.

Griffith, J. D., Comeau, L., Rosenfield, S., R. M. Stansel, Bianchi, A., Moss, H., and Lange, T. D. (1999). Mammalian telomeres end in a large duplex loop. *Cell* **97,** 503–514.

Hale, P., Woodward, R. S., and Lebowitz, J. (1980). *E. coli* RNA polymerase promoters on superhelical SV40 DNA are highly selective targets for chemical modification. *Nature* **284,** 640–644.

Hanvey, J. C., Klysik, J., and Wells, R. D. (1988). Influence of DNA sequence on the formation of non-B right-handed helices in oligopurine.oligopyrimidines inserts in plasmids. *J. Biol. Chem.* **263,** 7386–7396.

Hayatsu, H. (1976). Bisulfite modification of nucleic acids and their constituents. *Prog. Nucleic Acid Res. Mol. Biol.* **16,** 75–124.

Hayatsu, H., Wataya, Y., Kai, K., and Iida, S. (1970). Reaction of sodium bisulfite with uracil, cytosine and their derivatives. *Biochemistry* **9,** 2858–2865.

Hentschel, C. C. (1982). Homopolymer sequences in the spacer of a sea urchin histone gene repeat are sensitive to S1 nuclease. *Nature* **295,** 714–716.

Johnston, B. H. (1988). The S1-sensitive form of d(C-T)n.d(A-Gn): Chemical evidence for a three-stranded structure in plasmids. *Science* **241,** 1800–1804.

Kohwi, Y., and Kohwi-Shigematsu, T. (1988). Magnesium ion-dependent triple-helix structure formed by homopurine-homopyrimidine sequences in supercoiled plasmid DNA. *Proc. Natl. Acad. Sci. USA* **85,** 3781–3785.

Lee, J. S., Burkholder, G. D., Latimer, L., Haug, B., and Braun, R. P. (1987). A monoclonal antibody to triplex DNA binds to eukaryotic chromosomes. *Nucl. Acids Res.* **15,** 1046–1061.

Leonard, N. J., McDonald, J. J., Henderson, R. E. L., and Reichmann, M. E. (1971). Reaction of diethyl pyrocarbonate with nucleic acid components. Adenosine. *Biochemistry* **10**, 3335–3342.

Lilley, D. M. (1983). Structural perturbation in supercoiled DNA: Hypersensitivity to modification by a single-strand-selective chemical reagent conferred by inverted repeat sequences. *Nucleic Acids Res.* **11**, 3097–4012.

Lilley, D. M. J., and Dahlberg, J. E. (1992a). DNA Structures Part A: Synthesis and Physical Analysis of DNA. *In* "Methods in Enzymology" (J. N. Abelson and M. I. Simon, eds.), Vol. 211, pp. 3–506. Academic Press, Inc.

Lilley, D. M. J., and Dahlberg, J. E. (1992b). DNA Structures Part B: Chemical and Electrophoretic Analysis of DNA. *In* "Methods in Enzymology" (J. N. Abelson and M. I. Simon, eds.), Vol. 212, pp. 3–458. Academic Press, Inc.

Michelotti, G. A., Michelotti, E. F., Pullner, A., Duncan, R. C., Eick, D., and Levens, D. (1996). Multiple single-stranded cis elements are associated with activated chromatin of the human c-myc gene *in vivo*. *Mol. Cell. Biol.* **16**, 2656–2669.

Nordheim, A., Pardue, M. L., Lafer, E. M., Moller, A., Stollar, B. D., and Rich, A. (1981). Antibodies to left-handed Z DNA bind to interband regions of *Drosophila* polytene chromosomes. *Nature* **294**, 417–422.

Palecek, E. (1992). Probing DNA structure with osmium tetroxide complexes *in vitro*. *In* "Methods in Enzymology" (J. N. Abelson and M. I. Simon, eds.), Vol. 212. Academic Press, Inc.

Raghavan, S. C., Chastain, P., Lee, J. S., Hegde, B. G., Houston, S., Langen, R., Hsieh, C.-L., Haworth, I., and Lieber, M. R. (2005c). Evidence for a triplex DNA conformation at the bcl-2 Major breakpoint region of the t(14;18) translocation. *J. Biol. Chem.* **280**, 22749–22760.

Raghavan, S. C., Houston, S., Hegde, B. G., Langen, R., Haworth, I. S., and Lieber, M. R. (2004b). Stability and strand asymmetry in the non-B DNA structure at the bcl-2 major breakpoint region. *J. Biol. Chem.* **279**, 46213–46225.

Raghavan, S. C., Hsieh, C.-L., and Lieber, M. R. (2005b). Both V(D)J coding ends but neither signal end can recombine at the bcl-2 major breakpoint region and the rejoining is ligase IV dependent. *Mol. Cell. Biol.* **25**, 6475–6484.

Raghavan, S. C., and Lieber, M. R. (2004). Chromosomal translocations and non-B DNA structures in the human genome. *Cell Cycle* **3**, 762–768.

Raghavan, S. C., Swanson, P. C., Ma, Y., and Lieber, M. R. (2005a). Double-strand break formation by the RAG complex at the bcl-2 Mbr and at other non-B DNA structures *in vitro*. *Mol. Cell. Biol.* **25**, 5904–5919.

Raghavan, S. C., Swanson, P. C., Wu, X., Hsieh, C.-L., and Lieber, M. R. (2004a). A non-B-DNA structure at the Bcl-2 major break point region is cleaved by the RAG complex. *Nature* **428**, 88–93.

Rahmouni, A. R., and Wells, R. D. (1989). Stabilization of Z DNA *in vivo* by localized supercoiling. *Science* **246**, 358–363.

Romier, C., Dominguez, R., Lahm, A., Dahl, O., and Suck, D. (1998). Recognition of single-stranded DNA by nuclease P1: High resolution crystal structures of complexes with substrate analogues. *Proteins: Str. Fun. Genet.* **32**, 414–424.

Sakamoto, N., Chastian, P. D., Parniewski, P., Ohshima, K., Pandolfo, M., Griffith, J. D., and Wells, R. D. (1999). Sticky DNA: Self-association properties of long GAA.TTC repeats in R.R.Y. triplex structures from Friedreich's ataxia. *Mol. Cell* **3**, 465–475.

Sambrook, J., Fritsch, E. F., and Maniatis, T. (1989). "Molecular Cloning." Cold Spring Harbor Laboratory Press, New York.

Sasse-Dwight, S., and Gralla, J. D. (1989). KMnO4 as a probe for lac promoter DNA melting and mechanism *in vivo*. *J. Biol. Chem.* **264**, 8074–8081.

Secrist, R. J. A., Barrio, J. R., Leonard, N. J., Villar-Palasi, C., and Gilman, A. G. (1972). Structural perturbation in supercoiled DNA: Hypersensitivity to modification by a single-strand-selective chemical reagent conferred by inverted repeat sequences. *Science* **177,** 279–280.

Shapiro, R., Braverman, B., Louis, J. B., and Servis, R. E. (1973). Nucleic acid reactivity and conformation. II. Reaction of cytosine and uracil with sodium bisulfite. *J. Biol. Chem.* **248,** 4060–4064.

Shapiro, R., Cohen, B. I., and Servis, R. E. (1970). Specific deamination of RNA by sodium bisulphite. *Nature* **227,** 1047–1048.

Shapiro, R., DiFate, V., and Welcher, M. (1974). Deamination of cytosine derivatives by bisulfite: Mechanism of the action. *J. Am. Chem. Soc.* **96,** 906–912.

Sinden, R. R. (1994). "DNA Structure and Function." Academic Press, San Diego.

Soyfer, V. N., and Potaman, V. N. (1996). "Triple-Helical Nucleic Acids." Springer-Verlag, New York.

Ussery, D. W., and Sinden, R. R. (1993). Environmental influences on the *in vivo* level of intramoleular level of triplex DNA in *E. coli. Biochemistry* **32,** 6206–6213.

Yu, K., Chedin, F., Hsieh, C.-L., Wilson, T. E., and Lieber, M. R. (2003). R-loops at immunoglobulin class switch regions in the chromosomes of stimulated B cells. *Nature Immunol.* **4,** 442–451.

[18]   Detection and Structural Analysis of R-Loops

By Kefei Yu, Deepankar Roy, Feng-Ting Huang, and
Michael R. Lieber

Abstract

R-loops are structures where an RNA strand is base paired with one DNA strand of a DNA duplex, leaving the displaced DNA strand single-stranded. Stable R-loops exist *in vivo* at prokaryotic origins of replication, the mitochondrial origin of replication, and mammalian immunoglobulin (Ig) class switch regions in activated B lymphocytes. All of these R-loops arise upon generation of a G-rich RNA strand by an RNA polymerase upon transcription of a C-rich DNA template strand. These R-loops are of significant length. For example, the R-loop at the col E1 origin of replication appears to be about 140 bp. Our own lab has focused on class switch regions, where the R-loops can extend well over a kilobase in length. Here, methods are described for detection and analysis of R-loops *in vitro* and *in vivo*.

METHODS IN ENZYMOLOGY, VOL. 409   0076-6879/06 $35.00
                    DOI: 10.1016/S0076-6879(05)09018-X

Introduction

R-loops are structures where an RNA strand is base paired with one DNA strand of a DNA duplex, leaving the displaced DNA strand single-stranded. In all cases where R-loops have been described *in vivo*, the RNA strand is generated by an RNA polymerase transcribing a C-rich template, such that a G-rich RNA is generated. It is not yet clear in any of these cases whether the RNA:DNA duplex results as an extension of the usual 9 bp RNA:DNA hybrid of the RNA polymerase (called the extended RNA:DNA hybrid model) or arises by threading back of the RNA before the two strands of the DNA duplex reanneal (called the thread-back model).

Stable R-loops exist *in vivo* at prokaryotic origins of replication (Baker and Kornberg, 1988; Carles-Kinch and Kreuzer, 1997; Lee and Clayton, 1996; Masukata and Tomizawa, 1984, 1990), the mitochondrial origin of replication (Lee and Clayton, 1996), and mammalian immunoglobulin (Ig) class switch regions (in activated B lymphocytes) (Yu *et al.*, 2003). All of these R-loops arise upon generation of a G-rich strand by an RNA poly-merase upon transcription of a C-rich template strand. These R-loops are of significant length. For example, the R-loop at the col E1 origin of replication appears to be about 140 bp.

RNA:DNA associations are typically more stable than DNA:DNA associations for reasons that are not entirely clear (Roberts and Crothers, 1992). Among the possible reasons are the following. First, the RNA:DNA duplex is thought to adopt a structure that is closer to A-form than the B-form DNA duplex. A-form duplexes permit the $2'$-OH of residue i to form an extra hydrogen bond with the O4$'$ of residue i + 1. Second, the stacking within the RNA:DNA hybrid may permit a more stable structure. G-rich RNA strands that are complementary to C-rich DNA strands form particularly stable RNA:DNA hybrids relative to the corresponding DNA:DNA duplexes.

Association of RNA transcript with the DNA strand increases the molecular mass, as well as changes the conformation of the nucleic acid. If formed on supercoiled plasmid or short linear DNA (<3 kb), it can be readily detected by electrophoretic mobility shift (EMSA) (Daniels and Lieber, 1995; Yu *et al.*, 2003). This is a convenient way of assessing R-loop formation ability of a DNA fragment of interest. Recently, we adapted a chemical probing method called native sodium bisulfite sequencing that allows us to determine DNA secondary structures at sequence level reso-lution, as well as to detect rare R-loops on the chromosomes (Yu *et al.*, 2003). The principle of this method is to use the bisulfite anion to convert unpaired cytidine to uracil under nondenaturing conditions, followed by PCR amplification and sequencing (see Chapter 17 in this volume). DNA single-stranded regions are detected as sites of C to T changes in the final PCR product. For detection of rare R-loop structures in the chromosome, PCR amplification can be used to enrich for alleles in the R-loop structure. This was achieved by using a "converted" primer, which only anneals to bisulfite converted sites. We discuss the advantages and disadvantages of this PCR strategy in detail.

Procedures

Detection of R-Loop Formation Based on Gel Mobility

Procedure. Detection of R-loop formation on *in vitro* transcription. A DNA fragment of interest (e.g., a 2.2 kb class switch region, Sγ3) must be cloned onto a plasmid, and this region should be flanked by phage promoters (e.g., T7, T3, or SP6), which allow *in vitro* transcription by commercially available purified phage RNA polymerases. Plasmid DNA should be of high purity (e.g., prepared from a cesium chloride density gradient or a silica-based column). We typically transcribe 1 μg of plasmid in a 20 μl reaction mix containing 40 mM Tris-Cl, pH 8.0, 6 mM MgCl$_2$, 2 mM spermidine, 10 mM NaCl, 10 mM DTT, 0.5 mM each of rATP, rCTP, rGTP, and rTTP, and 15 $\sim$ 20 U of RNA polymerase (Promega, Madison, WI). The transcription mixture is incubated for 1 h at 37°. The transcription is terminated by incubation at 70° for 15 min. Free RNA is degraded by the addition of 1 μg of RNase A and a 30 min incubation at 37°. An identical reaction, except for addition of 0.5 U of RNase H (Promega) at the RNase A digestion step, was usually included as a control. Nucleic acid was purified by phenol and chloroform extraction and ethanol precipitation, and dissolved in 10 μl of TE buffer (10 mM Tris-Cl, pH 8.0, 1mM EDTA). Purified nucleic acid is then resolved on a 1% agarose gel run in 1$\times$ Tris-Borate-EDTA buffer (without ethidium bromide) for 2–3 h at 70 volts. The agarose gel is visualized on a UV-illuminator after staining with ethidium bromide (Fig. 1).

Precautions

1. Plasmids containing large G-rich or repetitive sequences are often unstable in the bacteria host, possibly due to the formation of R-loops upon prokaryotic transcription. Cloning using lacZ-based blue-and-white selection often only permits cloning of the insert in the nonphysiological orientation with regard to the transcription direction. Because of this, we deleted the lacZα promoter in pBlueScript-KS(+) by removing the Pvu II fg and reclosing the plasmid with a linker containing T7, T3

promoters and multiple cloning sites. We use this plasmid to clone all of our class switch sequences in either orientation.

2. It is important to use high quality plasmid DNA. Plasmid preparations containing large amounts of nicked or linear species complicate the assay, as R-loop formation is revealed as a change of plasmid mobility from the supercoiled to the nicked circular form.

3. It is critical that electrophoresis is carried out without any ethidium bromide. To remove trace amounts of ethidium bromide contamination, the buffer tank, gel tray, and comb can be washed thoroughly with detergent and soaked in detergent water overnight followed by extensive rinsing with de-ionized water.

4. Sometimes, the transcription mixture becomes turbid after heat inactivation of the RNA polymerase, especially with T7 RNA polymerase. A brief spin generates a small white pellet that contains most of the nucleic acid. This is usually caused by oxidized (old) DTT or contaminants in the water. Use of fresh DTT and milli-Q water is recommended.

Detection of R-loop formation on episomes harvested from the mammalian cell nucleus. The DNA fragment of interest can be cloned into an episome that replicates in mammalian cell nucleus. We have used an EBV-based episome, pREP4 (Invitrogen, Grand Island, NY), which can be

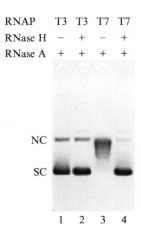

RNAP	T3	T3	T7	T7
RNase H	−	+	−	+
RNase A	+	+	+	+

NC

SC

1 2 3 4

Fig. 1. *In vitro* transcription of class switch sequences. A plasmid containing the full Sγ3 region was transcribed with phage RNA polymerases. Purified nucleic acid was resolved on a 1% agarose gel in the absence of ethidium bromide. The gel was stained with ethidium bromide after electrophoresis. SC, supercoiled position; NC, nicked circular position; RNAP, RNA polymerase. Other components of the reactions were indicated above the gel image.

stably maintained in the cell. The DNA fragment of interest (e.g., Sγ3) is cloned downstream of the RSV-LTR promoter in both orientations (one serves as a control). The episome is then transfected into 293/EBNA1 cells (available from ATCC or Invitrogen). We grow 293/EBNA1 cells in DMEM high glucose medium supplemented with 10% fetal bovine serum and 100 µg/ml of penicillin and streptomycin. The 293/EBNA1 are very easy to transfect. We routinely transfect 293/EBNA1 cells by calcium phosphate precipitation (10 µg plasmid/10 cm dish). Untransfected cells are killed by adding 200 µg/ml of hygromycin to the culture medium. Transfected cells are maintained in culture for about 7 days before Hirt harvest of the episome DNA. Briefly, cells are collected and washed once with 1× phosphate-based saline (PBS). About 2×10^7 cells are resuspended in 2.91 ml of 10 mM Tris-Cl and 10 mM EDTA (Hirt buffer) before the addition of 90 µl of 20% SDS (final concentration 0.5%). Cells are lysed by inverting the tube 7–10 times. One-quarter volume (750 µl) of 5 M NaCl is added to the cell lysate, and the tube is inverted immediately 7–10 times. White precipitates will form upon mixing with NaCl. The mixture is kept at 4° for at least 24 h. Cell debris and genomic DNA are removed as precipitates after centrifugation for 30 min at 10,000 rpm. Clear supernatant containing episomal DNA is transferred to a new tube and digested with proteinase K (final concentration = 100 µg/ml) for 1 h at 55° (or 37° overnight). The proteinase K digested supernatant is then extracted with phenol and chloroform, and the DNA is recovered by ethanol precipitation. Finally, the recovered DNA is resuspended in 50 µl of TE buffer, pH 8.0 (for DNA from $1–2 \times 10^7$ cells).

For detection of R-loop on the harvested episome, 10 µl of Hirt harvest (about 10 ng episomal DNA) is digested with appropriate restriction enzymes to release the DNA fragment of interest. Dpn I, which recognizes bacterial-specific adenine-methylated GA^mTC sites, is included in the digestion mixture to degrade unreplicated episomes. Digested DNA is then analyzed by Southern blot. In the case of our episomes containing the murine Sγ3 repeats (Fig. 2), digested DNA was resolved on a 0.8% agarose gel electrophoresed in TBE buffer at 70 volts for 2 h. DNA was transferred to a nylon membrane and probed with the radiolabeled full-length Sγ3 probe. Gels are exposed to phosphorimager screens, scanned on a Molecular Dynamics imager 445SI (Sunnyvale, CA), and analyzed with Image-Quant software version 5.0. The RNA::DNA hybrid (R-loop) is detected as a zone of smeary bands above the normal position of the double-stranded DNA fragment (Fig. 2).

Discussion. It is normal that the Hirt harvest is contaminated with a small amount of genomic DNA. Genomic DNA contamination can be reduced by increasing the incubation time after the addition of NaCl

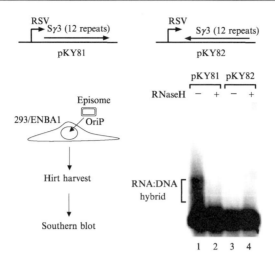

Fig. 2. RNA:DNA hybrid formation on a stable human mini-chromosome. Large arrows indicate orientation of the 12 Sγ3 repeats. Transcription initiates from the RSV promoter as indicated. Episomes harvested from 293/ENBA1 cells were digested and resolved on an agarose gel. The resolved DNA was then transferred to a nylon-membrane and probed with murine Sγ3. The arrow indicates the position of the unshifted switch fragment. The smear above the dsDNA band indicates the RNA:DNA hybrids.

(e.g., up to 3 days). However, the amount of genomic DNA present in the Hirt harvest does not interfere with this type of assay.

Structural Analysis of the R-Loop with Native Sodium Bisulfite Probing and DNA Sequencing

Sodium bisulfite sequencing is widely used in studying DNA methylation because methylated cytidine residue reacts very slowly with the bisulfite anion as compared to unmethylated cytidines. Unmethylated cytidines are converted to uracils via deamination during the sodium bisulfite treatment. To fully convert the unmethylated cytidines, the DNA template needs to be fully denatured because base paired cytidines are not accessible to bisulfite, which must carry out its nucleophilic attack from above or below the plane of the cytidine ring at the 5–6 C-C bond. We exploited this unique property of sodium bisulfite action on DNA to determine the structure (DNA base pairing status) of R-loops formed *in vitro* and *in vivo* in nondenaturing condition. For more detailed information about the chemistry and application of sodium bisulfite reaction, readers should refer to Chapter 17 in this volume.

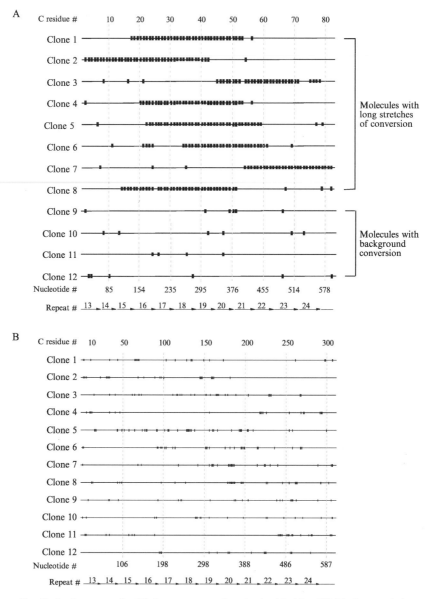

FIG. 3. *In vitro* transcribed Sγ3 sequences analyzed using bisulfite. (A) Single-strandedness of the G-rich nontemplate strand after *in vitro* transcription. A plasmid containing 12 repeats of Sγ3 was transcribed with T7 RNA polymerase *in vitro* (physiological transcriptional orientation) to form a stable RNA:DNA hybrid. The RNA:DNA hybrid was then treated with bisulfite followed by PCR amplification. The PCR products were cloned and fully

Procedure. One μg of plasmid (or transcribed plasmid) in 30 μl of distilled water is mixed with 12.5 μl of 20 mM hydroquinone and 457.5 μl of 2.5 M sodium bisulfite (pH 5.2). The mixture is sealed with mineral oil in a 500-μl microcentrifuge tube and incubated for 16 h at 37° in the dark. Bisulfite-treated DNA is then purified with the Wizard DNA clean-up system (Promega) according to the manufacturer's instructions. Briefly, 1 ml Wizard resin is used to bind DNA. The resin is then packed into a mini-column set on a vacuum apparatus. The resin is washed with 2 ml 80% isopropanol twice and dried by centrifuging the minicolumn inside an Eppendorf tube for 1 min at 14,000 rpm. DNA is eluted by adding 50 μl of prewarmed (55°) TE, pH 8.0, and centrifuged for 1 min in a new Eppendorf tube. Purified bisulfite-treated DNA is desulfonated with 0.3 M NaOH (by adding 86 μl of water and 24 μl of 2 M NaOH) at 37° for 15 min. Desulfonated DNA is recovered by ethanol precipitation and resuspended in 20 μl TE, pH 8.0.

PCR using bisulfite-modified DNA as the template is done with a pair of primers flanking the DNA fragment of interest. PCR products are resolved on agarose gels and the correct-size fragment is recovered (necessary only if multiple PCR bands are present). Purified PCR product is cloned using the TOPO-TA cloning kit (Invitrogen, Carlsbad, CA). Plasmid DNA from each clone is purified using the GenElute Plasmid Miniprep Kit (Sigma, St. Louis, MO). Sequencing reactions are carried out using a SequiTherm Excel II sequencing kit (Epicentre, Madison, WI) and MWG thermal cycler model primus 96 plus (MWG Biotech, High Point, NC). Automated sequencing is carried out using a Li-Cor DNA analyzer Model 4200 (Li-Cor, Lincoln, NE). A typical analysis is shown in Fig. 3.

Discussion. Sodium bisulfite treatment significantly reduces the amount of intact DNA, possibly due to damage by the action of free radicals (Hayatsu, 1976). Treated plasmid usually fails to transform *E. coli*. It is estimated that only 10–20% of the input DNA is recoverable after the treatment. Accordingly, the following PCR amplification requires seemingly more than the normal amount of DNA template.

---

sequenced. Each long horizontal line represents a single clone. A total of 81 C residues within the cloned switch region were evenly displayed along the horizontal line. The short bold vertical marks indicate those C residues that have been converted to U. C residue numbers are indicated on the top of the figure in increments of 10. Actual nucleotide positions are indicated below the line and are not to scale. Each horizontal arrow on the bottom represents a 49 bp Sγ3 repeat (repeats 13–24 of the 41 repeats in Sγ3). (B) The C-rich strand of the *in vitro* R-loop is reactive with bisulfite at only background levels. Twelve molecules derived from the C-rich strand are shown. All of the symbols are the same as in (A).

Cloned PCR product can be categorized into two groups, depending on which strand shows C to T conversion (the other strand G to A, accordingly). Each group is derived from either one of two DNA strands of the original template. This distinction is caused by noncomplementarity of the two DNA strands after sodium bisulfite treatment. Rare, sporadic C to T conversion can be found essentially in every cloned PCR product even from bisulfite-treated cannonical B-form DNA, reflecting a background level of sodium bisulfite reactivity. This is most likely due to the transient DNA breathing during the long incubation even at 37°. This background level of conversion sometimes facilitates the distinction between the two groups of molecules (top versus bottom strand origin). Occasionally, there are cloned PCR products showing C to T conversion on both strands. These molecules are called mosaic molecules that are most likely generated by template switching during the PCR amplification (Ford et al., 1994). These rare mosaic molecules (<10%) are excluded in our analysis.

As mentioned previously, methylated cytidines are protected against bisulfite-mediated deamination. If the plasmid DNA is prepared from a dcm^+ bacterial strain, the second cytidine residue at sites CC(A/T)GG may be methylated by the bacterial dcm methyltransferase. This, in some cases, may interfere with the structural analysis. Therefore, propagation of plasmid in dcm^- bacterial strain is strongly recommended.

Detection of R-Loops at Ig Class Switch Regions in Genomic DNA

Procedure. Induction of class switch recombination in cultured mouse naive spleen B cells. Mouse spleen B cells are purified by magnetic cell sorting (Miltenyi, Auburn, CA) according to the manufacturer's instructions. Briefly, a mouse spleen is removed aseptically and rinsed once in 10 ml RPMI1640 medium in a petri dish. The spleen is transferred to another petri dish with 10 ml fresh RPMI1640 medium and ground using the frosty regions between two sterile glass slides. The homogenate is transferred to a 15-ml conical tube on ice, and the big chunks are permitted to settle for 10 min. The single-cell suspension is transferred to a new 15-ml conical tube and centrifuged for 10 min at 500g at 4° to collect the cells. Cells are resuspended in 10 ml ACK red cell lysis buffer (150 mM NH$_4$Cl, 10 mM KHCO$_3$, 0.1 mM EDTA, pH 7.2–7.4) and incubated for 5 min at room temperature to lyse the red blood cells. Intact white blood cells are washed twice in 1× PBS containing 1% bovine serum albumin (BSA). Cells are resuspended in 900 μl of 1× PBS with 1% BSA and mixed with 100 μl anti-CD43 microbeads (Miltenyi, Auburn, CA) for 30 min at 4°. An LS25 column (Miltenyi) is attached to a magnetic stand (Miltenyi) and the column is balanced with 1× PBS plus 1% BSA. The column is washed three times with 3 ml 1× PBS plus 1% BSA, and the flow-through (CD43⁻) is collected. Cell numbers are determined by using a hemocytometer, and a fraction (e.g., 20%) of the cells are harvested as unstimulated B cells. Other CD43⁻ cells are resuspended in complete RPMI1640 supplemented with 10% FBS, 100 μg/ml Pen-Strep, 50 μM β-mercaptoethanol, and 20 μg/ml of lipopolysaccharide (LPS) at 2×10^5 cells/ml and cultured in a 37°, 5% CO_2 humidified cell incubator for 2 days.

Genomic DNA from mouse B cells is harvested by lysing the cells in TE buffer, pH 8.0 plus 0.5% sodium dodecyl sulfate (SDS) followed by proteinase K digestion overnight at 37°. Typically, 1×10^7 cells are harvested and resuspended in 9.65 ml TE, pH 8.0. Cells are lysed by adding 250 μl of 20% SDS and mixing by inverting the tube several times. One hundred microliters of 20 mg/ml proteinase K is added to the cell lysate, and the mixture is incubated overnight at 37°. The genomic DNA is then extracted with phenol and chloroform, and precipitated in ethanol. Air-dried genomic DNA pellets are dissolved in TE, pH 8.0 at about 1 mg/ml (a total of 10 μg genomic DNA can be expected from 10^6 cells). Before bisulfite treatment, genomic DNA is digested with *Eco*RI to reduce the size of the DNA. We typically digest 10 μg DNA in a 50 μl reaction incubated overnight at 37°. Spermidine (up to 4 mg/ml) can be added to improve the digestion. For RNase H-treated control, two units of RNase H (Promega) are added during the restriction enzymatic digestion. Digested genomic DNA is again extracted with phenol and chloroform, and precipitated in ethanol.

Sodium bisulfite treatment and PCR amplication of R-looped alleles is done as follows. After purification of the digested genomic DNA, 5 μg of DNA in 30 μl of water is treated with sodium bisulfite at 37° overnight as described above. Treated DNA is resuspended in 20 μl of TE, pH 8.0. For PCR amplification, 1 to 2 μl of DNA is used as template.

Due to a low percentage of alleles in the R-loop form in B cells stimulated with LPS and an overwhelming strand bias during amplication (>95% PCR clones were derived from the C-rich template strand), we developed a PCR strategy that can enrich the PCR products from template in the R-loop form. For amplification of the murine Sγ3 locus, PCR was carried out using a unconverted (regular) primer located about 500 bp upstream of the Sγ3 region and a converted primer located at the 19th repeat of Sγ3 (equivalent to the 23rd repeat for the Balb/c strain). A converted primer is a primer whose sequence matches the anticipated conversions due to deamination of C to yield U. A PCR product is expected only if the priming region for the converted oligonucleotide is single-stranded and is, therefore, converted by bisulfite. Regular Taq DNA polymerase

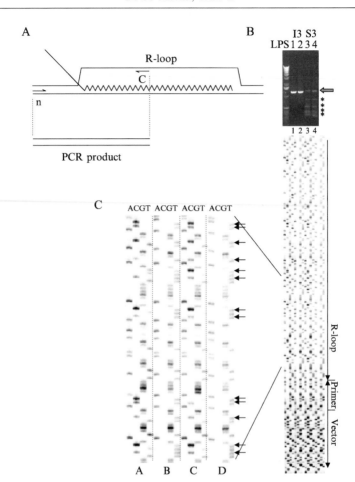

FIG. 4. *In vivo* structure of Sγ3 after LPS stimulation. (A) Scheme of G-rich strand amplification. The zig-zag line represents the G-rich RNA in the RNA:DNA hybrid at Sγ3 sequences. The primer located within the switch region ("c" for converted primer) is complementary to the bisulfite converted G-rich strand. The primer located outside of the switch region is complementary to the unconverted C-rich strand ("n" for native primer). Only templates that were accessible to bisulfite, and hence, single-stranded, were capable of converted primer annealing and exponential amplification. (B) PCR product using the converted/native primer pair. Bisulfite-treated DNA from unstimulated or stimulated mouse spleen B cells was PCR amplified. The PCR product of the correct size was cloned and sequenced. As a control for the overall template quality for PCR, a region upstream of Sγ3, including the Iγ3 exon, was also amplified. Lane 1, Control PCR using unstimulated B cell template. Lane 2, Control PCR using stimulated B cell template. Lane 3, Sγ3 G-rich strand specific PCR using unstimulated B cell template. Lane 4, Sγ3 G-rich strand specific PCR using stimulated B cell template. The block arrow indicates the expected size PCR product. Asterisks indicate PCR by-products due to mis-priming in the most similar repeats of the

failed to amplify the Sγ3 genomic region. We have successfully used AccuTaq LA polymerase (Sigma) in this case. Multiple bands of PCR product were evident, in our case, owing to the repetitiveness of the Sγ3 region. A typical PCR reaction (for amplifying the Sγ3 region with primer set KY267 and KY262) contains 1× AccuTaq LA buffer (Sigma), 400 μM of each dATP, dTTP, dGTP, and dCTP, 0.5 μM of native outside primer, 0.5 μM of converted internal primer, 2 μl of sodium-bisulfite treated DNA (~100 ng actual DNA), 0.5 M betaine (Sigma, B-0300), and 1 unit of AccuTaq LA DNA polymerase in a 20 μl reaction. Cycling is carried out on a Robocycler (Stratagene, La Jolla, CA) as follows: 95° for 3 min, 1 cycle; (94° for 30 sec; 60° for 40 sec, 68° for 5 min), 30 cycles; 68° for 10 min, 1 cycle. Only the PCR product of the correct size is purified and cloned for sequencing. CpG site are very rare in the switch regions. Therefore, mammalian CpG methylation has minimal effect on our analysis.

Discussion. Class switch regions are G-rich on the nontemplate DNA strand and C-rich on the template strand. Copying a G-rich template strand by thermostable DNA polymerases is much less efficient than that of a C-rich template strand, especially with a large amplicon. This results in a high bias in the final pool of PCR product from the bisulfite treated DNA. When amplifying the Sγ3 region, more than 95% of the cloned PCR product is derived from the C-rich template strands. This distinction is only specific to bisulfite-treated DNA template, as the two template strands are completely complementary in the normal PCR reaction. Because of the large degree of strand bias in the PCR, we were not able to obtain enough PCR clones derived from the G-rich strand (nontemplate) by using un-converted primers outside of the switch region. Instead, we have to use an enrichment method that selectively amplifies the G-rich strand in the R-loop form (single-stranded) based on its reactivity to bisulfite modification. This is achieved by using a primer within the switch region, and the sequence of that primer is modified such than it only anneals to the DNA template where all the C residues at that site are converted to U by sodium bisulfite. A limitation of this enrichment method is that the proportion of molecules that display long R-loops will not reflect the actual frequency of formation of this structure in the genome. A second limitation is that the size of the R-loops recovered after this enrichment will be underestimated.

---

switch region. (C) Extensive continuous C to U conversion on the G-rich strand of Sγ3 in stimulated mouse B cells. A portion of a sequencing gel shows the presence of long stretches of C to U (T) conversion on the G-rich strand of the switch region in stimulated B cells. Each sequencing lane was labeled with the complementary nucleotide to facilitate the reading of the G-rich strand sequence. Four sequenced samples are presented as examples.

The extent of detected R-loops will be limited to the region between the two priming sites. No information can be gained concerning the portion of Sγ3 outside of the amplification region, and this will lead to an underestimation of the sizes of the structures recovered from the genome. Similarly, any R-loop that is not positioned to encompass the converted primer site will fail to be detected by this method, even if it is within the amplified region. A third limitation is that it is not possible to make any conclusion about the base pairing status of the corresponding C-rich DNA strand. Even a large number of C-rich strand-derived PCR clones can be generated with a pair of native primers, the vast majority of these clones are simply derived from the normal duplex DNA. It is impossible to determine how many of these C-rich strand molecules are derived from the alleles in the R-loop form. Despite these limitations, recovery of any molecule displaying an extensive single-strandedness on the G-rich strand that is sensitive to RNase H treatment would be clear evidence that R-loop structures can form.

Ideally, PCR products should all come from alleles in the R-loop form using this strategy. In reality, amplification can occur if there are substantial background conversions at the priming site for the converted primer that allow the primer to anneal. Amplification from a genuine R-loop or B-form duplex accidentally converted at the converted primer binding site can be distinguished after sequencing the PCR product (with the obvious exception where the priming site happens to be on the edge of an R-loop). PCR products from real R-loops contain a long tract of continuous C to T conversion (up to 1 kb) starting from the converted primer site. These are striking on a sequencing gel because of the appearance of a long empty C lane (Fig. 4, molecule B and D). In contrast, PCR products derived from non-R-loop alleles with conversions at the priming site only show rare, sporadic C to T conversion across the entire amplicon, as was the case for PCR products generated from unstimulated B cell DNA.

References

Baker, T. A., and Kornberg, A. (1988). Transcriptional activation of initiation of replication from the *E. coli* chromosomal origin: An RNA-DNA hybrid near *oriC*. *Cell* **55**, 113–123.
Carles-Kinch, K., and Kreuzer, K. N. (1997). RNA/DNA hybrid formation at a bacteriophage T4 replication origin. *J. Mol. Biol.* **266**, 915–926.
Daniels, G. A., and Lieber, M. R. (1995). RNA:DNA complex formation upon transcription of immunoglobulin switch regions: Implications for the mechanism and regulation of class switch recombination. *Nucl. Acids Res.* **23**, 5006–5011.
Ford, J. E., McHeyzer-Williams, M. G., and Lieber, M. R. (1994). Chimeric molecules created by gene amplification interfere with the analysis of somatic hypermutation of murine immunoglobulin genes. *Gene* **142**, 279–283.

Hayatsu, H. (1976). Bisulfite modification of nucleic acids and their constituents. *Prog. Nucleic Acid Res. Mol. Biol.* **16**, 75–124.

Lee, D. Y., and Clayton, D. A. (1996). Properties of a primer RNA-DNA hybrid at the mouse mitochondrial DNA leading-strand origin of replication. *J. Biol. Chem.* **271**, 24262–24269.

Masukata, H., and Tomizawa, J. (1984). Effects of point mutations on formation and structure of the RNA primer for ColE1 DNA replication. *Cell* **36**, 513–522.

Masukata, H., and Tomizawa, J. (1990). A mechanism of formation of a persistent hybrid between elongating RNA and template DNA. *Cell* **62**, 331–338.

Roberts, R. W., and Crothers, D. M. (1992). Stability and properties of double and triple helices: Dramatic effects of RNA or DNA backbone composition. *Science* **258**, 1463–1466.

Yu, K., Chedin, F., Hsieh, C.-L., Wilson, T. E., and Lieber, M. R. (2003). R-loops at immunoglobulin class switch regions in the chromosomes of stimulated B cells. *Nature Immunol.* **4**, 442–451.

[19] The *Delitto Perfetto* Approach to *In Vivo* Site-Directed Mutagenesis and Chromosome Rearrangements with Synthetic Oligonucleotides in Yeast

By Francesca Storici and Michael A. Resnick

Abstract

In vivo genome manipulation through site-directed mutagenesis and chromosome rearrangements has been hindered by the difficulty in achieving high frequencies of targeting and the intensive labor required to create altered genomes that do not contain any heterologous sequence. Here we describe our approach, referred to as *delitto perfetto*, that combines the versatility of synthetic oligonucleotides for targeting with the practicality of a general selection system. It provides for an enormously wide variety of genome modifications via homologous recombination. Exceptional high frequencies of mutations are reached when a site-specific double-strand break (DSB) is induced within the locus targeted by the synthetic oligonucleotides. Presented in this chapter is an in-depth description of a series of applications of the *delitto perfetto* strategy for mutagenesis and chromosome modification both with and without the induction of a DSB, along with the procedures and materials.

Overview

Site-directed mutagenesis systems in which specific DNA sequences are targeted for alteration *in vitro* have been instrumental in dissecting genetic pathways and gene regulation and in understanding structure-function

METHODS IN ENZYMOLOGY, VOL. 409   0076-6879/06 $35.00
                                                   DOI: 10.1016/S0076-6879(05)09019-1

relationships in proteins. Often there is a need for direct *in vivo* modification. However, the modification of genomic DNA within cells, without retention of heterologous sequences, is generally an elaborate and inefficient process that includes multiple cloning steps and extensive DNA sequencing of each mutant allele that is created (Erdeniz *et al.*, 1997; Langle-Rouault and Jacobs, 1995; Scherer and Davis, 1979).

There has been a longstanding interest in developing systems for the effective *in vivo* modification of chromosome structure and chromosomal sequences. Oligonucleotides have been proposed for creating localized mutational changes and were initially shown to target homologous chromosome sequences in yeast by Moerschell *et al.* (1988). The approach was restricted to the generation of mutants with a selectable phenotype. Other oligonucleotide-based strategies for genome modification include triple-helix-forming oligonucleotides (TFOs) (Barre *et al.*, 2000; Vasquez *et al.*, 2000) and chimeric RNA/DNA oligonucleotides folded into a double hairpin conformation (Kmiec *et al.*, 2000; Liu *et al.*, 2001). However, since these oligonucleotide approaches are limited to the generation of point mutations and there is no genetic selection system, extensive screening is required. The *delitto perfetto* approach developed in the budding yeast *Saccharomyces cerevisiae* (Storici and Resnick, 2003; Storici *et al.*, 2001) is the only oligonucleotide-based genome targeting system that provides direct selection for almost any chromosomal modification. Presented in this chapter is the rationale for the original *delitto perfetto* approach followed by a detailed description of the break-mediated oligonucleotide targeting system (Storici *et al.*, 2003).

The *delitto perfetto* approach that was developed in yeast is a simple and rapid method for *in vivo* genome mutagenesis using synthetic oligonucleotides. It eliminates many of the steps required to produce multiple changes in the genome. The name ascribed to this versatile and accurate system for *in vivo* targeted mutagenesis derives from the idiomatic Italian term for perfect murder since there is complete elimination (i.e., perfect deletion) of the marker sequences that are used for selection. Mutagenesis is accomplished in two steps (Fig. 1), each of which requires the homologous recombination function of Rad52 (Storici *et al.*, 2003). The first step involves integration of a *CO*unterselectable *RE*porter (CORE) cassette that is integrated into a desired genomic locus using standard targeting procedures. The second step is accomplished by transformation with the appropriate targeting oligonucleotides such that the CORE cassette is lost and the desired change is generated. Since no heterologous sequence remains, mutagenesis is accomplished via a self-cloning process. Once the CORE cassette is inserted, the resulting strain provides many opportunities, as

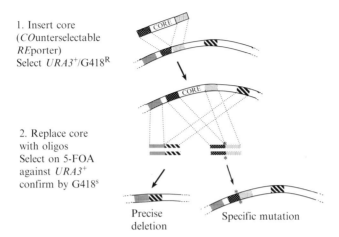

1. Insert core
(*CO*unterselectable
*RE*porter)
Select *URA3*⁺/G418ᴿ

2. Replace core
with oligos
Select on 5-FOA
against *URA3*⁺
confirm by G418ˢ

Precise   Specific mutation
deletion

FIG. 1. Schematic drawing of the *delitto perfetto* system that uses *I*ntegrative *R*ecombinant *O*ligonucleotides (IROs). This illustrates the deletion of a sequence and the creation of a specific point mutation. In Step 1, a CORE (*CO*unterselectable *RE*porter) cassette with *KlURA3* (counterselectable) and *kanMX4* (reporter) is inserted by standard DNA targeting procedures at a DNA sequence. The insertion site is anywhere in the sequence that has been chosen to be deleted or is close to a site where a specific mutation is to be created. In Step 2, cells containing the CORE cassette are transformed with IROs. This leads to loss of the CORE cassette and deletion of the desired region or introduction of the desired mutation (*). Generic DNA sequences are indicated as stippled or striped boxes. In this example the IROs have a short overlap. (This model is printed with permission from Storici *et al.*, 2001.)

described later, for mutation over a large distance from the CORE. This is particularly relevant to structure-function studies, since many different mutations can be targeted to sequences coding for key protein motifs.

Following the initial development of the *delitto perfetto* system, a subsequent approach, known as break-mediated *delitto perfetto*, took advantage of efficient recombinational repair of double-strand breaks in yeast (Storici *et al.*, 2003). The CORE cassette was modified to include both a galactose regulatable I-*Sce*I endonuclease and an I-*Sce*I cut site (Colleaux *et al.*, 1986; Plessis *et al.*, 1992). Generation of a DSB prior to oligonucleotide transformation (Fig. 2) increased oligonucleotide targeting more than 1000-fold, reaching efficiencies as high as 20% of all cells. This level is over 2 orders of magnitude higher than any reported DNA integration frequency in yeast. Surprisingly, we found that the DSB can also strongly stimulate recombination with single-strand DNA, suggesting new twists on present models of DSB repair (Storici *et al.*, 2003, submitted). The extremely high transformation frequencies and versatility of the break-mediated *delitto perfetto* system has

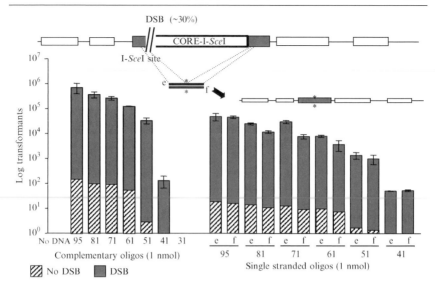

Fig. 2. Factors affecting IRO targeting efficiency: complementary pairs, single IRO molecules, size, and presence of a DSB. Presented are the frequencies of targeting by various IROs to a CORE-I-*Sce*I cassette inserted in the *TRP5* gene. The diagram above the graph shows the position of the DSB and the result of the oligonucleotide-targeting event. The IROs examined are referred to as "e" and "f", fully complementary with equal lengths of homology to both sides of DSB and from 31 to 95 nucleotides in length. The types of IROs analyzed are indicated below the bars. The amount of total IRO DNA used in each transformation is 1 nmole (~10–30 μg), unless otherwise indicated. The frequency of Trp$^+$ transformants are presented as the Log$_{10}$ Transformants, where the transformants correspond to the total transformants per 10^7 viable cells plated from the transformation mix. The vertical bars represent the average values from three to six determinations and include standard deviations. Dark-grey bars represent values from pairs of fully complementary IROs or single-strand IROs, and striped bars represent values from no-DSB controls (corresponding to cells not incubated in galactose media). The control value (no DNA) corresponds to no Trp$^+$ colonies.

resulted in powerful new tools for rapid genome modification as well as dissection of DNA repair mechanisms.

Presented later are protocols for the development and use of *delitto perfetto* systems that provide for targeted olignoucleotide recombination events. The first part describes the development of a CORE and transformation procedures. In the second part, emphasis is placed on the break-mediated *delitto perfetto* approach. Presented in Fig. 2 (striped bars) are frequencies of targeting with different sizes of complementary and single IROs using the CORE without a DSB generating system. The frequencies are much higher if targeting is mediated by a DSB (dark-grey bars in Fig. 2).

Creation and Utility of *Delitto Perfetto* Systems

The Delitto Perfetto *System Without a DSB*

CORE Cassettes. There are five CORE cassettes available (Fig. 3A). The original CORE cassette (Storici *et al.*, 2001), which contains the counterselectable *URA3* gene from *Kluyveromyces lactis* (*KlURA3*) and the reporter geneticin (G418) resistance gene *kanMX4*, is amplified as a 3.2 kb DNA fragment from pCORE using chimeric 70-mers consisting of 50 nucleotides homologous to the appropriate flanking region of the genomic target locus plus, as shown here, 20 nucleotides that allow for the amplification of the CORE-cassette:

P.1: 5'-...GAGCTCGTTTTCGACACTGG-3' for the *kanMX4* side
P.2: 5'-... TCCTTACCATTAAGTTGATC-3' for the *KlURA3* side-.

The genomic target site can be anywhere in a sequence that has been chosen to be deleted or within 100 bp of a site at which a specific mutation is to be created. Four additional CORE cassettes have been constructed in order to make the *delitto perfetto* approach applicable to a much wider set of yeast strains, including those that are URA$^+$ and G418 resistant. Two heterologous markers have been utilized: a reporter that provides resistance to hygromycin (*hyg*) (Goldstein and McCusker, 1999) and a new, counterselectable marker coding for the p53 mutant V122A (Storici and Resnick, 2003). High expression of this mutant p53 results in growth inhibition and toxicity (Inga and Resnick, 2001). These markers along with the previous markers (*KlURA3* and *kanMX4*) have been used in the construction of the following cassettes (see Fig. 3A): CORE-UK (*KlURA3* and *kanMX4*), CORE-UH (*KlURA3* and *hyg*), CORE-Kp53 (*kanMX4* and *GAL1/10*-p53), and CORE-Hp53 (*hyg* and *GAL1/10*-p53) (Storici and Resnick, 2003). The CORE-UK, -UH, -Kp53, and -Hp53 cassettes are amplified as 3.2, 3.5, 3.7, and 4.0 kb DNA fragments, respectively, from the corresponding vectors using chimeric 70-mers consisting of 50 nucleotides that are homologous to the flanking region of the genomic target locus plus the following 20 common nucleotide sequences that allow for the amplification of all four CORE cassettes (Note: these primers do not amplify the original CORE cassette from the pCORE plasmid):

P.I: 5'-...TTCGTACGCTGCAGGTCGAC-3' for the *KlURA3* side in pCORE-UK and pCORE-UH, and for the *kanMX4* or *hyg* side in pCORE-Kp53 and pCORE-Hp53
P.II: 5'-...CCGCGCGTTGGCCGATTCAT-3' for the *kanMX4* or *hyg* side in pCORE-UK and pCORE-UH, and for the *GAL1/10*-p53 side in pCORE-Kp53 and pCORE-Hp53.

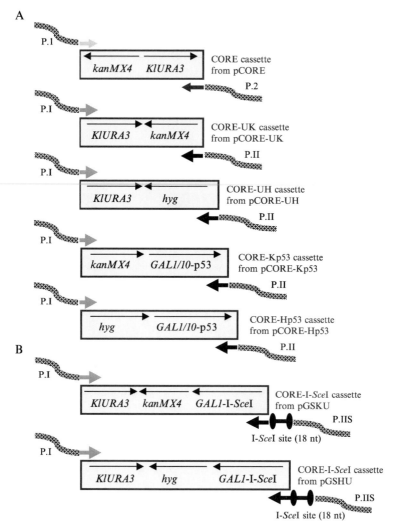

FIG. 3. CORE cassettes. (A) CORE cassette with *KlURA3*, or *GAL1/10*-p53 as a counterselectable marker and with *kanMX4*, or *hyg*, as reporter markers. (B) CORE-I-*Sce*I cassettes with *KlURA3* as the counterselectable marker, with *kanMX4*, or *hyg*, as reporter markers and with *GAL1*-I-*Sce*I for the induction of the DSB. Orientation of each gene in the cassettes is indicated by an arrow. Cassettes with *KlURA3* derived from CORE-UK, CORE-UH, and CORE-I-*Sce*I should have *KlURA3* oriented opposite to the transcription direction of the target gene (see text in the section CORE Cassettes). Primers used to amplify each cassette are also shown.

It should be noted that when inserting the CORE-UK, CORE-UH, and CORE-I-*Sce*I (see later) cassettes into a gene, *KlURA3* should be oriented opposite to the transcription direction of the gene. If *KlURA3* is in the same orientation as the targeted gene, transcription from the promoter of the targeted gene may interfere with the transcription of *KlURA3*. This may delay the growth on Ura⁻ media and may increase the background during selection of oligonucleotide transformants on 5-FOA. In such cases, to reduce the background during selection of oligonucleotide targeting events, we suggest replica plating the cells growing on 5-FOA on to a new 5-FOA plate once every day for a few days until the Ura⁻ (CORE-loss) clones are clearly distinguishable from the background.

PCR Reaction Conditions. PCR amplification of CORE cassettes from circular plasmids (about 0.1 µg) is performed with high yield in a final volume of 40 µl using Takara Ex Taq DNA polymerase (Takara Bio Inc., Temecula, CA), with 2 min at 94°, 32 cycles of 30 s at 94°, 30 s at 57°, and 4 min at 72°. For all PCRs of CORE cassette amplification a higher yield is obtained with Takara DNA polymerase than Taq DNA polymerase. The product of three PCR reactions is collected and precipitated with ethanol and resuspended in 20 µl of water. Ten µl are used for each transformation.

Steps in the Development of the Delitto Perfetto *System*

Integration of CORE Cassette

MEDIA AND COMPONENTS. Standard yeast media and growth conditions are described in Sherman *et al.* (1986). Transformation is done using the lithium acetate protocol described by Wach *et al.* (1994) with some modifications. Stock solutions are as follows: LiAc 1 *M*, filter sterilized; TE 10X (Tris 100 m*M* pH 7.5, EDTA 10 m*M*) pH 7.5, filter sterilized; PEG 4000 50%, sterilized by autoclave. Working solutions are Solution 1 (LiAc 0.1 *M*, TE 1X pH 7.5) and Solution 2 (LiAc 0.1 *M*, TE 1X pH 7.5 in PEG 4000 50%). Salmon sperm carrier DNA, in slight excess volume relative to the amount needed is denatured at 100° for 5 min and then put on ice.

TRANSFORMATION PROTOCOL. Below are the steps involved in transforming cells with the oligonucleotides.

1. Inoculate 5 ml of YPDA (standard rich media supplemented with adenine) medium with the chosen strain and shake at 30° over night (O/N).

2. Inoculate 50 ml of YPDA medium (volume of flask) with 1.5 ml of the O/N culture and shake vigorously at 30° for 3 h.

3. Transfer culture to a 50-ml tube and spin at 3000 rpm for 2 min.

4. Wash cells with 50 ml of sterile water and spin as stated previously.

5. Resuspend cells in 5 ml of Solution 1 and spin as stated previously.

6. Resuspend cells in 250 μl of Solution 1. This amount of cells is sufficient for about 7–8 transformations.

7. Aliquot 50 μl of the cell suspension in Eppendorf tubes and add 10 μl of concentrated CORE PCR product, 5 μl of denatured carrier DNA, and 300 μl of Solution 2 for each transformation reaction. Mix briefly by vortexing.

8. Incubate transformation reactions at 30° for 30 min with shaking.

9. Heat shock at 42° for 15 min.

10. Collect cells by centrifugation at 5000 rpm for 4 min.

11. Remove supernatant and resuspend cells in 100 μl of water.

12. Plate all cells from each transformation tube on one plate of synthetic media lacking uracil (SD-Ura⁻ plate) and incubate at 30° for 3 days or plate on one YPDA plate and incubate at 30° O/N and then replica-plate on YPDA media containing 200 μg/ml of G418 (Gibco BRL, Grand Island, NY) or 300 μg/ml of hygromycin B (Invitrogen, Carlsbad, CA) (Goldstein and McCusker, 1999), depending on the CORE cassette used.

13. Once transformant colonies appear (typically 5 to 30), streak a few colonies on YPDA to obtain single colony isolates.

14. Patch the single colony isolates (3 to 6) on YPDA and test for the presence of the other marker of the CORE, if applicable, and for various other phenotypes (according to the strain used) and on YP glycerol (YPG) to eliminate petite mutants.

COLONY PCR REACTION CONDITIONS TO IDENTIFY DESIRED ISOLATES. To identify clones with the correct CORE-cassette integration, colony PCR is performed using primers designed for annealing upstream and downstream of the integration locus and within the cassette and designed to give a band from 500 bp to 1500 bp, according to Wach *et al.* (1994) with modifications. Specifically, approximately 1 mm³ of cells from the single-colony derived patch is resuspended in 50 μl of water containing 1 unit of Lyticase (Sigma Chemical Co., St. Louis, MO) in an Eppendorf tube and incubated at room temperature (RT) for 10 min. Cells are then collected by centrifugation at 5000 rpm for 2 min. Supernatant is discarded and tubes with cell pellets are put in a 100° heat block for 5 min. Cell debris are finally resuspended in 40 μl of water. 10 μl of cell debris are used for each PCR reaction containing 5 μl of Buffer 10× with magnesium (Roche, Indianapolis, IN), 1 μl of dNTPs (Roche), 1 μl of each primer (from a 50 μM primer solution), 0.2 μl of Taq (Roche), and water to 50 μl. PCR is performed as follows: 2 min at 94°, 30 cycles of 30 s at 94°, 30 s at 55°, and 1 min at 72°.

Oligonucleotide Targeting and CORE Removal

OLIGONUCLEOTIDE DESIGN. The Integrative Recombinant Oligonucleotides (IROs) (Storici *et al.*, 2001) are desalted custom primers synthesized at the 50 nM scale (Invitrogen, Carlsbad, CA). They can be used in transformation experiments either singly or as pairs. Pairs of fully complementary oligonucleotides generally transform 5- to 10-fold more efficiently than corresponding single strand molecules (Fig. 2). Longer oligonucleotides also transform better than shorter (Fig. 2). Fully overlapping 80 to 100-mers are suggested for the introduction of point mutations, substitutions, or small insertions. Since the external 30 to 40 bases are used for efficient targeting, only the central sequence of the oligonucleotide pair can be designed for mutagenesis. The longer the oligonucleotide, the larger is the window available for modifications (Fig. 4). Therefore, if only a single point mutation is desired in a specific region, the CORE should be inserted at that site, and fully complementary IROs or single IROs can be used to introduce the mutation.

The window for mutagenesis can be expanded using IROs that share a short overlap of 20 nucleotides (nt) (Fig. 4). Using a pair of 100-mers that have a 20 nt complementary overlap at the 3′ end, the window available for mutagenesis can be increased to 100 nt (Fig. 4). To increase the efficiency of targeting of IRO pair sharing only a short overlap of 20 nt, we suggest annealing and extension of the IROs *in vitro* (this is not required when using the break-mediated *delitto perfetto* system, see later). In order to extend the sequence of 3′-overlapping IROs *in vitro*, 0.5 nmoles (15–20 μg) of each IRO is added to a 50-μl reaction mix containing 4 units Platinum Pfx (Gibco BRL), 5 μl 10× Buffer, 1 μl 50 mM MgSO4, and 2 μl 10 mM dNTPs (Roche). Extension is performed as follows: 1 min at 94° and 30 s to 3 min at 68°. Samples are ethanol precipitated and resuspended in 10–20 μl of water. For more details about oligonucleotide design, see Storici and Resnick (2003).

PROTOCOL FOR TRANSFORMATION WITH OLIGONUCLEOTIDE. Before adding IRO DNA to the cells, the oligonucleotides are denatured at 100° for 2 min and placed immediately on ice in order to eliminate possible secondary structures. Strains containing the CORE cassette are transformed with chosen oligonucleotides. Other steps in transformation are the same as previously except for the following differences in steps 7 and 12.

    Change in Step 7. To 50 μl of the cell suspension in Eppendorf tubes add 20 μl of 50 μM solution (i.e., 1 nmole) of oligonucleotides (1 nmole of a single oligonucleotide or for a pair of complementary

Mutagenesis region around the CORE ~200 bp

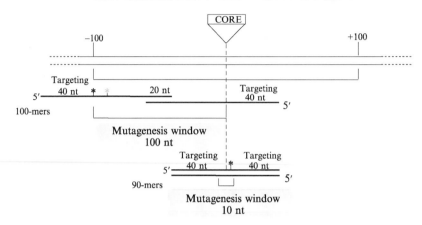

FIG. 4. Use of IROs for rapid production of a variety of mutations. Site-specific mutations can be accomplished at positions as far as 100 bp upstream or downstream of the CORE cassette integration point with different IRO pairs without changing position of CORE cassette. A pair of 100-mers, overlapping for 20 nt, is filled-in with Pfx and has a 100-nt region that can be modified. Within the 100-nt region, 40 nt on each side of the overlap can also be used for random mutagenesis. 40 nt at each end of the pair are designed to have identity upstream and downstream of the CORE cassette integration for efficient homologous integration. Complementary 90-mers have a central region of 10 nt, which can be used for site-directed mutations. 40 nt at each end must remain unchanged for efficient homologous integration. The figure shows examples of oligonucleotides designed to introduce different mutations (*) from close to the CORE site, up to 100 bp upstream of the CORE-cassette integration point. The efficiency of mutagenesis may decrease as the mutation position in the IRO sequence is placed closer to the 5′ end since the IRO recombination event leading to excision of the CORE could occur without inclusion of the mutation in the IRO (Storici et al., 2001).

IROs use 0.5 nmole of each) and 300 μl of Solution 2 for each transformation reaction. Vortex briefly. At this concentration of oligonucleotides, carrier DNA is not needed.

Change in Step 12. Plate all cells from each transformation tube on one YPDA plate and incubate O/N at 30°. The day after, if the counterselectable marker is *KlURA3*, replica-plate on media (synthetic complete) containing 60 mg/liter of uracil and 1 g/liter of 5-fluoroorotic acid (5-FOA) (Toronto Research Chemicals Inc., North York, ON, Canada). Three days later replica-plate colonies from 5-FOA to YPDA and YPDA containing G418 or hygromycin. If the counterselectable marker is *GAL1/10*-p53, replica-plate cells on synthetic complete media containing 2% galactose. Replica-plate every day for two days on a new plate of synthetic complete with 2% galactose. On the third day, replica-plate colonies from synthetic complete with 2% galactose to YPDA and YPDA containing G418 or hygromycin. In both cases, isolate colonies that are G418 or hygromycin sensitive.

IDENTIFICATION OF RELEVANT SINGLE COLONY ISOLATES. The procedure is similar to that described previously. Colony PCR products are sequenced after phenol-chloroform extraction and purification using the QIAquick PCR Purification Kit (Qiagen USA, Valencia, CA).

The Break–Mediated Delitto Perfetto *System*

Integration of CORE-I-SceI Cassette. The break–mediated *delitto perfetto* system utilizes a CORE that contains the gene for the double strand endonuclease I-*SceI* and the DSB target site. The components are diagrammed in Fig. 3B. The CORE-I-*SceI* cassette, starting from the 5′ end, contains *KlURA3, kanMX4* (or *hyg*), and GAL1-I-*SceI*. The cassette is amplified as a 4.6 kb or 4.8 kb DNA fragment from pGSKU, or pGSHU, respectively, using Takara Ex Taq DNA polymerase, with 2 min at 94°, 32 cycles of 30 s at 94°, 30 s at 57° and 5 min at 72°. For integration of the cassette into chromosomal loci, chimeric primers are used, consisting of 50 nucleotides homologous to the genomic target and an additional 20 nucleotides for the amplification of the CORE-I-*SceI* cassette:

P.I for the *KlURA3* side primer: 5′-TTCGTACGCTGCAGGTC-GAC-3′

P.IIS for the GAL1-I-*SceI* side that also contains 18-nt of the I-*SceI* target double-strand break site (italics): 5′-...*TAGGGA-TAA-CAGGGTAAT*-CCGCGCGTTGGCCGATTCAT-3′

Transformation of CORE-I-*SceI* cassettes and identification of clones with the correct cassette integration are performed as described previously.

Protocol for Targeting Oligonucleotides to a Region Containing a DSB. The procedures described previously (transformation protocol) are used with the following modifications:

Change in Step 2. Inoculate 50 ml of synthetic complete media containing 2% galactose with 1.5 ml of the O/N culture in YPDA and shake at 30° for 4 h to express *GAL1*-I-*SceI*, which will induce a DSB at the cloned target site. In experiments with strains having the CG379 background, about 30% of cells were cut by I-*SceI* at the *TRP5* locus on chromosome VII after 4-h induction of *GAL1*-I-*SceI* (Storici *et al.*, submitted).

Change in Step 7. Aliquot 50 μl of the cell suspension in Eppendorf tubes and add 20 μl of 50 μM solution (i.e., 1 nmole) of oligonucleotides (1 nmole of a single oligonucleotide or 0.5 nmoles of each strand of a pair of complementary oligonucleotides) and 300 μl of Solution 2 for each transformation reaction. Mix briefly by vortexing. At this concentration of oligonucleotides, carrier DNA is not needed.

Change in Step 12. Plate a dilution of the cells (over the range 10- to 10,000-fold, depending on oligonucleotide chosen) onto YPDA plates and incubate at 30° O/N. Replica-plate to media (synthetic complete) containing 60 mg/liter of uracil and 1 g/liter of 5-fluoroorotic acid (5-FOA). Three days later replica-plate cells from 5-FOA to YPDA and YPDA containing G418 or hygromycin depending on selective markers used. Isolate colonies that are G418 or hygromycin sensitive.

Frequencies of Targeting and Identification of Relevant Single Colony Isolates. The frequencies of targeting with different sizes of complementary and single IROs are shown in Fig. 2 (dark gray bars). In all transformation experiments 10^5-fold dilutions are plated directly to YPDA in order to determine viability of the cell population. Survival is normally ~25% after transformation with or without IROs, in glucose or in galactose (not shown). To directly measure events when there is a very high frequency of targeting, the 10^{-5}-fold diluted suspensions are plated to YPDA directly. Subsequently, the plates (containing 100–1000 colonies) are replica-plated to 5-FOA and to YPDA containing G418 or hygromycin.

The procedures to identify and isolate single colony isolates of transformant clones are the same as described previously. Colony PCR products are sequenced after phenol-chloroform extraction and purification using the QIAquick PCR Purification Kit.

Break-Mediated Delitto Perfetto with Oligonucleotides in Diploid Cells

The ability to apply the *delitto perfetto* approach to diploid cells has many utilities for gene and genome modification. Recent results have demonstrated that a pair of 95-mers (1 nmole) can be targeted to a single DSB in diploid yeast, in spite of the competition by an unbroken homologous chromosome (Storici *et al.*, 2003). Loss of the CORE-I-*SceI* cassette was primarily due to recombination with the homologous chromosome; however, approximately 4% of the loss events were due to direct oligonucleotide targeting. These results suggest that clones with the desired IRO generated mutation can be isolated by screening a pool of clones selected for CORE loss. For productive targeting in diploid cells, IROs used for the generation of point mutations are designed to also create a restriction site. That way the desired clones can be identified within a pool of samples by PCR analysis followed by restriction digestion. Mutations that generate several base deletions or insertions or rearrangements can be identified simply by PCR analysis of pooled samples.

High-Throughput Mutagenesis and Rapid Gene Diversification with the Delitto Perfetto *Systems*

Use of the CORE-I-*Sce*I cassette provides the opportunity for direct, highly efficient generation of many mutations within a gene. The cassette is placed in the middle of the desired region (<200 bp) of a gene. This provides a window for mutagenesis that is 100 bases upstream and downstream from the CORE insertion site. If the domain of interest is longer than 200 bases, or covers the entire gene, two or more CORE-I-*Sce*I cassettes are placed at appropriate positions in different strains at 150 nucleotide intervals across the region of interest. This leads to a panel of isogenic yeast strains, each containing one CORE-I-*Sce*I cassette, as described in Fig. 5. Placing the cassettes at intervals of 150 bases ensures some overlap between contiguous windows of mutagenesis. The chosen mutations are generated simply by replacement of the CORE cassette with IROs. The oligonucleotides can be specific for a single desired mutation, or the IROs could be degenerate in order to produce random mutations in the region of interest (Figs. 4 and 5).

Generation of Chromosomal Rearrangements

IROs can be targeted to sites of homology distant from a DSB, in fact, the distance can be quite large based on recent observations (Storici *et al.*, 2003). A pair of complementary 83-mers that were designed to delete 16.6 kb around a CORE-I-*Sce*I inserted at the *TRP5* locus on chromosome VII resulted in more than 1% deletion efficiency. They had 40-nt of homology at a position 4.7 kb on one side of the DSB and 11.9 kb on the other. Single-strand oligonucleotides also could be targeted to the sites distant from the DSB, although with 10-fold less efficiency. The maximum distance for targeting is currently under investigation.

The oligonucleotide-mediated process enables the production of a variety of rearrangements including circularization, interchromosomal fusion, and reciprocal translocations. Briefly, a CORE-I-*Sce*I cassette is integrated near the border of the proposed rearrangement. Oligonucleotides are designed that will repair the break, eliminate the cassette, and generate the desired genomic modifications.

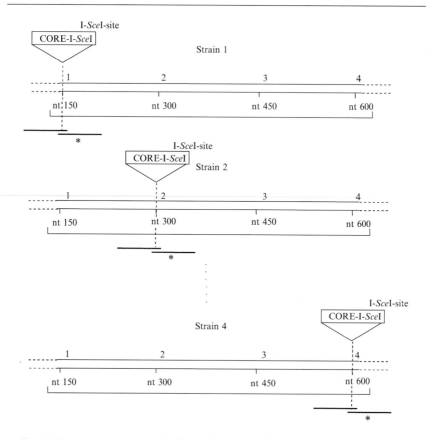

FIG. 5. Strategy for *in vivo*, site-directed mutagenesis in a large DNA region. In this example, four isogenic yeast strains are created each containing a CORE-I-*Sce*I cassette integrated at a different position (every 150 nucleotides) along the gene sequence. Mutations can be created simply by designing IROs that contain the desired change. Transformation with oligonucleotides, which are homologous to a region surrounding the CORE cassette, results in the elimination of the CORE cassette and creation of a site-specific change. A mutation can be designed 100 bases upstream and downstream of each CORE cassette insertion point (see Fig. 4). The specific scheme shows the introduction of a mutation upstream from a CORE cassette that was inserted at nucleotide positions 150, 300, and 600. (To decrease the likelihood that a distant mutation is not included during the targeting of the IRO, part of the sequence surrounding the site where the CORE is to be targeted can be deleted during the step of CORE insertion and reintroduced with IRO transformation.)

An example of redesigning chromosomes is presented in Fig. 6. To generate a reciprocal translocation between chromosome V and VII, a CORE-I-*Sce*I cassette was integrated on chromosome VII at the site of translocation. A DSB was induced and the break was repaired using two

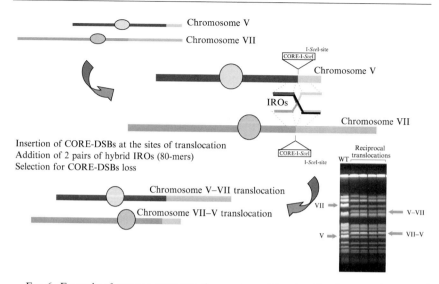

FIG. 6. Example of genome reconstruction: reciprocal translocation between the arms of chromosome V and VII. The first step involves the insertion of one CORE-I-*Sce*I cassette at one site of translocation. (Insertion of a CORE-I-*Sce*I cassette at each site of translocation on both chromosomes [using the cassette with *kanMX4* and the one with *hyg*, respectively] is predicted to enhance translocation frequency by IROs.) The second step involves transformation with two pairs of IROs, each containing sequence from both chromosomes adjacent to the translocation point. Selection of translocation events is accomplished by isolating 5-FOA resistant clones, which also have lost the reporter marker (or both reporter markers if using two cassettes). The picture on the right shows the karyotype display using pulse-field gel electrophoresis. Bands corresponding to chromosome V–VII and VII–V translocations are indicated. (See color insert.)

pairs of complementary 80-mers. One pair had half of its sequence identical to a region upstream of the translocation point on chromosome V and half of the sequence identical to a region downstream of the translocation point on chromosome VII. The other pair had the opposite configuration. Cells were selected for loss of the markers in the CORE and screened by colony PCR of small pools. Using this system, the frequency of events was approximately 10^{-6} (unpublished). The frequency of targeting might be higher if two CORE-I-*Sce*I cassettes are used, one with *KlURA3* and *kanMX4* markers (from pGSKU) and the other with *KlURA3* and *hyg* markers (from pGSHU) at the two translocation sites, since two DSBs would be generated. Transformant colonies would be selected for loss of *KlURA3* (loss of both cassettes) on media containing 5-FOA and checked for the loss of both *kanMX4* and *hyg* markers. This approach would be expected to eliminate most false positive clones (spontaneous *KlURA3* mutants; see Fig. 6). A similar strategy could be used to circularize or fuse chromosomes.

Conclusions

The protocols and examples described previously provide the essence of the *delitto perfetto* strategy and describe examples for the *in vivo* modification of genes and chromosomes in yeast. The efficiency of genetic modification is greatly enhanced when oligonucleotide targeting is stimulated by a DSB. Targeting frequencies as high as 20% of all cells can be obtained. More remarkable is the fact that the IROs can generate almost any kind of genetic modification *in vivo*, from a single-base mutation to a large deletion or a reciprocal chromosomal translocation. While developed in the yeast *S. cerevisiae*, the approach could be applied to other organisms where homologous recombination is efficient.

The *delitto perfetto* system has proven useful not only for the engineering of genetic material; studies on the mechanisms of DSB repair by single-strand oligonucleotides are revealing new features of DNA recombination and repair (Storici *et al.*, sumbitted). The information about repair and recombination mechanisms in yeast is expected to provide insights into gene/chromosome targeting and modification in other systems, including mammalian cells. We also anticipate that the oligonucleotide targeting procedure can be expanded to develop genome-wide analysis of spontaneous and induced DNA damage. Given the almost unlimited number of oligonucleotide variants that can be designed, there are vast numbers of genomic targets, substrates, and genomic modifications that can be investigated with the *delitto perfetto* approach using oligonucleotides.

References

Barre, F. X., Ait-Si-Ali, S., Giovannangeli, C., Luis, R., Robin, P., Pritchard, L. L., Helene, C., and Harel-Bellan, A. (2000). Unambiguous demonstration of triple-helix-directed gene modification. *Proc. Natl. Acad. Sci. USA* **97**, 3084–3088.

Colleaux, L., d'Auriol, L., Betermier, M., Cottarel, G., Jacquier, A., Galibert, F., and Dujon, B. (1986). Universal code equivalent of a yeast mitochondrial intron reading frame is expressed into *E. coli* as a specific double strand endonuclease. *Cell* **44**, 521–533.

Erdeniz, N., Mortensen, U. H., and Rothstein, R. (1997). Cloning-free PCR-based allele replacement methods. *Genome Res.* **7**, 1174–1183.

Goldstein, A. L., and McCusker, J. H. (1999). Three new dominant drug resistance cassettes for gene disruption in *Saccharomyces cerevisiae*. *Yeast* **15**, 1541–1553.

Inga, A., and Resnick, M. A. (2001). Novel human p53 mutations that are toxic to yeast can enhance transactivation of specific promoters and reactivate tumor p53 mutants. *Oncogene* **20**, 3409–3419.

Kmiec, E. B., Ye, S., and Peng, L. (2000). Targeted gene repair in mammalian cells using chimeric oligonucleotides. *In* "Genetic Engineering, Principle and Methods" (J. K. Setlow, ed.), Vol. 22, pp. 23–31. Kluwer Academic/Plenum Publisher, Upton, NY.

Langle-Rouault, F., and Jacobs, E. (1995). A method for performing precise alterations in the yeast genome using a recyclable selectable marker. *Nucleic Acids Res.* **23,** 3079–3081.

Liu, L., Rice, M. C., and Kmiec, E. B. (2001). *In vivo* gene repair of point and frameshift mutations directed by chimeric RNA/DNA oligonucleotides and modified single-stranded oligonucleotides. *Nucleic Acids Res.* **29,** 4238–4250.

Moerschell, R. P., Tsunasawa, S., and Sherman, F. (1988). Transformation of yeast with synthetic oligonucleotides. *Proc. Natl. Acad. Sci. USA* **85,** 524–528.

Plessis, A., Perrin, A., Haber, J. E., and Dujon, B. (1992). Site-specific recombination determined by I-*Sce*I, a mitochondrial group I intron-encoded endonuclease expressed in the yeast nucleus. *Genetics* **130,** 451–460.

Scherer, S., and Davis, R. W. (1979). Replacement of chromosome segments with altered DNA sequences constructed *in vitro*. *Proc. Natl. Acad. Sci. USA* **76,** 4951–4955.

Sherman, F., Fink, G. R., and Hicks, J. B. (1986). "Methods in Yeast Genetics." Cold Spring Harbor Laboratory Press, Cold Spring Harbor, NY.

Storici, F., Lewis, L. K., and Resnick, M. A. (2001). *In vivo* site-directed mutagenesis using oligonucleotides. *Nat. Biotechnol.* **19,** 773–776.

Storici, F., and Resnick, M. A. (2003). *Delitto perfetto* targeted mutagenesis in yeast with oligonucleotides. *In* "Genetic Engineering, Principle and Methods" (J. K. Setlow, ed.), Vol. 25, pp. 189–207. Kluwer Academic/Plenum Publisher, Upton, NY.

Storici, F., Durham, C. L., Gordenin, D. A., and Resnick, M. A. (2003). Chromosomal site-specific double-strand breaks are efficiently targeted for repair by oligonucleotides in yeast. *Proc. Natl. Acad. Sci. USA* **100,** 14994–14999.

Storici, F., Snipe, J. R., Chan, G. K., Gordenin, D. A., and Resnick, M. A. (submitted). Double-strand break repair by single-strand DNA via two steps of annealing.

Vasquez, K. M., Narayanan, L., and Glazer, P. M. (2000). Specific mutations induced by triplex-forming oligonucleotides in mice. *Science* **290,** 530–533.

Wach, A., Brachat, A., Pohlmann, R., and Philippsen, P. (1994). New heterologous modules for classical or PCR-based gene disruptions in *Saccharomyces cerevisiae*. *Yeast* **10,** 1793–1808.

[20] Assays for Transcriptional Mutagenesis in Active Genes

By Damien Brégeon and Paul W. Doetsch

Abstract

Cells exposed to DNA-damaging agents in their natural environment do not undergo continuous cycles of replication but are more frequently engaged in gene transcription. Despite the relatively high efficiency of the different DNA repair pathways, some lesions remain in DNA. During transcription, RNA polymerase can bypass DNA damage on the transcribed strand of an active gene. This bypass can be at the origin of the production of "mutated" mRNA because of the transcriptional miscoding

METHODS IN ENZYMOLOGY, VOL. 409
0076-6879/06 $35.00
DOI: 10.1016/S0076-6879(05)09020-8

(transcriptional mutagenesis) due to the altered pairing specificities of the lesion. *In vivo* consequences of transcriptional mutagenesis on normal cell physiology have not well been documented because of the lack of a robust system allowing for its study. We describe here a procedure that we developed using a plasmid-based luciferase reporter assay to analyze the transcriptional mutagenesis events induced by different types of DNA lesions. Introduction of the DNA lesion to be studied at a specific site on the plasmid is based on the synthesis of a complementary strand of a circular, single-stranded DNA (ssDNA) from a DNA lesion-containing oligonucleotide. Once obtained, this construct can be transformed into different *Escherichia coli* strains that can express the luciferase gene under nongrowth conditions. Quantification of luciferase activity and sequencing of luciferase cDNAs allow for the characterization of transcriptional mutagenesis both quantitatively and qualitatively.

Introduction

The genomes of all organisms are constantly exposed to chemical and physical agents that damage the DNA molecule. A large number of these lesions has been shown to be at the origin of an increased mutagenesis due to their altered pairing specificities during replication (for review see Friedberg *et al.*, 1995). In the literature, there are numerous reports dealing with the *in vitro* and *in vivo* effects of DNA damage on replication associated events. However, the consequences of these DNA lesions on transcription when present on the transcribed strand of an active gene is less well documented. Non-bulky DNA lesions do not generally block the transcriptional machinery and can be transcribed through by the RNA polymerase. As mentioned previously, DNA lesions often have altered pairing specificities thus resulting in the insertion of a noncomplementary nucleotide opposite the lesion during transcription. These misinsertion events at the lesion site can result in the generation of a population of erroneous transcripts, a process that was termed "transcriptional mutagenesis" (TM) (Bregeon *et al.*, 2003; Doetsch, 2002; Tornaletti and Hanawalt, 1999). Such erroneous transcripts can potentially lead to the production of a large amount of mutant proteins that may mediate deleterious outcomes for the cell (Doetsch, 2002; Holmquist, 2002). The *in vivo* consequences of TM have not been extensively investigated because of the lack of robust techniques allowing for its study.

To analyze *in vivo* consequences of TM, we developed a luciferase reporter assay that is engineered to reveal the level of TM induced by specific DNA lesions in nondividing *E. coli* cells (Bregeon and Doetsch, 2004; Bregeon *et al.*, 2003; Viswanathan *et al.*, 1999; You *et al.*, 2000). The

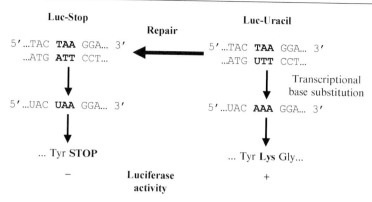

FIG. 1. Luciferase gene modification: example of a uracil construct. For TM measurement, constructs are engineered to contain a DNA lesion at the first base position of codon 445 (bold type) on the transcribed strand of the luciferase gene. In the example provided, the DNA lesion is uracil (U). Transcripts resulting from TM through uracil will result in the generation of a lysine for codon 445 and, therefore, will be translated into an active form of the luciferase protein. If DNA repair occurs, the uracil will be replaced by an adenine and thus convert codon 445 to a stop codon. This will result in the production of a truncated and inactive form of the luciferase protein.

system is based on the luciferase expression from the pBEST-f1 plasmid (Bregeon *et al.*, 2003) and takes advantage of the fact that C-terminal truncations of the firefly luciferase protein result in an inactive form of the enzyme (Sala-Newby and Campbell, 1994). With this system, the transcription template can be designed to contain specifically engineered base damage products within the luciferase gene transcribed strand. Bypass and miscoding by RNA polymerase at the site of the lesion will result in conversion of a premature stop codon into one that changes the resulting transcript into an mRNA encoding a full length, active luciferase (Fig. 1).

The critical step in this system is the production of a large quantity of the expression construct containing a DNA lesion at a specific location. This construct has to be generated *in vitro* as it cannot be propagated in cells via DNA replication-related events. For this purpose, we developed a procedure that is based on the extension by a DNA polymerase of 5′ phosphorylated DNA damage-containing oligonucleotide primer hybridized to a circular ssDNA (Fig. 2) (Bregeon and Doetsch, 2004). Once this construct is generated, it can be used to determine whether or not a particular type of DNA lesion can be bypassed by *E. coli* RNA polymerase *in vivo*. Furthermore, sequencing of luciferase cDNAs obtained from cells containing such constructs allows the specificities of RNA polymerase insertions opposite to the studied lesion to be determined (Fig. 3).

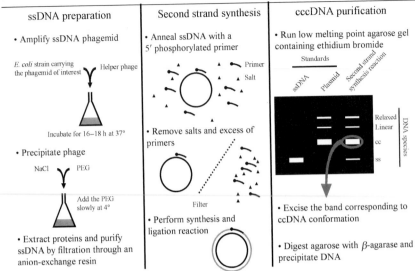

| ssDNA preparation | Second strand synthesis | cccDNA purification |

FIG. 2. Critical steps for the generation of double-stranded DNA vectors containing site-specific base modifications. The generation of closed circular (ccDNA) vectors containing site-specific base modifications is achieved in three steps, which include the production of ssDNA, the *in vitro* synthesis of the complementary strand containing a site-specific base modification, and the purification of the closed circular dsDNA. *E. coli, Escherichia coli*; PEG, polyethylene glycol. (Reprinted with permission from Brégeon, D., and Doetsch, P. W., 2004).

ssDNA Production

The protocol for ssDNA production is based on a method originally described by Kunkel (1987) that we have optimized to achieve higher yields. Unless otherwise specified, chemicals used for this procedure, and other methods described throughout this article were obtained from Sigma (St. Louis, MO) and bacterial culture media was purchased from Invitrogen (Carlsbad, CA).

For ssDNA production, a single colony of the bacterial strain DH12S (Φ80d*lac*ZM15, *mcr*A, Δ(*mrr-hsd*RMS-*mcr*BC), *ara*D139, Δ(*ara-leu*)7697, Δ*lac*X74, *gal*U, *gal*K, *rps*L, *deo*R, *nup*G, *rec*A1 [F′ *pro*AB⁺ *lac*I^q^ZΔM15]) carrying the appropriate phagemid is resuspended in 5 ml of 2YT (peptone 16 g/liter, yeast extract 10 g/liter, NaCl 5 g/liter) containing ampicillin (100 μg/ml) and grown for 16–18 h. This culture is diluted 2000-fold in 200 ml of 2YT containing ampicillin and incubated in a baffled flask at 37° to an OD_{600} of 0.1 with vigorous shaking (300 rpm) to maximize aeration of the culture. M13KO7 helper phage (Invitrogen) is then added to the media at a multiplicity of infection of 5 p.f.u. per cell. To achieve efficient phage infection, the culture is incubated at 37° for 30 min without shaking, and incubation is continued for an additional 1.5 h with vigorous shaking prior to the addition of kanamycin (50 μg/ml). The infected cells are then incubated at 37° overnight (16–18 h) with vigorous shaking.

Phage particles containing ssDNA are separated from bacteria by centrifuging the culture at 16,000g for 15 min at 4° and filtering the supernatant through a 0.2-μm filter to eliminate bacteria and to avoid potential contamination with phagemid dsDNA. Phage particles contained in the filtered supernatant are precipitated at 4° by adding 6 g of NaCl and 10 g of PEG-8000 (Sigma). The addition of PEG-8000 to the media is made one gram at a time at 4° with slow stirring and allowing for complete dissolution of PEG between each addition. Phage precipitate is centrifuged at 16,000g for 20 min at 4°, and the pellet is resuspended into 30 ml of buffer M2 (1% Triton X100, 500 mM Guanidine-HCl, 10 mM MOPS pH 6.5). Once the phage pellet is completely resuspended, it is incubated for 40 min at 80° to lyse the phage particles. During this incubation, the solution becomes turbid and for complete lysis of phage particles, it is important to completely clear the suspension by mixing it by inversion. The suspension is then cooled to room temperature.

To purify ssDNA, the DNA solution is passed through a Qiagen tip 500 following manufacturer's instructions. The final ssDNA precipitation is made with a Qiaprecipitator^{TM}, and ssDNA is resuspended into 1 ml of Tris-HCl 10 mM pH 8.5. With this method, the amount of ssDNA obtained from a 200-ml phage preparation ranges from 150 to 250 μg and should be of good quality based on an A_{260}/A_{280} ratio between 1.8 and 2.0.

Synthesis of the Transcribed Strand

Determination of Oligonucleotide Sequence

For the synthesis of the damage-containing transcribed strand, a phosphorylated oligonucleotide, containing a DNA lesion at a specific residue, is annealed to the circular ssDNA. This oligonucleotide has to be carefully chosen to obtain an efficient second strand synthesis. For this optimization, one can follow the strategy described by Piechocki and Hines (1994). We are routinely using 20–45 mers damage-containing oligonucleotides with the lesion located at least 8 bases away from both ends and roughly equal stabilities at the 3' and 5' end of approximately −8 kcal/mol as estimated with the Oligo 6 software.

Second Strand Synthesis

For the second strand synthesis, 40 μg of ssDNA is mixed with a 10-15 molar excess of the damage-containing oligonucleotide in 1× SSC (150 mM NaCl, 15 mM sodium citrate, pH 7.0). The tube containing this mixture is

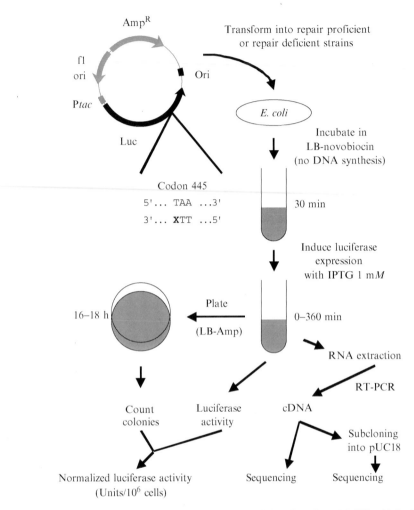

FIG. 3. Strategy for determining TM in bacterial cells. The pBESTluc-f1 luciferase reporter construct contains the firefly luciferase gene driven by the *E. coli tac* promoter. It also contains the ampicillin resistance gene, the f1 origin of replication for the production of ssDNA corresponding to the coding strand of the luciferase gene, and an origin of dsDNA replication (Ori) to allow for its propagation. Modifications of the luciferase gene are engineered so that the DNA lesion to be studied is placed at the first base of codon 445 on the transcribed strand of the luciferase gene. The constructs are electroporated into DNA repair-proficient and -deficient *E. coli* cells subsequently incubated in novobiocin-containing LB medium for 30 min. Luciferase gene expression is induced by the addition of IPTG in the medium. At different times following luciferase induction (0–360 min), aliquots of cells are plated onto LB-Amp medium to determine transformation efficiency, and luciferase activity of these cells is measured. Normalized luciferase activity is determined on the basis of transformation efficiency and total luciferase activity. Additionally, RNA is extracted from

then placed in a 1 liter beaker filled with 250 ml of boiling water and allowed to cool to 4° in a cold room.

At this step, the annealed DNA has to be purified as it was previously shown that an excess of monovalent cations may inhibit the subsequent polymerization/ligation reaction (Chiu *et al.*, 1982; Hayashi *et al.*, 1985) and that unannealed, phosphorylated oligonucleotides can also inhibit this reaction (Bregeon and Doetsch, 2004). In order to eliminate sodium ions and excess oligonucleotide, the mixture is filtered through Microcon Montage-PCR filter units. For this filtration step, it is necessary to divide the annealing reaction into four samples of equivalent volume in order to avoid saturation of the membrane. Samples are subsequently treated as follows:

1. Adjust volume of each aliquot to 400 μl with water and transfer each sample into separate filter units. Centrifuge the Montage PCR units at 1000g for 15 min. In the event that all of the solution has not completely passed through the membrane, one can increase the centrifugation time but not the relative centrifugal force.
2. To recover the purified DNA, add 20 μl of water onto the membrane and invert the filtration unit into a clean vial. Centrifuge at 1000g for 2 min.

The total amount of ssDNA/oligonucleotide hybrid (40 μg) is recovered in 80 μl of water and is used for polymerization/ligation reaction as follows.

The DNA polymerization/ligation reaction is carried out in a 300-μl reaction mixture containing the above 80 μl of ssDNA/oligonucleotide hybrid, 50 mM Tris-HCl pH 8.8, 15 mM (NH$_4$)$_2$SO$_4$, 7 mM MgCl$_2$, 0.1 mM EDTA, 10 mM 2-mercaptoethanol, 20 μg/ml bovine serum albumin, 1200 μM each of dATP, dCTP, dTTP, and dGTP, 1 mM ATP, 7.5% PEG-8000, 30 units of T4 DNA polymerase, and 50 units of T4 DNA ligase. To stabilize the initial duplex between the template DNA and the primer, the reaction is first incubated 5 min on ice and 5 min at room temperature. The polymerization/ligation reaction is then completed at 37° for 16–18 h.

Purification of Closed Circular DNA Molecules

Under the conditions described previously, the conversion of ssDNA to closed circular double-stranded DNA (ccDNA) is incomplete. Consequently, further purification of the ccDNA is required. Separation of ccDNA from other unwanted DNA species can easily be achieved by

cells and luciferase mRNA is converted to cDNA by RT-PCR, and the corresponding products are either directly sequenced or subcloned into pUC18 and subsequently isolated and sequenced. (Reprinted with permission from Brégeon, D., Doddridge, Z. A., You, H. J., Weiss, B., and Doetsch, P. W., 2003.)

separation of the polymerization/ligation products in an agarose gel containing 0.3 μg/ml ethidium bromide under conditions where ccDNA separates as supercoiled DNA. Once separated, the ccDNA can be extracted from the agarose. For this step, the most efficient and versatile procedure is the digestion of low melting point (LMP) agarose with β-agarase as previously described (Bregeon and Doetsch, 2004). Briefly:

1. Perform electrophoresis of the polymerization/ligation reaction in a 0.6% LMP agarose gel containing 0.3 μg/ml ethidium bromide.
2. UV light visualization and excision of the band corresponding to ccDNA conformation (Fig. 3).
3. Determine the weight of the agarose band and add 1/9 volume of 10× β-agarase buffer.
4. Agarose band melted (10 min at 70°) and cooled to 42°.
5. Addition of 1 unit of β-agarase per 100 mg of agarose and incubate 2 h at 42°.
6. Incubate 16–18 h at 4° to solidify any remaining, undigested agarose.
7. Centrifuge to pellet the undigested agarose and transfer the supernatant to a new tube.
8. Ethanol precipitation of the ccDNA and resuspension of the pellet into 50 μl of water.
9. Determination of DNA concentration by UV spectrophotometry (A_{260}).

The previous method is relatively efficient and can easily be scaled up or down. Furthermore, the extracted DNA is of good quality as revealed by measurement of the A_{260}/A_{280} ratio (Bregeon and Doetsch, 2004).

Following these steps, the presence of the lesion to be studied in the synthetic ccDNA needs to be confirmed. This can be achieved by treating the DNA with specific enzymes (e.g., DNA N-glycosylases, apurinic/apyrimidinic [AP] endonucleases, AP lyases) targeted towards the lesion of interest and verify that the ccDNA is then converted to the relaxed conformation following treatment. Once verified, the synthetic ccDNA can be used for transcriptional mutagenesis experiments as described later.

Quantification of Transcriptional Mutagenesis

Preparation of Competent E. coli Cells and Transformation Conditions

Transformation of synthetic ccDNA into competent *E. coli* cells can be achieved as follows:

1. Incubate a single colony of *E. coli* cells into 5 ml of LB (peptone 10 g/liter, yeast extract 5 g/liter, NaCl 5 g/liter). Grow for 16–18 h at 37° with shaking (200 rpm).
2. Dilute the preculture to 1/1000 into 500 ml LB medium in a 2-liter flask. Grow at 37° with shaking (200 rpm) to an OD_{600} of 0.4–0.5.
3. Chill cells in ice for 30 min and transfer to a prechilled centrifuge bottle. Centrifuge for 30 min at 5000g at 4°.
4. Resuspend pellet in 500 ml ice-cold water, mix, and centrifuge for 20 min at 5000g at 4°.
5. Repeat step 4 with 250 ml ice-cold water.
6. Repeat step 4 with 100 ml ice-cold water.
7. Resuspend pellet in 50 ml ice-cold 10% glycerol and transfer the suspension to a prechilled 50-ml polypropylene tube. Centrifuge 15 min at 5000g at 4°.
8. Estimate pellet volume and add an equivalent volume of ice-cold 10% glycerol to resuspend cells. Competent cells are aliquoted, frozen on dry ice, and stored at −80°.
9. Competent cells (50 μl) can then be electrotransformed at 2500 V with 100 ng of the synthetic ccDNA.

Inhibition of Replication in Transformed E. coli Cells

For transcriptional mutagenesis assays, it is important to inhibit DNA replication of the plasmid used for the experiment to avoid replication-error events that may cause false positive results. One way to block DNA synthesis in *E. coli* cells is to incubate cells in the presence of the DNA gyrase inhibitor novobiocin. This antibiotic effectively inhibits both chromosomal and plasmid replication (DeMarini and Lawrence, 1992) and does not induce the SOS DNA repair response in *E. coli* (Gellert *et al.*, 1976). In our experiments, the presence of 50 μM of novobiocin in the culture medium completely inhibits DNA synthesis over a 240 min period following isopropyl-β-D-1-thiogalactopyranoside (IPTG) induction of the luciferase gene (Viswanathan *et al.*, 1999). Importantly, RNA synthesis is not inhibited under these conditions (Viswanathan *et al.*, 1999). Inhibition of DNA synthesis and unaltered RNA synthesis can be monitored by using a specific protocol described previously (You *et al.*, 2000).

Luciferase Assay

After electroporation, bacterial cells are placed in 1 ml of LB containing 50 μM of novobiocin and incubated at 37° with shaking (200 rpm). Luciferase expression is induced 30 min after electroporation by the addition of IPTG (1 mM final concentration) into the medium. Luciferase activity can then be determined from 15 min to 360 min following IPTG induction as follows:

1. Remove an aliquot of culture and dilute it to 10^{-2}, 10^{-3}, and 10^{-4}. Plate 100 μl of each dilution onto LB plates containing ampicillin. Grow 16–18 hours at 37° and count Amp^R colonies to calculate the number of transformants in the culture.

2. Pellet 200 μl of the culture and resuspend bacterial cells into 50 μl of luciferase lysis buffer (25 mM Tris-phosphate pH 7.8, 2 mM DTT, 2 mM 1,2-diaminocyclohexane-N,N,N',N'-tetraacetic acid, 10% glycerol, 1% Triton X-100).

3. Lyse bacterial cells by freeze-thawing five times (dry ice in ethanol for 1 min, then 42° for 1 min).

4. Transfer 20 μl of bacterial cell extract into 100 μl of luciferase assay reagent (Promega) and mix by pipetting gently up and down.

5. Measure the luminescence level (RLU units) for 10 s by placing the reaction in a luminometer (e.g., Femtomaster Model FB12 from Zylux Corp., Maryville, TN).

6. The transformation efficiency will be variable from one electroporation to another, therefore luminescence levels have to be normalized to the number of transformants (AmpR cells calculated at step 1) for each measurement.

Determining the Transcriptional Mutagenesis Spectrum

RNA Extraction and RT-PCR

Measurement of luciferase activity is a good indicator for the evaluation of transcriptional mutagenesis. In order to identify the type of nucleotide that was introduced opposite to the lesion by the RNA polymerase, it is necessary to sequence luciferase messenger RNA. This determination is facilitated when performing this analysis with cells expressing the highest levels of luciferase activity following transformation with the lesion-containing, synthetic construct. At the time of luciferase activity measurement, total RNA can be extracted as follows:

1. Pellet 200 μl of the culture and resuspend bacterial cells into 1 ml of TRIzol (Invitrogen). Incubate samples for 5 min at room temperature.
2. Add 0.2 ml of chloroform and shake vigorously for 15 s and incubate for 3 min at room temperature. Centrifuge samples for 15 min at 12,000g at 4°.
3. Transfer the upper, colorless aqueous phase to a fresh tube and precipitate DNA by the addition of 0.5 ml of isopropanol and

incubation at room temperature for 10 min. Centrifuge samples for 15 min at 12,000g at 4°.

4. Wash the RNA pellet with 1 ml of ethanol 75% and air-dry the pellet. The RNA pellet can then be resuspended in 20 μl of DEPC-treated water.

At this step, RNA is not completely free of DNA contaminants as plasmid DNA is not completely removed with the TRIzol procedure. To remove any contaminating DNA, samples are treated twice with DNase I, and RNA is subsequently purified using DNA free RNA kit (Zymo Research, Orange, CA) as follows:

1. Digest the RNA sample (20 μl) with 2.5 units of DNase I in a final volume of 50 μl containing the manufacturer's supplied buffer and water. Components of the reaction must be mixed by pipeting as vortexing may inhibit the reaction. Incubate the reaction for 15 min at 37°.

2. When the reaction is completed, add 200 μl of RNA binding buffer and transfer the mixture into the provided columns. Centrifuge at high speed for 30 s and discard the flow-through.

3. Wash the RNA-containing column two times with 200 μl of RNA wash buffer and centrifuge 1 min at high speed.

4. Recover RNA by eluting it two times with 10 μl of RNase-free water and centrifuging 1 min at high speed.

5. Repeat steps 1 to 4 and quantify the RNA concentration by UV spectrophotometry (A_{260}).

The eluted DNA-free RNA can be used immediately for RT-PCR reactions or can be stored at −70°. For RT-PCR reactions, 50 ng of RNA is reverse transcribed with Sensiscript RT enzyme (Qiagen) and the cDNA is amplified with *Taq* DNA polymerase (Qiagen). When the DNA lesion is located at the first base position of codon 445 of the luciferase gene, one can use LBRT1 (TTGACTGGCGACGTAATCC) for reverse transcription and LBRT1 and LBRT2 (GACCAACGCCTTGATTGAC) for cDNA amplification, thus generating a 295-bp DNA fragment. RT-PCR products can be directly used for sequencing reactions or can be subcloned into pUC18 to determine the transcriptional mutagenesis spectrum within a population of transcripts induced by the DNA lesion (Bregeon *et al.*, 2003).

Sub-Cloning of RT-PCR Product

The subcloning of RT-PCR products is important to determine the RNA polymerase insertion specificity when encountering a DNA lesion on the transcribed strand of an active gene. Subcloning is accomplished by digesting the RT-PCR products with appropriated enzymes (e.g., BamHI and HincII when RT-PCR is made with LBRT1 and LBRT2). The digested product can then be ligated into the multiple-cloning site of pUC18 and subcloned by transforming DH5α competent cells. Transformed cells are plated onto LB plates complemented with IPTG (1 mM), ampicillin (100 μg/ml), and 5-bromo-4-chloro-3-indolyl-beta-D-galactopyranoside (X-gal; 50 μg/ml) and incubated for 16–18 h at 37°. Isolated white colonies can be selected with a toothpick and resuspended into 50 μl of water. The subcloned RT-PCR fragment can be amplified with Clo18U (GCTGCA-AGGCGATTAAGTT) and Clo18L (CGGCTCGTATGTTGTGTGG) and by using 2 μl of resuspended bacterial cells as a source of DNA template in a 50 μl (final) PCR reaction. The 375-bp amplification product can then be used directly for sequencing. It is recommended that 100 to 200 RT-PCR subclones should be sequenced to obtain the insertion specificities opposite to the studied lesion during transcription.

Conclusions

The ability of DNA lesions to cause transcriptional mutagenesis greatly depends on how rapidly they can be recognized and eliminated by various DNA repair pathways. Consequently, the level to which an RNA polymerase miscoding lesion is capable of inducing the production of erroneous proteins largely depends on the cell ability to repair this lesion. As we have shown, using the procedures described here, uracil and 8-oxoguanine are two DNA lesions that are able to induce transcriptional mutagenesis (Bregeon et al., 2003). It is important to note that the level of transcriptional mutagenesis was substantially elevated in bacterial cells compromised for the repair of those DNA lesions. This method was utilized to identify DNA repair pathways involved in the in vivo processing of the 8-oxoguanine lesion and allowed for the determination that such DNA damage can be eliminated from the transcribed strand of an active gene by Mfd-mediated transcription-coupled repair (TCR) (Bregeon et al., 2003).

Thus, the method described in this paper makes it possible to determine precisely which classes of DNA lesions cause transcriptional bypass and the nature of miscoding events, resulting in a phenotypic change in vivo. The ability to use this technique in bacterial strains with various genetic backgrounds renders it more valuable. This method can be used to ascertain the roles of different DNA repair pathways involved in the removal of transcriptional mutagenesis-inducing lesions. Using different vectors, this method can also be used to determine whether transcriptional mutagenesis spectra are different in eukaryotic systems and the influence of the DNA repair capacity of the cell on this process.

Acknowledgments

This work was supported by NIH grants CA73041 and ES011163. D. B. was supported by "Association pour la Recherche sur le Cancer." We thank all past and current members of the Doetsch laboratory for helpful discussions.

References

Bregeon, D., Doddridge, Z. A., You, H. J., Weiss, B., and Doetsch, P. W. (2003). Transcriptional mutagenesis induced by uracil and 8-oxoguanine in *Escherichia coli. Mol. Cell* **12**, 959–970.

Bregeon, D., and Doetsch, P. W. (2004). Reliable method for generating double-stranded DNA vectors containing site-specific base modifications. *Biotechniques* **37**, 760–766.

Chiu, C. S., Cook, K. S., and Greenberg, G. R. (1982). Characteristics of a bacteriophage T4-induced complex synthesizing deoxyribonucleotides. *J. Biol. Chem.* **257**, 15087–15097.

DeMarini, D. M., and Lawrence, B. K. (1992). Prophage induction by DNA topoisomerase II poisons and reactive-oxygen species: Role of DNA breaks. *Mutat. Res.* **267**, 1–17.

Doetsch, P. W. (2002). Translesion synthesis by RNA polymerases: Occurrence and biological implications for transcriptional mutagenesis. *Mutat. Res.* **510**, 131–140.

Friedberg, E. C., Walker, G. C., and Siede, W. (1995). "DNA Repair and Mutagenesis." ASM Press, Washington DC.

Gellert, M., O'Dea, M. H., Itoh, T., and Tomizawa, J. (1976). Novobiocin and coumermycin inhibit DNA supercoiling catalyzed by DNA gyrase. *Proc. Natl. Acad. Sci. USA.* **73**, 4474–4478.

Hayashi, K., Nakazawa, M., Ishizaki, Y., and Obayashi, A. (1985). Influence of monovalent cations on the activity of T4 DNA ligase in the presence of polyethylene glycol. *Nucleic Acids Res.* **13**, 3261–3271.

Holmquist, G. P. (2002). Cell-selfish modes of evolution and mutations directed after transcriptional bypass. *Mutat. Res.* **510**, 141–152.

Kunkel, T. A. (1987). Oligonucleotide-directed mutagenesis without phenotypic selection. *In* "Current Protocols in Molecular Biology" (F. M. Ausubel, R. Brent, D. D. Moore, J. G. Seidman, J. A. Smith, and K. Struhl, eds.), Vol. 1, pp. 8.1.1–8.1.6. Wiley Interscience, Boston.

Piechocki, M. P., and Hines, R. N. (1994). Oligonucleotide design and optimized protocol for site-directed mutagenesis. *Biotechniques* **16**, 702–707.

Sala-Newby, G. B., and Campbell, A. K. (1994). Stepwise removal of the C-terminal 12 amino acids of firefly luciferase results in graded loss of activity. *Biochim. Biophys. Acta* **1206**, 155–160.

Tornaletti, S., and Hanawalt, P. C. (1999). Effect of DNA lesions on transcription elongation. *Biochimie* **81**, 139–146.

Viswanathan, A., You, H. J., and Doetsch, P. W. (1999). Phenotypic change caused by transcriptional bypass of uracil in nondividing cells. *Science* **284**, 159–162.

You, H. J., Viswanathan, A., and Doetsch, P. W. (2000). *In vivo* technique for determining transcriptional mutagenesis. *Methods* **22**, 120–126.

[21] Methods for Studying Chromatin Assembly
Coupled to DNA Repair

By ANNABELLE GÉRARD,* SOPHIE E. POLO,*
DANIÈLE ROCHE, and GENEVIÈVE ALMOUZNI

Abstract

In the eukaryotic nucleus, the DNA repair machinery operates on chromatin-embedded DNA substrates. Currently, a favored model for DNA repair into chromatin involves the transient disruption of chromatin organization to facilitate access of the repair machinery to DNA lesions. Importantly, this model implies that, in addition to DNA repair, a subsequent step is necessary to restore a proper chromatin structure. To study this latter step, we describe here methods for simultaneously analyzing chromatin assembly and DNA repair both *in vitro* and *in vivo*. Several cell-free systems have been developed that reproduce both DNA repair and nucleosome assembly. These *in vitro* systems are based on the use of defined damaged DNA. Two complementary assays are routinely used: (i) with circular DNA molecules, one can monitor in a combined analysis both repair synthesis and plasmid supercoiling; (ii) with immobilized damaged DNA, one follows specific protein interactions including histone deposition. In addition, *in vivo* assays have been designed to monitor the recruitment of chromatin assembly factors onto damaged chromatin either at a global level over the whole cell nucleus or locally at sites of DNA damage. Combination of these approaches provides powerful tools to gain insights into the mechanism by which chromatin organization can be restored after repair of DNA lesions.

Introduction

A variety of environmental and intracellular genotoxic agents induce DNA damage that must be repaired to preserve genomic integrity. DNA repair occurs in the context of chromatin (Kornberg, 1977), a regularly repeated nucleoproteic structure, whose fundamental unit is the nucleosome. The nucleosome core particle consists of DNA wrapped around an octamer of histone proteins comprising one $(H3-H4)_2$ tetramer flanked by two H2A-H2B dimers. Histone proteins through their post-translational modifications and the existence of histone variants can contribute to an epigenetic marking

---

*A. G. and S. E. P. equally contributed to this work.

METHODS IN ENZYMOLOGY, VOL. 409   0076-6879/06 $35.00

which is currently intensively studied (Vaquero *et al.*, 2003). During their cellular metabolism, these small basic proteins are escorted by histone chaperones (Loyola and Almouzni, 2004) from their site of synthesis to their site of deposition onto DNA, where they are assembled into nucleosomes. The assembly of nucleosomes and their dynamics, in combination with histone modifications and variant incorporation, appear as key processes to control the perpetuation (or switch) of nucleosomal patterns. Indeed perturbations of factors involved in any of these processes have invariably led to genome function defects, including aberrant gene expression and genomic instability. Thus, it has become crucial to better understand how the dynamics of chromatin organization is regulated and coordinated with a variety of DNA dependent processes, among which we would like to focus on DNA repair.

In order to explain how DNA lesions can be repaired within chromatin and how epigenetic information can be maintained in these conditions, the Access, Repair, and Restore model has been proposed. Based initially on the observation of transient changes in the nuclease sensitivity of chromatin undergoing nucleotide excision repair (NER) (Smerdon and Lieberman, 1978), which is involved in UV-damage response, this model has been extended to other types of DNA damage (Green and Almouzni, 2002). It postulates that chromatin organization is first destabilized upon DNA damage, so that the repair machinery can get access to DNA lesions, and then restored after repair to preserve epigenetic as well as genomic integrity. This latest step is supposed to involve chromatin assembly factors, among which CAF-1 (chromatin assembly factor 1) represents an attractive candidate (Mello and Almouzni, 2001). This three-subunit complex (p150, p60, and p48 in human cells) is indeed unique in its ability to stimulate nucleosome assembly coupled to DNA synthesis. Such a coordination has been reported during DNA replication (Smith and Stillman, 1989; Stillman, 1986) but also during DNA repair in the context of NER (Gaillard *et al.*, 1996) and repair of single-strand breaks and gaps (Moggs *et al.*, 2000). A common molecular link in all these cases was found through the direct interaction of p150CAF-1 with proliferating cellular nuclear antigen (PCNA), a polymerase accessory factor (Moggs *et al.*, 2000; Shibahara and Stillman, 1999).

We describe here methods for analyzing respectively *in vitro* and *in vivo* both nucleosome assembly and DNA repair, with a special focus on NER and single-strand break repair. These methods enable us to address the role of specific histone chaperones as well as the nature of the epigenetic marks on histones (modifications, variants) that are incorporated during a specific repair process. Furthermore, they should also help to monitor the role of regulatory factors on both the repair and histone metabolic pathways.

Analysis of Chromatin Assembly Coupled to DNA Repair *In Vitro*

To study chromatin assembly *in vitro*, the systems so far have been developed based on a simplified approach using naked damaged DNA on which *de novo* nucleosome assembly can be followed. For these analyses, one needs to rely on powerful cell-free systems that can support both DNA repair and chromatin assembly. Several types of extracts have been used for such studies: HSE extract (high speed egg extract, derived from *Xenopus* eggs) (Almouzni, 1998) and *Drosophila* embryo extract (Bonte and Becker, 1999). These extracts contain all necessary compounds to ensure both reactions (Gaillard *et al.*, 1996, 1997). In contrast, cytosolic extracts initially developed to support DNA replication (Li and Kelly, 1984) were found capable to support chromatin assembly on replicated DNA only when supplemented with nuclear extracts (Stillman, 1986; Stillman and Gluzman, 1985). In this way, such extracts offer the interesting possibility of a complementation assay that has been exploited advantageously to identify critical components. These cytosolic extracts are proficient for NER and when complemented with CAF-1 can support a repair coupled chromatin assembly reaction (Gaillard *et al.*, 1996). They have also been used successfully to analyze changes in histone-associated complexes after replication fork arrest (Groth *et al.*, 2005). The basic principle for their preparation is to use hypotonic buffer for cell extraction. Resulting cytosolic extracts contain, in addition to cytoplasmic components, soluble nuclear proteins among which are many important ones for both repair and replication reactions. We will describe here two distinct ways of preparing such extracts: starting from cells in suspension to obtain the so-called S100 extract (Stillman and Gluzman, 1985) or from adherent cells (Krude *et al.*, 1997; Martini *et al.*, 1998). We will also highlight their respective advantages and distinct properties.

Preparation of Cell Extracts

All extracts are prepared at 4° in the presence of protease inhibitors (10 μg/ml pepstatin, 10 μg/ml leupeptin, 100 μM PhenylMethylSulfonyl-Fluoride) and 1 mM DiThioTreitol, freshly added.

Cytosolic Extract from Cells Grown in Suspension

S100 EXTRACT. Typical S100 extract is obtained from cells grown in suspension, which is convenient for large-scale production. A human cell line commonly used for these preparations (Stillman, 1986; Stillman and Gluzman, 1985) is the HEK-293 cell line (CRL-1573, ATCC) which grows easily in suspension. These S100 extracts have to be complemented with nuclear extract to support chromatin assembly coupled to DNA replication (Stillman, 1986). Complementation assays have shown that the only limiting factor provided by the nuclear extract is the largest subunit of CAF-1:p150. In these conditions, S100 extract can promote chromatin assembly reactions coupled to DNA replication (Kaufman *et al.*, 1995) and DNA repair (Quivy *et al.*, 2001). Exponentially growing cells (5–6.10^5 cells/ml, usually 1 liter minimum) are harvested by centrifugation at $1000g$ for 5 min. The cell pellet is washed twice with PBS (137 mM NaCl, 2.7 mM KCl, 1.4 mM NaH$_2$PO$_4$, 4.3 mM Na$_2$HPO$_4$ pH 7.4) and resuspended in 1/3 volume of ice-cold hypotonic buffer (20 mM Hepes-KOH pH 7.5, 5 mM KCl, 1.5 mM MgCl$_2$, 0.1 mM DTT). The cells are allowed to swell on ice for 10 min before disruption by Dounce homogenization (20 strokes of a B pestle). After 30 min incubation on ice, the lysate is centrifuged ($15,700g$ in rotor JS 13.1, Beckman, for 10 min at $4°$) to remove nuclear pellet. The supernatant is collected in ultra clear tubes (Beckman, Fullerton, CA) and NaCl concentration is adjusted to 100 mM. S100 extract is obtained by ultracentrifugation ($100,000g$ in SW55Ti rotor, Beckman, for 1 h at $4°$). Usually, 1.5 ml of S100 extract can be recovered from 1 liter of cell culture.

CYTOSOLIC EXTRACTS FROM ADHERENT CELLS GROWN ON PLATE. Another type of cytosolic extract can be obtained from adherent cells as initially described for HeLa B cells (Marheineke and Krude, 1998; Martini *et al.*, 1998). This protocol has been successfully applied to several other cell lines. These cytosolic extracts also need to be complemented with nuclear extract (to provide CAF-1 complex) in order to promote chromatin assembly coupled to DNA repair *in vitro*. However, contrary to S100 extract, which only requires p150CAF-1 addition, here both p60 and p150CAF-1 subunits are in limiting amounts (Martini *et al.*, 1998).

Cells on plates are washed twice with PBS and allowed to swell for 10 min in ice-cold hypotonic extraction buffer (20 mM Hepes-KOH pH 7.8, 5 mM potassium acetate, 0.5 mM MgCl$_2$, and 0.5 mM DTT). After removal of excess buffer, cells are scraped off the plate and subjected to Dounce homogenization (25 strokes with a loose-fitting pestle). Nuclei are pelleted by centrifugation ($1500g$ for 3 min at $4°$). The supernatant is centrifuged ($14,000g$ for 20 min at $4°$) to obtain the cytosolic extract.

NUCLEAR EXTRACTS. In brief, nuclear extracts are obtained using high salt buffer to extract the nuclear pellet both from cells grown in suspension, according to Dignam *et al.* (1983) and from adherent cells, as described in Martini *et al.* (1998).

Total protein concentration in the extracts is determined by Bradford assay (Bradford, 1976) and routinely ranges between 5 and 10 μg/μl for both cytosolic and nuclear extracts. Salt concentration in the extract is systematically measured using a conductivity meter (MeterLab CDM210; Radiometer Analytical, Lyon, France) and typical values are equivalent to 25 mM salt in cytosolic extracts prepared from adherent cells, 100 mM in

S100 extract, 300 mM in nuclear extract. When high amounts of extracts have to be used in reaction mixtures, dialysis is needed to maintain appropriate salt concentration. Indeed, high salt can inhibit both nucleosome assembly and DNA synthesis reactions. All extracts are aliquoted, snap-frozen in liquid nitrogen, and stored at $-80°$.

Plasmid Labeling and Supercoiling Assay: Nucleosome Assembly Associated with DNA Synthesis (Fig. 1)

This assay allows to follow both chromatin assembly and DNA repair by monitoring simultaneously the supercoiling of closed circular DNA molecules and DNA synthesis on the same molecules. A damaged circular plasmid is introduced in the cell-free system in the presence of a radioactive desoxyribonucleotide (dNTP) that can be incorporated during a repair-synthesis reaction. Given that nucleosome formation introduces negative superhelical turns into DNA, the accumulation of supercoiling on labeled DNA is thus indicative of nucleosome assembly associated with a repair process (Gaillard *et al.*, 1999). We use DH5α bacteria transformed with pBlueScript KS+ plasmid DNA (Stratagene, La Jolla, CA), to produce our stock of template DNA. This standard circular plasmid of 3.2 kb in size is convenient for supercoiling analysis. A 500-ml bacteria culture is prepared and plasmid DNA is subsequently isolated with the Qiagen plasmid purification kit. The purified DNA solution (usually 50 ng/μl) is then exposed to UV-C irradiation (254 nm at 500 J/m^2) in order to produce DNA damage, as described (Gaillard *et al.*, 1999). The UV fluence rate is adjusted to 5 J/m^2/s as measured with a Vilber Lourmat VLX-3W dosimeter. These conditions are estimated to give rise to about 16 pyrimidine dimer photoproducts per DNA molecule.

A standard labeling and supercoiling assay is performed for one reaction (one time point) in 25 μl under the following conditions to reach final concentrations of 6 μg/ml plasmid DNA, 40 mM Hepes-KOH pH 7.8, 5 mM MgCl$_2$, 0.5 mM DTT, 4 mM ATP, 20 μM of each dGTP, dATP, dTTP, 8 μM dCTP, 40 mM phosphocreatine, 2.5 μg creatine phosphokinase, 5 μCi of [α^{32}P]-dCTP (3000 Ci/mmol, MP Biomedicals, Irvine, CA) and 10 μl (100 μg) of either cytosolic or S100 extracts together with 1 μl (6 μg) of nuclear extract, the final volume is adjusted with water. The labeled precursor [α^{32}P]-dCTP allows to monitor DNA repair synthesis. In the assay, instead of nuclear extracts, one can use recombinant proteins corresponding to CAF-1 complex or its individual subunits. We usually found that 50 ng of recombinant purified p150CAF-1 effectively complemented S100 extracts. We set up the mix by adding 3 μl of plasmid stock at 50 ng/μl, 5 μl of 5× reaction buffer (200 mM Hepes-KOH pH 7.8, 25 mM MgCl$_2$, 2.5 mM DTT,

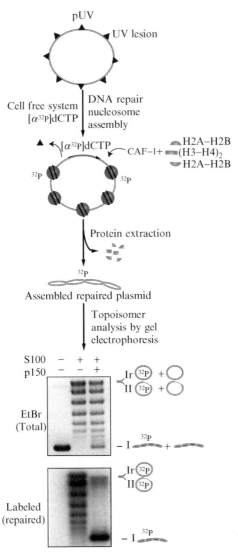

Fig. 1. *In vitro* analysis of chromatin assembly coupled to DNA repair by plasmid supercoiling assay. UV-damaged supercoiled plasmid (pUV) is incubated with radiolabeled desoxyribonucleotides and a cell-free system: here, cytosolic S100 extract complemented or not with recombinant purified p150CAF-1. The plasmid is repaired and assembled into chromatin by the concerted action of repair proteins and chromatin assembly factors. The final reaction product (after 3 h) is run on an agarose gel and visualized by intercalation of ethidium bromide (EtBr, total), repaired DNA is detected by autoradiography or phosphor-imaging (labeled). The extent of chromatin assembly coupled to DNA repair synthesis (CAF-1 dependent pathway) is given by the ratio of labeled supercoiled plasmid (I) over labeled relaxed (Ir) and nicked molecules (II).

100 μM of each dGTP, dATP, dTTP, 40 μM dCTP, 200 mM phosphocreatine), 1 μl ATP at 100 mM, 1μl creatine phosphokinase at 2.5 mg/ml, and
5 μCi of [α^{32}P]-dCTP. All compounds are stored at $-80°$ in small aliquots
(10–100 μl) and used only once to ensure reproducibility of the reaction.
Volume is adjusted with water, and the extracts added at last to start the
reaction. If necessary, the conductivity can be adjusted by adding KCl to
obtain identical salt conditions in each tube (usually around 35 mM salt).
Incubation is performed at 37° for various times usually ranging from 5 min
up to 3 h to achieve maximal supercoiling. Reactions are stopped by adding
one volume of stop buffer (30 mM EDTA, 0.7% SDS) and adjusted to
100 μl with H$_2$O before RNase treatment (10 μg of DNase free RNase A)
for 1 h 30 min at 37°. Subsequently, 20 μg of proteinase K is added
and incubation continued an extra hour. DNA is extracted using phenol-
chloroform-isoamyl alcohol (25:24:1, v/v) and then precipitated for 20 min
at $-80°$ with one volume of 5 M ammonium acetate and two volumes of
cold 100% ethanol in the presence of glycogen (1 μl at 20 mg/ml) as a
carrier. After centrifugation (20,000g at 4° for 30 min), DNA is recovered
in the pellet, which is then washed with 70% ethanol, air-dried, and
resuspended in 10 μl of loading buffer (10% glycerol, 0.084% bromophenol
blue in TE: 10 mM Tris-HCl pH 8, 1 mM EDTA pH 8). DNA samples are
subjected to analysis on 1% agarose gel in TAE buffer (40 mM Tris-acetate
pH 8, 1 mM EDTA pH 8) by electrophoresis at 1.5 V/cm for 20 h at 4°. The
gel has to be run in the absence of ethidium bromide to avoid changes in
DNA topology introduced by this intercalating drug. When gel migration is
completed, the gel is soaked for 20 min in TAE containing ethidium
bromide (0.5 μg/ml, BioRad, Hercules, CA) to stain DNA and an image
is taken under UV light to record the migration pattern of total DNA.
Then, the gel is dried (gel dryer, Fisher Bioblock Scientific, Illkirch,
France) for 3 h at 80° and subjected to autoradiography or phosphorimager
analysis to visualize the migration pattern corresponding to repaired DNA.

This assay, initially developed to follow nucleosome assembly using
UV-damaged DNA, as described here, was also extended to DNA molecules containing single-strand break and 1,3-intrastrand cisplatin-cross-
link (Gaillard et al., 1997; Moggs et al., 2000) which allows to address
chromatin assembly coupled to specific repair processes.

Histone Deposition on Immobilized DNA (Fig. 2)

Damaged DNA coupled to paramagnetic beads is a useful tool to
follow chromatin assembly coupled to DNA repair *in vitro*. Here we
describe first how to prepare, link, and damage DNA on beads and then

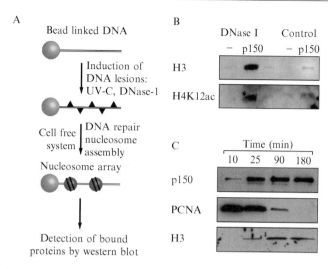

Fig. 2. *In vitro* analysis of chromatin assembly coupled to DNA repair using immobilized damaged DNA. (A) Scheme of the experiment: bead-linked DNA is damaged as indicated and incubated with a cell-free system competent for repair synthesis and nucleosome assembly. The recruitment of repair and chromatin assembly factors as well as histone deposition onto DNA are analyzed by SDS-PAGE and Western blot. (B) CAF-1 dependent histone deposition coupled to DNA repair synthesis. DNase-1 treated and control DNA are used as templates in a nucleosome assembly reaction with cytosolic S100 extract alone (−) or complemented with recombinant human p150CAF-1 (p150). Bound H3 and H4K12ac histones (the latter corresponding to newly synthesized histones associated with CAF-1 complex) are detected after 3 h at 37° for reactions complemented with p150 on damaged DNA. (C) Kinetics of repair and chromatin assembly are followed by PCNA, CAF-1, and histone H3 detection onto DNase-1 treated DNA: PCNA is recruited early and p150CAF-1 is recruited later on, concomitantly with histone H3 deposition. A.G., unpublished results.

how to use it as a template in a nucleosome assembly reaction (Mello *et al.*, 2004). In this assay, by comparing results on damaged and undamaged DNA, it is possible to distinguish events that are specific to the damaged template from nonspecific DNA binding events (background).

Plasmid DNA (pUC19) is purified from transformed *E. coli* DH5α bacteria (GibcoBRL, Gaithersburg, MD) using Qiagen Maxi Kit. The plasmid (40 μg) is linearized by two successive enzymatic digestions (EcoRI at 37° overnight and XmaI at 25° for 3 h, New England Biolabs, Ipswich, MA) in a final volume of 200 μl in TE pH 8.0. After phenol/chloroform extraction and ethanol precipitation, linearized DNA is resuspended in 40 μl of TE pH 8.0. Klenow exo-polymerase (large fragment of DNA polymerase I from *E. coli* lacking 3′-5′ and 5′-3′ exonuclease activities; 15 units) is used to fill in the 5' overhang with biotinylated ATP (0.4 mM) in the presence of 10 mM of the other dNTPs, in a 60-μl reaction volume for 2 h at 37°. After inactivation of the Klenow enzyme for 20 min at 65°, unincorporated dNTPs and small DNA fragments are removed using a Sephadex G-50 column (Roche, Indianapolis, IN) to isolate the biotinylated DNA. Recovery is assessed by running input and recovered samples on 1% agarose gel (TAE buffer). Biotinylated DNA (10 μg) is then coupled to Dynabeads M280-streptavidin (Dynal, 320 μl of stock solution) in 480 μl final volume of buffer A (2 M NaCl, 10 mM Tris-HCl pH 7.5, 1 mM EDTA) by incubation in siliconized tubes on a rotating wheel overnight at room temperature. These binding conditions, leading to a low amount of DNA per bead, are established to optimize induction of DNA lesions. We checked binding efficiency on a 1% agarose gel (TAE buffer) to ensure that no remaining DNA is detected in the supernatant at the end of the reaction. Usual yield is 99% efficiency. Bead-linked DNA is concentrated on a magnetic rack (Dynal, Carlsbad, CA), washed three times with buffer A, and resupended in this buffer at 100 ng/μl. Immobilized DNA can be stored for up to one month at 4° in buffer A before and after inducing damage using genotoxic agents such as UV-C light or DNase-1. UV-C irradiation (254 nm) is used to produce predominantly cyclobutane pyrimidine dimers and 6-4 photoproducts, which are repaired by the NER pathway. Alternatively, DNase-1 induces single-strand DNA (ssDNA) breaks that can be repaired by the base excision repair pathway. Immobilized DNA (about 100 μl of the stock solution in buffer A) is transferred on a siliconized Petri dish on ice before irradiating with a prewarmed UV lamp (254 nm, 6 J/cm^2). The UV fluence rate is adjusted to 25 J/m^2/s as measured with a Vilber Lourmat VLX-3W dosimeter. For DNase-1 treatment, 10 μg of bead-linked DNA is incubated in DNase-1 digestion buffer (10 mM Hepes-KOH pH 7.6, 50 mM KCl, 1.5 mM MgCl$_2$, 0.5 mM EGTA, 10% glycerol) with 7 units of DNase-1 (Roche) in a 650-μl reaction volume. After 5 min digestion at 25°, the reaction is stopped by adding EDTA (100 mM final). The efficiency of DNase-1 digestion is monitored on agarose gel after boiling the samples in Laemmli buffer (2% SDS, 10% glycerol, 60 mM Tris pH 6.8, 0.05% bromophenol blue) complemented with 100 mM DTT, to disrupt DNA-beads interactions. A smear is indicative of the effective digestion.

Damaged DNA immobilized on paramagnetic beads is used as a template for nucleosome assembly reaction under conditions as defined previously with plasmid DNA. For each reaction we use 5 μg/ml bead-linked DNA. Routinely, the assay is carried out in a 60-μl final volume for one reaction per time point. Bead linked DNA (3 μl at 100 ng/μl) is first washed in buffer B (150 mM NaCl, 20 mM Hepes, 5 mM MgCl$_2$, 0.025% NP-40) before incubation with reaction mix. We set up the reaction mix as above, taking into account the final volume and adjusting all components in proportion. The reaction is performed under constant agitation using a thermomixer (1100 rpm, 37°, Eppendorf, Hamburg, Germany). At different time points, usually from 2 min to 3 h (for a complete assembly reaction), the tubes are transferred on ice to stop the reaction and beads are concentrated on a magnetic rack. The unbound fraction is collected and beads are washed three times with buffer B. The stringency of the washings can be adjusted by increasing salt concentration. Finally, bound and unbound fractions are resuspended in Laemmli buffer complemented with 100 mM DTT, boiled for 5 min, and subjected to SDS-PAGE and Western blot analysis to monitor the binding of repair and chromatin assembly factors as well as histone deposition onto damaged DNA.

Controls

Several control experiments are required to draw valid conclusions in these *in vitro* nucleosome assembly assays. First, the use of untreated DNA is a necessary reference in each assay since unspecific nicks in the template can generate background signal, which would lead to misinterpretation of the results. The nucleosome assembly process in response to specific DNA lesions can also be assigned to a precise repair pathway. Indeed detection of excised damaged oligonucleotides and assessment of the length of the repair patch (around 30 nucleotides) enabled to identify NER in response to UV lesions (Gaillard *et al.*, 1996; Moggs and Almouzni, 1999). Second, to verify the quality of the final product of the assembly reaction, micrococcal nuclease (MNase) digestion is a suitable assay (Gaillard *et al.*, 1999). MNase cleavage indeed occurs preferentially between adjacent nucleosomes where DNA is the most accessible, generating oligonucleosome-sized fragments. A regular ladder is thus indicative of the proper spacing of the nucleosomes. Third, when performing plasmid supercoiling assays, it is particularly important to assess topoisomerase activity in the extracts (i.e., checking plasmid relaxation at early time point in the reaction—less than 5 min). In this assay, to be able to assess the efficiency of nucleosome assembly, it is also important to make a distinction between nonassembled plasmids (relaxed circular) and nicked plasmids (indicated Ir and II, respectively) which migrate at the same position under our electrophoresis conditions. This can be achieved by analyzing the final reaction product on an agarose gel in the presence of chloroquine (10 μg/ml). Chloroquine intercalation will change the topology of the closed molecule, which will then migrate faster as opposed to relaxed plasmids, which will not accumulate torsional stress.

Analysis of Chromatin Assembly Coupled to DNA Repair *In Vivo*

Global Recruitment of Repair and Chromatin Assembly Factors to Damaged Chromatin (Fig. 3)

Monitoring at a global level the recruitment of repair and chromatin assembly factors to damaged chromatin *in vivo* allows us to gain insight into the mechanisms whereby chromatin architecture can be maintained upon DNA damage in the cell nucleus. Importantly, such recruitment is

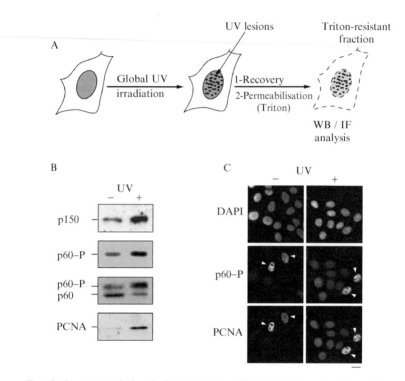

FIG. 3. *In vivo* analysis of chromatin assembly in response to global UV damage. (A) Experimental procedure: cells are subject to global UV-C irradiation (30 J/m^2). After recovery and removal of soluble proteins by detergent extraction (Triton), the recruitment of specific factors to UV damaged chromatin is analyzed by Western blot (WB) on cell extracts or by immunofluorescence (IF) on fixed cells after 2 h recovery post UV irradiation. (B) Recruitment of the chromatin assembly factor CAF-1 (p150 and phosphorylated p60 subunits) and PCNA to UV-damaged chromatin is detected by Western blot on Triton-treated extracts from G2 synchronized HeLa cells. Equal numbers of lysed cells were loaded in each lane (Reproduction with permission from Martini *et al.*, 1998). (C) Recruitment of the chromatin assembly factor CAF-1 (phosphorylated p60 subunit) and PCNA to UV-damaged chromatin is detected by immunofluorescence in asynchronous HeLa cells. S-phase cells, in which PCNA and CAF-1 are recruited to replication foci, are indicated by arrowheads. Scale bar, 10 μm.

more easily observed in cells that are synchronized out of S phase in order to exclude replication-associated processes. Induction of global DNA damage is carried out by irradiating cells (80–90% confluency) on plates in PBS with UV-C (254 nm). All kinds of plates can be used depending on the amount of material that you want to process. It is important though to cover the cell with PBS (5 mm in depth). The UV fluence is carefully determined using a dosimeter (1 $J/m^2/s$) in order to get in a short time period a final UV dose that is still compatible with cell survival (usually 30 J/m^2). The cap of the Petri dish needs to be removed during the irradiation procedure, which is carried out in PBS instead of culture medium so that UV light is not absorbed. Irradiated cells are allowed to recover in culture medium at 37° for various time periods (usually 1 to 2 h are required to reach a detectable signal) before treatment with detergent Triton X100 in order to remove soluble proteins. This is achieved by washing the cells in CytoSKeleton (CSK) buffer (10 mM PIPES pH 7.0, 100 mM NaCl, 300 mM sucrose, 3 mM MgCl$_2$) before subjecting them to detergent extraction in CSK-Triton 0.5% for 5 min at room temperature. Sucrose within CSK buffer helps to preserve "cell integrity" during the extraction procedure. Nevertheless, depending on the cell type, extraction conditions may need to be adjusted (duration, temperature, Triton concentration). Extraction is stopped by washing the cells twice in CSK buffer. Detergent-treated cells can be processed for immunofluorescence or Western blotting in order to monitor the recruitment of repair and chromatin assembly factors such as PCNA and CAF-1 onto UV damaged-chromatin (Martini et al., 1998). For immunofluorescence analysis, Triton-extracted cells, grown on glass coverslips, are fixed in 2% paraformaldehyde (freshly prepared by 1:1 dilution in PBS of a 4% paraformaldehyde solution that is stored at −20°) before immunodetection. For Western blot analysis, the extraction can be carried out as described previously either on cell plates or on cell pellets, and extracted cells are resuspended in equal volumes of CSK and Laemmli buffers to reach a final concentration of 8.10^4 cells/μl. These detergent-treated cell extracts, enriched in chromatin-associated proteins, are then analyzed by SDS-PAGE and Western blotting. The presence of repair and chromatin assembly factors is analyzed using appropriate antibodies, respectively, for the immunofluorescence or Western blot. In this way, one can estimate the proportion of proteins enriched on chromatin upon damage in a global manner compared to nondamaged cells.

Monitoring Chromatin Assembly Locally at Damage Sites (Fig. 4)

Simultaneous detection of repair and chromatin assembly factors on damaged chromatin is a first indication that chromatin re-assembly occurs concomitantly with DNA repair *in vivo*. More recently, a powerful technique

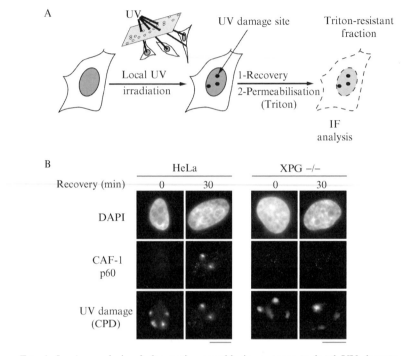

FIG. 4. *In vivo* analysis of chromatin assembly in response to local UV damage. (A) Experimental procedure: cells are subject to local UV-C irradiation (150 J/m^2, through 3 μm pore filters). After recovery and removal of soluble proteins by detergent extraction (Triton), the recruitment of specific factors to UV damage sites (CPD) is analyzed by immunofluorescence (IF). (B) Recruitment of the chromatin assembly factor p60CAF-1 to UV damage sites is efficiently detected 30 min after local UV irradiation but its recruitment is abolished in NER deficient cells (XPG$^{-/-}$). Scale bars, 10 μm.

was developed to visualize their retention locally at sites of DNA damage. For this purpose, local UV damage is induced by irradiating the cells with UV-C (254 nm, 100–150 J/m^2) through a 3-μm Isopore filter (Millipore, Billerica, MA) (Green and Almouzni, 2003; Mone *et al.*, 2001). Note that various presized filters can be used. Advantageously, this technique allows direct comparison between irradiated and nonirradiated areas within the same nucleus, thus providing a straightforward internal control. It is possible to resort to rather high UV doses since up to 98% of UV light is absorbed by these filters. UV fluence is adjusted to 5 J/m^2/s to limit the irradiation time to a maximum of 30 s, which is critical considering that cells are out of the culture medium during this procedure. Indeed, for local UV irradiation, cells are grown on glass coverslips, a piece of filter is laid on drained coverslips and removed after irradiation by adding PBS. The coverslips can be treated with 20 μg/ml collagen and 1 μg/ml fibronectin for 1 h at 37° prior to seeding the cells if these tend to detach during the irradiation process, which often occurs with primary cells. Irradiated cells are allowed to recover in culture medium at 37° for various time points: 5 min is enough to detect the recruitment of repair factors involved in lesion recognition, 30 min is required to reach the final steps of the repair process concomitant with chromatin assembly. Recruitment of repair and chromatin assembly factors to UV damage sites is monitored by immunofluorescence on Triton-extracted and paraformaldehyde-fixed cells. UV damage sites can be specifically detected with an anti-cyclobutane pyrimidine dimer (CPD) antibody (Kamiya Biomedicals, Seattle, WA) in preference to detection of 6-4 photoproducts since CPDs get repaired with slower kinetics (only half of them are removed in 24 h) and thus are more stable markers of UV irradiated regions. Notably, their immunodetection requires an antigen-unmasking step consisting of 5 min denaturation in 0.5 M NaOH at room temperature.

Controls

To check that the recruitment of factors detected at UV damage sites is indeed a UV-specific dynamic process, it is necessary to control that such factors are not already present at these sites in the absence of UV treatment and/or at a t_0 time point post-irradiation. In that respect, it is also interesting to induce local UV damage of broader size by using different filter sets (e.g., 8-μm pore filters), which should give rise to larger areas of recruited factors, reinforcing the UV-specificity of the observed process. Furthermore, it is possible to assign the *in vivo* response to a specific repair process using naturally deficient cell lines (e.g., cells from Xeroderma Pigmentosum patients, which are deficient in NER).

Conclusion and Perspectives

We describe here methods to analyze chromatin assembly coupled to DNA repair both *in vitro* and *in vivo*. These methods have been used successfully to document the involvement of the chromatin assembly factor CAF-1 in nucleosome assembly coupled to NER, as discussed extensively herein. Additionally, *in vitro* methods offer the attractive possibility to perform complementation analyses in order to address the role of other histone chaperones in such processes and to provide evidence of possible synergy between them. They have also been adapted to dissociate DNA synthesis-dependent and independent nucleosome assembly pathways by immunodepleting defined chromatin assembly factors within cell extracts prior to the assembly reaction (Ray-Gallet and Almouzni, 2004; Tagami *et al.*, 2004). As an alternative approach to immunodepletion, loss of function by RNA interference has also been coupled to plasmid super-coiling assay (Hoek and Stillman, 2003; Nabatiyan and Krude, 2004).

The *in vitro* systems that we presented here are dedicated to the analysis of *de novo* nucleosome assembly on naked, damaged DNA. In the future, new approaches should be developed to investigate histone dynamics by using a pre-assembled chromatin template. Different methods can be used for this purpose (e.g., nucleosome arrays reconstituted onto DNA *in vitro* by salt dialysis using purified histones incubated with dam-aged DNA; Brand *et al.*, 2001), see also Wassarman and Kornberg (1989) for other methods. Finally, *in vivo* methods can be extended to the study of chromatin assembly coupled to repair of various types of DNA damage since other groups have developed strategies to induce local single- (Okano *et al.*, 2003) or double-strand breaks (Lukas *et al.*, 2003; Rogakou *et al.*, 1999) in cultured cells. Both strategies rely on presensitization of the cells, either by expression of UV damage endonuclease in NER deficient cells prior to local UV-C irradiation or with halogenated thymidine analogues before exposure to a UV-A pulsed laser beam, respectively.

Importantly, the combination of such *in vitro* and *in vivo* analyses is useful in uncovering physiological regulatory mechanisms involved in the coupling between nucleosome assembly and DNA repair. These approaches are crucial to our understanding of the coordination in the maintenance of both genetic and epigenetic information.

Acknowledgments

We thank Dominique Ray-Gallet for critical reading. This work was supported by la Ligue Nationale contre le Cancer (Equipe labellisée la Ligue), Euratom (FIGH-CT-2002-00207), the CEA (LRC no. 26), Contract RTN (HPRN-CT-2002-00238), by a PIC Paramètres Epigénétiques, NoE Epigenome (LSHG-CT-2004-503433), ACI-DRAB (n° 04393) and Cancéropôle.

References

Almouzni, G. (1998). "Chromatin: A practical approach," pp. 195–218. Oxford University Press, New York.
Bonte, E., and Becker, P. B. (1999). Preparation of chromatin assembly extracts from preblastoderm Drosophila embryos. *Methods Mol. Biol.* **119,** 187–194.

Bradford, M. M. (1976). A rapid and sensitive method for the quantitation of microgram quantities of protein utilizing the principle of protein-dye binding. *Anal. Biochem.* **72,** 248–254.

Brand, M., Moggs, J. G., Oulad-Abdelghani, M., Lejeune, F., Dilworth, F. J., Stevenin, J., Almouzni, G., and Tora, L. (2001). UV-damaged DNA-binding protein in the TFTC complex links DNA damage recognition to nucleosome acetylation. *EMBO J.* **20,** 3187–3196.

Dignam, J. D., Lebovitz, R. M., and Roeder, R. G. (1983). Accurate transcription initiation by RNA polymerase II in a soluble extract from isolated mammalian nuclei. *Nucleic Acids Res.* **11,** 1475–1489.

Gaillard, P. H., Martini, E. M., Kaufman, P. D., Stillman, B., Moustacchi, E., and Almouzni, G. (1996). Chromatin assembly coupled to DNA repair: A new role for chromatin assembly factor I. *Cell* **86,** 887–896.

Gaillard, P. H., Moggs, J. G., Roche, D. M., Quivy, J. P., Becker, P. B., Wood, R. D., and Almouzni, G. (1997). Initiation and bidirectional propagation of chromatin assembly from a target site for nucleotide excision repair. *EMBO J.* **16,** 6281–6289.

Gaillard, P. H., Roche, D., and Almouzni, G. (1999). Nucleotide excision repair coupled to chromatin assembly. *Methods Mol. Biol.* **119,** 231–243.

Green, C. M., and Almouzni, G. (2002). When repair meets chromatin. First in series on chromatin dynamics. *EMBO Rep.* **3,** 28–33.

Green, C. M., and Almouzni, G. (2003). Local action of the chromatin assembly factor CAF-1 at sites of nucleotide excision repair *in vivo*. *EMBO J.* **22,** 5163–5174.

Groth, A., Ray-Gallet, D., Quivy, J. P., Lukas, J., Bartek, J., and Almouzni, G. (2005). Human Asf1 regulates the flow of S phase histones during replicational stress. *Mol. Cell.* **17,** 301–311.

Hoek, M., and Stillman, B. (2003). Chromatin assembly factor 1 is essential and couples chromatin assembly to DNA replication *in vivo*. *Proc. Natl. Acad. Sci. USA* **100,** 12183–12188.

Kaufman, P. D., Kobayashi, R., Kessler, N., and Stillman, B. (1995). The p150 and p60 subunits of chromatin assembly factor I: A molecular link between newly synthesized histones and DNA replication. *Cell* **81,** 1105–1114.

Kornberg, R. D. (1977). Structure of chromatin. *Annu. Rev. Biochem.* **46,** 931–954.

Krude, T., Jackman, M., Pines, J., and Laskey, R. A. (1997). Cyclin/Cdk-dependent initiation of DNA replication in a human cell-free system. *Cell* **88,** 109–119.

Li, J. J., and Kelly, T. J. (1984). Simian virus 40 DNA replication *in vitro*. *Proc. Natl. Acad. Sci. USA* **81,** 6973–6977.

Loyola, A., and Almouzni, G. (2004). Histone chaperones, a supporting role in the limelight. *Biochim. Biophys. Acta* **1677,** 3–11.

Lukas, C., Falck, J., Bartkova, J., Bartek, J., and Lukas, J. (2003). Distinct spatiotemporal dynamics of mammalian checkpoint regulators induced by DNA damage. *Nat. Cell. Biol.* **5,** 255–260.

Marheineke, K., and Krude, T. (1998). Nucleosome assembly activity and intracellular localization of human CAF-1 changes during the cell division cycle. *J. Biol. Chem.* **273,** 15279–15286.

Martini, E., Roche, D. M., Marheineke, K., Verreault, A., and Almouzni, G. (1998). Recruitment of phosphorylated chromatin assembly factor 1 to chromatin after UV irradiation of human cells. *J. Cell. Biol.* **143,** 563–575.

Mello, J. A., and Almouzni, G. (2001). The ins and outs of nucleosome assembly. *Curr. Opin. Genet. Dev.* **11,** 136–141.

Mello, J. A., Moggs, J. G., and Almouzni, G. (2004). Analysis of DNA repair and chromatin assembly *in vitro* using immobilized damaged DNA substrates. *Methods Mol. Biol.* **281,** 271–281.

Moggs, J. G., and Almouzni, G. (1999). Assays for chromatin remodeling during DNA repair. *Methods Enzymol.* **304,** 333–351.

Moggs, J. G., Grandi, P., Quivy, J. P., Jonsson, Z. O., Hubscher, U., Becker, P. B., and Almouzni, G. (2000). A CAF-1-PCNA-mediated chromatin assembly pathway triggered by sensing DNA damage. *Mol. Cell. Biol.* **20,** 1206–1218.

Mone, M. J., Volker, M., Nikaido, O., Mullenders, L. H., van Zeeland, A. A., Verschure, P. J., Manders, E. M., and van Driel, R. (2001). Local UV-induced DNA damage in cell nuclei results in local transcription inhibition. *EMBO Rep.* **2,** 1013–1017.

Nabatiyan, A., and Krude, T. (2004). Silencing of chromatin assembly factor 1 in human cells leads to cell death and loss of chromatin assembly during DNA synthesis. *Mol. Cell. Biol.* **24,** 2853–2862.

Okano, S., Lan, L., Caldecott, K. W., Mori, T., and Yasui, A. (2003). Spatial and temporal cellular responses to single-strand breaks in human cells. *Mol. Cell. Biol.* **23,** 3974–3981.

Quivy, J. P., Grandi, P., and Almouzni, G. (2001). Dimerization of the largest subunit of chromatin assembly factor 1: Importance *in vitro* and during Xenopus early development. *EMBO J.* **20,** 2015–2027.

Ray-Gallet, D., and Almouzni, G. (2004). DNA synthesis-dependent and -independent chromatin assembly pathways in Xenopus egg extracts. *Methods Enzymol.* **375,** 117–131.

Rogakou, E. P., Boon, C., Redon, C., and Bonner, W. M. (1999). Megabase chromatin domains involved in DNA double-strand breaks *in vivo*. *J. Cell. Biol.* **146,** 905–916.

Shibahara, K., and Stillman, B. (1999). Replication-dependent marking of DNA by PCNA facilitates CAF-1-coupled inheritance of chromatin. *Cell* **96,** 575–585.

Smerdon, M. J., and Lieberman, M. W. (1978). Nucleosome rearrangement in human chromatin during UV-induced DNA-repair synthesis. *Proc. Natl. Acad. Sci. USA* **75,** 4238–4241.

Smith, S., and Stillman, B. (1989). Purification and characterization of CAF-I, a human cell factor required for chromatin assembly during DNA replication *in vitro*. *Cell* **58,** 15–25.

Stillman, B. (1986). Chromatin assembly during SV40 DNA replication *in vitro*. *Cell* **45,** 555–565.

Stillman, B. W., and Gluzman, Y. (1985). Replication and supercoiling of simian virus 40 DNA in cell extracts from human cells. *Mol. Cell. Biol.* **5,** 2051–2060.

Tagami, H., Ray-Gallet, D., Almouzni, G., and Nakatani, Y. (2004). Histone h3.1 and h3.3 complexes mediate nucleosome assembly pathways dependent or independent of DNA synthesis. *Cell* **116,** 51–61.

Vaquero, A., Loyola, A., and Reinberg, D. (2003). The constantly changing face of chromatin. *Sci. Aging Knowledge Environ.* **2003,** RE4.

Wassarman, P. M., and Kornberg, R. D. (1989). "Nucleosomes." Academic Press, Inc., San Diego, CA.

[22] Structure–Function Analysis of SWI2/SNF2 Enzymes

By HARALD DÜRR and KARL-PETER HOPFNER

Abstract

Biochemical and structural progress over the last years has revealed that SWI2/SNF2 family chromatin remodeling or DNA repair enzymes are molecular motors that transport duplex DNA along a helicase-like domain using ATP-hydrolysis. The screw motion of DNA along the active site probably generates the force to disrupt chromatin or other protein:DNA complexes. In this chapter, we describe biochemical and structural approaches to study the molecular mechanism of SWI2/SNF2 enzymes. In particular, we describe assays to monitor DNA dependent ATPase activity, translocation on duplex DNA, and DNA distortion activity. We also describe recent progress in the crystallization and structure determination of SWI2/SNF2 enzymes in complex with duplex DNA.

Introduction

Dynamic remodeling of nucleosomes or disruption of other protein: DNA complexes is an important event to create accessible DNA in gene expression and genome maintenance. ATP-dependent chromatin remodelling by nucleosome sliding, nucleosome displacement and histone variant exchange is carried out by large multisubunit remodeling complexes such as SWI/SNF, ISWI, SWR1, and RSC (Becker and Horz, 2002; Vignali *et al.*, 2000). Several of the remodeling factors are specifically implicated in DNA repair. For instance, Rad54 is suggested to remodel chromatin or disrupt other protein:DNA complexes at the template sister-chromatid in homologous recombination (Alexeev *et al.*, 2003; Jaskelioff *et al.*, 2003). Other proven or putative remodeling factors are involved in DNA double-strand break repair (*S. cerevisiae* Ino80 and human Tip60 complexes), global nucleotide excision repair (*S. cerevisiae* NEF4 complex), transcription coupled nucleotide excision repair (human Cockayne syndrome protein B or *S. cerevisiae* Rad26), and postreplication repair (*S. cerevisiae* Rad5) (Morrison and Shen, 2005; Ulrich, 2003; van den Boom *et al.*, 2002). The overall composition and activity of remodeling factors is very diverse. However, they all contain a principal core ATPase that belongs to the SWI2/SNF2 family of proteins. SWI2/SNF2 enzymes possess a conserved

METHODS IN ENZYMOLOGY, VOL. 409
0076-6879/06 $35.00
DOI: 10.1016/S0076-6879(05)09022-1

ATPase catalytic core that shares sequence and structure homology to DExx box helicases (Durr et al., 2005; Tuteja and Tuteja, 2004). In contrast to DExx box helicases, SWI2/SNF2 enzymes do not display helicase activity in classical strand displacement assays. In the last decade, biochemical characterization of SWI2/SNF2 enzymes indicated that SWI2/SNF2 possess double-stranded DNA (dsDNA) stimulated ATPase activity, DNA translocation activity, and DNA distortion activity (Flaus and Owen-Hughes, 2001; Langst and Becker, 2004). Based on these biochemical and recent structural results, a picture emerges where the catalytic core of SWI2/SNF2 enzymes is a structure and sequence unspecific translocase that moves along the minor groove of duplex DNA without strand displacement. This minor groove tracking presumably translocates DNA along the SWI2/SNF2 ATPase active site cleft in a screw motion, a process that is well suited to translocate and distort DNA at nearby DNA:protein interfaces.

In this chapter, we will summarize biochemical and structural approaches to characterize the *in vitro* activity of SWI2/SNF2 family remodeling enzymes. In the first part, we briefly summarize protocols that have been developed for the analyses of DNA dependent ATP hydrolysis, DNA translocation, and DNA distortion activity. In the second part, we will address recent progress in the structural analysis of SWI2/SNF2 enzymes in complex with DNA. SWI2/SNF2 enzymes bind DNA preferentially in a sequence and structure independent manner. This property provides a formidable obstacle for crystallization of a specific protein:DNA complex, because the enzyme can bind to multiple sites on a DNA oligonucleotide duplex. The resulting heterogeneity severely interferes with crystallization. We will discuss strategies on how to overcome such problems.

Biochemical Characterization of SWI2/SNF2 ATPases

ATPase Activity

SWI2/SNF2 enzymes have a low intrinsic ATPase activity that is greatly stimulated by DNA. The optimal DNA substrate for stimulation, however, is variable and depends on the enzyme. Whereas some SWI2/SNF2 enzymes are stimulated by both dsDNA and single-stranded DNA (ssDNA), others are stimulated by dsDNA only, or by chromatin fragments. Therefore, the choice of the right DNA substrate is a critical aspect for assessing the ATPase activity. Additional important criteria are the nucleotide composition and length of the DNA substrates. It is a common procedure to use plasmids (e.g., ϕ174 or M13mp19) as dsDNA substrates, or their (+) or (−) strands as ssDNA substrates, respectively. However, these ssDNA substrates tend to form secondary structures that make a distinction of true ssDNA dependent activity difficult and can lead to false-positive results. A better choice to measure ssDNA stimulated ATPase activity would be to use poly(dT) or poly(dA) for ssDNA and a poly(dA):poly-(dT) duplex for dsDNA. Poly(dT) is not able to form secondary structures, representing a true ssDNA substrate. For instance, ϕ174 ssDNA resulted in substantial stimulation of the ATPase activity of yeast Rad54, whereas Rad54 is not stimulated by poly(dA) or poly(dT) ssDNA substrates (Petukhova *et al.*, 1998). The dsDNA substrate that is generated by annealing of poly(dA) to poly(dT), however, can result in a heterogenic mixture with variable single-stranded overhangs. In our opinion, good dsDNA substrate can be obtained from plasmid digestion with blunt-end restriction enzymes or by the use of complementary oligonucleotides that contain some partial non-A tracks.

The second criteria—the length of the DNA—is also very important. It has been shown for several SWI2/SNF2 enzymes that the ATPase activity depends on the length of the DNA fragment used. Maximum stimulation is reached with DNA of 60–100 bp, but is typically not further stimulated with longer DNA, at least when assessing the catalytic subunit of remodelers by itself. The length dependent stimulation of ATP-hydrolysis by DNA can have two reasons. First, DNA binding to an extended binding site or to multiple binding sites on large remodeling factors may be necessary for full stimulation. Second, length dependent stimulation is also a feature of processive translocation and the "diluting" of potential DNA end effects. This is due to the fact that in the presence of shorter DNA substrates, processive enzymes will fall off the DNA more rapidly. Until rebinding, these enzymes will not hydrolyze ATP, resulting in an overall reduction of ATPase activity in the reaction mixture. Two exemplary protocols for radioactive or nonradioactive monitoring of ATPase activity are as follows.

Example Protocols for DNA Dependent ATPase Assay

In a typical ATPase assay, reaction mixtures (20 μl) contain 20 mM Tris/HCl [pH = 7.5], 50 mM NaCl, 5 mM MgCl$_2$, 100 μg/ml BSA, 0.1 mM unlabeled ATP, 20 nM γ-^{32}P-ATP (3000 Ci/mmol), plus an excess (3–5-fold DNA concentration over protein concentration or saturating amounts in terms of DNA binding constants) of ds/ssDNA. The reaction should be kept on ice and is started by addition of typically 1–100 nM protein. The protein concentration used depends on the activity of the protein and has to be adapted in each case. The reaction is incubated at 30°–37° for 15–30 min. Care should be taken to make sure that at maximum 10% of ATP is hydrolyzed for the fastest reactions. Otherwise, substrate depletion and product inhibition can artificially slow down the reaction and lead to non-Michaelis-Menten like kinetics. The reactions are stopped by addition of 1 μl of 0.5 M EDTA to remove magnesium ions that are required for ATP hydrolysis. To analyze the amount of liberated phosphate, aliquots of 1 μl of the reaction mixture are spotted onto a thin layer chromatography polyethyleneimine-cellulose plate (Merck, Darmstadt, Germany). A height of about 7–10 cm of the TLC is sufficient to allow an efficient separation of ATP and phosphate, and the spots should be placed about 1 cm above the bottom of the plate. The TLC plate is air-dried and developed in 0.5 M LiCl/1.0 M formic acid. The liberated radioactive γ-phosphate almost runs with the front of the developing solution. After the front of the developing solution has reached 2/3 of the plate height, the TLC plate is removed from the container and air dried. The products are visualized by phosphoimager (e.g., the STORM system of Amersham/Pharmacia, Piscataway, NJ) and the spots are quantified by image analysis software. The upper spot corresponds to liberated γ-^{32}P and the lower visible spot to nonhydrolyzed ATP. The percentage of hydrolyzed ATP can be easily calculated after background and negative control correction according to the data by dividing the (amount of liberated phosphate)/(amount of unhydrolyzed ATP + liberated phosphate).

Non-Radioactive ATPase Assay as an Alternative

An easy to use, nonradioactive method uses Biomol green™ (Biomol Research Laboratories, Plymouth Meeting, PA) that sensitively detects liberated phosphate by forming a stoichiometric complex. The reaction mixture is the same as described previously, except nonradioactive ATP is used. After the reaction is stopped, the liberated phosphate can be detected by mixing an aliquot (e.g., 10 μl) of the reaction mixture with Biomol green (e.g., 90 μl; see also manufacturer's instruction). Incubate for 20 min at room temperature (RT) to allow development of the green color and measure the absorption at 620 nm. Biomol green forms a stoichiometric green complex with free phosphate that has an absorption maximum around 620 nm. Liberated phosphate can be determined by comparison with a standard curve. This method is also suitable for kinetic analysis using microwell plates and microtiter reader.

DNA Translocation Monitoring–Triplex Displacement Assay

Firman and Szczelkun (2000) developed a method to directly monitor translocation on duplex DNA. This approach uses the displacement of a triplex forming oligonucleotide (TFO) from a DNA duplex. Triplex

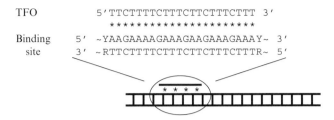

TFO   5′ TTCTTTTCTTTCTTCTTTCTTT 3′
                     *
Binding   5′ ~YAAGAAAAGAAAGAAGAAAGAAAY~ 3′
site   3′ ~RTTCTTTTCTTTCTTCTTTCTTTR~ 5′

FIG. 1. Schematic drawing of a triplex substrate and sequences of the triplex forming oligonucleotide (TFO) and the corresponding target sequence. The TFO binding is mediated by Hoogsten hydrogen bonds (depicted as asterisks).

substrates are generated by annealing a TFO to its target duplex DNA sequence (Fig. 1). The TFO interacts with the major groove of its target DNA duplex sequence, through the formation of a triple helical structure. Hereby, the pyrimidine rich (CT) TFO occupies the major groove parallel to a purine-rich (GA) strand and utilizes Hoogsten hydrogen bonds to form T_AT and C+_GC triplets (Gowers et al., 1999). For this interaction, the N3 position of cytosine has to be protonated, which demands a low pH. However, once the triplex is formed, the TFO remains stably associated with its target site, even if diluted into neutral buffer (up to pH 8.5). The stable association of the TFO with the DNA duplex requires Mg^{2+} ions that may stabilize the protonated form of cytosine in a C+_GC triplet. If the TFO is displaced by a DNA translocating enzyme, no reassociation is observed at neutral pH, due to rapid deprotonation of the cytosines. Thus, at the typical assay conditions, triplex disruption is an irreversible process. These properties make this method an ideal probe to test DNA transloca-tion. The triplex can be stored at pH 5.5 and 4° without dissociation of ^{32}P labeled TFO from the duplex for several days.

Preparation of the Triplex Substrate

TFO is ^{32}P labelled by T4 polynucleotide kinase and purified with commercially available oligonucleotide purification kits. To generate the triplex, equimolar concentration (100 nM) of NotI linearized pMJ5 plasmid (Greaves et al., 1985) and ^{32}P-labeled TFO are mixed in 25 mM MES [2-(N-Morpholino)ethanesulfonic acid] [pH 5.5] and 10 mM MgCl$_2$, incubated at 57° for 15 min and left to cool to 4° overnight. The triplex should be purified to remove free TFO. We established a quick and inexpensive procedure: as a small disposable "gel filtration" column we use a Pasteur pipette and seal it at the bottom with a glass bead or with cotton wool. The column is gravity packed with 2–3 ml degassed Superdex G50 fine media suspension (Amersham, Piscataway, NJ). The column is equilibrated with about 5 ml of 25 mM MES [pH 5.5] and 10 mM MgCl$_2$ before use. Load the triplex on the column and allow it to enter the column. Subsequently apply equilibration buffer carefully to the column and collect 0.1 to 0.2-ml fractions. The desired triplex will elute rapidly due to its large size and can be separated from free labeled TFO that migrates more slowly. The elution profile can be monitored using a scintillation counter. The efficiency of triplex formation and purity or homogeneity of the fraction should be checked by agarose gel electrophoresis (see later). Since triplex formation is never 100% complete, you will always have a mixture of linearized pMJ5 plasmid and triplex DNA. To minimize the amount of linearized pMJ5 plasmid you can use a slightly higher concentration of TFO over linearized pMJ5 for triplex formation. The free TFO is eliminated by the procedure described previously.

Triplex Displacement Assay

Prior to the translocase reaction, the triplex substrate is diluted with the desired buffer (e.g., 50 mM Hepes [pH 7.5], 50 mM NaCl, 10 mM MgCl$_2$, 100 μg/ml BSA, 1 mM DTT). 2.5 nM of triplex substrate is incubated at 30° for 30 min with typically 0–100 nM protein in Hepes [pH 7.5], 50 mM NaCl, 10 mM MgCl$_2$, 100 μg/ml BSA, 1 mM DTT, and 4 mM ATP. Similarly, one can also measure time dependence by using a fixed amount of protein and vary the incubation time. The reactions are quenched with GSMB buffer (15% [w/v] glucose, 3% SDS, 250 mM MES [pH 5.5], 0.1 mg/ml proteinase K and 0.4 mg/ml bromphenole blue) and analyzed with a 1.5% agarose gel (40 mM Tris-acetate [pH 5.5], 5 mM Na-acetate, 2.5 mM Mg-acetate) at 10 V/cm for 1.5 hour. Alternatively, a low percentage polyacrylamide gel can be used. The proportion of bound and free TFO can be determined using a phosphorimager and suitable image analysis software.

Detection of Superhelical Torsion–Cruciform Extrusion Assay

Havas et al. (2000) developed a cruciform extrusion assay that can measure superhelical tension in DNA. With this method, they demonstrated that SWI2/SNF2 enzymes can introduce negative superhelical torsion into linear DNA. Havas et al. used the pXG540 plasmid with an (AT)$_{34}$ inverted repeat as a cruciform forming element. Inverted (AT) repeats have been shown to adopt cruciform structures at moderate levels of supercoiling and without a significant kinetic barrier (Greaves et al., 1985). In negatively supercoiled DNA, the energy required for cruciform extrusion is compensated by the relaxation of the negative superhelical density (Lilley, 1980). The superhelical density required for cruciform extrusion hereby depends on the size of the inverted repeat. For instance a $(AT)_{34}$ inverted repeat incurs a free energy cost of 14 kcal/mol (McClellan *et al.*, 1990). In linearized DNA, the superhelical torsion has to be restrained by the enzyme to maintain the superhelical tension and to support the stable existence of a cruciform. This could be achieved by loop formation as suggested for large remodeling complexes (Bazett-Jones *et al.*, 1999). Havas *et al.* observed cruciform extrusion in the presence of large remodeling factors at a molar ratio of 1:1 (protein:DNA). For smaller SWI2/SNF2 enzymes like yeast Rad54, alteration of superhelicity has also been observed using this assay (Jaskelioff *et al.*, 2003). However, to detect cruciform extrusion by Rad54, a higher protein:pXG540 plasmid ratio is necessary. In contrast to multidomain remodeling factors, the restraint of the superhelical torsion by yeast Rad54 has been proposed to be a result of transient multimerization of the protein on DNA, since it is unlikely that the relatively small Rad54 can bind large DNA loops at multiple DNA binding sites.

The idea behind the cruciform extrusion assay is that SWI2/SNF2 enzymes are incubated with linearized pXG540 plasmid DNA that contains the $(AT)_{34}$ inverted repeat. In the presence of ATP, SWI2/SNF2 enzymes generate superhelical torsion that results in the formation of the cruciform. The cruciform structure can be specifically cleaved by the structure specific T4 endonuclease VII (ENDOVII) (Lilley and Kemper, 1984). EndoVII introduces bilateral cleavages and generates a double-stranded break in the DNA close to the junction. The resulting two DNA fragments can be analyzed by gel electrophoresis (Fig. 2). The amount of ENDOVII cleavage provides a direct measure for the DNA distortion activity of the remodeler. A typical protocol follows.

Protocol for Monitoring the Generation of Altered Superhelical Torsion by Cruciform Extrusion

30 ng (1.5 n*M*) AvaI linearized pXG540 plasmid [containing a $(AT)_{34}$ inverted repeat] is incubated in 20 m*M* Hepes [pH 7.5], 50 m*M* NaCl, 3 m*M* MgCl$_2$, 5% glycerol, and 0.1 m*M* DTT with appropriate quantities of protein in the presence of 1 m*M* ATP and 0.15 μg/ml endonuclease VII. For efficient remodeling factors, a 1:1 stoichiometric ratio of protein:plasmid DNA can be used. The reactions are incubated at 20° for 45 min and stopped by addition of 0.6% SDS, 20 m*M* EDTA, and 0.5 mg/ml Proteinase K followed by incubation at 50° for 15 min. The reactions are analyzed by separation on a 1.2% agarose gel and stained with the sensitive SYBR gold (Molecular Probes, Eugene, OR). The gels can be analyzed by fluorescence

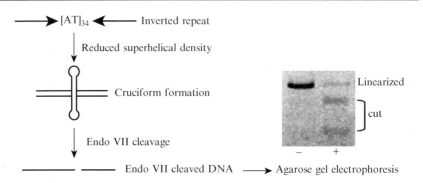

FIG. 2. Schematic illustration of the principle of the Cruciform extrusion assay. Generation of unconstrained superhelical torsion by SWI2/SNF2 enzymes leads to cruciform extrusion that is recognized by the structure specific endonuclease VII (ENDO VII). The cleavage products (cut) are analyzed by agarose gel electrophoresis (right). The gel shows an example of negative (−) and positive (+) cruciform extrusion assay obtained with *Sulfolobus solfataricus* Rad54 catalytic domain in the absence (−) and presence (+) of ATP. Figures adapted from Havas *et al.*, 2000.

excitation with maximal sensitivity at 300 nm. Hereby, a standard fluorescence visualization box for ethidium bromide DNA could be used, in conjunction with a gel documentation system. Better results are obtained with a multimode fluorescence imaging system.

Crystallization of Protein:DNA Complexes

For a full understanding of how SWI2/SNF2 enzymes modulate their protein:DNA substrates, it is important to examine the structural interplay between the enzyme and its DNA substrate. SWI2/SNF2 enzymes bind duplex DNA in a predominantly sequence and structure independent manner. This property provides a formidable obstacle for crystallization of a specific protein:DNA complex, since crystallization typically requires a homogenous population of the solution. In the following part, we will describe strategies to deal with this aspect.

Protein

For crystallization of a protein:DNA complex, the protein has to satisfy the same criteria as for crystallization of protein alone, for instance high purity, conformational homogeneity, and sufficient stability over the crystallization period. SWI2/SNF2 enzymes are often large, multidomain proteins that can interact with one or more partner proteins. These features often result in an inherently flexible molecule that hampers crystallization. To increase the probability to get crystals, a commonly used approach is to remove flexible regions that might hamper crystal packing [e.g., Rad54, (Thoma *et al.*, 2005)]. The boundaries of such protein domain(s) are best identified by limited proteolyses of the full-length protein and its DNA complex. Thereby the protein is incubated at 4° or 37° for 30 min with proteases (trypsin, chymotrypsin, proteinase K,...) at different concentrations (typically molar protein:protease ratio of 10:1 to 10^5:1) and the cleavage products are analyzed by SDS-PAGE. The obtained fragments can then be identified by mass-spectrometry or N-terminal Edmann sequencing. Proteolysis should be ideally performed in the presence of an appropriate DNA substrate. Often, flexible regions that are subject to protease digestion in the apoprotein become more resistant to digestion if they participate in DNA binding. These regions are therefore important for function and should be included in the protein fragment. Based on the identified fragments, new protein variants are expressed and used for crystallization.

DNA Design

If the protein specifically recognizes DNA sequences or structural features (hairpins, forks, bubbles, etc.), these elements greatly simplify the formation of a specific protein:DNA complex. Besides these elements, the length of DNA used is one of the most important criteria in crystallization of protein:DNA complexes. In most successful cases, the DNA contained 1–5 additional base pairs flanking the central recognized sequence. In practice, the minimal number of base pairs used to grow protein-duplex DNA complex crystals is about 10 bps, since shorter fragments might not be sufficiently stable at the crystallization temperature. A single helical turn, 10–11 bps, is often a good starting point for short DNA recognition sites. Subsequently, length increments (e.g., 2 bps) could be used. The DNA length might roughly match the diameter of the protein around the DNA binding site, with the chance that DNA ends might participate in crystal lattice contacts.

However, if the protein binds DNA in a predominantly sequence and structure independent manner, like SWI2/SNF2 enzymes, mixing of protein with DNA usually results in a quite heterogenous mixture of protein bound somehow to DNA. One strategy could be to employ DNA that has a more flexible internal region combined with stiffer ends. The concept of sequence-directed structural softness or flexibility has been introduced some years ago and may provide a physical basis for protein:DNA recognition. For instance, a slightly twisting, bending, or deformation of the

DNA for optimal DNA binding may be facilitated by the more flexible region of the DNA. Duplex DNA with a flexible central part and more inflexible ends may mimic a central "flexible wedge" that may promote specific DNA binding for crystallization by directing DNA binding to a specific DNA site. Analyses of sequence-dependent dynamics in duplex DNA revealed that AAAAAA tracts sequences or GGGCCC motifs have an increased stiffness compared to non A-tract sequences (Dlakic and Harrington, 1995). A number of specific sequences have been suggested to be more flexible. In particular, CA and TA dinucleotides have been proposed as candidates for regions of increased flexibility (Harrington and Winicov, 1994). The use of sequence-dependent flexibility in duplex DNA was important in crystallizing the *Sulfolobus solfataricus* Rad54 catalytic domain:DNA complex (Durr *et al.*, 2005). In our approach, we used a 13-bp internal sequence of predicted increased flexibility in combination with terminal poly(dA) runs of different lengths. In our experience increments of 2–3 bps should be reasonable for the initial screen, followed by smaller steps for refinement.

In the case of the *Sulfolobus solfataricus* Rad54 catalytic domain:DNA complex, the DNA binds alongside the entrance of the active site cleft by recognition of the two phosphate chains along the minor groove. Although the DNA bound to the protein is most similar to B-DNA, the enzymes induce a slight widening of the minor groove at the contact region and a slight bending of DNA. Both distortions, although small, may have helped to place the protein onto the DNA in a more homogeneous fashion.

Another important variable is the design of the DNA ends. In many crystals of protein:DNA complexes, it has been observed that the DNA ends are involved in crystal packing by contacting another DNA end. Most popular are complementary sticky ends with single or double base overhangs. Such ends, for instance a 5′-dA overhang combined with a 3′-dT overhang, can interact in a head-to-tail fashion to form a repetitive linear array that potentially facilitates crystal growth (Rice *et al.*, 1996). An alternative type of DNA:DNA contact in crystals are triple helices, although less regularly observed. For instance G:C and G (or C) or A:T and A (or T) can interact by forming triplexes. Combinations of overhanging G or A with C:G or A:T blunt ends, respectively, may help to generate crystal contacts by DNA base stacking. For instance, in the *Sulfolobus solfataricus* Rad54 catalytic domain:DNA complex crystals, the DNA ends have been involved in crystal lattice contacts. The two DNA duplex molecules form an unusual "head to head" crystal contact by utilizing base stacking and extrahelical hydrogen bonds of the T-overhang to the symmetry related DNA molecule, similar to a triplex structure (Fig. 3).

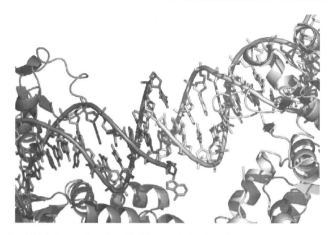

FIG. 3. In *Sulfolobus solfataricus* Rad54 catalytic domain:DNA complex crystals, crystal lattice contacts were to some extend mediated by head-to-head crystal contacts between two DNA molecules (grey and black stick models with backbone tube), which form a triple helix interface. Protein models are shown as black and grey ribbon model with highlighted secondary structure.

Preparation of Duplex DNA

We use HPLC purified oligonucleotides, desolved in 10 m*M* Tris/HCl (pH 8.0) at a concentration of 3 nmol/μl to generate duplex DNA substrate. The dsDNA is generated by mixing the two corresponding oligonucleotides in a molar ratio of exactly 1:1, heating up to 80° for 5 min followed by slow cooling to 4°. In our experience, the formation of duplex DNA is sufficiently homogenous, and we use the DNA sample directly for the formation of the protein:DNA complex without further purification. However, one can purify the obtained duplex DNA by polyacrylamide gel electrophoresis.

Preparation of the Protein:DNA Complex

The binary protein:DNA complex can be produced by incubating the protein with the DNA substrate at an effective protein:DNA molar ratio starting from 1:1.2 to 1:3. Incubate the protein and DNA for 1 h on ice. Typical buffer conditions for protein:DNA complex formation would be 20 m*M* Tris/HCl (pH 7.5), 50–150 m*M* NaCl, 1 m*M* EDTA, and 1 m*M* DTT and protein concentrations of 10–20 mg/ml. Depending on the solubility of the protein or the protein:DNA complex, respectively, DNA may be added at high protein concentrations. In some cases, this procedure can result in protein precipitation. In this case, DNA could be added at a low protein concentration. The complex is then concentrated using ultrafiltration devices. In our experience it was not necessary to further purify the protein: DNA complex prior to crystallization. Additional purification of the protein:DNA complex (e.g., by gel filtration) can be used if the protein:DNA complex is very stable (K_d values less than nanomolar). For crystallization, we use standard vapor diffusion or batch methods with commercially available crystallization screens. However, crystallization conditions that use salt as precipitants should be avoided, because under these conditions, the DNA:protein complex is presumably disrupted.

Trapping of Conformational States

In many cases, additional ATP-binding might be required to form a stable DNA:protein complex. To capture such a relevant enzyme-substrate complex, for instance a tertiary complex with protein, DNA, and ATP, one can apply different approaches. Often, slow- or nonhydrolyzable ATP analogues such as ATPγS, AMPPNP, or AMPPCP are used with great success. However, in several cases, this approach was not successful, because ATPγS might still be slowly hydrolyzed in the course of crystallization, while the nitrogen or carbon in AMPPNP or AMPPCP could interfere with formation of a fully assembled nucleotide bound active site. For instance, in many P-loop ATPases such as helicases, ABC ATPases, and AAA$^+$ ATPases, ATP is "embedded" in the active site. On one side, ATP is bound to Walker A and B motifs. On the other, side regulatory elements and allosteric activators (e.g., arginine fingers, the ABC ATPase signature motif) can interact with the oxygen atoms of the β- and γ-phosphates. These interactions can be sufficiently perturbed in the presence of ATP-analogs to prohibit formation of a fully assembled active site. In the case of several P-loop ATPases, the use of nonhydrolyzable ATP-analogs did not give satisfactory results. A good example is ABC ATPases, where ATP-binding promotes dimerization of two ABC domains by sandwiching ATP in the dimer interface. In our experience, a complementary and perhaps better approach is to use a particular mutant enzyme, at least for crystallographic studies. For many P-loop ATPases, like SWI2/SNF2 enzymes, the Walker B (hhhhDE, h: hydrophobic) or DExx box glutamic acid positions and polarizes the water molecule for nucleophilic attack on the γ-phosphate. Substitution of this glutamic acid with a glutamine abolishes ATP hydrolysis, but preserves ATP-binding. E $\rightarrow$ Q mutations in the Walker B motif have been used to trap ATP-bound state, where the use of nonhydrolyzable ATP analogs has failed (Lammens *et al.*, 2004). In those cases where crystallographic information is available, the resulting active sites are fully assembled, including a suitably positioned water molecule that is prone to nucleophilic attack (Lammens *et al.*, 2004). In the case of the SWI2/SNF2 enzyme SsoRad54, the use of the DExx box E → Q mutation results in an improved DNA binding activity, paired with a complete loss of ATP-hydrolysis. Such a stable trapped state is a good starting point for crystallization of ternary complexes.

Conclusions

Seminal contributions of several laboratories worldwide lead to the development of the described biochemical and structural methods. These methods form a powerful basis with which to unravel the complex and fascinating multisubunit machineries that remodel chromatin or disrupt other protein:DNA complexes in genome maintenance and gene expression. However, the methods are not limited to the described basic analysis of SWI2/SNF2 enzymes, but can be used as a powerful basis for the development of more specific assays to address special features of remodeling factors. For instance, these assays can be used in conjunction with nucleosome localization sites (Jaskelioff *et al.*, 2003), to study the effect of translocase or distortion activity in a more chromatin like template. Due to space, however, we could only focus on some of the available technologies and we apologize to our colleagues, whose key contributions could not be cited.

Acknowledgments

We thank the members of the Hopfner laboratory for discussions and Tom Owen-Hughes, Craig Peterson, and Börries Kemper for reagents. Work in K.P.H.s laboratory is funded by the German Research Council (DFG), the EU 6th framework program "DNA Repair," the EMBO Young Investigator Program. Harald Dürr thanks Boehringer Ingelheim Fonds for support.

References

Alexeev, A., Mazin, A., and Kowalczykowski, S. C. (2003). Rad54 protein possesses chromatin-remodeling activity stimulated by the Rad51-ssDNA nucleoprotein filament. *Nat. Struct. Biol.* **10**, 182–186.

Bazett-Jones, D. P., Cote, J., Landel, C. C., Peterson, C. L., and Workman, J. L. (1999). The SWI/SNF complex creates loop domains in DNA and polynucleosome arrays and can disrupt DNA-histone contacts within these domains. *Mol. Cell. Biol.* **19**, 1470–1478.

Becker, P. B., and Horz, W. (2002). ATP-dependent nucleosome remodeling. *Annu. Rev. Biochem.* **71**, 247–273.

Dlakic, M., and Harrington, R. E. (1995). Bending and torsional flexibility of G/C-rich sequences as determined by cyclization assays. *J. Biol. Chem.* **270**, 29945–29952.

Durr, H., Korner, C., Muller, M., Hickmann, V., and Hopfner, K. P. (2005). X-ray structures of the Sulfolobus solfataricus SWI2/SNF2 ATPase core and its complex with DNA. *Cell* **121**, 363–373.

Firman, K., and Szczelkun, M. D. (2000). Measuring motion on DNA by the type I restriction endonuclease EcoR124I using triplex displacement. *EMBO J.* **19**, 2094–2102.

Flaus, A., and Owen-Hughes, T. (2001). Mechanisms for ATP-dependent chromatin remodelling. *Curr. Opin. Genet. Dev.* **11**, 148–154.

Gowers, D. M., Bijapur, J., Brown, T., and Fox, K. R. (1999). DNA triple helix formation at target sites containing several pyrimidine interruptions: Stabilization by protonated cytosine or 5-(1-propargylamino)dU. *Biochemistry* **38**, 13747–13758.

Greaves, D. R., Patient, R. K., and Lilley, D. M. (1985). Facile cruciform formation by an (A-T)34 sequence from a Xenopus globin gene. *J. Mol. Biol.* **185**, 461–478.

Harrington, R. E., and Winicov, I. (1994). New concepts in protein-DNA recognition: Sequence-directed DNA bending and flexibility. *Prog. Nucleic Acid Res. Mol. Biol.* **47**, 195–270.

Havas, K., Flaus, A., Phelan, M., Kingston, R., Wade, P. A., Lilley, D. M., and Owen-Hughes, T. (2000). Generation of superhelical torsion by ATP-dependent chromatin remodeling activities. *Cell* **103**, 1133–1142.

Jaskelioff, M., Van Komen, S., Krebs, J. E., Sung, P., and Peterson, C. L. (2003). Rad54p is a chromatin remodeling enzyme required for heteroduplex DNA joint formation with chromatin. *J. Biol. Chem.* **278**, 9212–9218.

Lammens, A., Schele, A., and Hopfner, K. P. (2004). Structural biochemistry of ATP-driven dimerization and DNA-stimulated activation of SMC ATPases. *Curr. Biol.* **14**, 1778–1782.

Langst, G., and Becker, P. B. (2004). Nucleosome remodeling: One mechanism, many phenomena? *Biochim. Biophys. Acta* **1677**, 58–63.

Lilley, D. M. (1980). The inverted repeat as a recognizable structural feature in supercoiled DNA molecules. *Proc. Natl. Acad. Sci. USA* **77**, 6468–6472.

Lilley, D. M., and Kemper, B. (1984). Cruciform-resolvase interactions in supercoiled DNA. *Cell* **36**, 413–422.

McClellan, J. A., Boublikova, P., Palecek, E., and Lilley, D. M. (1990). Superhelical torsion in cellular DNA responds directly to environmental and genetic factors. *Proc. Natl. Acad. Sci. USA* **87**, 8373–8377.

Morrison, A. J., and Shen, X. (2005). DNA repair in the context of chromatin. *Cell Cycle* **4**, 568–571.

Petukhova, G., Stratton, S., and Sung, P. (1998). Catalysis of homologous DNA pairing by yeast Rad51 and Rad54 proteins. *Nature* **393**, 91–94.

Rice, P. A., Yang, S., Mizuuchi, K., and Nash, H. A. (1996). Crystal structure of an IHF-DNA complex: A protein-induced DNA U-turn. *Cell* **87**, 1295–1306.

Thoma, N. H., Czyzewski, B. K., Alexeev, A. A., Mazin, A. V., Kowalczykowski, S. C., and Pavletich, N. P. (2005). Structure of the SWI2/SNF2 chromatin-remodeling domain of eukaryotic Rad54. *Nat. Struct. Mol. Biol.* **12**, 350–356.

Tuteja, N., and Tuteja, R. (2004). Unraveling DNA helicases. Motif, structure, mechanism and function. *Eur. J. Biochem.* **271**, 1849–1863.

Ulrich, H. D. (2003). Protein-protein interactions within an E2-RING finger complex. Implications for ubiquitin-dependent DNA damage repair. *J. Biol. Chem.* **278**, 7051–7058.

van den Boom, V., Jaspers, N. G., and Vermeulen, W. (2002). When machines get stuck—obstructed RNA polymerase II: Displacement, degradation or suicide. *Bioessays* **24**, 780–784.

Vignali, M., Hassan, A. H., Neely, K. E., and Workman, J. L. (2000). ATP-dependent chromatin-remodeling complexes. *Mol. Cell. Biol.* **20**, 1899–1910.

[23] Genomic Approach for the Understanding of Dynamic Aspect of Chromosome Behavior

By YUKI KATOU, KIYOFUMI KANESHIRO,
HIROYUKI ABURATANI, and KATSUHIKO SHIRAHIGE

Abstract

Various functions are integrated into a single chromosome molecule. The genomic approach (ChIP-chip) we introduce here is a very powerful tool to study dynamic changes of the structure and function of the chromosome at the level of protein-DNA interaction as precisely as possible without prejudice or bias. This technology opens up the way to understand how local protein-protein or protein-DNA interactions lead to the dynamic changes of chromosome structure and how various chromosomal functions are connected to make a network for the faithful maintenance of the genome.

Introduction

In order to maintain the integrity of the genome, the precise duplication of all chromosomes in each round of the cell cycle is required. Errors in this process are known to cause mutations that can lead a cell to cancer and aging. To understand the molecular mechanism that guarantees the genome integrity, it is essential to study the process of chromosome dynamics (i.e., replication, recombination, repair, and partition) using a genomic approach. Genetic and biochemical approaches have so far identified hundreds of proteins that function in some aspects of chromosome dynamics. Now, genomic approaches are able to show us how these proteins are integrated in the process of whole chromosomal dynamics, that is, how each elemental process is connected to make a complex network in order to guarantee the faithful maintenance of genomes.

ChIP-chip (*Ch*romatin *I*mmuno-*P*recipitation combined with high resolution tiling DNA *chip*) is a powerful technique to explore the interplay of proteins on chromosomes in detail. In this chapter, first we will introduce the outline of ChIP-chip technique and next, we will describe the precise protocols for analyzing the location of proteins and newly synthesized DNA (replicated regions) on S-phase chromosome.

METHODS IN ENZYMOLOGY, VOL. 409   0076-6879/06 $35.00
    DOI: 10.1016/S0076-6879(05)09023-3

Outline of ChIP-Chip Technique

ChIP-chip technique was originally developed by Ren *et al.* (2000) (Fig. 1). The major difference between their method and the method described here is the use of high-resolution DNA tiling chip (oligonucleotide DNA chip) instead of DNA microarray in which 500–1000 bp DNA fragments amplified by PCR were used as probes (Katou *et al.*, 2003). Here, we use an oligonucleotide tiling DNA chip where 25 bp oligonucleotides are used as probes, because it is possible to design target sequences as densely and uniquely as possible (to avoid cross-hybridization of probes to target with similar sequences) (Fig. 2A). The chip we used in this section is a *S. cerevisiae* chromosome VI tiling chip provided by Affymetrix (Santa Clara, CA) (Katou *et al.*, 2003). 16 or 11 sets of unique 25 bp oligo probes (probe set) cover every 300 bp region of chromosome VI, which corresponds to one locus of analysis. A probe set is made up of probe pairs comprised of perfect match (PM) and mismatch (MM) probes. The intensities of each probe pair are the key ingredients used to measure the amount of each locus. This measurement is calculated for each probe set and is described in the form of quantitative values. The quantitative data of two chips can be compared to analyze the relative change in the amount of each locus (Wodicka *et al.*, 1997).

The ChIP-chip technique is a combination of a modified chromatin immunoprecipitation (ChIP) and DNA chip analysis (Fig. 1). Briefly cells were fixed with formaldehyde, harvested, and disrupted by sonication. The DNA fragments cross-linked to a protein of interest are enriched by immuno-precipitation with a specific antibody. In the case of yeast, strains expressing a specific epitope tagged version of the protein of interest are often used. After reversal of the cross-links, the enriched DNA is amplified and labeled by biotin with the use of random primed PCR method. A sample of DNA prepared from WCE (whole cell extract) is also subjected to random primed PCR, and each sample, immuno-precipitation (ChIP)-enriched and none-enriched (WCE) pools of labeled DNA are hybridized to a single DNA chip respectively. After hybridization and scan of DNA chips hybridized with ChIPed and WCE fraction, signal intensities of each locus on two chips are compared by the expression analysis protocol provided by Affymetrix to calculate protein binding profile. For the discrimination of positive and negative signals of the binding, the following three criteria are used. First, the reliability of the strength of the signal is judged by detection p-value. Second, reliability of the binding ratio is judged by change p-value. Third, clusters consisted of at least three contiguous loci that satisfied the previous two criteria are selected, because it is known that a single site of protein-DNA interaction results in immuno-precipitation of DNA fragments of the actual binding site but also to its

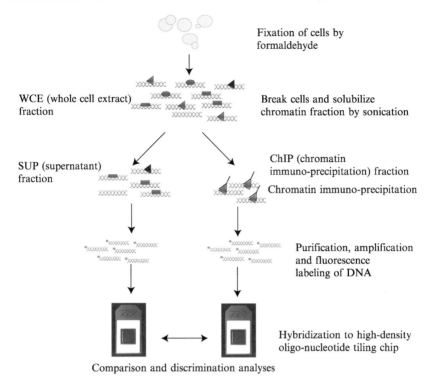

Fixation of cells by
formaldehyde

WCE (whole cell extract)
fraction

Break cells and solubilize
chromatin fraction by sonication

SUP (supernatant)
fraction

ChIP (chromatin
immuno-precipitation) fraction

Chromatin immuno-precipitation

Purification, amplification
and fluorescence
labeling of DNA

Hybridization to high-density
oligo-nucleotide tiling chip

Comparison and discrimination analyses

FIG. 1. Schematic presentation of ChIP-chip experiment (see text for details). (See color insert.)

neighbors. This third criterion is very important and only guaranteed by using an oligonucleotide tiling array (Katou *et al.*, 2003).

ChIP-Chip Protocol for the Analysis of Protein Binding Profile in S-Phase

In this section we describe the protocol for the location analysis of HA, PK, or FLAG-tagged DNA replication related protein. S-phase extracts can be prepared from cells that are arrested at G1 phase by α-factor, and then released into S-phase in the presence or absence of 200 mM HU (hydroxy urea). One can use yeast strain that expresses thymidine kinase of herpes simplex virus for the location analysis of BrdU labeled newly synthesized DNA (see next section). Analysis of the locations of various proteins—together with replicated regions—can provide us with a very detailed picture of the replicating chromosome.

Solutions for ChIP-chip Analysis

TBS: 20 mM Tris-Cl pH 7.5, 150 mM NaCl

Lysis buffer: 50 mM Hepes-KOH pH 7.5, 140 mM NaCl, 1 mM EDTA,
1% Triton-X100, 0.1% Na-deoxycholate (for cell breakage add
immediately before use of protease inhibitor Complete cocktail and
1 mM PMSF). Protease inhibitor cocktail can be dissolved in water
(25×) and kept frozen at −20°.

Wash buffer: 10 mM Tris-Cl pH 8, 250 mM LiCl, 0.5% NP-40, 0.5%
Na-deoxycholate, 1 mM EDTA

Elution buffer: 50 mM Tris-Cl pH 8, 10 mM EDTA, 1% SDS

TE: 10 mM Tris-Cl pH 8, 1 mM EDTA

For Western blot: 5% milk in PBST. α-HA (16B12 Babco, Richmond,
CA, 1 mg/ml) 1:500; α-PK (Serotec, Oxford, UK, MCA1360) 1 mg/ml
1:500; α-Flag (Sigma, St. Louis, MO, M2) 5 mg/ml 1:500.

Extract Preparation

Take 50–100 ml of cells at about 1 × 10^7/ml

Add formaldehyde to 1% final, incubate at room temperature (RT) for
30 min with gentle shaking. Eventually transfer to 4° and continue
incubation overnight with gentle shaking. At this stage prepare the
magnetic beads (see later) for chromatin immuno-precipitation.

Spin down cells for 3 min at 3000 rpm 4° and resuspend in 20 ml ice
cold TBS. Repeat this washing step 2 more times.

Resuspend cells in 0.4 ml of Lysis buffer (supplemented immediately
before use with 1 mM PMSF and antiproteolytic cocktail [Complete
protease inhibitor cocktail tablets, Roche 1697498, Indianapolis,
IN]), transfer to 2-ml tubes with screw caps and O-ring, add glass
beads (Sigma G-8772) up to the rim (1 mm below the meniscus is also
acceptable).

Break cells with a multibeads shocker (Yasui-kikai, Osaka, Japan)
(2700 rpm/min; total time 20 min; 5 cycles of 4 × [1 min shaking +
1 min pause at 4°]/each). Please make sure that more than 90% of
cells are broken by microscopy.

Wipe the tubes, puncture their bottom with a needle and transfer
them to a 15-ml Falcon tube. Spin at 3000 rpm for 1 min at 4°,
transfer the collected cell extracts to a 1.5-ml Eppendorf tube, spin
Falcon tube again at 3000 rpm for 1 min to collect all the remaining
extracts. Please make sure to collect all extracts.

Shear chromatin by sonicating the extracts (tune 1.5, 15″ for 5 cycles)
(Sonifier Branson 2508, Danbury, CT)

Spin at 15,000 rpm for 5 min at 4° and transfer the supernatant to a 1.7-ml Costar tube (Corning, Corning, NY, #3207). Add 5 μl of this WCE to 5 μl of 2× Laemmli buffer for SDS page later on (to be boiled before loading)

Add antibodies bound to magnetic beads and incubate on a wheel at 4° from 6 h to overnight.

Preparation of the Magnetic Beads

Usually we use magnetic beads for ChIP because background due to nonspecific DNA contamination was dramatically reduced when compared to sepharose beads. Use of Protein A or anti-IgG beads is highly recommended, because Protein G tends to give high background due to contamination of DNA.

For each immuno-precipitation, use 15 μl of Dynabeads (Dynal, Lake Success, NY, Dynabeads protein A, #100.02 or 100.05). Use 30 μl of beads for HA-tagged proteins.

Put dynabeads in a Costar tube and spin at 4° for 1 min at 5000 rpm. Amount of Dynabeads should be optimized according to the antibody used.

Aspirate supernatant with vacuum pump and resuspend in 0.5 ml cold PBS containing 5 mg/ml BSA

Repeat this washing step one more time

Resuspend in 15 μl of BSA/PBS, add 5 μg of antibody per IP (use 10 μg of antibody for HA-tagged proteins). Incubate with rotation at 4° from a few hours to overnight

Immediately before use wash 2 times with ice cold BSA/PBS and resuspend in the same initial volume of BSA/PBS.

Beads Wash After IP

Wash beads using magnetic stand; don't forget to keep the first supernatant, as this might be used later on for SDS page and for hybridization to the chip.

Washes

2× with 1 ml of ice cold Lysis buffer (w/o anti-proteolytics)

2× with 1 ml of ice cold Lysis buffer plus 360 mM NaCl

2× with 1 ml of ice cold wash buffer

1× with 1 ml of ice cold TE pH8

Spin 3 min at 3000 rpm and discard completely the supernatant.

Add 40–50 μl of elution buffer and incubate at 65° for 10 min. During this time flick the tubes 3 times.

The following steps can be done at RT.

Spin 1 min at 15,000 rpm at RT.

Transfer 5 μl to a new tube containing the same volume of 2×
Laemmli buffer and incubate at 95° for 20–30 min (this sample is for
loading on SDS page and Western blot).

Transfer the remaining 35–45 μl to a new tube and add 4 volumes
(i.e., 140–180 μl depending on the IP volume) of TE/1% SDS
(IP fraction). Also take 5 μl of the first supernatant of the IP (see
previously) and add 95 μl of TE/1% SDS (WCE fraction)

Incubate overnight at 65° in an oven to reverse the crosslink.

Check the amount of protein precipitated by Western blot. Due to the
cross-linking, IP efficiency is usually not high. Efficiency of IP will
be varied (5% to 50% of WCE) depending on the protein and
condition to be analyzed.

Clean-Up of DNA

Mix together:

For IP fraction: 175 μl IP fraction + 120.5 μl TE + 1.5 μl glycogen
(20 mg/ml) + 3 μl Proteinase K (50 mg/ml)

For WCE fraction: 100 μl WCE fraction + 97 μl TE + 1 μl glycogen
(20 mg/ml) + 2 μl Proteinase K (50 mg/ml)

Incubate at 37° for 2 h

Add NaCl to 200 mM final from a 5 M stock, mix, and extract 2× with
the same volume of phenol/chloroform/iso-amylalcohol

Add 2 volumes of cold EtOH, vortex, and incubate at −20° for at least
15 min

Spin at 4° at 12,000 rpm for 10 min

Discard supernatant and wash the pellet with 80% EtOH

Dry the pellet, resuspend it in 30 μl TE containing 10 μg RNaseA, and
incubate at 37° for 1 h

Purify by Qiagen PCR purification kit (Qiagen, Valencia, CA, Cat.
#28106). Elute with 50 μl of buffer EB

Precipitate DNA by adding 5 μl NaOAc 3 M, 1 μl 20 mg/ml glycogen,
and 2.5 volumes of cold EtOH. Incubate at −20° for 20 min,
centrifuge at max speed for 10 min, and wash the pellet with 70%
EtOH. Resuspend in 7 μl TE final volume.

DNA Amplification

DNA amplification is absolutely required for hybridization and detection, because more than 2 μg of DNA is required for the reliable detection
of the signal on the DNA chip. In the case of *S. cerevisiae*, random PCR

amplification method is used; however, the same method cannot be used for higher eukaryotes with more complex genome sequence. In the case of higher eukaryotes, LM-PCR (ligation mediated PCR) and IVT (*in vitro* transcription) amplification method (see alternative protocol section) are used and dramatically reduce the background signal caused by repetitive sequence (Liu *et al.*, 2003). The choice of DNA polymerase used for PCR is very important and affects the results seriously by biased amplification. Here we recommend using KOD-dash (King of DNA polymerase) from TOYOBO (Osaka, Japan) with proof reading activity and high processivity.

Round A

Sequenase Ver2.0 T7 DNA polymerase (USB #70775)

Primer A: GTT TCC CAG TCA CGA TCN NNN NNN NN

Set up

DNA	7 μl
5× sequenase buffer	2 μl
Primer A (40 mM)	1 μl
Total volume	10 μl

Reaction Mix

5× sequenase buffer	1 μl
dNTPs (3 mM each)	1.5 μl
DTT (0.1 M)	0.75 μl
BSA (0.5 mg/ml)	1.5 μl
Sequenase	0.3 μl

Incubate Round A. Set up in PCR machine at 94° for 2 min. Go to 10° and hold for 5 min while adding Reaction mix

Ramp to 37° over 8 min. Hold at 37° for 8 min and then go to 94° for 2 min

Go to 10° and hold for 5 min while adding 1.2 μl of diluted sequenase (diluted 1:4 with sequenase dilution buffer)

Ramp to 37° over 8 min

Increase the total volume to 60 μl by water and use one-fourth for Round B

Round B

Primer B: GTT TCC CAG TCA CGA TC

Template	15 μl
TOYOBO KOD-dash (LDP-101)	2 μl
10× KOD-Dash PCR buffer	10 μl
dNTP mixture (2.0 mM each)	10 μl
Primer B (100 μM)	1 μl
Water	62 μl
Total volume	100 μl

Perform amplification cycle as

    94° 5 min

    94° 20 s $\Rightarrow$ 40° 30 s $\Rightarrow$ 50° 30 s $\Rightarrow$ 74° 3 min (repeat for 32 cycles)

    74° 7 min

Run 2.5 μl of Round B on 1–1.2% agarose gel and check the size of amplified products, which should be in the range of 300–1000 bp.

Concentrate the volume down to less than 40 μl by Etoh precipitation. Alternatively Millipore (Bedford, MA) MICROCON YM10 can be used. Measure the concentration of DNA by A260. Usually yield is higher than 5 μg.

DNA Digestion

DNase I (Gibco BRL, Grand Island, NY, Amplification Grade #18068–015)

10× One-Phor-All-Buffer plus (Pharmacia, Piscataway, NJ, #27–0901–02)

DNase I Set up for 13 Samples

DNase I (1 U/μl)	2 μl
10× One-Phor-All-Buffer plus	2 μl
25 mM CoCl$_2$	1.2 μl
Water	14.8 μl

Reaction Mix

DNase I Set up	1.5 μl
10× One-Phor-All-Buffer plus	4.85 μl
25 mM CoCl$_2$	2.9 μl
DNA (2–5 ug) + Water	40.75 μl
Total volume	50 μl

Incubate at 37° for 2 min then transfer to 95° for 15 min to inactivate DNaseI.

Check the size of DNA by agarose gel electrophoresis (make sure the average size is less than 100 bp). Continue digestion until majority of the DNA reaches a size less than 100 bp.

DNA Labeling

Terminal transferase (Roche, Indianapolis, IN, #333 574)

Biotin-11-ddATP (NEN #NEL548)

Transfer DNA into 1.5-ml tube

Add 1 μl terminal transferase (25 U/μl) and 1 μl Biotin-N6-ddATP (1 nM/μl)Incubate at 37° for 1 h.

Hybridization. Steps after hybridization are essentially carried out following the protocols provided by Affymetrix.

Pre-hybridize chips (rikDACF, chromosome VI tiling array) with 190 μl prehybridization buffer. Incubate in Affymetrix oven (GeneChip hybrid oven 320, Affymetrix, CA) at 42° for 10–20 min

Prepare hybridization mix

DNA (2–5 μg)	50 μl
3 nM control Oligo B2	3.3 μl
20× eukaryotic hybridization control	10 μl
10 mg/ml Herring sperm DNA	2 μl
20× SSPE	60 μl
0.1% Triton-X100	10 μl
Water	64.7 μl
Total volume	200 μl

Boil at 100° for 10 min and transfer in ice. Spin down at max speed for 2–3 min

Remove prehybridization solution from chips. Add 150 μl of boiled probe and hybridize at 42° for 16 h in Affymetrix hybridization oven.

Washing and Staining of the Chips

Remove hybridization solution

Add 200 μl washing buffer/chip

Prepare 0.6 ml staining buffer/chip:

2× stain buffer	300 μl
Water	270 μl
Acetylated BSA (50 mg/ml)	24 μl
SAPE (1 mg/ml)	6 μl

Washing and staining protocol (Midi_euk1 ver2) provided by Affymetrix is performed automatically on a fluidics station (GeneChip fluidics station 450, Affymetrix, CA).

Scanning. Each DNA chip is scanned by Affymetrix GeneChip Scanner 3000 7G (Affymetrix). Grids are aligned to the scanned images using the known feature of the chip. After hybridization and scan of DNA chips hybridized with ChIPed and Sup fraction (or WCE fraction), signal intensities of each locus on two chips are compared by the GCOS expression analysis protocol.

Discrimination Analysis. Here, we briefly describe about Affymetrix statistical algorithms used in our analysis. For more detail, one can download sufficient information from Affymetrix Web site (http://www.affymetrix.com). It provides researchers with a mathematical concept behind quantitative measurements of DNA for either single chip or comparison analysis (profile analysis).

Analysis of each single chip provides the initial data required to perform comparison between ChIP and WCE fraction. This analysis generates signal intensity value and detection p-value for each locus (Fig. 2B). Detection p-value indicates whether each locus is reliably detected or not. Protein binding profile is obtained by comparison analysis of ChIP fraction with WCE fraction. Before comparing ChIP fraction with WCE fraction, a normalization method must be applied. The intensities of all of the probes from the ChIP and WCE fractions are summed, and the ratio of total ChIP/WCE intensity is set equal to one.

Two sets of algorithms are used to generate change significance and change quantity metrics for every probe set. A change algorithm (the Wilcoxon's signed-rank test is used) generates a change p-value. A second algorithm produces a quantitative estimate of the relative change in amount of DNA for each locus in the form of Signal Log Ratio (the computed ratio of corrected ChIP/WCE for each probe set). The change p-value ranges in scale from 0.0 to 1.0, and provides a measure of the likelihood of change and direction. Values close to 0.0 indicate likelihood for an enrichment of the locus in ChIP fraction compared to WCE, whereas values close to 1.0 indicate likelihood for a none enrichment. The Signal Log Ratio estimates the magnitude and direction of change of an abundance of each locus in ChIP fraction. It is calculated by comparing each probe pair of the ChIP fraction to the corresponding probe pair of the WCE fraction. This strategy can cancel out differences due to different probe binding coefficients and PCR amplification bias. The log scale used is base 2. Thus, a Signal Log Ratio of 1.0 indicates an enrichment of the locus by 2-fold. The average Signal Log Ratio of significantly none enriched locus is usually -1.0, which is a base line of protein binding profile.

For discrimination of positive and negative signals for the binding, we use three criteria as follows (Katou et al., 2003). First, the reliability of strength of signal is judged by detection p-value of each locus (p value ≤ 0.025). Secondly, reliability of binding ratio is judged by change p-value (p value ≤ 0.025). Thirdly, clusters consisting of at least three contiguous loci that satisfied the previous two criteria are selected, because it is known that a single site of protein:DNA interaction resulted in immuno-precipitation of DNA fragments that hybridized not only to the locus of the actual binding site but also to its neighbors. The software for the calculation of protein binding profile is available through our Web site (http://charlie.bio.titech.ac.jp/shirahigelab/tools/). When no significant binding of the protein of interest is observed, one has to be sure that whether it is because of no binding (due to high detection p-value) or no localized binding (due to high change p-value).

If the signal-to-noise ratio of ChIP fraction is high enough, the result of protein binding profile is not seriously affected by WCE fraction used for comparison. For example, using WCE fraction of a different cell cycle or a different strain background does not essentially change the profile. This is a simple but useful method of checking the justification of profile data.

ChIP-chip Protocol for the Detection of BrdU Incorporated (Replicated) Regions. For the analysis of BrdU incorporated regions, cells expressing thymidine kinase of herpes simplex virus are used (Katou *et al.*, 2003). Strain K5601(W303-1A GDP-thymidine kinase::URA3x7) is once arrested at G1 phase by α-factor for 2 h and 30 min then released into S phase in the presence or absence of HU. 5'-Bromo-2-Deoxyuridine (BrdU) is added to the culture (200 μg/ml final concentration) 30 min before the release (BrdU should be added in the dark). After labeling cells for the appropriate period of time, the cells are fixed by ice cold buffer containing Azide (equal volume of ice-cold 200 mM EDTA [pH 8.0]/20 mM Tris-HCl [pH 8.0]/0.2% NaH3 is added to the culture). Culture is kept on ice for 15 min then total DNA from 3 $\times$ 10^8 cells is purified using QIAGEN Genomic DNA purification Kit. After Etoh precipitation, DNA is resuspended in 250 μl of TE (DNA concentration should be 50–150 ng/μl). DNA is shared into the size of 400 bp by sonication (Power 1.5 15 s $\times$ 6 times) then heat denatured by heating 200 μl DNA solution at 100° for 10 min. Equal volume of 200 μl ice-cold 2$\times$ PBS (pH 7.4) is added and denatured DNA solution is kept on ice for 5 min. For immuno-precipitation of BrdU labeled DNA, equal volume of ice-cold immuno-precipitation buffer (400 μl of 2% B SA/1 $\times$ PBS/0.2% Tween 20) and 4 μg anti-BrdU monoclonal antibody (2B1D5F5H4E2, MBL, Nagoya, Japan) bound to 20 μl of Dynabeads (M-450 Rat anti-mouse IgG1) is added. Incubation with antibody is continued for 6 h at 4°. Washing and elution of antibody bound DNA is carried out essentially following the protocol for ChIP. After eluting antibody bound DNA in 50 μl elution buffer, 1 μl of Proteinase K (50 mg/ml) is added, and eluted DNA solution is incubated at 37° for 2 h. DNA is further purified by Qiagen PCR Purification Kit and eluted by 50 μl TE. After Etoh precipitation, DNA is resuspended in 7 μl of water and subsequently used for amplification and labeling exactly following the protocol for ChIPed DNA. Concentration of BrdU in precipitates is confirmed by Western blotting and further by conventional ChIP PCR method. Comparative analysis is carried out with DNA prepared from cells at G1 phase.

Alternative Protocol for DNA Amplification (IVT Amplification Method). As we mentioned in the previous section, PCR-based amplification protocol is highly susceptible to bias and is not suitable for amplification of DNA of ?tlsb?> higher eukaryotes or of a small amount. The protocol introduced here is mainly based on the amplification protocol developed by Liu *et al.* (2003). First, terminal transferase is used to add poly-T-tails to the ends of DNA fragments. Second, strand synthesis is carried out using a T7-polyA primer adapter to produce double stranded templates suitable for *in vitro* transcription (IVT). After amplification by IVT, RNA is transformed into cDNA and then used for labeling and hybridization (Fig. 3).

Materials

Ambion MEGAscript T7 kit #1333 (Ambion, Austin, TX)
Ambion RNase Cocktail #2286

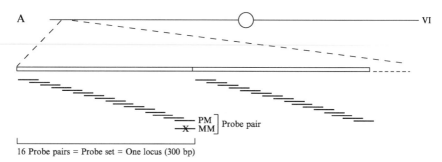

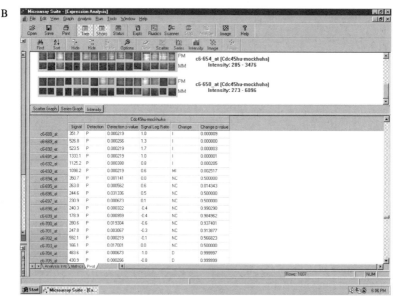

FIG. 2. (A) Design of high-density oligonucleotide array of chromosome VI. White box corresponds to one locus for analysis, which is covered by one probe set. One probe set comprised of 16 probe pairs, 16 perfect match and mismatch oligos. (B) Desktop image of the data obtained by ChIP-chip. At the top, actual scanned image of 16 probe pairs from two loci are shown. Below is the formatted data output by analytical software.

Ambion RNase H #2293
Invitrogen SuperScript II RT #18064-071 (Invitrogen, Carlsbad, CA)
Invitrogen RNase OUT #10777-019
Invitrogen T7 RNA polymerase #18033-100
Invitrogen Second Strand Buffer #10812-014
Invitrogen DNA polymerase I #18010-025
Invitrogen Random Primer #48190-011
Invitrogen DEPC-treated water #10813-012
Invitrogen NTP mix #18109-017
Invitrogen *E. coli.* DNA Ligase #18052-019
Invitrogen RNase H #18021-071
Invitrogen DNase I #18068-015
NEB Alkaline Phosphatase, Calf Intestinal (CIP) #M0290S (NEB, Beverly, MA)
NEB T4 DNA polymerase #M0203S
Perkin Elmer Biotin-N6-ddATP #NEL508 (Perkin Elmer, Norwalk, CT)
Pharmacia One-Phor-All Buffer #27–0901-02 (Pharmacia, Piscataway, NJ)
Roche Terminal Transferase, recombinant #3333566 (Roche, Indianapolis, IN)
TAKARA ddCTP #4033 (TAKARA, Shiga, Japan)
TAKARA dTTP #4028
QIAGEN MinElute Reaction Cleanup Kit #28204 (QIAGEN, Valencia, CA)
QIAGEN RNeasy Mini Kit #74104
T7-PolyA primer
　　5′-GCATTAGCGGCCGCGAAATTAATACGACTCACTA-
　　TAGGGAGAAAAAAAAAAAAAAAAAAA[C or T or G]-3′

CIP Treatment
Reaction Mix

ChIP'd DNA (10–100 ng)	17.5 μl
10× NEB Buffer 3	2 μl
CIP (10 U/μl)	0.5 μl
Total volume	20 μl

Incubate for 60 min at 37°
Cleanup with MinElute Reaction Cleanup Kit (see manufacturer's instruction)
Elute into 20 μl EB buffer
Poly-dT Tailing
Reaction Mix

CIP-treated DNA   20 μl
5 × TdT reaction buffer   8 μl
25 mM CoCl2   1.2 μl
100 μM dTTP   1.84 μl
10 μM ddCTP   1.6 μl
Terminal transferase (400 U/μl)   2 μl
Water   5.4 μl
Total volume   40 μl
Incubate for 20 min at 37°
Add 0.5 M EDTA 8 μl to stop the reaction
Cleanup with MinElute Reaction Cleanup Kit (see manufacturer's
    instruction)
Elute into 12.3 μl EB buffer
T7-primer Annealing and 2nd Strand Synthesis

Reaction Mix
Poly-dT tailed DNA   12.3 μl
25 μM T7-PolyA primer   0.12 μl
5× 2nd strand buffer   4 μl
2.5 mM dNTPs   1.6 μl
Total volume   18 μl
Perform following cycle in the Thermal Cycler
    94°, 2 min
    Ramp down to 35° (1°/s)
    35°, 2 min
    Ramp down to 25° (0.5°/s)
    Hold
    Add 1/5 diluted DNA polymerase 2 μl
    37°, 90 min
To stop the reaction, add 0.5 M EDTA 5 μl
Cleanup with MinElute Reaction Cleanup Kit (see manufacturer's
    instruction)
Elute into 10 μl RNase-free water.
1st IVT Amplification

Reaction Mix
NTP mix   2.5 μl × 4
10× Reaction Buffer   2.5 μl
Template DNA   10 μl
Enzyme Mix   2.5 μl
T7 RNA polymerase (200 U/μl)   1.2 μl
Total volume   26.2 μl
Incubate for 16 h at 37°

Cleanup with RNeasy Mini Kit (see manufacturer's instruction)

Elute into 40 μl RNase-free water

Measure the absorbance at 260 nm to determine cRNA yield. Typically, the amount of cRNA synthesized should be 5 to 50 μg (Optional) Check the size of synthesized cRNA by Bioanalyzer (Agilent Technologies, Polo Alto, CA). The average size should be similar to that of starting ChIP'd DNA. For optimization, increase/decrease the amount of T7 RNA polymerase in the 1st IVT amplification reaction. Also, when a peak of small size fraction originated from a self amplification of T7 primer is observed, reduce the amount of T7 primer in the 2nd strand synthesis reaction or go to the next 2nd IVT amplification step in which a portion of self-amplified primer disappears (Fig. 4).

1st Strand cDNA Synthesis

Set-up Mix

cRNA (600 ng)	10 μl
Random Primer (0.4 ug/μl)	1 μl
Total volume	11 μl

Perform following step in the Thermal Cycler

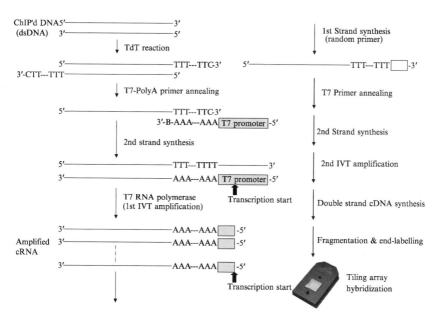

FIG. 3. Schematic presentation of IVT amplification method (see text for details).

70°, 10 min
4°, hold
Add followings to the set-up mix on ice
5× 1st strand buffer	4 μl
100 mM DTT	2 μl
RNase OUT	1 μl
10 mM dNTP mix	1 μl
SuperScript II RT (200 U/μl)	1 μl
Total volume	9 μl

Perform following steps in the Thermal Cycler
42°, 60 min
4°, 5 min
Add RNase H 1 μl
37°, 20 min
95°, 5 min
4°, 5 min

T7-primer Annealing and 2nd Strand cDNA Synthesis

Set-up Mix
1st strand cDNA	21 μl
2.5 μM T7-PolyA primer	4 μl
Total volume	25 μl

Perform following steps in the Thermal Cycler
70°, 6 min
4°, hold

Add followings to the set-up mix on ice
Water	88 μl
5× 2nd strand buffer	30 μl
10 mM dNTP mix	3 μl
DNA polymerase (10 U/μl)	4 μl
Total volume	125 μl

Perform following steps in the Thermal Cycler
16°, 120 min
Add T4 DNA polymerase (3 U/μl)   2 μl
16°, 10 min
4°, hold

Cleanup with MinElute Reaction Cleanup Kit (see manufacturer's instruction)
Elute into 10 μl RNase-free water
Measure the absorbance at 260 nm to determine the cDNA yield.
    Typically, the amount of cDNA synthesized should be 550 to 650 ng

2nd IVT Amplification

Reaction Mix

NTP mix	2 μl $\times$ 4
10$\times$ Reaction Buffer	2 μl
DNA template	8 μl
Enzyme mix	2 μl
Total volume	20 μl

Incubate for 16 h at 37°

Cleanup with RNeasy Mini Kit (see manufacturer's instruction)

Elute into 40 μl RNase-free water

Measure the absorbance at 260 nm to determine the cRNA yield.

Typically, the amount of synthesized cRNA should be 80 to 100 μg

1st Strand cDNA Synthesis for Double-Strand cDNA Synthesis

Set-up Mix

cRNA	11 μg
Random primer (3 μg/μl)	2 μl
DEPC-treated water	Up to 18 μl
Total volume	18 μl

Perform following steps in the Thermal Cycler

    70°, 10 min

    15°, 30 min (20 min ramp)

Add followings as soon as temperature of set-up mix reaches 15°

5$\times$ 1st strand buffer	8 μl
100 mM DTT	4 μl
10 mM dNTP mix	2 μl
Total volume	14 μl

Perform following steps in the Thermal Cycler

    Just before ramping up to 42°, add 8 μl of SuperScript II RT

        (200 U/μl)

    42°, 60 min (20 min ramp)

    Add SuperScript II RT 2 μl (200 U/μl)

    42°, 60 min

    70°, 15 min

    4°, hold

2nd Strand cDNA Synthesis

Reaction Mix

1st strand cDNA	42 μl
Water	180 μl
5$\times$ 2nd strand buffer	60 μl
10 mM dNTP mix	6 μl
E. coli DNA Ligase (10 U/μl)	2 μl

DNA polymerase (10 U/μl, Invitrogen)	8 μl
RNase H (Invitrogen)	2 μl
Total volume	300 μl

Incubate for 2 h at 16°
Incubate for 15 min at 70°
4° hold
RNA Removal (Optional)
Reaction Mix

2nd strand cDNA	300 μl
RNase H (10 U/μl)	3 μl
RNase Cocktail (A: 0.5 U/μl, T1: 20 U/μl)	3 μl
Total volume	306 μl

Incubate for 20 min at 37°

Add 306 μl phenol:chloroform:isoamyl alcohol (25:24:1), vortex thoroughly, and centrifuge at RT for 5 min at 14,000g to separate the aqueous phase (Approximately 280 μl of aqueous phase can be retrieved.)

Add 140 μl of 7.5 M NH$_4$OAc, add 1 ml of ice-cold ethanol, vortex the mixture thoroughly, and centrifuge immediately at 25° for 20 min at 14,000g

Wash with 500 μl of ice-cold 70% ethanol, centrifuge for 2 min at 14,000g

Dry the cDNA at 37° for 10 min, and dissolve the pellet with 11 μl water

Measure the absorbance at 260 nm to determine the concentration of DNA. Typically, the amount of synthesized dsDNA should be 10 to 11 μg.

Usually 10 μg of DNA is used for labeling and hybridization, essentially following the protocol of the previous section. Time of DNaseI treatment can be extended up to 8–9 min.

DNA Fragmentation
DNase I Setup

DNase I (1 U/μl)	4 μl
10× One-Phor-All buffer	2 μl
25 mM CoCl$_2$	1.2 μl
Water	12.8 μl
Total volume	20 μl

Reaction Mix

Amplified DNA (10 μg)	20.5 μl
10× One-Phor-All buffer	2.44 μl
25 mM CoCl$_2$	1.4 μl

DNase I setup   0.6 μl
Total volume   25 μl

Incubate for 8 min at 37°
Incubate for 15 min at 95°
4° hold
Check the size of fragmented DNA, by running 2 μl of reaction mix
    on 2% agarose gel. The majority of DNA should distribute at
    50–100 bp.
TdT Labeling
Reaction Mix
Fragmented DNA   22.4 μl
5× TdT buffer   7 μl
25 mM CoCl$_2$ 3.5 μl
1 mM biotin-N6-ddATP   2.5 μl
Terminal transferase (400 U/μl)   0.125 μl
Total volume   35 μl
Incubate for 2 h at 37°

For the inactivation of Terminal transferase, incubate the reaction mix
    at 95° for 10 min.

Location Analyses of DNA Replication Proteins and Replicated Regions

In Fig. 5, there are two sets of protein binding profiles data obtained by ChIP-chip analyses. In late G1, Mcm proteins (DNA helicase complex required for pre-RC formation) and Cdc45 (their associated factor) bound to all origins of DNA replication. Dbf4, a subunit of regulatory kinase Cdc7, was also found at early origins of DNA replication. In S phase (arrested by adding 200 mM HU), elongation proteins started to move away, and BrdU incorporated regions appeared only on early replicating regions, ARS607, 606, 605, and 603.5. These data clearly show that data obtained by ChIP-chip is useful not only for the detection of localization of the protein of interest, but by comparing various protein binding profiles, they can help us identify protein complex and regulatory mechanisms.

The ChIP-chip method using a high-density oligonucleotide array can give us a continuous and detailed picture of a chromosome that we have never experienced with other methods. It is no longer a dream to know the structure of an eukaryotic chromosome at several bp resolutions by this new technology.

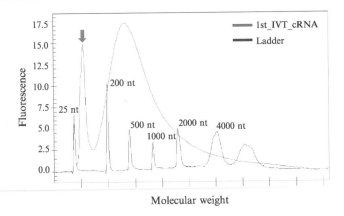

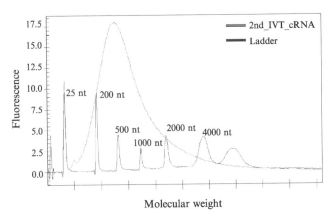

FIG. 4. Product size analysis of cRNA by bioanalyzer (Agilent Technologies). Top; product analysis after first IVT amplification. A peak of small size fraction originated from a self amplification of T7 primer is pointed by red arrow. Bottom; product analysis after 2nd amplification. A portion of self-amplified primer disappears. Vertical axis represents fluorescence intensity (amount of DNA) and horizontal axis represents molecular weight. Red line; size distribution of cRNA amplified by IVT. Blue line; size distribution of the ladder marker. Size of each peak is shown. (See color insert.)

However there are several problems we should mention with this method. First, ChIP-chip needs internal control to fairly compare strain-to-strain or protein-to-protein difference of chromosome behavior. Second, as we already mentioned, repetitive sequences are sources of biased amplification and could not be fairly evaluated. For those sequences, evaluation by other methods such as quantitative PCR will be required.

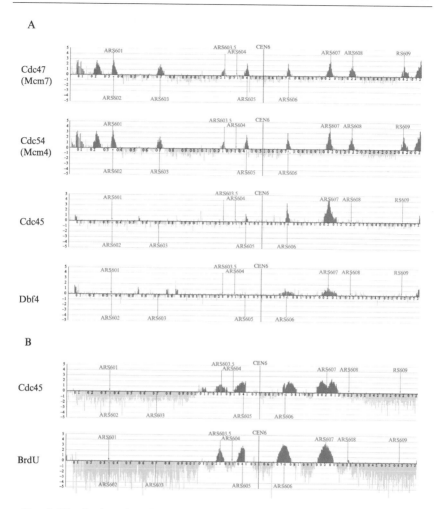

FIG. 5. Distribution of replication proteins during G1 and S on chromosome VI of *S. cerevisiae*. (A) Distribution of Cdc54, 47, 45 and Dbf4 in G1 phase arrested by α-factor. (B) Distribution of Cdc45 and BrdU incorporated regions in S phase challenged by HU. The blue shading represents the binding ratio of loci that show significant enrichment in the chromatin immuno-precipitated fraction. The yellow dotted line indicates the average signal ratio of loci that are not enriched in the ChIP fraction. The scale of the vertical axis is log2. The horizontal axis shows kilobase units (kb). (See color insert.)

Third, current algorithms for discrimination analysis cannot catch up with the protein that shows wide distribution across the chromosome. For example, for the analyses of DNA replication elongation proteins in the late stage of S phase, when replication forks get more asynchronously distributed, it is hard to get a good signal-to-noise ratio and would be hard to discriminate between the positive and negative signal for binding.

Acknowledgments

We thank Dr. Yoshikawa for the comments on the manuscript. We thank all the laboratory members of Shirahige and Aburatani laboratories. This work was supported in part by a grant of the Genome Network Project from the Ministry of Education, Culture, Sports, Science and Technology, Japan.

Reference

Katou, Y., Kanoh, Y., Bando, M., Noguchi, H., Tanaka, H., Ashikari, T., Sugimoto, K., and Shirahige, K. (2003). S-phase checkpoint proteins Tof1 and Mrc1 form a stable replication-pausing complex. *Nature* **424,** 1078–1083.

Liu, C. L., Schreiber, S. L., and Bernstein, B. E. (2003). Development and validation of a T7 based linear amplification for genomic DNA. *BMC Genomics* **4,** 19.

Ren, B., Robert, F., Wyrick, J. J., Aparicio, O., Jennings, E. G., Simon, I., Zeitlinger, J., Schreiber, J., Hannett, N., Kanin, E., Volkert, T. L., Wilson, C. J., Bell, S. P., and Young, R. A. (2000). Genome-wide location and function of DNA binding proteins. *Science* **290,** 2306–2309.

Wodicka, L., Dong, H., Mittmann, M., Ho, M. H., and Lockhart, D. J. (1997). Genome-wide expression monitoring in *Saccharomyces cerevisiae. Nat Biotechnol.* **15,** 1359–1367.

[24] Measurement of Chromosomal DNA Single-Strand Breaks and Replication Fork Progression Rates

By Claire Breslin, Paula M. Clements, Sherif F. El-Khamisy, Eva Petermann, Natasha Iles, and Keith W. Caldecott

Abstract

Chromosomal single-strand breaks (SSBs) are the most common lesions arising in cells, but are normally rapidly repaired by multiprotein complexes centered around the scaffold protein, XRCC1. Here, we describe protocols to measure chromosomal SSBs in cells and for recovering and identifying novel components of SSBR complexes *in vitro* and *in vivo*. We also describe an assay we employ to measure the rate of replication fork progression in mammalian/vertebrate cells in the presence or absence of DNA damage.

METHODS IN ENZYMOLOGY, VOL. 409   0076-6879/06 $35.00

Measurement of Chromosomal Single-Strand Breaks by Alkaline
  Single-Cell Agarose-Gel Electrophoresis (Alkaline Comet Assay)

Introduction and Overview

The alkaline comet assay is a widely used and well-established technique for measuring chromosomal DNA strand breakage, and readers are
directed to one of the many reviews (Fairbairn *et al.*, 1995; Olive, 1999,
2002; Wojewodzka *et al.*, 2002) and Web sites (e.g., http://cometassay.com)
for exhaustive coverage. Here, we describe the basic protocol that we
employ and tips we have found useful concerning the application of this
assay for measurement of chromosomal SSB repair. In this assay, cells are
embedded in low melting point agarose on microscope slides, lysed in an
alkaline buffer containing detergent and high salt, and then subjected to
electrophoresis at alkaline pH to facilitate migration of denatured and
broken DNA out of the nucleus, thereby forming a comet-like "tail" that
can be stained and visualized using fluorescence microscopy. The amount
of DNA strand breakage can be quantified by various parameters. The one
we employ routinely is the "tail moment," a unit that reflects the product
of the fraction of chromosomal DNA that has exited the nucleus to form
the comet tail and the distance migrated (tail length). However, it should
be noted that there are other useful parameters, such as tail intensity. In
our laboratory, quantification of tail moments and other parameters is
automated using Comet Assay III software (Perceptive Instruments, UK).
Alternatively, comets can be classified visually for their level of damage
(e.g., comet type 1 = low damage; comet type 2 = medium damage; comet
type 3 = high damage, etc.).

General Considerations

It cannot be overemphasized that this assay detects not only SSBs but
also alkali labile sites and double-strand breaks (DSBs). For our purposes,
we have not worried too much about alkali-labile sites and so do not
discriminate between these and proper SSBs. However, if desired, judi-
cious choice of pH during the protocol can help distinguish alkali labile
sites from true SSBs (Horvathova *et al.*, 1998). Similarly, discrimination
between SSBs and DSBs can be achieved by careful choice of genotoxin.
For example, hydrogen peroxide and ionizing radiation induce ∼2000 SSBs
and ∼40 SSBs for every DSB, respectively, and so the chromosomal strand
breaks initially induced by these agents are primarily (>95%) SSBs. Similarly, most if not all DNA strand breaks initially induced by base damaging
agents (we tend to use the alkylating agent MMS) are SSBs. Careful
consideration is needed when conducting repair experiments, however,

because replicating cells within an asynchronous population may convert SSBs to DSBs. This is generally not a problem for short repair periods of up to 1 h (which is often sufficient for repair of most SSBs in wild-type cells) when only a small percentage (<5%) of SSBs are likely to encounter a replication fork in mammalian cells, assuming that the breaks are randomly distributed across cell cycle phases and across the genome. Additionally, the use of synchronized cell populations or selective labeling of S-phase cells within an asynchronous population are excellent ways of assessing or circumventing the impact of DNA replication on DNA strand breakage. Note that when employing a genotoxin to induce DNA strand breakage, it is necessary to first conduct a dose response curve to examine the linearity of the relationship between tail moment and drug concentration and to identify concentrations that do not saturate the assay. Tail moment values may then be converted into strand break frequencies using γ-ray or X-ray dose response curves (for which the relationship between dose and DNA strand break frequency is known).

As a final note, we find that H_2O_2 and γ-rays are also useful for calibrating neutral versions of the comet assay [essentially, we use the assay described by Wojewodzka et al. (2002)] which primarily measures DSBs at levels of strand breakage above 2 Gy equivalent of X-rays/γ-rays. Note that SSBs up to this level ($\sim$2000 SSBs/cell, or a tail moment of $\sim$2 in our hands) do register in the neutral assay, due to their impact on the relaxation of chromatin loops. However, above this level, increases in tail moment appear to be specific for DSBs, in our hands at least. For example, whereas 150 μM H_2O_2 (10 min in PBS on ice) and 20 Gy of γ-rays induce similar tail moments ($\sim$15) in the alkaline assay in our hands (both inducing $\sim$20,000 total strand breaks), in a neutral assay conducted in parallel, the tail moments increase from a background level of $\sim$1 to $\sim$2 or $\sim$7, respectively, in the neutral assay.

Materials

    Frosted microscope slides (VWR)

    Coverslips

    Horizontal electrophoresis tank (ours is "home-made" and made of black plastic to prevent light contamination, but commercial ones are available from various sources including Fisher Scientific, Owl Scientific, and Thistle Scientific)

    Slide holder and blacked-out glass box (we use a slide tray and glass box covered with black tape)

    Fluorescence microscope

    Automated comet quantification software (we use Comet Assay III, Perceptive Instruments, but others are available)

Cell lysis buffer
>   2.5 M NaCl
>   100 mM EDTA, pH 8.0
>   10 mM Tris-HCl, pH 10 (pH to 10 with 10 M NaOH)
>   Less than 20 min before use: add 1/100 volume each DMSO
>   and Triton X-100 while stirring vigorously on magnetic
>   stirrer.
Electrophoresis buffer
>   1 mM EDTA, pH 8.0
>   50 mM NaOH (Note: some protocols employ 200 mM
>   NaOH, resulting in a pH > 13, to increase sensitivity)
>   Less than 20 min before use: add 1/100 volume DMSO while
>   stirring vigorously on magnetic stirrer.
Low gelling temperature agarose (Sigma, St. Louis, MO, Type VII)
Standard gelling temperature agarose (Invitrogen, Grand Island, NY,
>   ultrapure electrophoresis grade)
SYBR green (or other suitable DNA stain) (Sigma)

Protocol

The lysis and electrophoresis buffers are prepared on the day before or morning of the experiment and prechilled at 4° along with PBS and distilled water (1–2 liter). To prepare precoated frosted slides, weigh out sufficient standard gelling temperature agarose for 0.5 ml of a 0.6% solution per slide and dissolve thoroughly in PBS in a microwave. To coat slides, pipette 150 μl of the agarose down the underside of a cover slip that is resting on the slide at an angle of $\sim$45°, and then lower the coverslip onto the slide slowly, using the pipette tip, without forming bubbles. We generally prepare two cover slips per slide in order to provide duplicate samples per data point. Once coated, the slides are refrigerated for at least 1 h before use. The cover slips can be left on the slides at this stage, or can be removed 2 min (as soon as set) after coating and the slides labeled as appropriate with a pencil. For some users, removal at this point can reduce the risk of the agarose subsequently floating off. An appropriate quantity of 1.2% low melting point agarose in PBS is prepared and incubated at 42° until needed. When the cells of interest have been treated as desired and are ready for analysis, cell samples are washed and resuspended in ice cold PBS (typically 2×10^5 cells per data point in 1 ml PBS). A 0.2 ml sample of each cell suspension is then aliquot into 1.5-ml centrifuge tubes and stored on ice, in the dark, until ready for spreading. Cover slips are gently removed from the precoated slides (if not done so already) and, taking each 0.2-ml cell sample in turn, 0.2 ml of 1.2% low gelling temperature agarose (at 42°) is added to the cell sample, mixed quickly by gentle pipetting, and 150 μl aliquots (two per slide) pipetted down the underside of each of two fresh cover slips, as described previously, onto the precoated agarose. The cover slip is gently lowered onto the agarose/cell suspension and slides placed in the dark at 4° for at least 30 min, to allow the agarose to set. Once again, cover slips can be removed as soon as set, if desired (~2 min).

Note that the remaining steps are conducted at 4° and shielded from white light as much as possible (to minimize the introduction of additional damage). Once the LMP-agarose/cell suspensions have set, cover slips are gently removed and slides placed into the slide rack and lowered gently into lysis buffer in a blacked-out glass box and incubated for 1 h. The slides are then gently rinsed three times in distilled H_2O and finally aligned in the same orientation in a horizontal electrophoresis tank containing electrophoresis buffer. Note that slides should always be added gently to solutions, and not the other way round, to minimize the risk of agarose loss. The slides are allowed to equilibrate in the buffer for 45 min to facilitate DNA denaturation, prior to electrophoresis at 0.6 V/cm (25 V, 80 mA for our tank) for 25 min. Following electrophoresis, slides are neutralized in 0.4 M Tris pH 7 for 1 h and then stained with 1 ml SYBR green per slide (1:10,000 dilution in PBS). Cover slips are then replaced prior to microscopy.

Data Acquisition and Analysis

Microscope fields are chosen randomly across the whole of the slide, with care taken to avoid the edges, and fields omitted from scoring if cells are overlapping. Results from a single experiment can be plotted as the mean tail moment for each data point (we routinely score 100 cells per data point). In our hands, this gives a reliable indication of repair capacity, and ultimately we typically plot each data point as the average (+/–SEM) of the mean tail moment value from at least three independent experiments. Within each experiment, we also find it extremely important and useful to examine the distribution of tail moments for each data point. We routinely do this both by grouping individual tail moments into classes (e.g., 0–2, 2–5, 5–10, etc.) and by plotting histograms of these classes (X-axis) versus percentage of cells in that class. It is also advisable and very useful to plot the raw data as scatter graphs. For scatter plots, it is important to ensure that the individual cells remain numbered in the Excel sheet in the same sequence in which they were scored (e.g., 1–100) and that they are then plotted in this same order along the X-axis (with their individual tail moment value plotted on the Y-axis). In addition to revealing the level of homogeneity/heterogeneity in tail moments within the population of cells scored for a particular data point, this approach can highlight spurious data points and experimental artefacts such as variations in tail moment values that result from subpopulations of cells in a particular area of the slide. Such subpopulations will appear as a cluster in the scatter graph.

Identification of Novel SSBR Polypeptides by Yeast 2-Hybrid Library Screens

Introduction and Overview

The Gal 4-based yeast 2-hybrid assay is a well-established system for identifying and studying protein-protein interactions. In general, the protocols we use are based on those in the Clontech Yeast Protocols Handbook (PT3024-1 March 2001 [http://www.clontech.com/clontech/tech-info/manuals/PDF/PT3024-1.pdf]), and readers are directed there for details of materials, strains, and background information. Modifications to these protocols and tips that we have found useful when screening cDNA libraries for novel SSBR polypeptides are outlined here. Although we have employed these protocols for identifying novel SSBR-associated factors, they are generally applicable to other types of protein bait.

General Considerations

In general, library screens can be conducted in two ways. First, the cDNA library (e.g., in pACT) can be transformed into a suitable recipient strain (e.g., Y190) that has already been transformed with the bait (e.g., in pAS1CYH2 or pGBKT7). This is our favorite protocol because a single clone that expresses a desired level of bait can be selected and identified by immunoblotting. Moreover, the efficiency of the screen (the total number of independent screened library clones) is limited only by the efficiency of transformation of the library. However, it is possible to screen by cotransforming both bait and library plasmids, simultaneously. This approach has the disadvantage that the screening efficiency is limited by the transformation efficiency of both the library and bait plasmids. However, on occasion, we find that for some baits we achieve better overall transformation efficiencies with the cotransformation approach (perhaps stable expression of these baits has a detrimental impact on yeast transformation or growth). Whichever transformation protocol is adopted, "small-scale" trial transformations should be attempted and repeated on independent occasions to ensure that 1×10^4 transformants/μg library or greater can reproducibly be obtained before attempting a "large-scale" screen (we typically ensure this occurs on at least three consecutive occasions, using the same stock solutions that will be employed for the large scale screen). For large-scale screening we conduct 18–20 "small-scale" transformations in parallel. We find this retains the high transformation efficiency observed in the trial experiments more reproducibly than by conducting a single large transformation.

Protocol

Transformations: Our protocols/solutions are based on the Clontech Yeast Protocols Handbook, and so here we have focused on minor modifications and tips that we have found useful. We dilute a fresh overnight culture of Y190 cells in yeast extract/peptone/dextrose complete media (YPD) to an OD_{600} of ~0.05 in 40 ml YPD medium per small scale transformation, and incubate at 30° with shaking until the OD_{600} has doubled twice. The cells are then pelleted, washed in TE buffer, and resuspended in 200 μl fresh TE/LiAc (1× solution of TE and LiAc; see Handbook, mentioned previously) per 40 ml culture and kept on ice until needed (use within 1 h). For each small-scale transformation, 1 μg library DNA (and 1 μg of bait DNA if cotransformation is used) is added to a 1.5 ml microfuge tube containing 4 μl of 5× TE and 200 μg of sonicated salmon sperm DNA (freshly denatured at 90° and cooled rapidly on ice for several min). 200 μl of fresh competent cells are then added to the tube followed by 1.2 ml of PEG/LiAc (see Handbook) and 140 μl of DMSO. The cells are then heat-shocked (42°, 15 min) and chilled on ice for 2 min. For the large-scale screen, the cells from the 18–20 tubes can be pooled at this point, pelleted, and resuspended in 9.5–10.0 ml of 1× TE. 500 μl of the cell suspension is then spread on each of 18–20 15 cm plates containing minimal media (glucose and yeast nitrogen base w/o amino acids; Becton, Dickinson, Oxford, UK) +Ade, +25 mM 3-aminotriazole (3-AT). 100 μl of each of a 1:10, 1:100, and 1:1000 dilution series is also spread onto 10 cm minimal media +Ade + His plates to calculate the transformation efficiency of the experiment. Plates are then incubated at 30° for ~3–5 days for determining transformation efficiency and for up to 14 days for the library screen. Colonies are picked from the library plates as they appear, resuspended in 150 μl H$_2$O, and a 50-μl aliquot restreaked onto two fresh 10-cm plates containing minimal media +Ade + His for one stock plate and for one plate for a filter lift assay to examine for activation of the β galactosidase reporter gene, and also one plate containing minimal media +Ade, +25 mM 3-AT to confirm activation of the His3 reporter gene.

Isolation and identification of recovered pACT clones: Cells from 1–5 ml of minimal media overnight culture of each yeast colony recovered from the library screen are pelleted and resuspended in 100 μl lysis buffer (2% v/v Triton-X-100, 1% w/v SDS, 100 mM NaCl, 100 mM Tris-HCl pH 8.0, 1 mM EDTA). 0.1 ml of phenol/chloroform (1:1) and 0.1 ml acid-washed glass beads are added and the suspension vortexed (safety glasses required) for 2 min. The suspension is then pelleted at full speed in a microfuge for 5 min. The aqueous phase is recovered, and the DNA ethanol precipitated and resuspended in 10 μl ddH$_2$O. 1 μl of the extracted DNA solution is then transformed into electrocompetent bacteria. Several bacterial clones (typically five) are recovered from each bacterial transformation and mini-preparations of DNA digested with a frequent-cutting restriction enzyme (we use *Rsa*I). The pattern of *Rsa*I restriction fragments provides a useful diagnostic test to determine whether single or multiple different pACT library cDNA clones are present in each isolated yeast colony. A sample of each independent pACT cDNA recovered is then identified by DNA sequencing. The interaction of the encoded gene product with the bait protein can then be confirmed by standard analyses (further yeast/mammalian 2-hybrid analyses, coimmunoprecipitation, etc.). Note that when quantifying the level of interaction observed by 2-hybrid analysis by liquid β-galactosidase assays, we tend to use 10 ml of yeast culture for each sample rather than the 1.5 ml suggested in the Clontech Yeast Protocols Handbook (not forgetting to take this into account in the final calculation of β-galactosidase units). Also note that the level of interaction quantified in the liquid β-gal assay is extremely dependent on the levels of expression of the binding domain (BD) and activation domain (AD) fusion proteins. Therefore, we always take an aliquot of the cells/cell extract used for liquid assays for immunoblotting. We find that the anti-GAL4 BD (Upstate, #06–21, NY), anti-Myc (Cell Signaling, clone 9B11, Danvers, MA) and anti-GAL4 AD (Upstate, #06–283) work well.

Recovery of XRCC1 Protein Complexes from Cell Extract by Immobilized Metal-Chelate Chromatography (IMAC)

Introduction and Overview

Although this protocol is designed for recovery of XRCC1 protein complexes, it is generally applicable to other protein complexes, if an appropriately tagged polypeptide of interest is employed. The CHO cell line EM9 lacks XRCC1 and so provides a useful model system for the analysis of XRCC1 function and SSBR protein complexes. EM9 cells can be stably transfected with expression constructs that encode histidine-tagged XRCC1. To date, we have only employed human XRCC1 for correction of EM9 cells. We have not encountered any major problems in employing cross-species complementation, but this caveat should always be considered when interpreting results. For example, if expression levels of the recombinant protein are an issue, levels similar to the endogenous protein are difficult to achieve and validate with strong promoters and cross-species approaches. In particular, most if not all anti-XRCC1 antibodies will have different affinities for the human and hamster protein, rendering accurate comparisons of the level of recombinant and endogenous proteins invalid. If cross-species complementation becomes an issue, then the use of hamster XRCC1 cDNA is a possibility. Alternatively, with the emergence of RNAi technology, the use of human cells with depleted levels of endogenous XRCC1 as the recipient of an expression construct is another alternative. Of course if overexpression is not an issue, or is even desirable, then almost any cell line may be appropriate. In the latter case, transient overexpression may be best, since higher levels of recombinant protein can be achieved.

General Considerations

Affinity purification of XRCC1 complexes could be conducted using an automated chromatography system if desired, though we find that the manual purification procedure is better suited to processing multiple cell lines in parallel. For example, we find it very important to conduct a parallel chromatography on a negative control cell line, because many mammalian proteins copurify "non-specifically" with histidine-tagged proteins on nickel agarose. This control can be matched cells that do not express histidine-tagged XRCC1, or cells that express an unrelated histidine-tagged polypeptide. The specific buffer composition can of course be altered to suit the needs of the experiment. Attempting the purification with different salt concentrations is particularly useful in determining the stability of particular interactions. As a starting point, we typically use $\sim$300 mM. Similarly, it is useful to determine empirically the optimum concentration of imidazole in the column washes, though we typically use 25 mM as a starting point. Note that, with the obvious exception of imidazole concentrations, we keep buffer compositions (and particularly the chosen salt concentration) the same throughout any single experiment, to minimize the risk of protein precipitation. We have observed this phenomenon, which can be diagnosed by heating a sample of the NTA-agarose from the column in SDS-PAGE loading buffer and including it on the immunoblot. We thus find it better to change buffer composition gradually, at the end of the experiment, by dialysis of the eluate.

Materials

Equipment
Chilled bench-top centrifuge
Microfuge
Low pressure/gravity chromatography columns + tubing (we use disposable 5 ml Plastic Poly-prep® columns and tubing from Bio-Rad, Hercules, CA)
Clamps and column stand
Nickel-agarose (we use Nickel NTA-agarose from Qiagen, Valencia, CA)
Solutions
Lysis Buffer
       25 mM Hepes-NaOH, pH 8.0
       150–500 mM NaCl
       0.5% (v/v) TritonX-100
       10% (v/v) glycerol
       1 mM DTT
       25 mM imidazole, pH 8.0 (optional)
       1:100 (v/v) protease inhibitor cocktail
Wash Buffer
       25 mM Hepes-NaOH pH 8.0
       150–500 mM NaCl
       0.5% (v/v) TritonX-100
       10% (v/v) glycerol
       1 mM DTT
       25–40 mM imidazole, pH 8.0
Elution Buffer
       25 mM Hepes-NaOH, pH 8.0
       100–500 mM NaCl
       0.5% (v/v) TritonX-100
       10% (v/v) glycerol
       1 mM DTT
       250 mM imidazole

Protocol

Cells ($0.5–1.0 \times 10^7$) are harvested from spinner flasks/roller tubes by centrifugation or from subconfluent flasks or dishes with a rubber policeman, washed with ice-cold PBS ($2\times$). Adherent cells can also be recovered with trypsin, if preferred, but if so we include one wash in complete medium (i.e., with serum) prior to washing with PBS. Pellets can be used immediately or frozen at $-80°$ until use. When ready, frozen pellets are thawed on ice and resuspended in complete lysis buffer at $\sim 6 \times 10^6$ cells/ml.

Cell suspensions are incubated on ice for 15 min and genomic DNA sheared gently by passage through a needle and syringe (ten times, 0.4 mm diameter needle). The cell extracts are then clarified by centrifugation at ~8700g for 10 min at 4° and the supernatant retained. At this point we remove 0.1–0.2 ml as a sample of the column input or load, for immunoblotting. The remaining supernatant (0.8–1.0 ml) is then added to 0.5 ml bed volume of Nickel NTA agarose (prewashed 2× with H_2O and 1× with lysis buffer) in a 1.5-ml microfuge tube and the suspension incubated on ice for 20–30 min with gentle but frequent mixing (~every 5 min). The suspension is then added to the chromatography column in a cold room/cabinet and the flow rate adjusted to ~0.5ml/minute. The flow-through is retained (unbound material) and the column washed sequentially with 2 × 5 ml (2× 10-bed volumes), with the flow-through again retained (wash samples). Finally, bound proteins are eluted with 10 bed volumes (5 ml) of elution buffer and the flow-through (column eluate) collected as 0.5 ml fractions. The recovery of XRCC1 and associated polypeptides can then be examined by immunoblotting. If desired, the fractions containing XRCC1 complexes can be pooled, dialyzed, and further purified by other techniques for further analysis, or can be used for enzymatic assays.

Quantification of Chromosomal Replication Fork Rates on Damaged DNA by DNA Fiber Labeling

Introduction and Overview

This technique is essentially that of Jackson and Pombo (1998). Examples of our experiments employing fiber assays to measure fork progression rates in the presence or absence of cisplatin or UV damage are described in Henry-Mowatt *et al.* (2003). In this assay, each replication fork is first labeled with two halogenated nucleosides (we use BrdUrd and IdUrd), with DNA damage introduced (if desired) before or during the second label. The first pulse label serves to normalize for intrinsic differences in the rate of progression of individual forks. The second pulse label can be conducted during, or following, treatment with a genotoxin of interest, to examine the impact of DNA damage on fork progression. Differential immunostaining is then employed to measure the two pulse-labeled replication tracks.

General Considerations

For immunostaining, we use a combination of rat primary anti-BrU Mab that only recognizes the first pulse label (BrdUrd) and a sheep anti-BrU polyclonal that also detects the second (IdUrd). However, there can be some unwanted cross-reaction of the rat anti-BrU antibody with IdU, and the sheep anti-BrU antibody has a lower affinity for IdU than for BrU. Both factors impact on the level of differential staining. The concentration of the rat antibody should thus be as low, and the incubation time as short, as possible in order to prevent nonspecific staining. In contrast, the conditions for the sheep antibody should be optimized for maximum staining. The following protocol is a good starting point, but ideally the staining conditions should be determined empirically. Similarly, the duration of the HCl denaturation step, prior to immunostaining, should ideally be optimized empirically, though the following protocol is a good starting point. When analyzing stained slides, the rat anti-BrU antibody is not very stable and the antibody dissociates from DNA with time. Paraformaldehyde fixing minimizes this problem, but it is advisable to keep stained slides at 4°, or on ice, at all times, and to perform the microscope analyses as soon as possible. Do not store the samples more than one day. Also, note that the antifading reagent p-phenylenediamine used in Vectashield mounting medium can destroy cyanine dyes such as Cy3 with time.

Materials

Thymidine analogues and DNA damaging agents: 5-Bromodeoxyuridine (BrdU, Sigma) and 5-Iododeoxyuridine (IdU, Sigma) each separately at 2.5 mM in growth medium. BrdU stocks can be made at up to 100 mM, but IdU is only soluble up to 2.5 mM. Dissolve IdU by warming in a water bath (up to 60°) and vortexing vigorously for several minutes. Stock solutions can be stored at $-20°$ and freeze-thawed at least three times. Stock of DNA damaging drug at suitable concentration (dissolve drug in growth medium if possible).

Solutions

Spreading buffer (200 mM Tris-HCl pH 7.5, 50 mM EDTA, 0.5% SDS), methanol/acetic acid (3:1), 2.5 M HCl, blocking solution (PBS + 1% BSA + 0.1% Tween 20), 4% paraformaldehyde in PBS, mounting medium containing antifading reagent (e.g., Vectashield, Vector Laboratories, Burlingame, CA), nail polish.

Antibodies and DNA stains: For detecting only the first pulse label we use a rat anti-BrdU Mab that does not recognize IdU [Clone BU 1/75 (ICR1) from Immunologicals Direct (cat. no. OBT0030CX, Immunologicals Direct, Oxfordshire, UK)] and a corresponding Alexa Fluor 488-conjugated donkey anti-rat IgG secondary antibody from Molecular Probes (cat. no. A-21208, Molecular Probes, Eugene, OR). For detecting both pulse labels we use a sheep anti-BrdU polyclonal antibody that cross reacts with IdU (Biodesign International, Saco, ME cat. no. M20105S) and a corresponding Cy3-conjugated donkey anti-sheep IgG secondary antibody (Jackson Immunoresearch, West Grove, PA, cat. no. 713-165-147). For staining total DNA in fibers (optional) we use YOYO-1 iodide (Invitrogen, cat. no. Y3601).

Microscopy equipment: fully precleaned microscope slides (recommended: Menzel-Gläser Superfrost, Scientific Laboratory Supplies, Nottingham, UK, cat. no. MIC3021. Spreads may behave differently using other brands). Rectangular coverslips (thickness no. 1, 22×50 or 22×64 mm) confocal laser scanning microscope or, alternatively, high performance fluorescence microscope with CCD camera, equipped with $100\times$ oil immersion objective microscope image analysis software that allows measuring of distances.

Protocol

Labeling and Spreading. Cells must be subconfluent at the time of labeling. Have one T75 flask or 10 ml of suspension culture per sample (cell line, drug dose, etc.). For each sample, prepare 10 ml of medium containing 25 μM BrdUrd and 10 ml of medium containing 250 μM IdUrd. Split the IdUrd medium in two and, if appropriate, add genotoxin to one half. Warm the labeling media in a $37°$ water bath or CO_2 incubator. If using suspension cells, conduct experiment in centrifugation tubes and perform media changes by centrifugation for 3 min. To resuspend cells, flick the tubes or gently pipette cells up and down, once. Perform manipulations outside the incubator as quickly as possible to prevent cells from cooling. To label cells, add 10 ml of BrdUrd medium per sample, incubate the cells in a CO_2 incubator for 20 min, remove the medium, add 10 ml of IdUrd medium (with or without genotoxin, as desired), and incubate for another 20 min. If preferred, to minimize mechanical perturbation of the replicating cells, BrdU/IdUrd can be added directly from the stock tubes to the cell samples. Differential staining of replication fork movement during the first and second pulse label is still observed because the IdUrd is added at 10-fold higher concentration than the BrdUrd, though some increase in overlap may be observed. Following both pulse labels, remove the medium and wash with 2×10 ml ice-cold PBS and (following trypsinization if necessary) resuspend in 0.5–2.0 ml ice-cold PBS at a concentration of $0.5–1.0 \times 10^6$ cells/ml (will need to count cells for this step). Keep the cells on ice for the following steps. To obtain fiber spreads, 2 μl of cell suspension is spread per slide. The amount of cells used for spreading depends on cell size; for large cells such as HeLa use ~1000 cells per slide (i.e., 2 μl at 0.5 × 10^6 cells/ml). For smaller cells such as CHO or DT40, use ~1500–2000 cells per slide (i.e., 2 μl at 0.75–1.0 × 10^6 cells/ml). In order to obtain a better separation of labeled fibers, labeled cells should be diluted with unlabeled cells before spreading (e.g., try 1:2–1:10 dilutions, to find one that works best for you). It is advisable to do the first experiments without dilution, however, until you become used to finding labeled fibers under the microscope. To spread, place a drop of 2 μl of cells in the center of the top quarter of a microscope slide. Do this for 5–10 slides per sample. Let the drops dry at RT for 5–10 min until they start drying out at the edges, but do not let them dry out completely. Add 7 μl of spreading buffer to each drop and immediately mix by gently stirring with the pipette tip, and then incubate for a further 2–3 min. Tilt the slides slightly and disrupt the surface tension of the drops with a pipette tip. Place the top ends of the slides on the edge of a support (e.g., a tray) so that the slides remain tilted at 10–20°. Let the drops run down the slides slowly. They should reach the bottom edges after 3–5 min. If they run too slow or fast, vary the angles of the slides. The running speed is influenced by the extent of drying before running, which depends on the previous incubation times and on the room temperature. If the drops run too fast, extend incubation times, and if they run too slow, shorten incubation times. Air-dry the spreads, fix in methanol/acetic acid 3:1 for 10 min, and allow to dry. Fixed spreads can be stored at 4° for weeks.

Immunostaining. All steps are conducted at room temperature, unless indicated otherwise. Slides are washed 2× with H$_2$O for 5 min, 1× with 2.5 M HCl, denatured with 2.5 M HCl for 1 h, rinsed 2× with PBS, and then washed with blocking solution for 2 × 5 min and 1 × 1 h. Next, taking one slide at a time, discard blocking solution, drain using a paper towel to remove excess fluid, and add 100 μl of rat anti-BrdU antibody (1:1000 in blocking solution). Cover with a coverslip. Then, incubate all of the slides for 1 h, rinse 3× with PBS, fix for 10 min with 0.5 ml 4% paraformaldehyde/slide, rinse 3× with PBS, and wash with blocking solution for 3 × 5 min. Then, with the slides protected from light during subsequent incubations, to each slide add 100 μl of Alexa Fluor 488-conjugated anti-Rat antibody (1:500 in blocking solution) and incubate for 1.5–2 h. Rinse 2× with PBS, wash with blocking solution for 3 × 5 min, and add 100 μl of sheep anti-BrdU antibody (1:1000 in blocking solution) and incubate overnight at 4°. Rinse 2× with PBS, wash with blocking solution for 3 × 5 min, and add 100 μl of Cy3-conjugated anti-sheep antibody (1:500 in blocking solution) for 1.5–2 h. Finally, rinse 2× with PBS, wash with blocking solution for 3 × 5 min, and rinse 2× with PBS. Finally, drain slides and mount in 1 drop of mounting medium, seal with nail polish. Store slides in fridge. Optionally, DNA can be counterstained with YOYO-1 iodide that will, albeit weakly, stain single fibers. Note that YOYO-1 is green and thus not suitable for use with green fluorophores such as Alexa Fluor 488. To stain with YOYO-1, add 1 ml of YOYO-1 solution per slide (diluted 1:10,000 in PBS), for 10–20 min, before mounting. Rinse 3× with PBS and then mount the slides in PBS/glycerol (1:1), because YOYO-1 can produce high green background with other mounting media.

Image Collection. Immunostained fibers can be analyzed using a confocal laser scanning microscope (e.g., Zeiss LSM 510) or a CCD camera-equipped fluorescence microscope with high resolution. A LSM allows for much faster and easier image collection and analysis, and also provides higher resolution than a conventional fluorescence microscope. Matching image analysis software must also be available (see later), to measure fiber lengths. Using a 100× oil immersion objective, look for suitable fibers using the eyepiece of the microscope. The most difficult part is finding the fibers on the slide. The fluorescence intensity of the samples is much weaker than that of DAPI, for example. If the slide has frosted writing areas or letters then first focus on those, or try focusing on the edges of the slide or the coverslip. Usually, you will find fibers in or slightly below the area of the slide on which the initial drop was applied. You can then follow the spread down the slide. Both spreading and staining are usually very heterogeneous (i.e., you may find areas where the fibers are tangled) or where one or both antibodies have stained poorly. Search the entire slide to find other areas where the fibers are usable, and if necessary examine a duplicate slide. For this reason, it is advisable to stain more than one slide per sample. By switching between the red and the green filters, search for areas with good differential staining (i.e., where parts of the fibers are only stained by the antibody that recognizes both BrU and IdU). Collect images of as many fibers as possible, making sure that you collect only single fibers and not tangles or bundles. The latter are extremely long, unevenly thick, and very brightly stained.

Image Analysis. After collecting the images, measure fiber track lengths using suitable image analysis software. For the Zeiss LSM, this is the LSM Image Browser 3.2, which can be downloaded from the Zeiss website (www.zeiss.com). For measuring fiber lengths with the LSM Image Browser, use the line and ruler tools in the "overlay" section, which will measure the fiber lengths in μm. The conversion factor to Kb is ~2.5 μm/kb. To obtain a precise conversion factor for your spreads, apply DNA fragments of a defined length in control spreads (e.g., adenovirus DNA), and stain the spreads with YOYO-1.

References

Fairbairn, D. W., Olive, P. L., and O'Neill, K. L. (1995). The comet assay: A comprehensive review. *Mutat. Res.* **339**, 37–59.

Henry-Mowatt, J., Jackson, D., Masson, J. Y., Johnson, P. A., Clements, P. M., Benson, F. E., Thompson, L. H., Takeda, S., West, S. C., and Caldecott, K. W. (2003). XRCC3 and Rad51 modulate replication fork progression on damaged vertebrate chromosomes. *Mol. Cell* **11**, 1109–1117.

Horvathova, E., Slamenova, D., Hlincikova, L., Mandal, T. K., Gabelova, A., and Collins, A. R. (1998). The nature and origin of DNA single-strand breaks determined with the comet assay. *Mutat. Res.* **409**, 163–171.

Jackson, D. A., and Pombo, A. (1998). Replicon clusters are stable units of chromosome structure: Evidence that nuclear organization contributes to the efficient activation and propagation of S phase in human cells. *J. Cell Biol.* **140**, 1285–1295.

Olive, P. L. (1999). DNA damage and repair in individual cells: Applications of the comet assay in radiobiology. *Int. J. Radiat. Biol.* **75**, 395–405.

Olive, P. L. (2002). The comet assay. An overview of techniques. *Methods Mol. Biol.* **203**, 179–194.

Wojewodzka, M., Buraczewska, I., and Kruszewski, M. (2002). A modified neutral comet assay: Elimination of lysis at high temperature and validation of the assay with anti-single-stranded DNA antibody. *Mutat. Res.* **518**, 9–20.

[25]   Monitoring DNA Replication Following UV-Induced Damage in *Escherichia coli*

By CHARMAIN T. COURCELLE and JUSTIN COURCELLE

Abstract

The question of how the replication machinery accurately copies the genomic template in the presence of DNA damage has been intensely studied for more than forty years. A large number of genes has been characterized that, when mutated, are known to impair the ability of the cell to replicate in the presence of DNA damage. This chapter describes three techniques that can be used to monitor the progression, degradation, and structural properties of replication forks following UV-induced DNA damage in *Escherichia coli*.

Introduction

The failure to accurately replicate the genomic template in the presence of DNA damage, whether spontaneous or induced, is thought to produce most of the genetic instability and mutagenesis observed in cells of all

METHODS IN ENZYMOLOGY, VOL. 409
0076-6879/06 $35.00
DOI: 10.1016/S0076-6879(05)09025-7

types. DNA damage encountered during replication produces genomic rearrangements when it resumes from the wrong place, mutagenesis when the incorrect base is incorporated opposite to the lesion, or even cell death when the block to replication cannot be overcome. Several genetic disorders clearly demonstrate the severe consequences that occur when damaged templates are inappropriately processed during replication. Cells from patients with classical xeroderma pigmentosum (XP) exhibit high frequencies of chromosomal rearrangements, mutagenesis, and lethality due to an inability to repair DNA lesions that block replication, rendering patients extremely sensitive to UV and prone to skin cancers (Cleaver et al., 1975, 1999; De Weerd-Kastelein et al., 1977; Tsujimura et al., 1990). These same phenotypes are also observed in the variant form of xeroderma pigmentosum (XPV) but are instead produced specifically by the loss of a polymerase that replicates through blocking DNA lesions (Cordeiro-Stone et al., 1997; Griffiths and Ling, 1991; Lehmann and Kirk-Bell, 1978; Masutani et al., 1999; Svoboda et al., 1998). Abnormal replication patterns and high rates of chromosomal exchanges are also observed in cells from Bloom's syndrome and Werner's syndrome patients, other genetic disorders characterized by cancer predisposition and premature aging, and can be traced to the loss of a RecQ-like DNA helicase (Ellis et al., 1995; Epstein and Motulsky, 1996; Fukuchi et al., 1989; Giannelli et al., 1977; Gray et al., 1997; Hanaoka et al., 1985; Karow et al., 2000; Kuhn and Therman, 1986; Langlois et al., 1989; Lonn et al., 1990; Mamada et al., 1989; Shiraishi, 1990; Yamagata et al., 1998). In E. coli, RecQ processes the nascent DNA at UV-induced blocked replication forks prior to their resumption and is needed to suppress illegitimate recombination (Courcelle and Hanawalt, 1999; Courcelle et al., 2003; Hanada et al., 1997). These genetic disorders clearly indicate that inaccurate replication in the presence of DNA damage contributes significantly to the incidence of cancer and aging in humans. Considering the severe consequences that result from the improper processing of damaged DNA, the molecular events that normally allow replication to accurately duplicate damaged genomic templates have been intensely studied over the years. This has resulted in the identification of a large number of candidate genes in both prokaryotes and eukaryotes which, when mutated, are known to impair the accuracy of replication in the presence of DNA damage. A remaining challenge has been to determine the precise roles that these gene products play in the recovery process. The DNA replication machinery and its associated proteins, like RecQ, are highly conserved among evolutionary diverged organisms, making E. coli an extremely valuable and appropriate system for dissecting the mechanism by which replication recovers from DNA damage.

In this chapter, we describe three cellular assays that are designed to help monitor and elucidate the events that occur at replication forks that encounter DNA damage. Each assay is designed to focus on a different aspect of replication following UV irradiation, and we try to discuss the advantages and shortcomings of each approach. The first assay measures the rate of DNA synthesis, the second measures degradation that occurs at the replication fork, and the third examines the structural properties of the replication fork DNA. We typically have utilized UV-induced DNA damage as our model lesion, but these methods should be adaptable to other forms of DNA damage or treatments that disrupt the replication machinery.

Description of the Methods and Technical Comments

General Considerations for Cell Culture and UV Irradiation

The parental strain utilized most frequently in our lab has been SR108, a *thyA deoC* derivative of W3110 (Mellon and Hanawalt, 1989). However, other backgrounds that contain *thy deo* mutations have also been used. When working with thymine auxotrophs, we always add 10 μg/ml thymine to all growth media and plates. Lower concentrations of thymine can result in impaired growth and cell filamentation (unpublished observations). The *thy* mutation is required for labeling with thymine, which in our hands gives a linear incorporation during long labeling periods. Thymidine in our media works efficiently for short (pulse) labeling periods, but is not incorporated linearly over extended time periods (Ann Ganesan, personal communication and unpublished observations).

For all these assays, frozen cultures are typically struck on a fresh Luria Bertani (LB) plate, supplemented with 10 μg/ml thymine, and incubated overnight at 37°. The following day, a single colony is used to inoculate 2 ml of Difco Minimal Davis Broth, a phosphate-buffered minimal medium, supplemented with 0.4% glucose, 0.2% casamino acids, and 10 μg/ml thymine (DGCthy medium). Davis broth contains 7 g dipotassium phosphate, 2 g monopotassium phosphate, 0.5 g sodium citrate, 0.1 g magnesium sulfate, and 1 g ammonium sulfate per liter of water, pH 7.0. Cultures are then grown overnight at 37° with vigorous shaking in test tubes with loose-fitting lids to allow adequate gas exchange.

UV-irradiation experiments are carried out under yellow lighting (>500 nm) or in the dark to prevent photoreactivation of cyclobutane pyrimidine dimers (CPD) by the enzyme photolyase. For yellow lights, we use F40/GO 40W Gold from GE. For UV irradiation, we use a Sylvania 15-watt germicidal lamp, which emits primarily 254-nm light. During UV irradiation, care should be taken to ensure that the culture is agitated and that the irradiation time is sufficient to provide for a uniform exposure of the entire culture. The incident fluence should be determined with a UV photometer prior to use and the effective dose (CPD induced per kb of DNA) can be determined using T4 endonuclease V as previously described (Mellon and Hanawalt, 1989).

Measurement of DNA Synthesis Following UV-Induced Damage

The progress of the replication fork is impeded by UV-induced DNA damage. Over the years, several assays have been developed to monitor replication fork progression in the presence of DNA damage. The assay described here is a modification of one originally described by Khidhir *et al.* (1985). The protocol utilizes a dual, $[^{14}C]$-thymine and $[^{3}H]$-thymidine label to simultaneously monitor the overall DNA accumulation and the rate of DNA synthesis at specific times in thymine auxotrophs of *E. coli*.

To begin the assay, dilute the overnight culture 1:100 in 50-ml DGCthy medium supplemented with 0.1 μCi/ml $[2-^{14}C]$thymine, 50–60 mCi/mmol (Moravek Biochemicals, Brea, CA). Grow cells in a 37° water bath with vigorous shaking, monitoring cell growth using absorbance at 600 nm.

Once cells have reached an OD_{600} of precisely 0.3, begin sample collection as described later. When cultures above an OD_{600} of 0.3 are used, we observe that the rate of DNA synthesis begins to decrease before the 90-min time course ends, presumably because the culture begins to exit log-phase growth. Comparatively, cell densities that are significantly below 0.3 give highly variable results from experiment to experiment if not all the cells in the culture have entered log-phase growth.

UV Treatment and Recovery Assay

Pulse label solution:
Mix   2.97 ml DGCthy medium
          30 μl [methyl-^{3}H]-Thymidine, 1 mCi/ml, 78 Ci/mmol stock
          (MP Biomedicals, Irvine, CA)
As the culture approaches OD_{600} of 0.3, pipet 50 μl of the pulse label solution into 60 5-ml polypropylene, round-bottom tubes. A typical time course involves 15 time points, taken in duplicate, and always includes both UV- and mock-irradiated treatments that are run in parallel.

Beginning 20 min before irradiation (time = −20 min), remove duplicate 0.5-ml aliquots of culture and add each aliquot to a tube containing the pulse label solution. Vortex the tubes for 2 s to mix the culture and ^{3}H label together and immediately place them into a 37° shaking water bath. After exactly 2 min, remove the tubes from the water bath and add ice-cold 5%

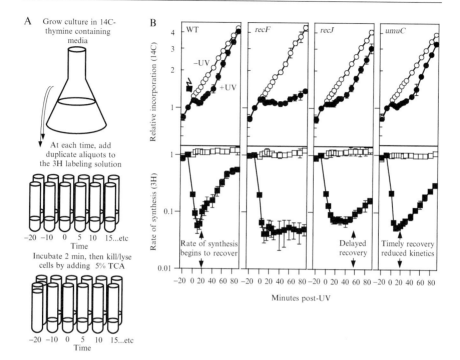

FIG. 1. DNA synthesis following UV irradiation. (A) Schematic of the radiolabeling and sampling process used in this assay. The steps involved in the technique are described in the text. (B) The rate of DNA synthesis and total DNA accumulation was measured for wild-type cells, *recF*, *recJ*, and *umuC* mutants at the times indicated following either UV irradiation (filled symbols) or mock treatment (open symbols). The relative amount of total DNA, ^{14}C (O), and DNA synthesis/2min, ^{3}H (□), is plotted. Typical initial values for ^{3}H and ^{14}C are between 7000–12,000 and 1500–2500 cpm, respectively, for all experiments.

trichloroacetic acid (TCA) to fill each tube. Figure 1A outlines the process used for sample collection. Repeat sampling process at time = −10 min. These two time points prior to treatment ensure that the cells are dividing and growing appropriately and provide a baseline measurement for DNA accumulation and DNA synthesis.

For UV treatment at time = 0, place 22 ml of the culture in a flat-bottomed container (e.g., the top of a plastic 90-mm Petri dish) and UV irradiate the culture with gentle agitation for the amount of time required for the desired dose. We typically irradiate at an incident dose of 1 J/m^2/s and expose the culture to 30 J/m^2. This dose induces a strong SOS response and generates approximately 1 CPD lesion per 8 kb DNA (Mellon and Hanawalt, 1989), but does not significantly reduce the survival of our parental strain. Immediately after irradiation, transfer the culture to a fresh flask, prewarmed to 37°, and immediately add duplicate 0.5-ml aliquots of the culture to each of two tubes containing the pulse label, vortex, incubate at 37° for 2 min, and add ice-cold 5% TCA as before.

For the mock-treated sample, place 22 ml of the culture in a flat-bottomed container and gently agitate it without irradiation for the same amount of time as used in the UV-treated culture. Transfer mock-irradiated cells to a fresh flask, prewarmed to 37°, and immediately remove duplicate 0.5-ml aliquots of culture for pulse labeling as described previously. We have found that staggering the UV- and mock-treatment time course by 2 min facilitates accurate and timely sampling of cultures.

Continue to incubate both UV- and mock-treated cultures in a shaking 37° water bath for the reminder of the time course. At 5, 10, 15, 20, 25, 30, 40, 50, 60, 70, 80, and 90 min post-treatment, remove duplicate 0.5-ml aliquots of culture and add them to tubes containing the pulse label solution as before. At the end of each 2-min pulse period, add ice-cold 5% TCA to fill the tube. Keep collected samples at 4° until all time points have been taken.

Sample Preparation and Analysis of Recovery. The addition of 5% TCA serves to lyse the cells and precipitate DNA fragments longer than ~12 bp. To collect the acid-precipitable DNA from each sample and determine how much ^{3}H and ^{14}C was incorporated into each sample, we filter the samples through a vacuum manifold onto Whatman glass fiber filters. The empty sample tubes are then filled with 95% ethanol and again poured over each respective glass fiber filter to wash the remaining traces of precipitate from the tube and wash through any TCA that remains on the filter. The washed glass fiber filters are then rinsed a second time with 95% ethanol and the filters are allowed to dry in a 55° incubator (2 h is more than sufficient). Each filter is then placed into a scintillation vial, scintillation fluid is added, and the amount of radioactivity in each sample is determined. The windows or program on the scintillation counter must be set to exclude any overlap between the ^{3}H- and the ^{14}C-detection profiles.

Average the ^{3}H and ^{14}C counts from duplicate sample time points. Then, using the counts from the time $= -10$ min sample as a reference, determine the relative rate of DNA synthesis (based on ^{3}H counts) and total amount of DNA accumulation (based on ^{14}C counts) for each time point and treatment.

$$\text{Relative } ^3\text{H at time } X = {}^3\text{H}_{\text{time } X}/{}^3\text{H}_{\text{time} - 10}$$
$$\text{Relative } ^{14}\text{C at time } X = {}^{14}\text{C}_{\text{time } X}/{}^{14}\text{C}_{\text{time} - 10}$$

Plot the relative rate of DNA synthesis for UV- and mock-irradiated samples over time, and the total amount of DNA accumulation for UV- and mock-treated samples over time. In the case of mock-treated cultures, the ^{14}C-DNA continues to increase while the relative rate of DNA synthesis (^{3}H-DNA/2min) should either remain constant, or increase slightly, over the time course (Fig. 1B). Use of the dual label and mock-irradiation treatment provides several levels of internal controls that should help indicate the quality of the experimental data that is obtained. Significant fluctuations between time points in the amount of ^{14}C labeled-DNA could be indicative of pipetting errors or problems with filtering the samples. Large fluctuations in the ^{3}H-labeled DNA of mock-irradiated samples may also suggest problems in sampling or culturing techniques, and care should be taken when interpreting these results. A significant decrease in the rate of synthesis in mock-irradiated cultures suggests that the culture growth began to decrease before the end of the experiment.

Typical results for UV-irradiated wild-type cultures should produce a marked decrease in the relative rate of DNA synthesis immediately following the induction of DNA damage. After a 30 J/m^2 dose, we normally observe a drop in rate of about 90%. It is not clear what the remaining 10% of DNA synthesis reflects. uvrA mutants exhibit a similar decrease, suggesting that the synthesis does not represent residual repair replication. dnaB^{ts} mutants also exhibit a similar decrease when shifted to the restrictive temperature suggesting that it does not represent continued synthesis by the holoenzyme at a reduced rate (not shown). It is possible that this may represent radiolabeled nucleotides that are bound to proteins or lipids in our samples, which precipitate upon the addition of the TCA. The time at which the ^{3}H-DNA first begins to increase after UV irradiation is the time that we interpret DNA synthesis to begin to recover. For our parental or wild-type strains, this occurs between the 15- and 20-min time point after a 30 J/m^2 dose (Fig. 1B). The efficiency of replication recovery following UV-induced DNA damage can be established from the slope of the graphs.

In the majority of cases, the graph of total DNA accumulation reflects the trends observed for DNA synthesis rates, with a period of little to no DNA accumulation corresponding to times prior to when the rate of synthesis begins to recover. As shown in Fig. 1B, mutants can be identified by this assay that exhibit a failure to recover, a delayed recovery, and timely recovery but with reduced kinetics.

DNA Degradation at Arrested Replication Forks

Arrested replication forks are subject to enzymatic processing and degradation. To help characterize the enzymes that may process arrested replication forks, we developed the simple assay described later to examine the degradation that occurs at the replication fork following arrest. In previous work this assay was used to show that following UV-irradiation, the nascent DNA of the replication fork is maintained and protected by the RecF-O-R proteins and partially degraded by RecQ, a 3′-5′ helicase, and RecJ, a 5′ single-strand nuclease (Courcelle and Hanawalt, 1999; Courcelle *et al.*, 1997, 1999, 2003). Our recent experiments suggest that the nucleases and helicases which process the ends of the replication fork depend upon the nature of the impediment that blocked or disrupted the replication fork.

Dilute a fresh overnight culture 1:100 in 10-ml DGCthy medium supplemented with 0.1 μCi/ml [^{14}C]thymine. A 50-ml conical tube works well for growing this volume of culture. Grow cells at 37° with vigorous shaking, monitoring cell growth using absorbance at 600 nm.

While the culture is growing, prewarm 10-ml nonradioactive DGCthy medium in a 50-ml tube along with an empty 50-ml tube to 37°. In addition, set up 23 5-ml polypropylene, round-bottom tubes for sample collection, two for every time point, except for time 0, which serves as a reference for all other time points and is collected in triplicate.

UV Treatment and Degradation Assay

1× NET rinse buffer:
100 mM NaCl
10 mM EDTA, pH 8
10 mM Tris-HCl, pH 8.0

Prior to UV irradiation, the room should be set up to work under yellow light conditions as before. When the culture reaches an OD$_{600}$ of 0.4, prefill three of the 5-ml polypropylene collection tubes with ice-cold 5% TCA, set up a vacuum filter holder and flask with a 0.45-μm general filtration membrane, and turn on the vacuum. Add 1 μCi/ml [^{3}H]thymidine to the culture for a 10-s pulse. Then, collect the cells by pouring the culture onto the 0.45-μm pore membrane filter. Once all the liquid has been sucked through, rinse the filter twice with 3-ml cold 1× NET. The EDTA in the NET buffer makes the cells more permeable and allows more of the [^{3}H] thymidine in the pulse label to be washed away. Figure 2A depicts the differential labeling of total and nascent DNA with ^{14}C and ^{3}H, respectively, resulting from this method.

Immediately resuspend the cells in the prewarmed, nonradioactive DGCthy medium by placing the filter into the conical tube and vortexing for about 5 s. Then, pour the culture into a flat-bottomed container and UV irradiate with gentle shaking for the desired dose, 30 J/m^2 in the example shown. After irradiation, transfer the culture to a fresh, warmed 50-ml

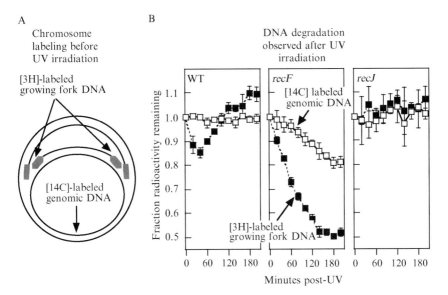

Fig. 2. Measurement of the amount of DNA degradation following UV-induced DNA damage. (A) Schematic depicting the differential labeling of total and newly synthesized DNA with [^{14}C]thymine and [^{3}H]thymidine, respectively. (B) The fraction of radioactive nucleotides remaining in the DNA from wild-type, *recF*, and *recJ* cells is plotted over time. Typical initial values for ^{3}H and ^{14}C are between 2500–4000 and 1200–1700 cpm, respectively, for all experiments. Total DNA (^{14}C, □); nascent DNA (^{3}H, ■).

tube. Remove triplicate 0.2-ml aliquots of the irradiated culture and place each aliquot in a tube with ice-cold 5% TCA (time = 0). Clearly, consistency and timing are important for this experiment when comparing the relative amount of degradation between strains. We find that about 20 s is required for rinsing and resuspending cells prior to irradiation, and with our UV apparatus, a 30 J/m^2 dose is delivered in 30 s. Therefore, to ensure consistency in the reference time point between strains and experiments, we typically remove the time = 0 aliquot 60 s after the time at which UV irradiation began.

Incubate the irradiated culture at 37° with vigorous shaking. Immediately before taking each time point, fill two tubes with ice-cold 5% TCA. Remove duplicate 0.2-ml aliquots of culture and place each into a tube with ice-cold 5% TCA at 20, 40, 60, 80, 100, 120, 140, 160, 180, and 200 min postirradiation.

Sample Preparation and Analysis of Degradation. Once the last time point is taken, collect the acid-precipitable DNA from each sample onto Whatman glass fiber filters and determine how much ^{3}H and ^{14}C was incorporated into each sample as described in the previous assay. Average the ^3H and ^{14}C counts that are obtained from the duplicate (triplicate in the case of time = 0) time points. Then, using the counts from the time = 0 min sample as a reference, determine the relative amount of nascent DNA (based on ^3H counts) and total DNA (based on ^{14}C counts) that remains at each time. The loss of radioactivity represents the amount of degradation in total DNA and DNA made at replication forks immediately prior to UV irradiation.

$$\text{Relative } ^3H \text{ at time } X = {}^3H_{time\ X}/{}^3H_{time\ 0}$$
$$\text{Relative } ^{14}C \text{ at time } X = {}^{14}C_{time\ X}/{}^{14}C_{time\ 0}$$

Plot the relative amount of nascent DNA and total DNA remaining after UV irradiation over time. In the case of the total DNA for our parental cultures, we typically see little to no variation in the amount of DNA remaining over the time course (Fig. 2B). The lack of degradation in the total DNA provides an internal control and serves as a baseline upon which to compare the amount of degradation that occurs specifically in the nascent DNA following UV irradiation. For our parental cells, we typically observe that 10–20% of the nascent DNA is degraded at times prior to when replication resumes. In cells that are able to recover replication, the observed degradation of the nascent DNA ceases at the time when replication resumes (Fig. 2B). However, in mutants that fail to resume DNA synthesis, the nascent DNA degradation continues and is much more extensive (Fig. 2B). Once robust replication resumes, the assay is no longer able to effectively detect degradation and an increase in acid-precipitable counts is sometimes observed in the nascent DNA, presumably due to the reincorporation of the remaining intracellular pools of radiolabeled nucleotides. As shown in Fig. 2B, mutants have been identified that exhibit more extensive degradation than wild-type cells, suggesting that they have a role in protecting the arrested fork. Alternatively, other mutants like RecJ exhibit less nascent DNA degradation than wild-type cells, suggesting it acts to degrade the DNA at the fork following arrest.

Plasmid Replication Intermediates Observed by 2D N/N Gel Analysis

Following UV irradiation, the structural properties of the DNA molecule can be observed on replicating plasmids in *E. coli* using two-dimensional (2D) agarose gel electrophoresis. The technique can be used to observe UV-induced intermediates associated with both recombination and replication following arrest (Courcelle *et al.*, 2003). We use a 2D agarose gel technique adapted almost directly from Friedman and Brewer (1995). However, the method for preparing total genomic DNA for this analysis is somewhat unusual in that it does not involve any procedures to enrich the samples for single-stranded fragments, nor does it involve any DNA precipitations that are often used in purifying DNA for 2D gel analysis. We have noticed that some DNA structural intermediates are sensitive to ethanol precipitation while others appear to form with higher frequencies. Thus the procedure was developed with the idea of keeping the manipulations during lysis and purification to a minimum.

An overnight culture of cells previously transformed with the plasmid pBR322 is grown at 37° with vigorous shaking in 2-ml DGCthy medium supplemented with 100 μg/ml ampicillin to maintain the plasmid.

To begin the experiment, 200 μl of the overnight culture is pelleted for 30 s at 12,000g in a microfuge tube. The cell pellet is resuspended in 200-μl fresh DGCthy and used to inoculate 20-ml DGCthy medium. This step is necessary to remove the ampicillin from the media as this antibiotic absorbs light strongly in the UV region of the spectrum and can significantly reduce the effective dose of irradiation to the culture. Grow cells in a shaking incubator at 37° without antibiotic selection.

UV Irradiation and DNA Isolation

2× NET:
>   200 mM NaCl
>   20 mM EDTA, pH 8
>   20 mM Tris-HCl, pH 8.0
Lysis Buffer:
>   1 mg/ml lysozyme
>   0.2 mg/ml RNase A in
>   TE (10 mM Tris-HCl, 1 mM EDTA, pH 8.0)
6× gel loading dye:
>   0.25% (w/v) bromophenol blue
>   0.25% (w/v) xylene cyanol FF
>   30% (v/v) glycerol in H_2O

While cells are growing, warm an empty flask at 37°. Set up six 2-ml microfuge tubes with 0.75 ml of 2× NET in an ice bucket, one for each time point. Make up lysis buffer (150 μl for each time point to be taken) and store on ice.

Prior to irradiation, the room should be set up to work under yellow light conditions as before. Once cultures have reached an OD_{600} of 0.5, place the culture in a flat-bottomed container and UV irradiate with gentle agitation. For studies using the plasmid pBR322, we typically irradiate with 50 J/m^2 because this dose generates, on average, 1 lesion per plasmid yet greater than 90% of our parental cells still survive to form colonies at this dose. Following irradiation, transfer the culture to a fresh, prewarmed flask. Then, immediately remove a 0.75-ml aliquot of the culture and place it into one of the prechilled tubes containing 2× NET (time = 0). The cold temperature and EDTA in the NET buffer effectively stop further replication and repair events from occurring.

Centrifuge the sample at 14,000 rpm for 90 s in a microfuge. Decant the supernatant taking care to remove all the liquid and resuspend the pellet in 150-μl lysis buffer. Store samples on ice for the duration of the time course.

Continue to incubate the irradiated culture in a 37° shaking incubator. At 15, 30, 45, 60, and 90 min, remove 0.75-ml aliquots of the culture and add them to the prechilled tubes containing the 2× NET buffer and process these samples as described for the initial time point.

After the last time point, place the tubes in a 37° water bath for 30 min to lyse the cells. Then, add 10-μl 10 mg/ml proteinase K and 10-μl 20% Sarkosyl to the samples, mix gently, and then incubate at 50° for 1 h. Samples should be clear after this incubation.

To minimize shearing of the DNA, pipetting in the following steps should be done gently using wide-bore micropipet tips (we typically enlarge the holes of standard tips by cutting the ends off with a razor blade). Add 2 volumes of phenol to samples and mix for 5 min on an orbital platform. Then add 2 volumes of chloroform/isoamyl alcohol (24/1) and mix gently by inversion. Extract the aqueous phase and repeat the extraction with 4 volumes of phenol/chloroform/isoamyl alcohol (25/24/1) if necessary. Finally, extract the samples once with 4 volumes chloroform/isoamyl alcohol (24/1).

The samples are then dialyzed for 3 h on Whatman 0.05-μm pore disks that float in a 250-ml beaker filled with 2 mM Tris-HCl, 1 mM EDTA, pH 8.0. Note, with careful pipetting, each disk can support up to three 150-μl aliquots of sample. Larger volumes than this will cause the disk to sink. Cover the beaker with plastic wrap to prevent excessive evaporation.

Following dialysis, the DNA samples are digested with the desired restriction enzyme, in this example PvuII was used, following the manufacturer's instructions. We typically make up a 5× master mix of the enzyme in the buffer, then add 20 μl of the enzyme mix to 80 μl of each genomic DNA sample. PvuII restricts pBR322 once just downstream of the origin, however the choice of restriction enzyme to use will depend on the plasmid being examined and the type of structural information being sought. A more thorough discussion of this and other aspects of neutral/neutral 2D gel electrophoresis can be found elsewhere (Brewer and Fangman, 1987; Dijkwel and Hamlin, 1997).

Following digestion, add 1 volume of chloroform/isoamyl alcohol (24/1) to the samples to denature the restriction enzyme prior to electrophoresis and then add 20 μl of 6× gel loading dye directly to each sample. Since no precipitation steps are involved, and the cell pellets from each sample are resuspended and lysed in an equal volume, the DNA concentration in each sample is typically consistent throughout the time course and yields about 20 ng/μl of genomic DNA.

Two-Dimensional Agarose Gel Electrophoresis

10× TBE: Per liter, dissolve
108 g Tris base
55 g boric acid
40 ml 0.5 *M* EDTA (pH 8.0)
add H_2O to 1 liter

Cast a 200-ml 0.4% agarose gel, 1× TBE in a 13 × 17-cm tray using a 20-well comb (approximate capacity 40 μl). Once gel is solidified at room temperature, pour 1× TBE buffer over the gel. Remove comb gently and fill the electrophoresis rig with enough 1× TBE to submerge the gel completely.

Load a lambda-Hind III size marker in the first lane, then skipping lanes, load 30 μl of each DNA sample into the wells. Loading every other lane makes it easier to cleanly cut the gel lanes for casting in the second-dimension gel. Extra gel lanes may be used to load a second set of samples to check the quality and quantity of the DNA in your samples. Run the gel at 1 V/cm until the (lower) bromophenol blue marker dye has migrated 6.5 cm (~12–14-h in our gel rigs). Note, the gel may be run longer depending on the time points and size of your gel in the second dimension. This migration distance was selected because it yields sufficient resolution and allows us to fit six samples on our gel for the second dimension.

For the second dimension, prepare a 500-ml 1% agarose gel solution in 1× TBE and allow it to cool to between 45 and 50°.

Slice the gel evenly between lanes. We use a large butcher knife to cut our gels, which we find makes it easy to slice even segments. The first lane containing the size markers can be stained with ethidium bromide and visualized. If a second set of lanes was loaded, these too can be stained and visualized. Once the migration distance of your linear fragment has been determined, trim the gel lanes to a length that comfortably spans this size. Under the gel conditions described here, linear pBR322 runs above the bromophenol blue marker. Rotate each gel lane 90° and place the first three lanes lengthwise along the top of a leveled 20 × 25-cm electrophoresis tray. The next three lanes are then placed across the middle of the gel tray. Pipette a small amount of the 1% agarose gel solution around each gel slice to set it in place. This small amount of agarose should set in 1–2 min. Then, pour the remaining 1% agarose solution into the gel tray, allowing it to hit the surface of the tray first and taking care to pour a smooth, even layer. The agar should completely cover the sliced lanes from the previous gel. If the agarose solution is too hot when it is poured, it could partially melt the agarose slices from the first dimension or affect fragile DNA structures.

Once the gel has solidified, place it into an electrophoresis rig and fill with enough $1 \times$ TBE to submerge the gel. Run the second-dimension gel at 6.5 V/cm until the xylene cyanol dye front has migrated about 10.5 cm ($\sim$7 h in our gel rigs).

After electrophoresis, transfer the DNA to a positively charged nylon membrane (e.g., Hybond N+) using standard procedures (Spivak and Hanawalt, 1995).

Prepare a ^{32}P-labeled pBR322 probe using nick translation. We use the protocol included with the nick translation kit from Roche with good success.

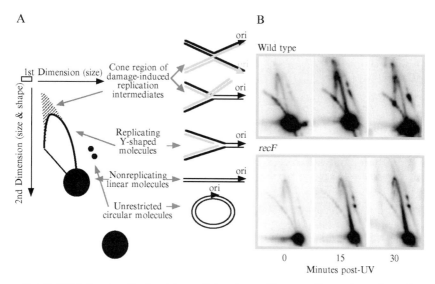

FIG. 3. UV-induced replication intermediates observed by neutral/neutral two-dimensional gel electrophoresis. (A) Predicted migration pattern for PvuII-digested pBR322 plasmid after UV treatment using 2D analysis. Nonreplicating plasmids run as a linear 4.4-kb fragment. Normal replicating plasmids form Y-shaped structures and migrate more slowly due to their increased size and nonlinear shape, moving as an arc that extends from the linear fragment. Double Y- and X-shaped intermediates migrate in the cone region. (B) Blocked replication fork and cone region molecules accumulate transiently in wild-type cells after UV irradiation. RecF mutants do not accumulate cone region intermediates.

Hybridize the membrane with the radiolabeled probe and visualize and quantitate the radioactivity on a phosphorimager. The predicted migration pattern for PvuII-linearized pBR322 is shown in Fig. 3A. It should be noted that this method can be easily adapted to examine other plasmids or used with alternative restriction enzymes. In wild-type cells following UV-induced damage, we typically see a transient increase in the amount of replicating Y-shaped intermediates due to an accumulation of blocked replication forks at UV lesions (Fig. 3B). In addition, double Y- or X-shaped intermediates also accumulate transiently in the cone region, peaking around 30 min following UV irradiation before waning at a time that correlates with the repair of the DNA lesions and the recovery of robust replication (Courcelle *et al.*, 2003). In contrast, cone region intermediates are not observed in *recF* mutants.

Taken together, the three assays described previously can be applied to various *E. coli* mutants to help characterize the potential functional role of those gene products *in vivo*. In the case of *recF* we believe that these assays are consistent with the idea that RecF is needed to protect and maintain the structural integrity of replication forks arrested at UV-induced damage.

Concluding Remarks

The three assays described previously each focus on a different aspect of the replication fork following DNA damage. It is critical to keep in mind, however, that while the first assay is designed to quantify the amount of DNA synthesis at the fork and the second to measure DNA degradation, both processes are clearly occurring simultaneously in the cell. While the use of both assays provides a more comprehensive picture of the events occurring at the replication fork, each process is likely to partially interfere with the measurement of the other. For instance, if significant amounts of degradation are occurring in the nascent DNA, then the observed amount of newly DNA synthesized in our assay may be less than that which is actually occurring. The limitations of these assays and determining precisely what they can measure are important factors that should be considered when interpreting these assays and when trying to develop new methods to observe and tease apart the biochemical reactions that occur in living cells.

Acknowledgments

These studies are supported by CAREER award MCB-0448315 from the National Science Foundation. C. T. C. is supported by award F32 GM068566 from the NIH-NIGMS.

References

Brewer, B. J., and Fangman, W. L. (1987). The localization of replication origins on ARS plasmids in S. cerevisiae. *Cell* **51**, 463–471.

Cleaver, J. E., Bootsma, D., and Friedberg, E. (1975). Human diseases with genetically altered DNA repair processes. *Genetics* **79**(Suppl.), 215–225.

Cleaver, J. E., Thompson, L. H., Richardson, A. S., and States, J. C. (1999). A summary of mutations in the UV-sensitive disorders: Xeroderma pigmentosum, Cockayne syndrome, and trichothiodystrophy. *Hum. Mutat.* **14**, 9–22.

Cordeiro-Stone, M., Zaritskaya, L. S., Price, L. K., and Kaufmann, W. K. (1997). Replication fork bypass of a pyrimidine dimer blocking leading strand DNA synthesis. *J. Biol. Chem.* **272**, 13945–13954.

Courcelle, J., Carswell-Crumpton, C., and Hanawalt, P. C. (1997). recF and recR are required for the resumption of replication at DNA replication forks in Escherichia coli. *Proc. Natl. Acad. Sci. USA* **94**, 3714–3719.

Courcelle, J., Crowley, D. J., and Hanawalt, P. C. (1999). Recovery of DNA replication in UV-irradiated Escherichia coli requires both excision repair and recF protein function. *J. Bacteriol.* **181**, 916–922.

Courcelle, J., Donaldson, J. R., Chow, K. H., and Courcelle, C. T. (2003). DNA damage-induced replication fork regression and processing in Escherichia coli. *Science* **299**, 1064–1067.

Courcelle, J., and Hanawalt, P. C. (1999). RecQ and RecJ process blocked replication forks prior to the resumption of replication in UV-irradiated Escherichia coli. *Mol. Gen. Genet.* **262**, 543–551.

De Weerd-Kastelein, E. A., Keijzer, W., Rainaldi, G., and Bootsma, D. (1977). Induction of sister chromatid exchanges in xeroderma pigmentosum cells after exposure to ultraviolet light. *Mutat. Res.* **45**, 253–261.

Dijkwel, P. A., and Hamlin, J. L. (1997). Mapping replication origins by neutral/neutral two-dimensional gel electrophoresis. *Methods* **13**, 235–245.

Ellis, N. A., Groden, J., Ye, T. Z., Straughen, J., Lennon, D. J., Ciocci, S., Proytcheva, M., and German, J. (1995). The Bloom's syndrome gene product is homologous to RecQ helicases. *Cell* **83**, 655–666.

Epstein, C. J., and Motulsky, A. G. (1996). Werner syndrome: Entering the helicase era. *Bioessays* **18**, 1025–1027.

Friedman, K. L., and Brewer, B. J. (1995). Analysis of replication intermediates by two-dimensional agarose gel electrophoresis. *Methods Enzymol.* **262**, 613–627.

Fukuchi, K., Martin, G. M., and Monnat, R. J., Jr. (1989). Mutator phenotype of Werner syndrome is characterized by extensive deletions. [published erratum appears in *Proc. Natl. Acad. Sci. USA* [Oct; 86(20):7994]. *Proc. Natl. Acad. Sci. USA* **86**, 5893–5897.

Giannelli, F., Benson, P. F., Pawsey, S. A., and Polani, P. E. (1977). Ultraviolet light sensitivity and delayed DNA-chain maturation in Bloom's syndrome fibroblasts. *Nature* **265**, 466–469.

Gray, M. D., Shen, J. C., Kamath-Loeb, A. S., Blank, A., Sopher, B. L., Martin, G. M., Oshima, J., and Loeb, L. A. (1997). The Werner syndrome protein is a DNA helicase. *Nat. Genet.* **17**, 100–103.

Griffiths, T. D., and Ling, S. Y. (1991). Effect of UV light on DNA chain growth and replicon initiation in xeroderma pigmentosum variant cells. *Mutagenesis* **6**, 247–251.

Hanada, K., Ukita, T., Kohno, Y., Saito, K., Kato, J., and Ikeda, H. (1997). RecQ DNA helicase is a suppressor of illegitimate recombination in Escherichia coli. *Proc. Natl. Acad. Sci. USA* **94**, 3860–3865.

Hanaoka, F., Yamada, M., Takeuchi, F., Goto, M., Miyamoto, T., and Hori, T. (1985). Autoradiographic studies of DNA replication in Werner's syndrome cells. *Adv. Exp. Med. Biol.* **190,** 439–457.

Karow, J. K., Wu, L., and Hickson, I. D. (2000). RecQ family helicases: Roles in cancer and aging. *Curr. Opin. Genet. Dev.* **10,** 32–38.

Khidhir, M. A., Casaregola, S., and Holland, I. B. (1985). Mechanism of transient inhibition of DNA synthesis in ultraviolet-irradiated *E. coli*: Inhibition is independent of recA whilst recovery requires RecA protein itself and an additional, inducible SOS function. *Mol. Gen. Genet.* **199,** 133–140.

Kuhn, E. M., and Therman, E. (1986). Cytogenetics of Bloom's syndrome. *Cancer Genet. Cytogenet.* **22,** 1–18.

Langlois, R. G., Bigbee, W. L., Jensen, R. H., and German, J. (1989). Evidence for increased *in vivo* mutation and somatic recombination in Bloom's syndrome. *Proc. Natl. Acad. Sci. USA* **86,** 670–674.

Lehmann, A. R., and Kirk-Bell, S. (1978). Pyrimidine dimer sites associated with the daughter DNA strands in UV-irradiated human fibroblasts. *Photochem. Photobiol.* **27,** 297–307.

Lonn, U., Lonn, S., Nylen, U., Winblad, G., and German, J. (1990). An abnormal profile of DNA replication intermediates in Bloom's syndrome. *Cancer Res.* **50,** 3141–3145.

Mamada, A., Kondo, S., and Satoh, Y. (1989). Different sensitivities to ultraviolet light-induced cytotoxicity and sister chromatid exchanges in xeroderma pigmentosum and Bloom's syndrome fibroblasts. *Photodermatol.* **6,** 124–130.

Masutani, C., Kusumoto, R., Yamada, A., Dohmae, N., Yokoi, M., Yuasa, M., Araki, M., Iwai, S., Takio, K., and Hanaoka, F. (1999). The XPV (xeroderma pigmentosum variant) gene encodes human DNA polymerase eta [see comments]. *Nature* **399,** 700–704.

Mellon, I., and Hanawalt, P. C. (1989). Induction of the *Escherichia coli* lactose operon selectively increases repair of its transcribed DNA strand. *Nature* **342,** 95–98.

Shiraishi, Y. (1990). Nature and role of high sister chromatid exchanges in Bloom syndrome cells. Some cytogenetic and immunological aspects. *Cancer Genet. Cytogenet.* **50,** 175–187.

Spivak, G., and Hanawalt, P. (1995). Determination of damage and repair in specific DNA sequences. *In* "Methods: A Companion to Methods in Enzymology." **7,** pp. 147–161. Academic Press, Inc., Burlington, MA.

Svoboda, D. L., Briley, L. P., and Vos, J. M. (1998). Defective bypass replication of a leading strand cyclobutane thymine dimer in xeroderma pigmentosum variant cell extracts. *Cancer Res.* **58,** 2445–2448.

Tsujimura, T., Maher, V. M., Godwin, A. R., Liskay, R. M., and McCormick, J. J. (1990). Frequency of intrachromosomal homologous recombination induced by UV radiation in normally repairing and excision repair-deficient human cells. *Proc. Natl. Acad. Sci. USA* **87,** 1566–1570.

Yamagata, K., Kato, J., Shimamoto, A., Goto, M., Furuichi, Y., and Ikeda, H. (1998). Bloom's and Werner's syndrome genes suppress hyperrecombination in yeast sgs1 mutant: Implication for genomic instability in human diseases. *Proc. Natl. Acad. Sci. USA* **95,** 8733–8738.

[26] Methods to Study Replication Fork Collapse in Budding Yeast

By GIORDANO LIBERI, CECILIA COTTA-RAMUSINO,
MASSIMO LOPES, JOSE' SOGO, CHIARA CONTI,
AARON BENSIMON, and MARCO FOIANI

Abstract

Replication of the eukaryotic genome is a difficult task, as cells must coordinate chromosome replication with chromatin remodeling, DNA recombination, DNA repair, transcription, cell cycle progression, and sister chromatid cohesion. Yet, DNA replication is a potentially genotoxic process, particularly when replication forks encounter a bulge in the template: forks under these conditions may stall and restart or even break down leading to fork collapse. It is now clear that fork collapse stimulates chromosomal rearrangements and therefore represents a potential source of DNA damage. Hence, the comprehension of the mechanisms that preserve replication fork integrity or that promote fork collapse are extremely relevant for the understanding of the cellular processes controlling genome stability.

Here we describe some experimental approaches that can be used to physically visualize the quality of replication forks in the yeast *S. cerevisiae* and to distinguish between stalled and collapsed forks.

Introduction

DNA replication forks frequently slow down or even stall when encountering obstacles on their way, such as repetitive DNA sequences, specialized protein-DNA complexes, heavily transcribed regions, or a damaged template (Rothstein *et al.*, 2000).

Most times replication forks can easily resume DNA synthesis after the block; nevertheless, in certain pathological situations (for instance in the absence of key factors required for fork integrity), they collapse, losing the replisome (Cobb *et al.*, 2003; Lucca *et al.*, 2004) and accumulate abnormal DNA intermediates such as four branched DNA molecules, long stretches of ssDNA, or DNA breaks (Cha *et al.*, 2002; Cotta-Ramusino *et al.*, 2005; Lopes *et al.*, 2001; Sogo *et al.*, 2002). These structures may prevent fork restart and cause cell lethality and/or genome instability. Occasionally, in normal cells it is also possible that forks collapse as a consequence of intrinsic DNA damage or at certain genomic locations.

METHODS IN ENZYMOLOGY, VOL. 409
Copyright 2006, Elsevier Inc. All rights reserved.
0076-6879/06 $35.00
DOI: 10.1016/S0076-6879(05)09026-9

Several genetic studies indicate that homologous recombination pathways can be used to rescue collapsed forks (Haber and Heyer, 2001); however, it should be stressed that the recruitment of recombination factors at forks might be an unscheduled rather than a programmed event, as too much recombination during replication may contribute to genome instability.

In general, several approaches can be used to visualize stalled and collapsed forks and to distinguish between these two situations; budding yeast certainly represents a powerful tool to study the mechanisms causing and preventing fork collapse due to the possibility to use genetic and biochemical strategies.

The two-dimensional gel electrophoresis (2D) technique (Brewer and Fangman, 1987) has been widely used to directly analyze nascent replication intermediates at specific chromosomal locations, including origins of replication or replication-risk regions.

Another approach to directly visualize DNA structures at stalled and collapsed forks is represented by electron microscopy combined with psoralen-crosslinking (Sogo et al., 2002), although, so far, this analysis has been carried out on the whole genome rather than at specific DNA locations.

Further, using the Molecular Combing technique (Bensimon et al., 1994) it is possible to analyze the size of the replicons, the rate of fork progression, and the inter-origin space.

Other complementary approaches, such as chromatin IP, can be helpful to analyze replisome-fork association or the recruitment of recombination factors at forks at certain genomic regions (Hecht and Grunstein, 1999).

Methods

Analysis of Stalled and Collapsed Replication Forks Using 2D Gel Electrophoresis

The classical 2D gel method allows the analysis of replication intermediates at the forks. Recently, using modified versions of the original protocols (Brewer and Fangman, 1987), it has been possible to visualize other type of intermediates such as reversed forks (Cotta-Ramusino et al., 2005; Lopes et al., 2001) and sister chromatid junctions (Liberi et al., 2005; Lopes et al., 2003). These branched structures represent important parameters to evaluate the integrity of stalled and collapsed replication forks (Fig. 1).

The DNA extraction procedure employed is crucial to preserve the different species of replication intermediates. Here we describe two different methods.

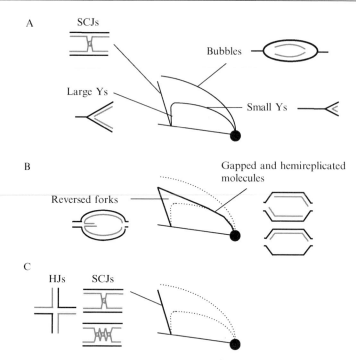

FIG. 1. 2D gel pattern of replication and recombination intermediates. Classical replication and recombination intermediates migrate with characteristic arcs (A) following 2D gel analysis. This is the typical pattern observed under untreated conditions in wild-type cells at a region containing an origin of replication. Bubbles, Y and X structures are clearly visible. The X molecules represent sister chromatid junctions (SCJs) resembling hemi-catenanes. Forks stalling in close proximity of the origin of replication exhibit a similar 2D profile. Unconventional replication intermediates, including reversed forks, gapped, and hemireplicated molecules (B) result from fork collapse and further processing of bubbles and SCJ (Cotta-Ramusino et al., 2005; Lopes et al., 2001; Sogo et al., 2002). Reversed forks seem to accumulate as a cone signal, while hemireplicated and gapped molecules as small Ys (Lopes et al., 2001). (C) SCJs migrate like Holliday junctions (HJs) (Zou and Rothstein, 1997) on 2D gels but the two species of branched cruciform structures can be differentiated by further analysis (Liberi et al., 2005; Lopes et al., 2003) (see text for details).

The first method (A) is a modification of the procedure described by Allers and Lichten (2000) and utilizes the cationic detergent CTAB and chloroform/isoamylalcohol for DNA extraction.

The second method (B) is based on the isolation of yeast nuclei and utilizes a modified version of the *Qiagen Genomic DNA Handbook* protocol of the Qiagen DNA Isolation Kit (Valencia, CA).

The CTAB method has the advantage that it is a rapid and efficient DNA extraction protocol compared to the other procedures. Moreover the

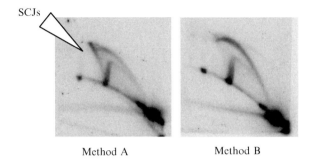

Method A   Method B

FIG. 2. Sister chromatid junctions are sensitive to certain DNA extraction procedures. Replication intermediates, such as bubbles and Ys molecules can be efficiently purified using both DNA extraction methods A and B. Conversely, the SCJs that resemble hemicatenanes (indicated by a white triangle) are preferentially enriched using method A.

CTAB procedure protects the integrity of branched molecules that migrate with an X-shape and that resemble hemicatenanes (Fig. 2). Conversely, Method B, which uses columns to enrich replication intermediates and glass beads to break the cells, minimizes the detection of the X-structures (Fig. 2), although it has been used to visualize reversed forks (Lopes *et al.*, 2001). On the other hand, with Method B the signals for bubbles and Ys are usually sharper, compared to the CTAB method (Fig. 2).

Cell Growth

Yeast strains are grown under appropriate conditions until they reach the logarithmical phase. Cells are then synchronized using either α-factor (2 μg/ml) or nocodazole (5 μg/ml) treatment. Samples are taken at different time points. Usually each sample requires 2×10^9–4×10^9 total cells (i.e., 200 ml of 1–2×10^7 cells/ml).

We strongly recommend that yeast cells be presynchronized as otherwise it would be difficult to visualize replication intermediates as they are not abundant. Further, sometimes it is necessary to treat cells with DNA damaging agents such as hydroxyurea or methyl methane sulfonate (usually, at the concentration of 0.2 M or 0.033%, respectively).

DNA Extraction Methods

Method A: Isolation of Total Genomic DNA Using CTAB Procedure

MATERIALS AND SOLUTIONS

• Sodium azide 10% store at 4°
• 10 mg/ml zymolyase stock (1000 U/ml)

- Spheroplasting buffer: 1 M sorbitol, 100 mM EDTA pH 8.0, 0.1% β-mercapto ethanol
- Solution I: 2% w/v CTAB (FLUKA-cetyltrimethylammonium bromide), 1.4 M NaCl, 100 mM Tris HCl pH 7.6, 25 mM EDTA pH 8.0
- 10 mg/ml RNase (DNasi free)
- 20 mg/ml Proteinase K
- 24:1 chloroform/isoamylalcohol
- Corex glass tubes
- Solution II: 1% w/v CTAB, 50 mM Tris HCl pH 7.6, 10 mM EDTA
- Solution III: 1.4 M NaCl, 10 mM Tris HCl pH 7.6, 1 mM EDTA
- Isopropanol
- 70% ethanol
- 10 mM Tris-HCl pH 8.0.

Procedure

1. Harvest cell samples in ice and treat them with 0.1% sodium azide (final concentration).

2. Collect the cells by centrifugation at 6000–8000 rpm 5–10 min, wash one time with cold water, and resuspend in 50-ml Falcon tube with 5 ml of spheroplasting buffer.

3. Place the cell suspension at 30° until spheroplasts are visible under microscope. Usually this step takes 45–60 min, but the appropriate time has to be calculated based on the stock of zymolyase and mainly on the cell growth conditions, which influence the dimension of the cell wall.

4. Collect the spheroplasts by centrifugation at 4000 rpm for 10 min at 4°; carefully remove the supernatant and replace it with 2 ml of water. Vigorously resuspend the spheroplasts on vortex and subsequentially add 2.5 ml of Solution I; kindly mix the suspension and place it at 50° with 200 μl of 10 mg/ml RNasi for 30 min.

5. Add 200 μl of 20 mg/ml Proteinase K and protract the incubation for further 1.5 h at 50°. Note that at this step the solution has to become clear, with no visible aggregates of cellular component; if necessary, incubate overnight at 30° with additional 100 μl of Proteinase K.

6. Separate the solution by centrifugation at 4000 rpm for 10 min at room temperature and process separately the obtained supernatant and the pellet as indicated in the following sections.

Supernatant

1. Transfer the supernatant into a 15-ml Falcon tube and add 2.5 ml chloroform/isoamylalcohol 24:1.

2. Mix vigorously 6 times and separate the two phases by centrifugation at 4000 rpm for 10 min.

3. Carefully transfer the clear upper phase into a Corex glass tube with a pipette and add two volumes (10 ml) of Solution II. Note that at this step the prolonged incubation (1–2 h) with Solution II might help DNA precipitation in the next step.
4. Separate the solution by centrifugation at 8500 rpm for 10 min in a swing out rotor, discard the supernatant, and resuspend the pellet in 2.5 ml of Solution III. Briefly incubate at 37° to help the dissolution of the pellet.

Pellet

1. Energically resuspend the pellet into 2 ml of Solution III and incubate 1 h at 50°.
2. Transfer the solution into a 15-ml Falcon tube and extract with 1 ml of chloroform/isoamylalcohol 24:1. Separate the two phases by centrifugation at 4000 rpm for 10 min at full speed in an appropriate centrifuge.
3. Carefully transfer the clear upper phase (Solution III) into the Corex glass tube containing Solution III obtained from the treatment of the supernatant (see treatment of "supernatant" step 4).
4. Precipitate the DNA with 1 volume (10 ml) of isopropanol at 8500 rpm in swing out rotor for 10 min.
5. Briefly wash the pellet with 2 ml of ethanol 70%.

After centrifugation (2 min, 8500 rpm), carefully remove the ethanol as much as possible with a pipette and dissolve the DNA into 250 μl of 10 mM Tris/HCl pH 8.

Method B: Isolation of Total Genomic DNA from Isolated Yeast Nuclei

MATERIALS AND SOLUTIONS

- 1 M spermidine stock, store −20°
- 0.5 M spermine stock, store 4°
- Nuclei isolating buffer (NIB) pH 7.2, store at 4°: 17% glycerol, 50 mM MOPS, 150 mM Potassium acetate, 2 MgCl$_2$, 500 μM spermidine, 150 μM spermine
- RNase (10 mg/ml)
- Proteinase K (20 mg/ml)
- Glass beads
- Qiagen Genomic-Tip 100/G
- Corex glass tube
- Tris-HCl 10 mM; EDTA 1 mM pH 8 (TE) 1×
- Isopropanol.

Procedure

1. Harvest cells in ice and treat them with sodium azide 0.1% final concentration, to freeze the RI. Collect cells by centrifugation of 6000 g for 5-10 min.
2. Wash the cells with 20 ml of water and pellet them by centrifugation at 7000 rpm, 4°, for 5 min.
3. Resuspend cells in 5 ml of NIB buffer, keeping them on ice
4. Add an equal volume of glass beads.
5. Vortex 30 s at the higher speed and subsequently chill out the tube in ice and water for 30 s. Repeat this step for 15 cycles. This allows cells breaking.
6. Carefully recover the supernatant in a new tube with a Pasteur pipette. Wash the glass beads two times with 5 ml of NIB buffer and recover the supernatant in the same tube.
7. Centrifuge the cell extract (around 15 ml) at 8000 rpm, 4°, for 10 min
8. Discard the supernatant and very carefully resuspend the pellet (made only from the nuclei of the cells) in 5 ml of G2 buffer of the Qiagen kit. Resuspend the pellet with the aid of the tip of the Pasteur pipette.
9. Add 100 μl of 10 mg/ml RNasi and incubate the tube for 30 min at 37°
10. Add 100 ml of Proteinase K and incubate for 1 h at 37°.
11. Rescue the supernatant by centrifugation at 5000 rpm, 4°, for 5 min
12. Equilibrate the Genomic tip 100/G with 4 ml of QBT buffer of the Qiagen kit.
13. Gently mix the supernatant with 5 ml of QBT and apply it to the equilibrated Genomic tip 100/G.
14. Wash the columns 2 times with 7.5 ml of QC buffer of the Qiagen kit.
15. Elute the DNA into a Corex tube with 5 ml of QF buffer of the Qiagen kit prewarmed at 50°.
16. Add 3.5 ml of isopropanol to precipitate the DNA.
17. Centrifuge for 25 min at 8000 rpm 4° in a proper swing out rotor.
18. Rescue the supernatant in a new Corex tube and incubate overnight at −20°.
19. Place the Corex containing the pellet upside down and allow air-dry of the pellet for 15 min.
20. Resuspend overnight the dry pellet in 150 μl of 1× TE.
21. The day after, centrifuge for 25 min at 8000 rpm 4° the supernatant left at −20° over-night (see step 18).
22. Air-dry the pellet and resuspend in 150 μl of 1× TE.
23. After at least 1 h collect the DNA-TE solution and make a pool with the one of the day before (see step 20).

After preparation of DNA samples, 1–2 μl of DNA preps are quantified using a DNA fluorimeter or using standard gel electrophoresis. An aliquot of sample, corresponding to 10 μg of total DNA, is digested with the appropriate restriction enzyme and subjected to neutral-neutral 2D gel electrophoresis.

First and Second Dimension Gel Electrophoresis and Southern Hybridization

For more details on 2D gel procedure see Brewer and Fangman, 1987. Prepare a 0.35% agarose gel without ethidium bromide (US Biological-LOW EEO, USA) in fresh TBE 1× and fill an appropriate gel tray in cold room (we routinely use apparatus W × L = 20 × 25). Wait 30 min, and put the gel in the box at room temperature containing a suitable volume of TBE 1×. Handle the gel very carefully because it is very fragile. Load the DNA samples and a molecular weight DNA marker, leaving one empty well in between each sample and run the gel at constant low voltage (50–60 V, c.a. 1 V/cm). The length of the run will be determined based on the dimension of the DNA fragment of interest (see later). Stain the gel in TBE 1× with 0.3 μg/μl ethidium bromide for 30 min. Use a big knife to cut out the gel lanes under a UV trans-illuminator with the aid of a ruler. If the fragment of interest is, for example, an origin of replication contained in a 5 kb restriction fragment, the gel slice will contain all DNA molecules ranging from 5 kb up to 10 kb. The intermediates with a complex shape, such as bubbles and joint molecules, migrate up to 10 kb; hence, we recommend to cut a piece larger than is expected based on the molecular weight of the fragment; generally, we manage slices from 9.5 to 6.5 cm for DNA fragment ranging from 3–6 kb. With the aid of a flexible piece of plastic, rotate gel slices by 90° before putting them in the second dimension gel tray. It is possible to use an apparatus of the same size of the one used for first dimension and set 4 to 6 slices, depending on their dimension. At the same time, prepare a 0.9% agarose gel in TBE 1×, with ethidium bromide 0.3 μg/μl. Pour the gel, this time at room temperature, around the gel slices and wait 20 min for solidification. Put the tray in gel box with an appropriate volume of TBE 1× containing 0.3 μg/μl ethidium bromide and run at constant high voltage in a cold room (180–250 V, c.a. 2–3 V/cm). During the run, linear DNA molecules will distribute along a characteristic arc that is visible under a UV lamp. When DNA molecules with lower molecular weight reach the bottom of the gel, stop the run and, using the knife, cut gel pieces 10 × 20 cm containing 2 or 3 DNA samples.

Depurinate the gel for 10 min in 0.25 N HCl, denaturate 30 min in 0.5 M NaOH, 1.5 M NaCl and finally neutralize for 30 min in 1 M AcNH$_4$, 0.02 M

NaOH. Transfer the gel in standard Southern blot conditions using Gene Screen (Perkin Elmer, USA) membrane in SSC 10× and leave overnight. Remove the membrane from the gel and cross-link the DNA to the membrane by UV.

The membranes are subjected to hybridization with a radiolabeled probe of interest. Different protocols can be employed at this step; here we propose the following rapid and efficient procedures: 50 ng of purified DNA probe is labeled with 50 μCi of ^{32}P dCTP using a random prime kit (Redi prime kit from Amersham Bioscience, USA). The reaction is stopped by adding 2 μl 0.5 M EDTA pH 8 and passed through Sepadex G50 (Amersham Bioscience, USA) to remove the non-incorporated nucleotides. During the preparation of the radiolabeled probe, the membranes are rinsed with SSC 6× and prehybridized with 20 or 30 ml of Hybridization Solution 1× (Sigma, St. Louis, MO) for at least 30 min at 65° in a rotating tube. The probe is boiled 10 min at 100° and added to prehybridization mix. The hybridization is prolonged at 65° overnight or at least 4–5 h. The filters are washed two times 15 min each with 500 ml 2× SSC, 1% SDS at 65°, and two times 15 min each with 500 ml 0.1× SSC, 0.1% SDS at 42°. The hybridized membranes are briefly air dried. The signals are analyzed using PhosphoImager Molecular Storm 820 and quantified using ImageQuant program. The membrane can be rehybridized with other probes of interest 3–4 times and can be stripped with boiling solution 0.015 M NaCl, 0.1% SSC, 1% SDS for 30 min.

In Vitro *Analysis of Replication Intermediates*

Sometimes, to gain more insight into the molecular nature of certain branched replication intermediates, it might be necessary to carry out an additional *in vitro* characterization.

Here we propose some analyses that can be performed on the joint molecules that might help to establish whether they represent classical recombination intermediates such as Holliday junctions or rather, other cruciform structures such as hemicatenanes.

Branch Migration of Joint Molecules. Cruciform structures have the capability to branch migrate *in vitro* leading to the formation of linear molecules in a reaction that is stimulated by high temperatures (Panyutin and Hsieh, 1994). This molecular transition can be assayed using the 2D gel technique, since these joint molecules will be converted into a characteristic spot following branch migration (Fig. 3A). In the presence of divalent cations, (Mg^{2+}), however, Holliday junctions are stabilized in a molecular conformation that prevents such transition (Panyutin and Hsieh, 1994).

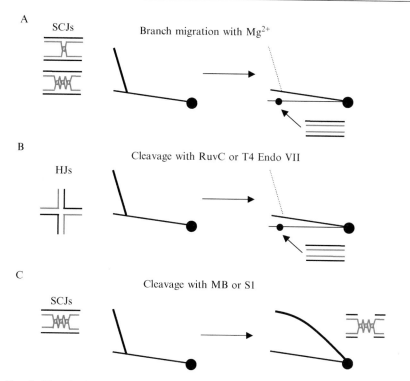

Fig. 3. 2D gel migration patterns of replication and recombination intermediates after different *in vitro* tests. (A) Differently from Holliday junctions (HJs) (Zou and Rothstein, 1997), the SCJs (Liberi *et al.*, 2005; Lopes *et al.*, 2003) that resemble hemicatenanes when subjected to *in vitro* branch migration in presence of Mg^{2+}, are converted into linear molecules that migrate as a spot. (B) HJs, but not SCJs, are cleaved into linear products after treatment with RuvC or T4 Endo VII resolvases (Zou and Rothstein, 1997). (C) In certain mutant backgrounds the SCJs resembling hemicatenanes are converted into recombination structures that are sensitive to treatment with ssDNA nuclease, such as Mung Bean or S1. These recombination structures resemble double HJs and likely result from an extension of the hemicatenated portion (see Liberi *et al.*, 2005 for further details).

Branch migration is performed in the following conditions: the slices of agarose, cut from a standard first dimension run in pulsed-field gel grade agarose (for example Seakem Gold Agarose from Cambrex Bioscence, Walkersville, MD), are incubated in small boxes with a suitable volume of branch migration TNM buffer (50 mM NaCl, 10 mM MgCl$_2$ 10 mM Tris-HCl pH 8, 0.1 mM EDTA) or TNE buffer without Mg^{2+} (100 mM NaCl, 10 mM Tris-HCl pH 8, 0.1 mM EDTA) as a control, at room temperature for 20 min. The gel slices are incubated with fresh aliquots of buffers without shaking at 65° for 5 h. Finally the gel slices are washed again with TBE 1× and than subjected to a standard second dimension.

Enzymatic Treatment of Joint Molecules. Holliday junctions can be cleaved and thus resolved into linear intermediates that can be visualized using 2D gel (Fig. 3B) by prokaryotic resolvase/nuclease, including RuvC and T4 EndoVII (Lilley and White, 2001; Zou and Rothstein, 1997). Although these enzymes cut efficiently synthetic Holliday junctions *in vitro*, it should be pointed out that it has been also reported that both enzymes can recognize and cleave other branched DNA intermediates (Bénard *et al.*, 2001; Gruber *et al.*, 2000). In addition, Holliday junctions and other four-branched molecules, such as hemicatenanes or replication intermediates, are expected to be differentially sensitive to the treatment with single strand nucleases, such as Mung Bean and S1 (Liberi *et al.*, 2005; Wellinger *et al.*, 2003).

The replication intermediates run in pulsed-field gel grade agarose are subjected to digestion using the different enzymes in the following conditions: gel slices cut from first dimension are first pre-equilibrated with an appropriate volume of reaction buffer in small boxes at room temperature for 20 min. The buffer is then totally removed and a suitable amount of enzyme (usually c.a. 50 U of T4 Endo VII from USB Corporation, Cleveland, OH, and 50 μg of purified RuvC) is spread on the surface of the slice, which is then incubated at 37° for 5 h or even overnight. After 2.5 h of incubation, a new aliquot of enzyme can be reapplied to the slices. Finally, the gel slices are washed again with TBE 1× and subjected to a standard second dimension.

Enzymatic treatments that are expected to change the mass together with the shapes of replication intermediates can also be performed just before first dimension in the suitable reaction buffers (Fig. 3C).

Preparation of In Vivo-Crosslinked DNA Samples for Electron Microscopy Analysis

Electron microscopy performed on cross-linked DNA has been used to analyze chromatin architecture (Sogo and Thoma, 1989). DNA packaged into nucleosomes is not accessible to cross-linking agents and appears as single-stranded bubbles under denaturing conditions. *In vivo* psoralen DNA cross-linking and electron microscopy has been also used to analyze DNA intermediates at replication forks. Particularly, extensive single-stranded DNA (ssDNA) regions, arising from unscheduled processing of forks, can be visualized also under nondenaturing conditions as gaps on replicating template (Cotta-Ramusino *et al.*, 2005; Sogo *et al.*, 2002).

Here we describe the *in vivo* DNA cross-linking procedure by Psoralen along with the two-step purification of replication intermediates that can be prepared using the CTAB extraction previously described. A comprehensive description of sample preparation for electron microscopy has been already described (Sogo and Thoma, 1989).

In Vivo *Psoralen-DNA Cross-Linking*

The Psoralen DNA cross-linking is performed on living cells at the time of their collection. To assure that the *in vivo* cross-linking is carried out efficiently, a crucial ratio between concentration of cells and amount of cross-linking reagent is required.

1. Collect 400 ml of 1.2×10^7 cells/ml for each time point.
2. Harvest cells by centrifugation, wash one time with cold water, and resuspend the pellet in 20 ml of cold water (including the volume occupied by the cells). Make sure to keep them on ice during all the following steps.
3. Transfer the cells in a standard petri dish, add 1 ml of psoralen (Sigma-TMP, dissolved in Et OH 100% at 200 μg/ml), mix with the aid of a blue tip, and incubate for 5 min.
4. Put the Petri dish under the 366 nm UV light source for 5 min (prewarm the lamp for 5 min during the first incubation with Psoralen). Note that the appropriate distance between the lamp and the sample surface has to be determined for each light source employed.
5. Repeat the incubation with Psoralen and the irradiation steps (3 and 4) another 4 times.
6. Transfer the cross-linked cells to a new Falcon tube. Wash twice the Petri dish with 1 ml of water and collect all the cells by centrifugation 5 min at full speed.

The CTAB-extraction of total DNA is carried out as described in the previous section. Additional purification steps of DNA samples are required before the digestion. The restriction enzyme usually employed is PvuI (use 10 μg of total DNA).

Purification of Replication Intermediates by BND Cellulose Column Chromatography

Before electron microscopy analysis, the replication intermediates, which constitute only a small percentage of total DNA, are enriched by BND cellulose column chromatography. BND cellulose has the capability to retain both dsDNA and ssDNA in low salt conditions. Subsequently, dsDNA will be eluted with high salt, while ssDNA, mainly represented by replication intermediates, will be eluted with high salt plus caffeine.

All the following steps are done at room temperature.

1. Apply 1 ml of BND cellulose (Sigma-0.1 g/ml resuspended in Tris-HCl 10 mM pH 8, NaCl 300 mM) to a proper chromatography column and allow gravity flow of the buffer.
2. Prewash the column 6 times with 1 ml Tris-HCl 10 mM pH 8, NaCl 800 mM.
3. Equilibrate the column 6 times with 1 ml Tris-HCl 10 mM pH 8, NaCl 300 mM.
4. Load the sample pre-equilibrated at NaCl 300 mM final concentration and increase the volume to 600 μl. Incubate for 30 min, gently resuspending the BND cellulose every 10 min. Collect the flow-through to verify later that all DNA has been absorbed by BND cellulose.
5. Wash two times with 1 ml Tris-HCl 10 mM pH 8, NaCl 800 mM and collect the two salt elution fractions containing dsDNA.
6. Add to the column 600 μl of caffeine 1.8%, Tris-HCl 10 mM pH 8, NaCl 1 M and incubate for 10 min. Collect the caffeine elution fraction containing the RI.

Aliquots of the different fractions can be checked using standard gel electrophoresis conditions to ensure that the purification of replication intermediates has been worked properly. Note that most of the DNA loaded into the column will be eluted in the salt elution and a smaller amount in the caffeine elution.

Purification of Replication Intermediates by Size-Exclusion Column

The final step of purification of replication intermediates is performed using a size-exclusion column (Microcon YM-100, Millipore, USA).

1. Load the caffeine elution fraction, obtained from BND cellulose purification, into the size-exclusion column. Spin 5 min at 5000 rpm in Eppendorf (maximum speed). Collect the flow-through to verify later that all DNA has been absorbed by the column.
2. Wash the column 2–3 times with c.a. 200 μl TE buffer. Spin the column for few minutes at no more than 3000 rpm in Eppendorf.
3. Recover the DNA in c.a. 10 μl of TE.

Check the concentration of purified replication intermediates in a standard mini-gel electrophoresis and adjust the final volume with TE if necessary.

Yeast Molecular Combing Protocol

The molecular combing represents a powerful method to study DNA replication at the level of single molecules. Using this approach, the DNA fibers are stretched and aligned on a glass surface by the force exerted by a receding air/water interface and then directly visualized using immunofluorescence techniques. Since a relevant number of DNA molecules are treated at the same time on a single coverslip, reliable measurements can be rapidly obtained (Bensimon *et al.*, 1994).

Molecular combing has been firstly used for the high resolution mapping of genetic alteration in the human genome, but recently it has also revealed an attractive system to study DNA replication (Herrick and Bensimon, 1999; Herrick *et al.*, 2000). In fact, also in yeast, the direct labeling of replicating DNA sequence, using modified nucleotides such as BrdU, allows origins of DNA replication to be visualized and mapped at a genome wide level (Fig. 4). Further, the size of replication bubbles and the interorigin space can be also determined and compared in different genetic backgrounds.

Cell Background

Yeast cells are unable to incorporate exogenous nucleosides into DNA for two main reasons. First yeast cells lack the enzyme required to phosphorylate deoxyribonucleosides, converting them to $5'$ deoxyribonucleosides monophosphates. Second they are unable to uptake exogenous deoxyribonucleosides through the cell wall. The inability to phosphorylate

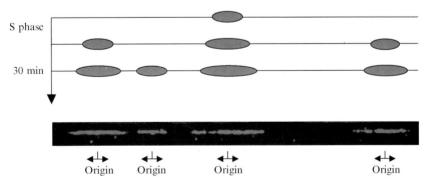

Fig. 4. Visualization of *S. cerevisiae* replicons by molecular combing. Yeast DNA samples, treated as described in the text, were collected 30 min after the release from G1 block. The schematic diagram shows the different timing of origin firing during S phase. (See color insert.)

can be bypassed by expressing the thymidine kinase (TK) from herpes simplex virus, while the uptake problem can be circumvented by expressing the human nucleoside transporter (hENT) (Lengronne *et al.*, 2001; Vernis *et al.*, 2003). For this reason before starting you need to be sure that your background is appropriately modified.

All the labeling procedures described here are performed using cells expressing both TK and hENT from the Gal1 promoter on multicopy plasmids (Vernis *et al.*, 2003). Note that in order to maintain the plasmids and to promote the expression of the genes cells must grow in selective medium in the presence of galactose as the carbon source.

Cell Growth

Exponentially growing cells must be used to efficiently label the replicons. You can directly label logarithmically growing cells or presynchronize the cells in G1 and then release them from the G1 block. This last condition is more likely indicated if you wish to enrich for clear replicon signals. We suggest to use α-factor to synchronize the cells. Moreover, it is important to slow down S-phase progression by lowering the temperature to 25° to avoid merged signal between the two forks. In the procedure described here cells were grown and presynchronized in G1 at 25° in untreated condition.

2.5×10^7 cells for each plug are recommended: this is the optimal concentration to have the right amount of DNA fibers on one coverslip. Ideally, at least 5 plugs for each sample should be used (hence, for each time point, 5 ml of culture at a concentration of 2.5×10^7 cells/ml).

DNA Labeling

This procedure gives information on the replicon size, the origin position, the timing of origin activation, and the inter-origin distance (Lebofsky and Bensimon, 2006). It is important to note that if you wish to compare the replicon size between two different strains, for example, it is crucial that both strains activate the origins at the same time. This is sometimes a problem with certain replication mutants that exhibit large cells and, once presynchronized by α-factor, activate the origins of replication slightly earlier than wild-type cells. Hence it might be important to check the timing of origin firing by 2D gels prior to the molecular combing analysis. If the mutant background exhibits a premature origin firing after the release from the G1 block, it would be better to presynchronize cells using other procedures, such as nocodazole treatment, which has the advantage of minimizing the difference in cell size of wild-type and mutant cells.

Details on nocodazole synchronization procedure have been already described (see for example Day *et al.*, 2004).

Solutions

- 10 m*M* CldU stock, dissolve the powder in 1× PBS and store small aliquots (2 ml) at −20°
- 10 m*M* IdU stock dissolve the powder in 1× PBS/ 2.4% DMSO and store a 4°
- 50 m*M* EDTA pH 8.0 chilled out before use.

Procedure

1. Before releasing the cells from the G1 block, add 100 μM of IdU (or CldU), so that the cells have already incorporated the base analogues before entering into S phase.
2. Release the cells in fresh medium containing 100 μM of IdU.
3. After 30–35 min collect the cells into prechilled falcon tubes and treat them with sodium azide 0.1% final concentration to prevent further movement of replication forks.
4. Spin down the cells by centrifugation at 4000 rpm for 5 min at 4°.
5. Resuspend cells in ice with 1 ml of cold 50 m*M* EDTA pH 8.0 and transfer them in 1.5-ml Eppendorf tube.
6. Spin down the cells by centrifugation at 14,000 rpm for 3 min at 4°.
7. Wash one more time with 1 ml of cold 50 m*M* EDTA pH 8.0.
8. Spin down the cells by centrifugation at 14,000 rpm for 3 min at 4° and be sure to remove all the EDTA with a pipette, leave the pellet in ice.

Note that the time of the labeling can be changed depending on the relative S-phase progression of your strain and on your experimental conditions. Since the technical resolution of the replicon is around 3 kb, shorter pulses are not recommended.

DNA Isolation in Agarose Plugs

Materials and Solutions

- 1% low melting point agarose (LMP) in 0.5 *M* EDTA pH 8.0 store at RT
- 10 mg/ml zymolyase stock (1000 U/ml)
- 20% sarkosyl stock store RT
- 10 mg/ml proteinase K stock
- SCE Solution: 1 *M* sorbitol, 0.1 *M* sodium citrate, 0.06 *M* EDTA pH 8.0

- Solution I: SCE, 0.2% β-mercapto ethanol, 1 mg/ml zymolyase (100 U/ml)
- Solution II: 0.5 M EDTA pH 8.0, 1% sarkosyl, 1 mg/ml proteinase K
- 1× TE pH 8.0.
- Plugs cast from Bio-Rad.

Procedure

1. Melt 1% LMP agar and store in a bath at 50°.
2. Resuspend the cells in Solution I (50 μl for each plug).
3. Add an equal volume of 50° molten LMP agarose and mix with a pipette. This step needs to be done quickly by putting the Eppendorf in a water bath at 50°; in this way the agarose does not solidify.
4. Cover the bottom of the plug cast with tape.
5. Fill plug cast with cell/agarose mix (approximately 90 μl per plug).
6. Put the cast at 4° for 20–30 min to allow the solidification of the plugs.
7. Eject plugs in a 50-ml Falcon tube and cover them with Solution I; generally, calculate around 0.5 ml for plugs.
8. Leave at 37° for 1 h.
9. Gently remove Solution I and wash the plugs with an abundant volume of 0.5 M EDTA pH 8.0.
10. Resuspend the plugs within Solution II (0.5 ml/plug).
11. Leave overnight at 37°.
12. Discard Solution II and wash 3 times with an abundant volume of 1× TE pH 8.0: you can fill the falcon tube with 1× TE pH 8.0 and wash the plugs directly in there. It is important to wash very well to eliminate the detergent and cell debris that might interfere with the combing.
13. Transfer the plugs that you wish to analyze in a new 50-ml Falcon tube and wash for 2 h with 1× TE pH 8 at RT on a rotating wheel. Since labeled DNA is sensitive to light wrap the tube in aluminum foil during the washing steps and to store the samples.
14. Transfer the plugs that you do not analyze immediately in 15-ml Falcon tubes and cover them with 0.5 M EDTA pH 8.0. Blocks can be stored indefinitely at 4°.

DNA Combing

Materials and Solutions

- 0.5 MES (2-morpholinoethanesulfonic acid) pH 6.5 stock, autoclave and store at 4°
- β-agarase (New England Biolabs, Ipswich, MA)

- YOYO-1 from Molecular Probes (Eugene, OR). Dissolve 2 μl of YOYO-1 into 2 ml of 0.1 *M* RES pH 5.5.
- Teflon reservoirs and 22 mm by 22 mm silanized coverslips are made at the Institute Pasteur (contact Aaron Bensimon at abensim@pasteur.fr).
- The combing machine is made at the Institute Pasteur, (contact Aaron Bensimon abensim@pasteur.tr).

Procedure

1. After washing the plugs with 1× TE, transfer the blocks into a sterile 2 ml Eppendorf tube. Carefully remove eventual drops of TE and resuspend the plugs into 0.1 *M* MES pH 6.5 solution to a final volume of 1.5 ml (consider that each plug is 100 μl).

2. Heat the block at 72° for 20 min to melt the LMT agarose.

3. Transfer the tube at 42° in a water bath for 10 min.

4. To digest agarose add 2 μl of β-agarase for each plug and incubate overnight at 42°.

5. Gently pour the DNA solution into a Teflon combing reservoir; you can store the DNA solution at RT in the dark up to one month but always cover the reservoirs with parafilm to avoid evaporation.

6. Using the combing machine insert the silanized coverslip into the reservoir containing the DNA. The machine is set up to remove the coverslip after approximately 5 min. During removal of the coverslip from the reservoir, the meniscus moving along the hydrophobic surface is combing the DNA. The DNA fibers are then stretched on the silanized coverslip at a speed of 300 μm/sec, for more details see Lebofsky and Bensimon (2005).

7. After combing the DNA on the silanized coverslip repeat step 6 but this time lower the coverslip into a reservoir containing YOYO-1 in order to stain the DNA and have a quality check of the DNA preparation.

8. Use superglue to stick the coverslip to a slide and visualize the YOYO-1 stained DNA by the microscope.

9. If the DNA is well stretched, at the right concentration with enough fibers and no bundles, you can proceed combing more coverslips in order to detect the DNA replication.

10. Heat the combed DNA at 65° for 1 h.

11. Use superglue to stick the coverslip to a slide and store at −20°, avoid touching the surface with the DNA.

Note that the DNA in solution is very fragile, hence avoid physical agitation to preserve the integrity of the DNA fibers.

Fluorescent Detection of DNA Replication

Materials and Solutions

- NaOH 0.5 *M*
- 1× PBS
- Detection buffer (DB): Blocking reagent 1% in 1× PBS, 0.05% Tween20, autoclave, and store 1 ml aliquots at −20°
- Primary antibody: IdU→FITC conjugated Mouse anti BrdU (from Becton-Dickinson, USA), CldU→Rat anti BrdU (from Sera Laboratories, UK)
- Secondary antibody: IdU→Alexa 488 Goat anti-mouse (from Molecular Probes), CldU→Alexa 594 Donkey anti-rat (from Molecular Probes)
- Vectalshield from Vector Laboratories.

Procedure

1. Remove the slides to be used from −20° and leave at RT for 1 h
2. Place the slides in 0.5 *M* NaOH for 20 min to denature the DNA
3. Rinse 3 times with 1× PBS each for 3 min, to neutralize the pH
4. While rinsing, prepare a mix of primary antibody in the detection buffer (DB). Each slide will be covered with 25 μl of antibody/DB mix; the dilution of the antibodies are 2.7 for the FITC conjugated mouse anti BrdU and 1:20 for the rat anti BrdU.
5. Place the antibody on the slide and cover with a glass coverslip.
6. Immediately put the slide in a humid chamber and leave at RT for 1 h.
7. Dip the slide into a Becker containing 1× PBS in order to gently remove the glass coverslip.
8. Wash the slide 3 times in 1× PBS for 3 min each.
9. While washing, prepare the secondary antibody mix in DB. Again, each slide will be covered with 25 μl of antibody/DB mix, the dilution for each antibody is 1:50.
10. Place the antibody on the slide, cover with a glass coverslip, and put the slides into the humid chamber at 37° for 20 min.
11. Remove the coverslip as described previously and wash 3 times with 1× PBS for 3 min each. Mount slides in Slowfade light antifade mounting media (Vector Laboratories, USA) and seal with nail polish.

At this point the slides are ready for microscope analysis.

Fluorescent signals were examined with a Zeiss Axioplan 2 microscope equipped with a 75-W xenon arc lamp; photos were taken using Smartcapture

2 software (Digital Scientific, UK) and a Photometrics HQ Coolsnap CCD camera.

All the replicon length measurements are done using cartographiX, a software developed in Aaron Bensimon's lab.

Acknowledgments

Work in M. F.s lab is supported by the Associazione Italiana per la Ricerca sul Cancro. C. C. R. was a recipient of an EMBO short-term fellowship. We thank all members of our laboratories for useful discussion and criticism.

References

Allers, T., and Lichten, M. (2000). A method for preparing genomic DNA that restrains branch migration of Holliday junctions. *Nucleic Acids Res.* **28**, e6.

Bénard, M., Maric, C., and Pierron, G. (2001). DNA replication-dependent formation of joint DNA molecules in *Physarum polycephalum*. *Mol. Cell* **7**, 971–980.

Bensimon, A., Simon, A., Chiffaudel, A., Croquette, V., Heslot, F., and Bensimon, D. (1994). Alignment and sensitive detection of DNA by a moving interface. *Science* **265**, 2096–2098.

Brewer, B. J., and Fangman, W. L. (1987). The localization of replication origins on ARS plasmids in *S. cerevisiae*. *Cell* **51**, 463–471.

Cha, R. S., and Kleckner, N. (2002). ATR homolog Mec1 promotes fork progression, thus averting breaks in replication slow zones. *Science* **297**, 602–606.

Cobb, J. A., Bjergbaek, L., Shimada, K., Frei, C., and Gasser, S. M. (2003). DNA polymerase stabilization at stalled replication forks requires Mec1 and the RecQ helicase Sgs1. *EMBO J.* **22**, 4325–4336.

Cotta-Ramusino, C., Fachinetti, D., Lucca, C., Doksani, Y., Lopes, M., Sogo, J., and Foiani, M. (2005). Exo1 processes stalled replication forks and counteracts fork reversal in checkpoint-defective cells. *Mol Cell.* **17**, 153–159.

Day, A., Schneider, C., and Schneider., B. L. (2004). Yeast cell synchronization. *Methods Mol. Biol.* **241**, 55–76.

Gruber, M., Wellinger, R. E., and Sogo, J. M. (2000). Architecture of the replication fork stalled at the 3′ end of yeast ribosomal genes. *Mol. Cell. Biol.* **20**, 5777–5787.

Haber, J. E., and Heyer, W. D. (2001). The fuss about Mus81. *Cell* **107**, 551–554.

Hecht, A., and Grunstein, M. (1999). Mapping DNA interaction sites of chromosomal proteins using immunoprecipitation and polymerase chain reaction. *Methods Enzymol.* **304**, 399–414.

Herrick, J., and Bensimon, A. (1999). Single molecule analysis of DNA replication. *Biochimie* **81**, 859–871.

Herrick, J., Stanislawski, P., Hyrien, O., and Bensimon, A. (2000). Replication fork density increases during DNA synthesis in *X. laevis* egg extracts. *J. Mol. Biol.* **300**, 1133–1142.

Lebofsky, R., and Bensimon, A. (2006). Fluorescent visualisation of genomic structure and DNA replication at the single molecule level. *In* "Cell Biology: A Laboratory Handbook," (J. E. Celis, ed.), pp. 429–439. Elsevier Science (USA).

Lengronne, A., Pasero, P., Bensimon, A., and Schwob, E. (2001). Monitoring S phase progression globally and locally using BrdU incorporation in TK(+) yeast strains. *Nucleic Acids Res.* **29**, 1433–1442.

Liberi, G., Maffioletti, G., Lucca, C., Chiolo, I., Baryshnikova, A., Cotta-Ramusino, C., Lopes, M., Pellicioli, A., Haber, J. E., and Foiani, M. (2005). Rad51-dependent DNA structures accumulate at damaged replication forks in sgs1 mutants defective in the yeast ortholog of BLM RecQ helicase. *Genes Dev.* **19,** 339–350.

Lilley, D. M. J., and White, M. F. (2001). The junction-resolving enzymes. *Nat. Rev. Mol. Cell. Biol.* **2,** 433–443.

Lopes, M., Cotta-Ramusino, C., Pellicioli, A., Liberi, G., Plevani, P., Muzi-Falconi, M., Newlon, C. S., and Foiani, M. (2001). The DNA replication checkpoint response stabilizes stalled replication forks. *Nature* **412,** 557–561.

Lopes, M., Cotta-Ramusino, C., Liberi, G., and Foiani, M. (2003). Branch migrating sister chromatid junctions form at replication origins through Rad51/Rad52-independent mechanisms. *Mol. Cell* **12,** 1499–1510.

Lucca, C., Vanoli, F., Cotta-Ramusino, C., Pellicioli, A., Liberi, G., Haber, J., and Foiani, M. (2004). Checkpoint-mediated control of replisome-fork association and signalling in response to replication pausing. *Oncogene* **23,** 1206–1213.

Panyutin, I. G., and Hsieh, P. (1994). The kinetics of spontaneous DNA branch migration. *Proc. Natl. Acad. Sci. USA* **91,** 2021–2025.

Rothstein, R., Michel, B., and Gangloff, S. (2000). Replication fork pausing and recombination or "gimme a break" *Genes Dev.* **14,** 1–10.

Sogo, J. M., Lopes, M., and Foiani, M. (2002). Fork reversal and ssDNA accumulation at stalled replication forks owing to checkpoint defects. *Science* **297,** 599–602.

Sogo, J. M., and Thoma, F. (1989). Electron microscopy of chromatin. *Methods Enzymol.* **170,** 142–165.

Vernis, L., Piskur, J., and Diffley, J. F. (2003). Reconstitution of an efficient thymidine salvage pathway in *Saccharomyces cerevisiae*. *Nucleic Acids Res.* **31,** e120.

Wellinger, R. E., Schär, P., and Sogo, J. M. (2003). Rad52-independent accumulation of joint circular minichromosomes during S phase in *Saccharomyces cerevisiae*. *Mol. Cell. Biol.* **23,** 6363–6372.

Zou, H., and Rothstein, R. (1997). Holliday junctions accumulate in replication mutants via a RecA homolog-independent mechanism. *Cell* **90,** 87–96.

[27] Analysis of Gross-Chromosomal Rearrangements in *Saccharomyces cerevisiae*

By Kristina H. Schmidt, Vincent Pennaneach, Christopher D. Putnam, and Richard D. Kolodner

Abstract

Cells utilize numerous DNA metabolic pathways and cell-cycle checkpoints to maintain the integrity of their genome. Failure of these mechanisms can lead to genome instability, abnormal cell proliferation, and cell death. This chapter describes a method for the measurement of the rate of accumulating gross-chromosomal rearrangements (GCRs) in haploid cells of the yeast *Saccharomyces cerevisiae*. The isolation of cells with GCRs relies on the simultaneous loss of two counterselectable markers, *CAN1*

METHODS IN ENZYMOLOGY, VOL. 409
0076-6879/06 $35.00
DOI: 10.1016/S0076-6879(05)09027-0

and *URA3*, within a nonessential region on the left arm of chromosome V. Healing of DNA breaks by *de novo* telomere addition, translocations, large interstitial deletions, and chromosome fusion has been detected using a PCR-based procedure for the mapping and amplification of breakpoint junctions, which is also described in detail here. This GCR analysis provides an effective tool for the assessment of the contribution by multiple cellular mechanisms to the maintenance of genome integrity.

Introduction

Stable inheritance of genetic information is essential for cell survival and normal growth. Recent genetic studies of the yeast *S. cerevisiae* have implicated numerous genes in the maintenance of genome integrity, including those that function in cell-cycle checkpoints, DNA replication, DNA mismatch repair, recombination, oxidant defense systems, and telomere maintenance (Huang and Kolodner, 2005; Myung and Kolodner, 2002; Myung *et al.*, 2001a, b; Pennaneach and Kolodner, 2004). Cells with defects in these pathways can generally be divided into those that have suffered small changes of only one or a few base pairs, such as point mutations and frameshifts, and those that exhibit gross chromosomal rearrangements (GCRs), such as translocations, chromosome fusions, duplications, large interstitial deletions, or terminal deletions with *de novo* telomere additions (reviewed in Kolodner *et al.*, 2002).

Mutations that disrupt the replication checkpoint (*rfc5–1*, *dpb11–1*) significantly increase the rate of accumulating GCRs (200-fold) (Myung *et al.*, 2001b). Mutations that inactivate either one of the two branches of the intra-S checkpoint, the *SGS1*-dependent branch or the *RAD24/MEC3*-dependent branch, cause moderate increases in the genome rearrangement rate (10- to 20-fold), whereas simultaneous inactivation of both branches causes a higher increase (100-fold) (Myung and Kolodner, 2002). An even larger increase in the GCR rate (1000- to 12,000-fold) is seen upon inactivation of both the intra-S checkpoint and the replication checkpoint, indicating extensive redundancy in cell-cycle checkpoints with regard to the maintenance of genome stability (Myung and Kolodner, 2002).

Yeast cells that have a reduced ability to inactivate reactive oxygen species (Δ*tsa1*) exhibit an increased GCR rate (50-fold), which is further elevated by disabling the *SGS1*-dependent branch of the intra-S checkpoint (300-fold). Similarly, cells that lack Ogg1, a purine-specific enzyme that plays an important role in repairing oxidized bases in DNA (7,8-dihydro-8-oxoguanine), have an increased GCR rate and combining *ogg1* and *tsa1* mutations results in a synergistic increase in the GCR rate (Huang and Kolodner, 2005; Huang *et al.*, 2003).

In the absence of a functional Tel1-dependent checkpoint telomerase-inactivating mutations, such as Δ*tlc1* or Δ*est2,* also lead to elevated GCR rates (100-fold), suggesting that Tel1-dependent recognition of shortening telomeres and the activation of appropriate alternative telomere maintenance pathways may prevent GCRs (Pennaneach and Kolodner, 2004). Tel1 may also be important for the suppression of error-prone repair pathways, such as nonhomologous endjoining (NHEJ), which may lead to GCRs in cells with defective telomeres.

Deletion of homologous recombination genes (*RAD51, RAD54, RAD55, RAD57, RAD59*) causes only a small GCR rate increase (5- to 30-fold) (Chen and Kolodner, 1999), whereas combinations of these mutations, growth at a lower temperature (Δ*rad55,* Δ*rad57*) or deletion of *RAD52* causes a larger GCR rate increase (100- to 300-fold) (Myung *et al.,* 2001a). Yeast strains with mutations that disrupt the NHEJ pathway (Δ*yku70,* Δ*yku80,* Δ*lig4*) exhibit a very low GCR rate similar to that seen in

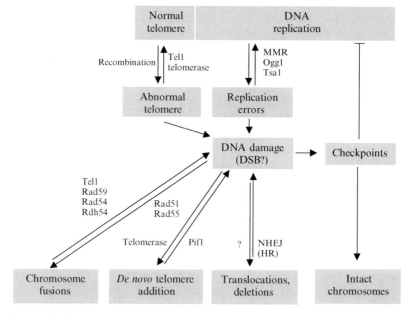

FIG. 1. Multiple pathways function to maintain genome stability in *S. cerevisiae.* Replication errors and telomere degradation lead to DNA damage and activation of checkpoint sensors, which in turn activate nonmutagenic responses that suppress DNA instability. Mutagenic DNA repair pathways can lead to translocations and large deletions (nonhomologous endjoining, homologous recombination), chromosome fusions (homologous recombination), or *de novo* addition of telomeric sequences (telomerase); the latter type of GCR is suppressed by Pif1.

wild-type cells (Chen and Kolodner, 1999). Deletion of *MRE11*, *XRS2*, or *RAD50*, which encode the three subunits of the MRX-DNA-repair complex with functions not only in DNA recombination, but also in cell-cycle checkpoints and telomere maintenance, leads to one of the strongest GCR rate increases observed thus far for a single gene (600-fold) (Chen and Kolodner, 1999). Taken together, these findings reveal that normal cells have an extensive network of redundant pathways at their disposal to suppress GCRs (Fig. 1).

Selection of Yeast Cells with Gross–Chromosomal Rearrangements

Several assays exist in *S. cerevisiae* to study genome rearrangements between a naturally occurring DNA sequence and an extra (artificial) copy of this sequence that has been inserted into an unrelated chromosome. Genome rearrangements detected in these assays are promoted almost exclusively by mitotic recombination and the homology between the two distant sequences. Recently, this laboratory has developed an assay that allows detection of a broader spectrum of genome rearrangements, particularly those that are not necessarily generated by homologous recombination between extended regions of homology. This gross-chromosomal rearrangement (GCR) assay involves measuring the rate of rearrangements of the nonessential portion of the left arm of chromosome V in haploid cells, which was modified by replacing the nonessential gene *HXT13* located distal to the *CAN1* gene with a second selectable marker, the *URA3* gene (Fig. 2). This modified yeast strain, RDKY3615 (*ura3–52, leu2Δ1, trp1Δ63, his3Δ200, lys2ΔBgl, hom3–10, ade2Δ1, ade8, hxt13:: URA3*), and its derivatives are available upon request. Cells expressing

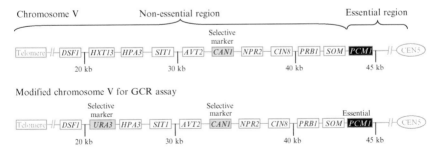

FIG. 2. Chromosome V modifications for the GCR assay. The nonessential *HXT13* gene on the left arm of chromosome V is replaced with the *URA3* cassette, resulting in a chromosome V with two counterselectable markers, *CAN1* and *URA3*, on the same chromosome arm. The location of the first essential gene, *PCM1*, is indicated by a black box.

the *CAN1* gene are sensitive to the arginine analog L-canavanine, which is found in the legume *Dioclea megacarpa* and may be converted to L-canaline, which inhibits the Krebs cycle. Yeast cells that express functional orotidine-5'-phosphate decarboxylase from the *URA3* gene are sensitive to 5-fluorotic acid (5-FOA) due to its conversion to toxic 5-fluorouracil. Thus, cells in which both, *CAN1* and *URA3*, have been inactivated can be obtained by selection on agar (GCR) plates containing 60 mg/liter L-canavanine and 1 g/liter 5-FOA (US Biological, Swampscott, MA). In addition, the selection plates contain 7 g/liter amino acid free yeast nitrogen base (Becton Dickenson, Franklin Lakes, NJ), 20 g/liter glucose, 50 mg/liter uracil, and 2 g/liter synthetic dropout mix without uracil and arginine (US Biological, Swampscott, MA). Prior to selection on solid media, liquid media consisting of 20 g/liter glucose, 10 g/liter yeast extract (Fisher Scientific, Hampton, NH), and 20 g/liter Bacto-peptone (Becton Dickenson, Franklin Lakes, NJ) is inoculated with an entire single colony ($\sim 10^7$ cells) of the yeast strain of interest and incubated with vigorous shaking at 30°. Typically, culture volumes of 5 ml or 10 ml are used, but the volume may have to be increased to 15–100 ml if the GCR rate of the strain is low. Once the culture has reached saturation (2 to 3 days depending on the doubling time of the yeast strain) all cells are harvested by centrifugation, resuspended in 0.5–2 ml of sterile, distilled H_2O, and spread onto GCR plates (150 × 15 mm), using aliquots of approximately 10^9 cells per plate. Yeast cells with chromosome V rearrangements can be harvested from GCR plates after incubation at 30° for 4 days. A variation of this assay using markers on chromosome VII has also been described (Myung *et al.*, 2001b); note that many yeast chromosomes have long, up to 80 kb, nonessential termini that can in principal be exploited for these kinds of assays.

Determining the Rate of Accumulating Gross–Chromosomal Rearrangements

In the absence of selective pressures, every cell has a non-zero chance of accumulating a mutation during its lifetime. Each of these cells will give rise to progeny that also possess this mutation; however, the randomness of when mutations occur and the dramatic impact of mutations that occur early in the culture cause the resulting number of mutants per culture to be quite variable. Together these factors make the trivial measurements of mutation frequencies (the number of mutants in a population divided by the size of the population) and any statistical measures based on averages inaccurate. What one would like to know is the mutation rate: that is the number of mutational events (m) divided by the total number of cells (N). Mutation rates are normally expressed in units of events per cell per generation. The most common method of determining rates is fluctuation analysis in which multiple parallel cultures are grown from clones of a single cell. The total number of both mutants (r) and cells (N) are measured in each culture. From this distribution, estimates of m, and hence the mutation rate, can be calculated.

The original estimate for m was from (Luria and Delbrück, 1943), who derived $m = \ln(C/Z)$ from the Poisson distribution, where C is the total number of cultures and Z is the number of cultures that lack colonies. Importantly, this method of determining m (the P_o method) is only appropriate when about half of the cultures lack any mutation. Lea and Coulson (1949) extended the analysis for experiments in which most or all cultures had mutations by empirically determining that the probability of cultures having r mutations or fewer per culture was 50% (e.g., the median) when Eq. (1) held.

$$m(1.24 + \ln[m]) - r = 0 \qquad (1)$$

Determining m is a simple problem in finding the root of a one-dimensional nonlinear equation. The equation itself is well-behaved and can be solved by a variety of iterative methods including the bisection and Newton-Raphson methods, which are easily implemented as macros for spreadsheets or independent programs. Other, more modern calculations for m are also available (reviewed in Rosche and Foster, 2000), but none of these are as heavily used as the Lea and Coulson calculation.

Thus, in the classical Lea and Coulson treatment, median m (m_{med}) is calculated from the median number of mutant colonies (r_{med}). Additionally, it is assumed that the ending population size for each culture is equal, so the number of colonies per culture is averaged from all of the cultures ($<N>$). The rate is then simply $m_{med}/<N>$. To calculate statistical distributions, such as 95% confidence intervals (CI), confidence intervals are calculated on the number of observed colonies (r) and then m values and rates are calculated from the r values defining the lower and upper bounds of the 95% CI.

In practice, we use a slightly modified Lea and Coulson method, as the original method makes it difficult to use many statistical tests to establish the significance of differences in rates between different isolates or mutant backgrounds. Thus, we determine rates from each individual culture (calculating a m from each culture's r and dividing by the number of cells in that culture) and use the median rate from all of the cultures. Then 95% CI can be calculated from the rates themselves, and statistical significances of any potential differences can be estimated by nonparametric tests such as the Mann-Whitney test. If all cultures are grown to identical densities, this method is identical to the classical Lea and Coulson treatment; therefore, it is important to ensure that no cultures have substantially different final density as the validity of calculating m has not been well established under these conditions.

Mapping and Sequencing of Chromosome V Breakpoints

Principle

Breakpoints that lead to inactivation of *CAN1* and *URA3* have to occur within the 12.1 kb region between (or within) *CAN1* and *PCM1*, the first essential gene on the left arm of chromosome V (Fig. 2). Before the exact nucleotide sequence of the breakpoint can be determined by sequencing, a set of 22 primer pairs, each designed to amplify overlapping fragments of about 500–600 bp, is utilized to determine the approximate location of the breakpoint within the 12.1 kb region of chromosome V in each Can^r 5-FOA^r clone. Since all mapping primers are specific to a small region of the left arm of chromosome V, the DNA fragment containing the break-point (and all fragments telomeric of it) will not yield a PCR product (Fig. 3), narrowing down the breakpoint location to within 500–600 bp of chromosome V. In the next step, the breakpoint junction is amplified by 2 rounds of nested, arbitrary-primed PCR using oligonucleotides that anneal to the region of the last successful breakpoint mapping PCR reaction. Nested PCR reaction products are analyzed by agarose gel elec-trophoresis and PCR products specific to a rearranged chromosome V are then purified and sequenced. Note that an alternate method for mapping GCRs has also been published (Smith *et al.*, 2004).

Procedure

DNA Extraction. Genomic DNA from a 1-ml YPD culture of a Can^r 5-FOA^r clone of interest ($\sim$1.5 $\times$ 10^8 cells) is purified using the Puregene DNA Isolation Kit (Gentra Systems, Minneapolis, MN). Routinely 5 μg of genomic DNA is purified using this protocol. The DNA is rehydrated overnight in 150 μl of Tris–EDTA (10 mM Tris, 1 mM EDTA, pH 8) and supplemented with RNase A to final concentration of 0.7 units/ml.

Breakpoint Mapping. For each clone 22 PCR reactions are performed to amplify the entire region of chromosome V between nucleotides 31759 and 43877, according to the *Saccharomyces* Genome Database (SGD) coordinates. Each 12.5 μl PCR reaction contains 1.25 μl of 10$\times$ PCR buffer PC2 (final concentration 50 mM Tris-HCl pH 9.1, 16 mM ammonium sulfate, 3.5 mM MgCl$_2$, 150 μg/ml BSA) from Ab-peptide Inc. (St. Louis, MO), 0.125 μl dNTP mix (final concentration at 20 μM for each

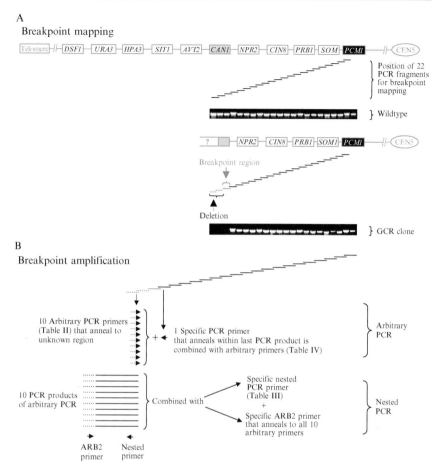

FIG. 3. Procedure for mapping and amplification of breakpoint junctions on chromosome V of *S. cerevisiae*. (A) Breakpoint mapping; 22 PCR reactions are used to amplify the 12-kb region between PCM1 and CAN1. All 22 PCR products are obtained in a cell without chromosome V rearrangements (wild-type), whereas PCR products are not obtained if a rearrangement has occurred (GCR clone), indicating the approximate location of the breakpoint. (B) Breakpoint amplification; a specific primer complementary to the last amplifiable PCR fragment is combined in 10 separate PCR reactions with 10 primers containing short arbitrary sequences (Arbitrary PCR). An aliquot of every PCR reaction is then used as template for a second round of PCR amplification (Nested PCR) using the ARB2 primer, which anneals to the 10 arbitrary primers, and an appropriate nested primer, which anneals to the last amplifiable mapping PCR product (see Table III for primer combinations). PCR products are analyzed on an agarose gel and GCR clone specific PCR products are purified and sequenced.

nucleotide), 2 μl genomic DNA ($\sim$60 ng), 0.0625 μl of Klen Taq DNA polymerase (25 units/μl) from Ab-peptide Inc., and 0.125 μl of each of two breakpoint mapping primers (Table I). PCR amplification is carried out in a MJ Research DNA Engine Tetrad® Thermal Cycler (Biorad, Hercules, CA) in 96-well plates, starting with an initial denaturation at 95° for 5 min and followed by 35 cycles of 30 s at 95°, 30 s at 55°, and 1 min at 68°. Three microliters of each of the 22 PCR reactions is analyzed on a 2% TAE agarose gel with Ethidium Bromide, run at 120 volts for 12 min.

Breakpoint Junction Amplification and Sequencing. Breakpoint junctions are amplified by a first round of arbitrary-primed PCR using a set of 10 arbitrary primers (Table II) in combination with a chromosome V specific primer that anneals to the region of the last successful mapping PCR reaction (Tables III and IV). Each PCR reaction is carried out in a total volume of 25 μl containing 2.5 μl of 10× PC2 buffer, 2.5 μl dNTP mix (final concentration at 20 μM for each nucleotide), 4 μl genomic DNA ($\sim$200–300 ng), 0.125 μl of Klen Taq DNA polymerase (25 units/μl), 0.2 μM of one arbitrary primer, and 0.2 μM of the chromosome V specific primer as determined by the breakpoint mapping PCR. Amplifications are carried out in a MJ research DNA engine Tetrad® Thermal cycler in 96-well plates. An initial denaturation step at 95° for 5 min is followed by 5 cycles of 30 s at 95°, 30 s at 25°, 2.5 min at 68°, and followed by 30 cycles of 30 s at 95°, 30 s at 50°, and 2.5 min at 68°.

To increase specificity and enhance yield, a second round of nested PCR is performed for each one of the 10 arbitrary PCRs using primer ARB2, which is complementary to the 5' tail of all arbitrary primers used in the first round of PCR, in combination with a nested, chromosome V specific primer that anneals proximal to the breakpoint relative to the chromosome V primer used in the arbitrary PCR reaction (Tables III and IV). The DNA sequence surrounding the breakpoint junction is amplified by adding 1 μl of PCR product from one of the 10 arbitrary PCRs to a 24 μl PCR mix containing 2.5 μl of 10× PC2 buffer, 20 μM of each dNTP, 3.125 U of Klen Taq DNA polymerase, 0.2 μM ARB2 primer, and 0.2 μM of the chromosome V specific primer. PCR amplification is performed using an initial denaturation step of 5 min at 95° and 35 cycles of 30 s at 95°, 30 s at 50° and 2.5 min at 68°. Of each PCR reaction, 1/8 is loaded on a 2% TAE agarose gel with Ethidium Bromide and electrophoresis is carried out at 120 volts for 20 min. Both PCR reactions, arbitrary and nested, are performed with genomic DNA prepared from the Can^r 5-FOA^r clone of interest as well as from the wild-type strain so that nested PCR reaction products from the wild-type strain and the CAN^r 5-FOA^r clone can be compared side-by-side on the agarose gel and only PCR products that are

TABLE I
SEQUENCES OF MAPPING PRIMERS

Primer name	Primer sequence
63A5	5′-GCCTCAATGTCTCTTCTATCGGAAT-3′
63A3	5′-AACACAATCTACTTCCTACGTTTCT-3′
63B5	5′-GCGGCAGAAATAATGGTTGTTAAGA-3′
63B3	5′-TTACATGTATTGGTTTTCTTGGGCA-3′
63C5	5′-ACCCAAAAAATACTAATCCATGCCG-3′
63C3	5′-CAGACTTCTTAACTCCTGTAAAAACA-3′
63D5	5′-GAAGCTTCAACGTCGTGAAAGAGGG-3′
63SEQ-KJ1	5′-AATGCATAAAAATAGTTACAATTAA-3′
63E5	5′-CCTATTGAAAATTCGGACACTTTAG-3′
63E3–2	5′-CATACTTAAAAGTGATCAACTTGTG-3′
62A5	5′-GTGTTCCATCCTACAGAAGGTTCTA-3′
62A3–2	5′-CAGGGGCACATCTTCTAATGAGACAC-3′
62B5	5′-CAAGATTTATTGATGAGAATTTTCCA-3′
62B3	5′-GCAGGAGGTTTGTGATTGCCTCCAG-3′
62C5	5′-TCAAACCCGGATTCTAGAACTACAT-3′
62DELA*	5′-TGATGGACAATGAATTTCTCTAAT-3′
62D5	5′-TTGGCAACAGGCAAAAGTACTATGC-3′
62D3	5′-ACGCTACCAAAGACGAAATGATCGA-3′
62E5	5′-TGGGCGATATTACACTTTGCACTGA-3′
62E3	5′-TGAATCGTCATGTGGATTCTACTTA-3′
61A5	5′-AGTTGTCAGATTCTGCATTATTAGT-3′
61A3	5′-ATCATTGAAAACTACCATAGATCATTT-3′
61B5	5′-GGGCCATTTGCATTACCTCAGTCAA-3′
61B3	5′-GAGCATTAAATCAACGTGCCAAAGA-3′
61C5	5′-TTTCGATTCACGGAAAGGTATATGG-3′
61C3	5′-TTGATTCAAGCACAGCAAATAATAC-3′
61D5	5′-TTAATAGAGCAGCTTTAGGTGTTAG-3′
61D3	5′-CCAACAGCATCAGTAAAAATGGCAA -3′
61E5	5′-TCATACTAATTTCCCTTTCATTCCTT-3′
61E3	5′-CTTCATTTAGAAAAATTTCAGCTGCT-3′
61F5	5′-GAAATTCAGATTGAAAATAATGGGG-3′
61F3	5′-TTGTAATACCTCCCCAGCTTCTGCT-3′
60A5	5′-AAACCTGGGGCGAAAACGTCGACAC-3′
60A3	5′-TACAGAATTTGACACTCAAAATAGCG-3′
60B5	5′-ACACCGCGACCGGCATCATCATCGT-3′
60B3	5′-AAAGGGCATGAAGCCTAAGCATGAA-3′
60C5	5′-CCTTTCATCTTCTTCTCCTCCACCT-3′
60C3	5′-GGAAATTTAGGTGACTTGTTGAAAAAG-3′
60D5	5′-TCGCGCTCTTTATCGCGGGTGTGTT-3′
60D3	5′-ACAGTAGATAAGACAGATAGACAGA-3′
59A5	5′-GCGCCATGAGGGTTGATTGTCTTAT-3′
58DELA	5′-CACCCCCACATACTGACCCTGAA-3′
59B5	5′-GATTAAGTAAGAACAAAAAACCGTTAA-3′

TABLE II
SEQUENCES OF PRIMERS FOR ARBITRARY PCR

Primer name	Primer sequence[a]
ARB 1	GGCCACGCGTCGACTAGTAC-N_{10}-GATAT
ARB 4	GGCCACGCGTCGACTAGTAC-N_{10}-TATAG
ARB 5	GGCCACGCGTCGACTAGTAC-N_{10}-TATAC
ARB 6	GGCCACGCGTCGACTAGTAC-N_{10}-TATAA
ARB G	GGCCACGCGTCGACTAGTAC-N_{10}-GGGGG
ARB A	GGCCACGCGTCGACTAGTAC-N_{10}-AAAAA
ARB T	GGCCACGCGTCGACTAGTAC-N_{10}-TTTTT
ARB C	GGCCACGCGTCGACTAGTAC-N_{10}-CCCCC
ARB T1	GGCCACGCGTCGACTAGTAC-N_{10}-TGTGT
ARB T2	GGCCACGCGTCGACTAGTAC-N_{10}-ACACA

[a] N_{10} indicates a random sequence of 10 nucleotides composed of any of the 4 nucleotides.

unique to the CAN^r 5-FOA^r clone are purified and sequenced. Selected PCR products are purified by adding 2 U of Shrimp Alkaline Phosphatase (USB, Swampscott, MA), 0.1 U Exonuclease I (USB, Swampscott, MA), and 2 μl of 10× PCR buffer (200 mM Tris-HCl pH 8.0, 100 mM $MgCl_2$), supplied by Perkin Elmer (Wellesley, MA), to 16 μl of the PCR product in a total reaction volume of 20 μl. Incubation at 37° for 20 min is followed by heat inactivation of the enzymes at 80° for 20 min prior to sequencing. Dideoxy-sequencing is carried out using the BigDye® Terminator v3.1 Cycle Sequencing Kit (Applied Biosystems, Foster City, CA) and the chromosome-specific primer used for the nested PCR, followed by analysis on an ABI 3730 × 1 DNA Analyzer (Applied Biosystems, Foster City, CA). To determine the nucleotide coordinates of the chromosome V breakpoint and to identify the rearrangement target, sequences are analyzed by performing BLAST searches against SGD.

Chromosome V Rearrangement Types

The nucleotide sequence of the nested PCR products is used for a BLAST search at SGD to identify the exact nucleotide coordinates of the chromosome V breakpoint and origin of the unrelated DNA fragment fused to this breakpoint.

Many GCRs appear to result from healing of a broken chromosome V by adding a telomeric sequence, either by telomerase-mediated addition of poly [TG_{1-3}] repeats at a short 4-bp seed sequence (*de novo* telomere addition) or by translocation to the subtelomeric or telomeric sequence

TABLE III

SEQUENCES OF PRIMERS FOR ARBITRARY PCR AND NESTED PCR

Primer name	Primer sequence
GCR2R	5'-TGCAAGCTTTGAAATACCGTGG-3'
GCR11R	5'-CTCACAATCATATGGGAATAC-3'
GCR14R	5'-GATGACATGAATATTTGAACTCG-3'
GCR24R	5'-GAGAAATTGAAAGTTTGACATCG-3'
59DELB	5'-CATGACGCACAAGCAAGCAAACAG-3'
GCR1R	5'-TGTTTGGATCTTATTTCAATGC-3'
GCR4R	5'-CTTTATTATTGCTATTGAGAACTCTGG-3'
GCR5R	5'-CCTGGAACTTAGTGTAGTTGGC-3'
GCR6R	5'-ACAAATTCAAAAGAAGACGCCGAC-3'
GCR9R	5'-CAAGAACTTATTTAGAACACG-3'
GCR12R	5'-CTTTATGTCTACTGCATTTCCC-3'
GCR16R	5'-TTATCAAATGTTTCGGCATTCC-3'
GCR19R	5'-CCAAATCTAAGTAAGAAAATGCCG-3'
GCR21R	5'-AATATTGAGAAGAGAACAAACG-3'
GCR25R	5'-CAACCAAAGTCTATTGACGC-3'
GCR27R	5'-CAACAGTGCTAGTAGTTCCAGG-3'
GCR30R	5'-AGGTTGGATGTCACACTGTTCC-3'
GCR32R	5'-CTTTCCTTAGGCGTCTCGAGG-3'
GCR34R	5'-TCGGAGCTTCCACGTTGAGC-3'
GCR36R	5'-ATGGGGGTTGGCCCGTATTTCC-3'
GCR38R	5'-GATGATAAGAAGAAGAAGCCTCACC-3'
GCR40R	5'-CAAAATTTGGGCTCAGTAATGCC-3'
GCR42R	5'-TGGGAGTGCCAATTCAAAGG-3'
GCR44R	5'-CGTCTCTAAGCTTACTGCTGACC-3'

of an unrelated chromosome (telomere capture). Translocations to an unrelated DNA sequence of chromosome V (isochromosomal transloca-tion) or of an unrelated chromosome (interchromosomal translocation) are also commonly observed. Translocations were found by PCR to be nonreciprocal, suggesting that they are generated by mechanisms that copy a portion of the target chromosome onto the end of a broken chromosome V (nonreciprocal translocations); these translocations can exist in both monocentric and dicentric orientations. Chromosome V rearrangements in which the partner chromosome fused to the breakpoint in inverted orientation is a portion of chromosome V are called isoduplications, where-as fusion of the chromosome V breakpoint to telomeric poly $[C_{1-3}A/TG_{1-3}]$ repeats or subtelomeric sequences of an unrelated chromosome, which are oriented towards the centromere of that chromosome, are called chromo-some fusions. Interstitial deletions that span the *CAN1* and *URA3* genes but leave the chromosome V telomere intact have also been found.

TABLE IV
PCR PRIMER COMBINATIONS FOR MAPPING AND AMPLIFICATION OF
CHROMOSOME V BREAKPOINTS

| Fragment number | Breakpoint mapping | | Breakpoint amplification | |
	Mapping primer pair	Arbitrary PCR primer	Nested PCR primer
22	63A5–63A3	GCR2R	GCR1R
21	63B5–63B3	63A3	GCR4R
20	63C5–63C3	63B3	GCR5R
19	63D5–63SEQKJ1 (CAN DEL4)	63C3	GCR6R
18	63E5–63E3–2	63SEQKJ1	GCR9R
17	62A5–62A3–2	GCR11R	63E3–2
16	62B5–62B3	62A3–2	GCR12R
15	62C5–62DELA*	GCR14R	62B3
14	62D5–62D3	62DELA*	GCR16R
13	62E5–62E3	62D3	GCR19R
12	61A5–61A3	62E3	GCR21R
11	61B5–61B3	GCR24R	61A3
10	61C5–61C3	61B3	GCR25R
9	61D5–61D3	61C3	GCR27R
8	61E5–61E3	61D3	GCR30R
7	61F5–61F3	61E3	GCR32R
6	60A5–60A3	61F3	GCR34R
5	60B5–60B3	60A3	GCR36R
4	60C5–60C3	60B3	GCR38R
3	60D5–60D3	60C3	GCR40R
2	59A5–58DELA	59DELB	GCR42R
1	59B5–59B3	58DELA	GCR44R

For every rearrangement the chromosome V fragment distal to the breakpoint is assumed to be intact and therefore to contain a centromere. Depending on whether or not the fused DNA fragment is predicted to contain a centromere, the rearranged chromosome V is assumed to be dicentric or monocentric. Rearrangements predicted to be monocentric are *de novo* telomere additions and interstitial deletions, rearrangements predicted to be dicentric are chromosome fusions and rearrangements that can be monocentric or dicentric are nonreciprocal translocations and isoduplications (Fig. 4).

Conclusion

The simple measurement of rates of accumulating GCRs in different mutant backgrounds of *S. cerevisiae* and the ability to determine the exact DNA sequence at the breakpoint junction, allowing prediction of GCR

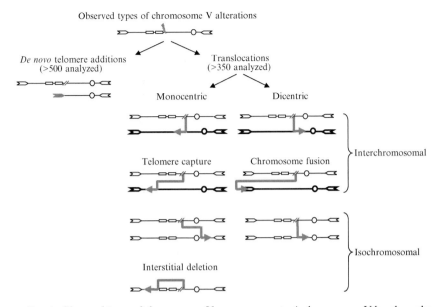

FIG. 4. Observed types of chromosome V rearrangements. A chromosome V break can be healed by recombining with a distant DNA sequence (translocation) or, in telomerase-proficient cells, by adding poly [TG_{1-3}] repeats (*de novo* telomere addition). Translocations that target chromosome V (isochromosomal) or an unrelated chromosome (interchromosomal translocation) are predicted to be dicentric or monocentric depending on whether the target sequence contains a centromere or no centromere, respectively. Translocations to the telomere or subtelomeric elements of an unrelated chromosome are also observed (telomere capture) as are fusions of chromosome V to telomeric poly [$C_{1-3}A/TG$] repeats, which are oriented towards the centromere of that chromosome (chromosome fusions) and thus are predicted to be dicentric. Rearrangements that lead to a partial deletion of the left arm of chromosome V and encompass the *CAN1* and *URA3* genes are always predicted to be monocentric (interstitial deletion).

structures, provides an effective tool for the analysis of global DNA metabolic pathways and their interaction with respect to their role in the maintenance of genome integrity. The methods described here are subject to some limitations because the assays require the GCRs to involve chromosome V as one partner and limit one of the GCR breakpoints to fall within a 12-kb region of chromosome V while GCRs elsewhere in the genome remain undetected. To obtain information about the extent of genome-wide DNA instability in mutants of interest, the breakpoint analysis described here may be combined with methods that allow genome-wide chromosomal analysis, such as array comparative genome hybridization (aCGH) and/or pulsed-field gel electrophoresis (PFGE). Nonetheless, these assays have shed considerable light on the pathways that both maintain and promote GCRs.

References

Chen, C., and Kolodner, R. D. (1999). Gross chromosomal rearrangements in *Saccharomyces cerevisiae* replication and recombination defective mutants. *Nat. Genet.* **23**, 81–85.

Huang, M. E., and Kolodner, R. D. (2005). A biological network in *Saccharomyces cerevisiae* prevents the deleterious effects of endogenous oxidative DNA damage. *Mol. Cell.* **17**, 709–720.

Huang, M. E., Rio, A. G., Nicolas, A., and Kolodner, R. D. (2003). A genomewide screen in *Saccharomyces cerevisiae* for genes that suppress the accumulation of mutations. *Proc. Natl. Acad. Sci. USA* **100**, 11529–11534.

Kolodner, R. D., Putnam, C. D., and Myung, K. (2002). Maintenance of genome stability in *Saccharomyces cerevisiae*. *Science* **297**, 552–557.

Lea, D. E., and Coulson, C. A. (1949). The distribution of the number of mutants in bacterial populations. *J. Genet.* **49**, 264–285.

Luria, S. E., and Delbrück, M. (1943). Mutations of bacteria from virus sensitivity to virus resistance. *Genetics* **28**, 491–511.

Myung, K., Chen, C., and Kolodner, R. D. (2001a). Multiple pathways cooperate in the suppression of genome instability in *Saccharomyces cerevisiae*. *Nature* **411**, 1073–1076.

Myung, K., Datta, A., and Kolodner, R. D. (2001b). Suppression of spontaneous chromosomal rearrangements by S phase checkpoint functions in *Saccharomyces cerevisiae*. *Cell* **104**, 397–408.

Myung, K., and Kolodner, R. D. (2002). Suppression of genome instability by redundant S-phase checkpoint pathways in *Saccharomyces cerevisiae*. *Proc. Natl. Acad. Sci. USA* **99**, 4500–4507.

Pennaneach, V., and Kolodner, R. D. (2004). Recombination and the Tel1 and Mec1 checkpoints differentially effect genome rearrangements driven by telomere dysfunction in yeast. *Nat. Genet.* **36**, 612–617.

Rosche, W. A., and Foster, P. L. (2000). Determining mutation rates in bacterial populations. *Methods* **20**, 4–17.

Smith, S., Hwang, J. Y., Banerjee, S., Majeed, A., Gupta, A., and Myung, K. (2004). Mutator genes for suppression of gross chromosomal rearrangements identified by a genome-wide screening in *Saccharomyces cerevisiae*. *Proc. Natl. Acad. Sci. USA* **101**, 9039–9044.

[28]   Formation and Processing of Stalled Replication Forks—Utility of Two-Dimensional Agarose Gels

By JENNIFER REINEKE POHLHAUS and KENNETH N. KREUZER

Abstract

Replication forks can be stalled by tightly bound proteins, DNA damage, nucleotide deprivation, or defects in the replication machinery. It is now appreciated that processing of stalled replication forks is critical for completion of DNA replication and maintenance of genome stability. In this chapter, we detail the use of two-dimensional (2D) agarose gels with Southern hybridization for the detection and analysis of blocked replication forks *in vivo*. This kind of 2D gel electrophoresis has been used extensively for analysis of replication initiation mechanisms for many years, and more recently has become a valuable tool for analysis of fork stalling. Although the method can provide valuable information when forks are stalled in random locations (e.g., after UV damage or nucleotide deprivation), it is even more informative with site-specific fork blockage, for example, blocks caused by tightly bound replication terminator proteins or by drug-stabilized topoisomerase cleavage complexes.

Introduction

Although replication is generally very processive, replication forks can become stalled or blocked due to nucleotide deprivation, template damage, or tightly bound proteins (for recent reviews, see Bussiere and Bastia, 1999; Kuzminov, 1999; Michel, 2000). Even in the absence of exogenous DNA damaging agents, forks stall in a significant fraction of the cell division cycles in both prokaryotic and eukaryotic cells. Replication fork stalling and disassembly of the replisome is likely a programmed response to problems with the DNA template, allowing access of repair, lesion bypass, and recombination machinery to damaged template DNA. It is now clear that cells have multiple pathways for processing stalled or blocked replication forks (for recent reviews, see Cox, 2001; Cox *et al.*, 2000; Kreuzer, 2005; Marians, 2000, 2004; McGlynn and Lloyd, 2002). In some cases, the blocked fork is cleaved by recombination nucleases, and the newly broken DNA must be processed by homologous recombination pathways to ensure cell viability (Hong and Kreuzer, 2003; Seigneur *et al.*, 1998). Stalled forks

METHODS IN ENZYMOLOGY, VOL. 409
0076-6879/06 $35.00
DOI: 10.1016/S0076-6879(05)09028-2

can also be restarted directly without fork cleavage, and there is evidence for at least two pathways of direct fork restart in *Escherichia coli* (Heller and Marians, 2005; Sandler, 2000). We have found the technique of two-dimensional (2D) neutral-neutral agarose gel electrophoresis to be extremely useful for analyzing the formation and resolution of blocked forks (Hong and Kreuzer, 2000, 2003; Pohlhaus and Kreuzer, 2005). This technique has been used widely in replication studies, since it is capable of identifying a variety of replication intermediates, including replication bubbles that form during initiation at origins, simple-Y structures from ongoing replication forks, and double-Y structures that form when two forks approach each other during termination of replication (Brewer and Fangman, 1987; Friedman and Brewer, 1995). Two dimensional gel electrophoresis utilizes a first dimension that separates DNA by size only and a second dimension that separates DNA by both size and shape. The first dimension gel contains a relatively low agarose concentration and is run slowly, allowing branched and circular DNA molecules to migrate at a mobility similar to the same-sized linear DNA. A gel lane is then excised and cast across the top of a second-dimension gel. The second dimension gel is at a higher agarose concentration, contains ethidium bromide to stiffen the DNA, and is run at a higher voltage. These conditions result in markedly reduced migration for DNA containing branches, and related families of branched molecules follow predictable arcs that deviate from the (roughly) diagonal arc of linear DNA molecules. In most cases, the DNA is precut with a suitable restriction endonuclease, although undigested plasmid DNA can be very informative as well (Lucas *et al.*, 2001; Martín-Parras *et al.*, 1998; Pohlhaus and Kreuzer, 2005). The earliest studies of fork blockage involved natural replication terminator systems, including the Tus-Ter system of *E. coli* and rDNA fork blockage in *Saccharomyces cerevisiae* (Brewer and Fangman, 1988; Bussiere and Bastia, 1999). In both cases, the cells encode site-specific binding proteins that block replication forks at particular DNA sites, and 2D gel electrophoresis was instrumental in analyzing this fork blockage. 2D gels revealed strong spots along the arcs of replication intermediates, which could be abolished by mutating either the binding site or the terminator protein. Likewise, in our studies of type II topoisomerase inhibitors, we observed an accumulation of spots along bubble arcs and Y-arcs that depended on the topoisomerase binding site, the inhibitor that stabilizes the enzyme at that site, and the topoisomerase itself (Hong and Kreuzer, 2000, 2003; Pohlhaus and Kreuzer, 2005). Two dimensional gel electrophoresis has also proven useful in studies of randomly located DNA damage, for example UV-induced damage (Courcelle *et al.*, 2003). In these cases, entire arcs or regions containing large numbers of branched forms are increased by the

DNA damage, causing mechanistic interpretation to be somewhat more difficult.

Protocol

While replication intermediates can be detected in either plasmid or genomic DNA, they are generally easier to detect and analyze in plasmids because of higher copy numbers. Isolation of plasmid and genomic DNA must be done in a manner that does not disrupt the replication intermediates of interest, which are often branched and fragile. In the following protocol, DNA is extracted from cells, associated proteins are removed by treatment with Proteinase K, and the DNA is then purified by phenol/chloroform extraction and dialysis. The DNA can be digested with a restriction enzyme before electrophoresis, or it can be analyzed without digestion. Finally, a Southern blot is used to identify the DNA molecules of interest.

Cell Manipulations

When studying plasmid replication, it is important to use cells that contain predominately monomeric plasmid, to simplify interpretation of the data. We routinely prepare multiple aliquots of frozen cells, after verifying that they contain predominately monomeric plasmid. For each particular experiment, we use one aliquot as an inoculant for growth to log-phase. In many of our experiments, we treat the cells with an inhibitor that stabilizes the covalent topoisomerase-DNA "cleavage complex." In our studies of replication in uninfected *E. coli*, we treat cells with the topoisomerase inhibitor norfloxacin for 6 min (Pohlhaus and Kreuzer, 2005). When analyzing phage T4-directed replication in an *E. coli* host, we allow a short period of time (e.g., 6 min) for adsorption of the bacteriophage and early gene expression, before adding the topoisomerase inhibitor *m*-AMSA (4′-[9-acridinylamino]methanesulfon-*m*-anisidide) for 6 to 18 min (Hong and Kreuzer, 2000, 2003). The unique nature of the topoisomerase cleavage complex led us to develop two different conditions for extracting DNA from the drug-treated cells. In one case, the cells (1–2 ml) are pelleted and resuspended with 300 μl Resealing Lysis Buffer (50 mM Tris-HCl [pH 7.8], 10 mM EDTA, 1% Triton X-100, lysozyme at 1.8 mg/ml), and incubated for 1 h at 65°. During this extraction, the latent breaks within topoisomerase cleavage complexes are resealed, resulting in intact DNA wherever a cleavage complex had been located. Proteinase K and SDS are then added (to 0.5 mg/ml and 0.2%, respectively) and incubation is continued for an additional hour. A somewhat different extraction method reveals the latent DNA breaks within cleavage complexes. In this case, we resuspend the cells in Cleavage Lysis Buffer (50 mM Tris-HCl [pH 7.8], 10 mM EDTA, 0.2% SDS). The SDS treatment disrupts the topoisomerase molecule in the cleavage complex without allowing for DNA resealing. Subsequently, a 1-h treatment at 65° with Proteinase K (0.5 mg/ml) removes any attached protein from the DNA. The differences between 2D gel patterns with DNA isolated under cleavage and resealing conditions have been very useful in interpreting the fork blockage events at topoisomerase cleavage complexes (Hong and Kreuzer, 2000; Pohlhaus and Kreuzer, 2005). For either of the two lysis methods described previously, total nucleic acids are extracted after the proteinase treatment by sequential treatment with phenol, phenol/chloroform/isoamyl alcohol, and chloroform/isoamyl alcohol. Nucleic acids are then further purified by dialysis into TE buffer (10 mM Tris-HCl [pH 7.8], 1 mM EDTA). The dialysis step removes SDS that could interfere with subsequent restriction digestion.

Neutral/Neutral 2D Gel Electrophoresis

In most experiments, the isolated DNA is cleaved with a restriction endonuclease; alternatively, the undigested DNA can be loaded directly onto agarose gels. The 2D gel electrophoresis method, described by Friedman and Brewer (1995), allows for the identification of branched DNA forms that are undergoing replication and/or recombination. After running the first dimension gel in the absence of ethidium bromide (typically 0.4% agarose in 0.5× TBE [1× TBE = 90 mM Tris borate, 2 mM EDTA] for 29 h at 1.2 V/cm), the gel is stained with ethidium bromide (0.3 μg/ml) and the sample lanes are excised. To minimize breakage of the fragile replication intermediates, we do not expose the sample lanes to ultraviolet light (UV). Instead, we excise the sample lanes based on the migration of bands in nearby marker lanes that have been exposed to UV. The excised sample lanes are then cast across the top of a second dimension gel (typically 1% agarose in 0.5× TBE), which is run at 4° in the presence of ethidium bromide (0.3 μg/ml) for 16 h at 6 V/cm with buffer recirculation. Depending on the size of the gel box, several first-dimension gel slices can be analyzed in a single second-dimension gel by carefully placing the slices so that the resulting patterns do not overlap with each other.

Southern Blot, Using Sponge Downward Transfer

We routinely use the sponge downward transfer method for Southern blotting (Hong and Kreuzer, 2000; Pohlhaus and Kreuzer, 2005; adapted from Koetsier *et al.*, 1993 and Zhou *et al.*, 1994). Downward transfer is preferred to the conventional upward transfer method because (1) the time for complete DNA transfer is decreased, and (2) the bottom of the gel is regular and flat, rendering inversion of the gel unnecessary (recommended in some upward transfer protocols). The sponge downward transfer protocol is recommended over the downward transfer protocol using a wick and a reservoir (see pages 2.9.7–2.9.8 in Ausubel *et al.*, 2005) for reasons of convenience.

In this method, the second-dimension gel is treated with 0.25 N HCl for 15–20 min, followed by 3 M NaCl, 0.4 M NaOH for at least 30 min. Next, the following items are layered on a tray from bottom to top: 2–3 inches of dry paper towels, 4 sheets of dry thick electrophoresis blotting paper (Ahlstrom, grade 320 filter paper), four sheets of dry 3 MM Chr Whatman paper (VWR, Westchester, PA) one sheet of 3 MM Chr Whatman paper (wet with transfer buffer [1.5 M NaCl, 0.4 M NaOH]), one sheet of Schleicher & Schuell Nytran Supercharge nylon membrane (Keene, NH) (wet with transfer buffer), agarose gel, three sheets of 3 MM Chr Whatman paper (wet with transfer buffer), and finally a cellulose sponge (Plato, Amarillo, TX, Sponge Sheet CS-17) saturated with transfer buffer. Capillary action from top to bottom serves to transfer DNA onto the nylon membrane. Complete transfer of DNA from 1-cm thick gels occurs in as little as 3.5 h and thinner gels presumably transfer even faster. After transfer, the nylon membrane is washed in 4× SSC (1× SSC = 0.15 M NaCl, 17 mM sodium citrate), briefly dried, and the DNA is crosslinked to the membrane with a 120 mJ/cm^2 UV exposure. Alternatively, the DNA can be fixed to the membrane by baking at 80° for 2–3 h. The membrane is then blocked in prehybridization buffer (6× SSC, 5× Denhardt's Solution [1× Denhardt's Solution = 0.2 mg/ml Ficoll, 0.2 mg/ml polyvinylpyrrolidone, 0.2 mg/ml bovine serum albumin], 1% SDS, 50% formamide, sonicated salmon sperm DNA at 0.1 mg/ml) before addition of a radiolabeled probe. Hybridization of the labeled probe is then allowed at 42° in a rolling glass tube incubator (Robbins Scientific, Sunnyvale, CA, Model 400) for 12–20 h. The nylon membrane is washed twice with Wash 1 (0.6% SDS, 1× SSC, at 42°), and twice with Wash 2 (0.6% SDS, 0.1× SSC, at 70°), for 5–10 min each wash. Finally, the membrane is briefly dried, wrapped in Saran Wrap, and exposed to X-ray film or a phosphorimager screen.

Interpretation

This section will focus on interpretation of unidirectional pBR322 plasmid replication (as shown in Fig. 1A), although similar patterns can be seen with bidirectional genomic or plasmid replication. The 2D patterns of both restriction endonuclease-linearized and undigested plasmid DNA

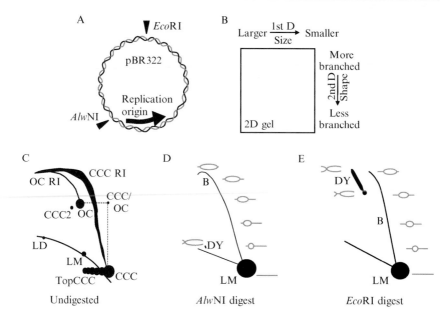

FIG. 1. 2D gel schematic of replicating plasmid molecules. (A) Plasmid pBR322 (4361 base pairs) replicates unidirectionally, as shown by the arrowhead at position 2535. The locations of the recognition sites for *Alw*NI (2890) and *Eco*RI (4361) are shown. (B) The setup of the 2D gel is shown, where the first dimension separates primarily by size and the second dimension separates molecules of equal size by differences in shape. (C) Plasmids undergoing normal replication exhibit a characteristic 2D gel pattern when undigested, where CCC = covalently closed circular monomers, TopCCC = topoisomers of CCC, OC = open circular monomers, CCC/OC = CCC in the first dimension and OC in the second dimension, CCC RI = CCC replication intermediates, OC RI = OC replication intermediates, CCC2 = covalently closed circular dimers, LM = linear monomer, and LD = linear dimer. (D) The molecules formed after *Alw*NI digestion include bubble (B) molecules, sized between 1.0× and 1.9×, and double-Y (DY) molecules, sized between 1.9× and 2.0×. The bubble-to-double-Y transition occurs at 1.9× with passage through a simple-Y (Y) intermediate. The simple-Y arc is not shown in these drawings. Linear monomer (LM) molecules are indicated by the dark circle. The shapes of the molecules formed by *Alw*NI digestion are shown in gray next to their expected position in the 2D gel pattern. (E) The molecules formed after *Eco*RI digestion include bubble (B) molecules, sized between 1.0× and 1.6×, and double-Y (DY) molecules, sized between 1.6× and 2.0×. In this case, the bubble-to-double-Y transition occurs at 1.6× with passage through a simple-Y (Y) intermediate. Linear monomer (LM) molecules are indicated by the dark circle. The shapes of the molecules formed by *Eco*RI digestion are shown in gray next to their expected position in the 2D gel pattern.

will be considered. In both cases, an arc of linears, which corresponds to all the nonreplicating DNA fragments of the plasmid and genome, can be visualized by UV excitation of the ethidium bromide-soaked second dimension gel.

Patterns Seen with Restriction Endonuclease–Linearized DNA

Normal Plasmid Replication. After Southern hybridization, the most intense spot is the linear monomer spot, which corresponds to nonreplicating monomeric plasmid. A schematic drawing of the arc of linears and the monomer spot (LM) after digestion with *Alw*NI or *Eco*RI is presented in Fig. 1D and E, respectively (also see Friedman and Brewer, 1995). Plasmids undergoing unidirectional replication exhibit a characteristic "bubble to double-Y" pattern after linearization with a restriction endonuclease. Replicating molecules in which the fork has not yet reached the restriction endonuclease site migrate as a bubble (B) arc. Replicating molecules in which the active fork has just barely passed the restriction site result in a Y-shaped molecule (after restriction digestion) that falls at one particular spot (the bubble-to-double-Y transition) on the simple-Y (Y) arc (not shown in Figure 1). Further replication results in plasmid molecules that have the double-Y (DY) shape, with two Y-shaped segments on either end connected by a linear segment. As replication nears completion, the resulting molecules approach $2.0\times$ size and convert from a double-Y (DY) molecule to an X-shaped molecule (Fig. 1D and E).

Plasmid Replication in the Presence of Topoisomerase Inhibitors. When cells are treated with topoisomerase inhibitors that stabilize the cleavage complex, characteristic arcs can be detected, including the asymmetric-Y (AY) arc, and part of the simple-Y (Y) arc, from the monomer spot to the bubble-to-double-Y transition (Pohlhaus and Kreuzer, 2005). We presented evidence that these breaks can reflect an induced DNA break that occurs *in vivo* after fork blockage (Pohlhaus and Kreuzer, 2005). The asymmetric-Y (AY) arc and part of the simple-Y (Y) arc are created by a break at either the fixed branch (at the origin, as in Fig. 2, arrow 2) or the replicating fork of a bubble (Fig. 2, arrow 1), and subsequent digestion by a restriction endonuclease. These kinds of molecules were previously attributed to branch breakage during harsh DNA isolation conditions (Martín-Parras *et al.*, 1992, 1998). However, in our study of topoisomerase inhibitors, these arcs were only detected after drug treatment (Pohlhaus and Kreuzer, 2005). These arcs were detected in both cleavage and resealing conditions, indicating that formation of the molecules is not dependent on topoisomerase cleavage.

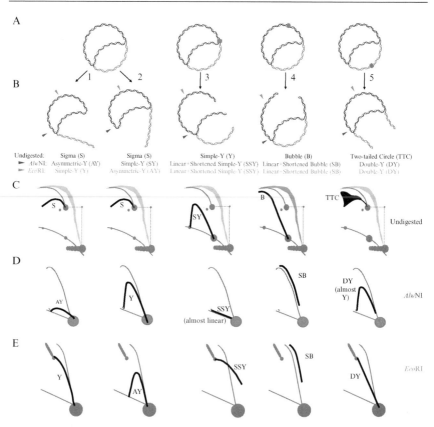

Fig. 2. 2D gel schematic of molecules formed during discontinuous replication. (A and B) Double-stranded pBR322 molecules are shown as undigested molecules; the locations of the recognition sites for *Alw*NI and *Eco*RI are indicated by the dark gray and light gray arrowheads, respectively. The plasmid considered in this figure is partially replicated, to the location of the major gyrase binding site, centered at 990 bps. (C, D, E) The patterns resulting from normal plasmid replication, as shown in Fig. 1, are indicated in gray. The black arc on each panel indicates the expected position for the molecule illustrated in Panel B when undigested (C), digested with *Alw*NI (D), or digested with *Eco*RI (E). Between Panels A and B, arrows 1 and 2 show that a covalently closed circular replication intermediate (CCC RI) can be converted into a sigma (S) molecule by breaking a single strand at either the active replication fork (arrow 1) or the fixed branch (arrow 2). The undigested sigma (S) molecules can be converted into asymmetric-Y (AY) molecules or simple-Y (Y) molecules by digestion with *Alw*NI or *Eco*RI; the 2D patterns of these molecules are shown in Panels C (undigested), D (*Alw*NI), and E (*Eco*RI). Arrow 3 shows that a replication intermediate blocked at a gyrase dimer (gray overlapping ovals) can be converted to a simple-Y (Y) molecule when cells are lysed with cleavage conditions. The undigested simple-Y (Y) molecule can be digested into a shortened simple-Y (SSY) molecule by either *Alw*NI or *Eco*RI; the 2D patterns of these molecules are shown in Panels C (undigested), D (*Alw*NI), and E (*Eco*RI). Arrow 4 shows that a CCC replication intermediate with a gyrase dimer bound ahead of the replication

In contrast, the shortened simple-Y (SSY) arc was detected only in cleavage conditions, indicating that one of the DNA ends was created by topoisomerase-induced cleavage. Shortened simple-Y (SSY) molecules can be created by topoisomerase-induced cleavage of replication intermediates at the replication fork (Fig. 2, arrow 3) and subsequent digestion by a restriction endonuclease away from the fork.

Patterns Seen with Undigested DNA

Normal Plasmid Replication. The topological state of the DNA markedly affects the 2D pattern of undigested DNA. Nonreplicating plasmids are detected in both the open circular (OC) form and the covalently closed circular (CCC), or supercoiled form, while replication intermediates extend up from the CCC or OC spots (Fig. 1C). The arcs described previously for digested DNA molecules undergoing normal plasmid replication (bubble, bubble-to-double-Y transition through a simple-Y intermediate, and double-Y) correspond to the smear of molecules migrating as CCC replication intermediates (CCC RI) without digestion. The CCC replication intermediates correspond to a smear instead of a defined arc because each plasmid has replicated to a different point and contains a variable level of supercoiling (Martín-Parras *et al.*, 1998). Circular replication intermediates containing a nick appear as open circular replication intermediates (Fig. 1C; OC RI).

Plasmid Replication in the Presence of Topoisomerase Inhibitors. The asymmetric-Y (AY) arcs and simple-Y (Y) arcs seen in 2D gel patterns of restriction endonuclease-linearized DNA correspond to sigma-shaped molecules when undigested (which form an eyebrow-shaped arc). Sigma-shaped molecules have one duplex DNA end (Fig. 2, arrows 1, 2); we discussed the genesis of this DNA break previously when we considered the DNA forms after restriction endonuclease digestion. In experiments where cells are lysed in cleavage lysis buffer, the latent DNA breaks within

---

fork can be converted to a bubble (B) molecule when cells are lysed with cleavage conditions. The undigested bubble (B) molecule can be digested into a shortened bubble (SB) molecule by either *Alw*NI or *Eco*RI; the 2D patterns of these molecules are shown in Panels C (undigested), D (*Alw*NI), and E (*Eco*RI). Arrow 5 shows that a CCC replication intermediate with a gyrase dimer bound behind the replication fork can be converted to a two-tailed circle (TTC) molecule when cells are lysed with cleavage conditions. The undigested two-tailed circle (TTC) molecule can be digested into a double-Y (DY) molecule by either *Alw*NI or *Eco*RI; the 2D patterns of these molecules are shown in panels C (undigested), D (*Alw*NI), and E (*Eco*RI).

the topoisomerase cleavage complexes are revealed, generating DNA molecules with a double-stranded break. Thus, even though there is no restriction enzyme digestion, we detect simple-Y (Y) arcs, bubble (B) arcs, and two-tailed circle (TTC) arcs, all corresponding to molecules with two or more double-stranded ends. These molecules result when the cleavage complex is located at, ahead, or behind the replication fork, creating a simple-Y (Y) arc, a bubble (B) arc, or a two-tailed circle (TTC) arc, respectively (Fig. 2, arrows, 3, 4, 5; see also Lucas *et al.*, 2001; Pohlhaus and Kreuzer, 2005). Note that restriction endonuclease digestion of these molecules results in shortened simple-Y (SSY) molecules, shortened bubble (SB) molecules, and double-Y (DY) molecules, respectively (Fig. 2, arrows 3, 4, 5, and gel patterns in panels C, D, and E).

Examples of 2D Gels with Blocked Replication Forks. Based on our studies with topoisomerase inhibitors, we present several typical 2D gel patterns in Fig. 3A, C, and E (corresponding schematics in Fig. 3B, D, and F). In each case, the 2D gel in the left panel contains DNA from drug-free control cells while the pattern in the right panel is from cells treated with the gyrase inhibitor norfloxacin (Pohlhaus and Kreuzer, 2005). A replication block from a drug-stabilized gyrase cleavage complex causes an accumulation of molecules of a particular size, resulting in a spot appearing on an arc.

Considering the undigested DNA (Fig. 3A), simple-Y (Y) molecules are created when a topoisomerase cleavage complex is located at the replication fork and DNA is isolated under cleavage conditions (which reveals the latent DNA break in the cleavage complex; see Fig. 2, arrow 3). The 2D gel in Fig. 3A (undigested DNA), right panel, shows the entire simple-Y (Y) arc composed of discrete spots, representing the accumulation of replication forks stalled at various locations. This gel also shows the bubble (B) and two-tailed circle (TTC) arcs discussed previously (see previous section, Plasmid Replication in the Presence of Topoisomerase Inhibitors).

In restriction enzyme-digested samples, the bubble (B) arc can only be composed of replicative molecules (see Fig. 1). The 2D gels in the right panels of Fig. 3C (*Alw*NI-digested DNA) and E (*Eco*RI-digested DNA) display discrete spots along the bubble arc, again corresponding to forks blocked at drug-stabilized gyrase cleavage complexes (but in this case, the DNA was isolated under resealing conditions, and so any latent DNA breaks were resealed upon lysis). The spots are dependent on the presence of the inhibitor and the presence of a drug-sensitive DNA gyrase.

We have analyzed the appearance of the most intense spot on the simple-Y (Y) arc of undigested DNA and the bubble (B) arc of digested DNA (arrow in Fig. 3A, C, and E; see also Pohlhaus and Kreuzer, 2005).

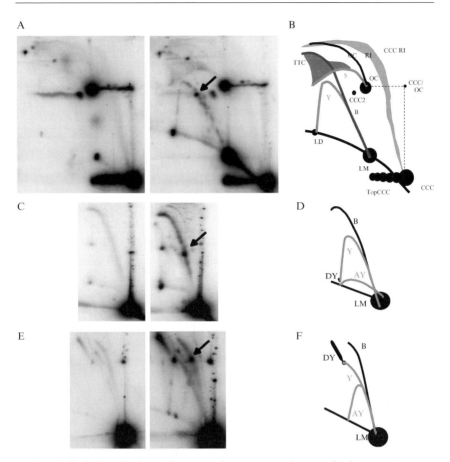

Fig. 3. Stalled replication forks accumulate as spots along replicative arcs. Arcs are labeled as in Figs. 1 and 2. Data panels were previously published in Pohlhaus and Kreuzer (2005). (A) Isolation of DNA under cleavage conditions results in the accumulation of a spot (indicated by the black arrow) along the simple-Y (Y) arc of undigested DNA after norfloxacin treatment (compare right panel, norfloxacin [1 μg/ml], with left panel, untreated). (B) The schematic serves as a guide for interpretation of Panel A. (C) Isolation of DNA under resealing conditions results in the accumulation of a spot (indicated by the black arrow) along the bubble (B) arc of AlwNI-digested DNA after norfloxacin treatment (compare right panel, norfloxacin [1 μg/ml], with left panel, untreated). (D) The schematic serves as a guide for interpretation of Panel B. (E) Isolation of DNA under resealing conditions results in the accumulation of a spot (indicated by the black arrow) along the bubble (B) arc of EcoRI-digested DNA after norfloxacin treatment (compare right panel, norfloxacin [1μg/ml], with left panel, untreated). (F) The schematic serves as a guide for interpretation of Panel B.

This intense spot was abolished when the major gyrase binding site in pBR322 was mutationally inactivated, providing the strongest evidence that gyrase cleavage complexes block replication forks *in vivo* (Pohlhaus and Kreuzer, 2005).

In Vivo *Processing of Blocked Replication Forks*

Detecting replication fork blockage by 2D gels is relatively straightforward, and the exact nature of the blocked fork can be carefully elucidated. However, it is more difficult to analyze the pathways by which the blocked forks are processed or restarted. In the case of forks blocked *in vivo* by drug-stabilized T4 topoisomerase cleavage complexes, we detect a rapid loss of the intense spots on 2D gels after drug washout *in vivo* (DT Long and KN Kreuzer, unpublished data). Loss of the spots implies that the forks are being processed for destruction or restart, but by itself is not informative of the actual pathway. We are analyzing various mutant infections, hoping to identify the gene products necessary for processing these topoisomerase-blocked forks. As mentioned in the Introduction, one fate of blocked forks is enzymatic cleavage by recombination nucleases. In the phage T4 system, we found that T4 endonuclease VII can cleave drug-stabilized topoisomerase-blocked forks *in vitro*, and that blocked forks accumulate to higher levels in endonuclease VII-deficient infections (Hong and Kreuzer, 2003). These results argue that endonuclease VII cleaves a significant fraction of blocked forks during a wild-type T4 infection, and led us to propose that endonuclease cleavage might be important in the cytotoxicity of topoisomerase inhibitors. More recently, our experiments in the *E. coli* system revealed replication fork breakage at the sites of norfloxacin-stabilized DNA gyrase cleavage complexes *in vivo* (see above; Pohlhaus and Kreuzer, 2005). In this case, the identity of the enzyme(s) that cleave(s) the blocked fork is not currently known. It is worth pointing out that unique patterns in 2D gels were crucial in identifying the broken replication forks. In this case, the 2D gels revealed both the fork blockage and the subsequent cleavage event (see above; Pohlhaus and Kreuzer, 2005).

The initiation of replication from T4 origins involves a two-stage bidirectional mechanism with a significant delay between the start of the first and second replication fork (Belanger and Kreuzer, 1998). This leads to the accumulation of branched Y-shaped molecules in which only the first fork has fired (and exited the restriction fragment). These molecules are somewhat analogous to blocked replication forks, and the assembly of the replication machinery for the second replication fork may occur by the same mechanism as replication restart from bona fide blocked forks. We have recently analyzed this assembly pathway and uncovered a unique role for the replicative helicase loader protein (gp59; Dudas and Kreuzer, 2005). In this study, the combination of 2D neutral-neutral and neutral-alkaline gels allowed us to identify the extent of both leading and lagging strand synthesis within the Y-shaped DNA molecules (Dudas and Kreuzer, 2005).

Closing Comments

Other methods can provide additional important evidence related to fork stalling and restart pathways, either alone or in combination with 2D gels. For example, genetic approaches can uncover the physiological importance of restart pathways or reveal recombination reactions that result from disturbances at the replication fork (e.g., see Sandler, 2000). Measurements of overall DNA synthesis rates have an important place in analyzing fork stalling and restart—an elegant series of studies used pulse-labeling to follow the process of replication restart after UV damage (Courcelle *et al.*, 1997, 1999; Rangarajan *et al.*, 2002). As mentioned previously, comparing the results of neutral-neutral and neutral-alkaline electrophoresis provides detailed information about branched molecules, particularly when combined with strand-specific probing (Dudas and Kreuzer, 2005). This general idea is incorporated into two related methods for performing "3-dimensional" gel electrophoresis, with an alkaline third dimension (Kalejta and Hamlin, 1996; Kalejta *et al.*, 1996; Liang and Gerbi, 1994).

One significant limitation of 2D gel electrophoresis is that more than one family of replication intermediates can migrate in similar locations, leading to uncertainty or misidentification of particular intermediates. For example, the "cone" region between the larger Y-form DNA and X-form DNA has been proposed to contain regressed replication forks, double-Y molecules, and Y-molecules with partially single-stranded regions (Courcelle *et al.*, 2003; Dudas and Kreuzer, 2005; Lopes *et al.*, 2001). Also, as described previously, asymmetric Y-molecules with a shortened arm can migrate in a pattern that is very similar to bona fide Y-forms. It is therefore important to validate interpretations of 2D gel patterns carefully. Performing multiple digests and/or analyzing undigested DNA can solidify conclusions (Pohlhaus and Kreuzer, 2005), as can additional analyses with alkaline electrophoresis (see paragraph above) or electron microscopy (EM) (Kuzminov *et al.*, 1997; Sogo *et al.*, 2002).

Another interpretational pitfall worth considering is the temptation to equate the intensity of an arc or spot on a 2D gel with the importance of that form or the fraction of molecules that follow a particular pathway. The intensity of an arc can be influenced by many factors, including the rate of formation, rate of processing into another form, and overall stability of the branched DNA molecule. When multiple pathways process a stalled fork, a very intense arc or spot could be generated from a relatively minor processing pathway, simply because the intermediate in that pathway has a much longer half-life than an intermediate in a more prominent pathway.

One also needs to be cognizant of the possibility that some branched intermediates may simply be lost and therefore are undetectable in a 2D gel. Holliday junctions are prone to branch migration under certain conditions, but can be stabilized either by crosslinking or by isolating DNA under certain conditions (Allers and Lichten, 2000). Many scientists routinely use crosslinking to stabilize a variety of branched DNA forms, and it is good practice to at least compare the results with and without crosslinking before any major conclusion is reached. Some studies utilize BND (benzoylated naphthoylated DEAE) cellulose column chromatography to concentrate replication intermediates, but it seems possible that these columns result in preferential recovery of a subset of branched DNA forms.

Finally, there has been only very limited analysis of DNA forms that contain extensive single-stranded regions, such as forks with leading-strand only synthesis (Belanger *et al.*, 1996). Several important issues arise with DNA containing single-stranded regions: (1) migration in the 2D gels may be different than predominantly duplex DNA forms, (2) 3-stranded branch migration may be more facile than 4-stranded branch migration, (3) most restriction endonucleases do not cleave their sequences in a single-stranded region, and (4) other restriction endonucleases cleave single-stranded DNA with less specificity than duplex DNA (possibly due to contaminating nucleases in some cases). In general, when considering the results of 2D gel electrophoresis, one should think carefully about the various DNA forms that might exist in the population but be undetectable for one reason or another.

In spite of these limitations and possible pitfalls, 2D gel electrophoresis is a powerful method to identify and analyze a large variety of replication intermediates. This method can uncover short-lived intermediates in replication pathways, and the migration pattern alone provides strong hints and sometimes essentially conclusive evidence about the exact structure of a branched molecule. The use of site-specific damage significantly increases the utility of 2D gel electrophoresis, since fork blockage can be definitively assigned to events at a single site (e.g., a protein binding site that can be mutationally altered in a control experiment). 2D gel electrophoresis is relatively straightforward and highly sensitive, can be used with a wide variety of biological systems, and will continue to be used extensively to elucidate important aspects of replication fork stalling, fork cleavage, and fork restart pathways.

Acknowledgments

We thank Jody Plank for teaching us how to create figures depicting double-stranded DNA. Our research is supported by grants GM34622 and GM072089 from the National Institutes of Health.

References

Allers, T., and Lichten, M. (2000). A method for preparing genomic DNA that restrains branch migration of Holliday junctions. *Nucleic Acids Res.* **28,** e6.

Ausubel, F. M., Brent, R., Kingston, R. E., Moore, D. D., Seidman, J. G., Smith, J. A., and Struhl, K. (2005). *Curr. Protocols in Molecular Biology* John Wiley & Sons, Inc, New York, NY.

Belanger, K. G., and Kreuzer, K. N. (1998). Bacteriophage T4 initiates bidirectional DNA replication through a two-step process. *Mol. Cell* **2,** 693–701.

Belanger, K. G., Mirzayan, C., Kreuzer, H. E., Alberts, B. M., and Kreuzer, K. N. (1996). Two-dimensional gel analysis of rolling circle replication in the presence and absence of bacteriophage T4 primase. *Nucl. Acids Res.* **24,** 2166–2175.

Brewer, B. J., and Fangman, W. L. (1987). The localization of replication origins on ARS plasmids in *S. cerevisiae. Cell* **51,** 463–471.

Brewer, B. J., and Fangman, W. L. (1988). A replication fork barrier at the 3′ end of yeast ribosomal RNA genes. *Cell* **55,** 637–643.

Bussiere, D. E., and Bastia, D. (1999). Termination of DNA replication of bacterial and plasmid chromosomes. *Mol. Microbiol.* **31,** 1611–1618.

Courcelle, J., Carswell-Crumpton, C., and Hanawalt, P. C. (1997). *recF* and *recR* are required for the resumption of replication at DNA replication forks in *Escherichia coli. Proc. Natl. Acad. Sci. USA* **94,** 3714–3719.

Courcelle, J., Crowley, D. J., and Hanawalt, P. C. (1999). Recovery of DNA replication in UV-irradiated *Escherichia coli* requires both excision repair and RecF protein function. *J. Bacteriol.* **181,** 916–922.

Courcelle, J., Donaldson, J. R., Chow, K. H., and Courcelle, C. T. (2003). DNA damage-induced replication fork regression and processing in *Escherichia coli. Science* **299,** 1064–1067.

Cox, M. M. (2001). Recombinational DNA repair of damaged replication forks in *Escherichia coli*: Questions. *Annu. Rev. Genet.* **35,** 53–82.

Cox, M. M., Goodman, M. F., Kreuzer, K. N., Sherratt, D. J., Sandler, S. J., and Marians, K. J. (2000). Re-establishment of inactivated replication forks as a bacterial housekeeping function. *Nature* **404,** 37–41.

Dudas, K. C., and Kreuzer, K. N. (2005). Bacteriophage T4 helicase loader protein gp59 functions as gatekeeper in origin-dependent replication *in vivo. J. Biol. Chem.* **280,** 21561–21569.

Friedman, K. L., and Brewer, B. J. (1995). Analysis of replication intermediates by two-dimensional agarose gel electrophoresis. *Methods Enzymol.* **262,** 613–627.

Heller, R. C., and Marians, K. J. (2005). The disposition of nascent strands at stalled replication forks dictates the pathway of replisome loading during restart. *Mol. Cell* **17**, 733–743.

Hong, G., and Kreuzer, K. N. (2000). An antitumor drug-induced topoisomerase cleavage complex blocks a bacteriophage T4 replication fork *in vivo*. *Mol. Cell. Biol.* **20**, 594–603.

Hong, G., and Kreuzer, K. N. (2003). Endonuclease cleavage of blocked replication forks: An indirect pathway of DNA damage from antitumor drug-topoisomerase complexes. *Proc. Natl. Acad. Sci. USA* **100**, 5046–5051.

Kalejta, R. F., and Hamlin, J. L. (1996). Composite patterns in neutral/neutral two-dimensional gels demonstrate inefficient replication origin usage. *Mol. Cell. Biol.* **16**, 4915–4922.

Kalejta, R. F., Lin, H. B., Dijkwel, P. A., and Hamlin, J. L. (1996). Characterizing replication intermediates in the amplified CHO dihydrofolate reductase domain by two novel gel electrophoretic techniques. *Mol. Cell. Biol.* **16**, 4923–4931.

Koetsier, P. A., Schorr, J., and Doerfler, W. (1993). A rapid optimized protocol for downward alkaline Southern blotting of DNA. *BioTechniques* **15**, 260–262.

Kreuzer, K. N. (2005). Interplay between DNA replication and recombination in prokaryotes. *Annu. Rev. Microbiol.* **59**, 43–67.

Kuzminov, A. (1999). Recombinational repair of DNA damage in *Escherichia coli* and bacteriophage lambda. *Microbiol. Mol. Biol. Rev.* **63**, 751–813.

Kuzminov, A., Schabtach, E., and Stahl, F. W. (1997). Study of plasmid replication in *Escherichia coli* with a combination of 2D gel electrophoresis and electron microscopy. *J. Mol. Biol.* **268**, 1–7.

Liang, C., and Gerbi, S. A. (1994). Analysis of an origin of DNA amplification in *Sciara coprophila* by a novel three-dimensional gel method. *Mol. Cell. Biol.* **14**, 1520–1529.

Lopes, M., Cotta-Ramusino, C., Pellicioli, A., Liberi, G., Plevani, P., Muzi-Falconi, M., Newlon, C. S., and Foiani, M. (2001). The DNA replication checkpoint response stabilizes stalled replication forks. *Nature* **412**, 557–561.

Lucas, I., Germe, T., Chevrier-Miller, M., and Hyrien, O. (2001). Topoisomerase II can unlink replicating DNA by precatenane removal. *EMBO J.* **20**, 6509–6519.

Marians, K. J. (2000). PriA-directed replication fork restart in *Escherichia coli*. *Trends Biochem. Sci.* **25**, 185–189.

Marians, K. J. (2004). Mechanisms of replication fork restart in *Escherichia coli*. *Philos. Trans. R. Soc. Lond. B Biol. Sci.* **359**, 71–77.

Martín-Parras, L., Hernández, P., Martínez-Robles, M. L., and Schvartzman, J. B. (1992). Initiation of DNA replication in ColE1 plasmids containing multiple potential origins of replication. *J. Biol. Chem.* **267**, 22496–22505.

Martín-Parras, L., Lucas, I., Martínez-Robles, M. L., Hernández, P., Krimer, D. B., Hyrien, O., and Schvartzman, J. B. (1998). Topological complexity of different populations of pBR322 as visualized by two-dimensional agarose gel electrophoresis. *Nucleic Acids Res.* **26**, 3424–3432.

McGlynn, P., and Lloyd, R. G. (2002). Recombinational repair and restart of damaged replication forks. *Nat. Rev. Mol. Cell. Biol.* **3**, 859–870.

Michel, B. (2000). Replication fork arrest and DNA recombination. *Trends Biochem. Sci.* **25**, 173–178.

Pohlhaus, J. R., and Kreuzer, K. N. (2005). Norfloxacin-induced DNA gyrase cleavage complexes block *Escherichia coli* replication forks, causing double-stranded breaks *in vivo*. *Mol. Microbiol.* **56**, 1416–1429.

Rangarajan, S., Woodgate, R., and Goodman, M. F. (2002). Replication restart in UV-irradiated *Escherichia coli* involving pols II, III, V, PriA, RecA and RecFOR proteins. *Mol. Microbiol.* **43**, 617–628.

Sandler, S. J. (2000). Multiple genetic pathways for restarting DNA replication forks in *Escherichia coli* K-12. *Genetics* **155**, 487–497.

Seigneur, M., Bidnenko, V., Ehrlich, S. D., and Michel, B. (1998). RuvAB acts at arrested replication forks. *Cell* **95**, 419–430.

Sogo, J. M., Lopes, M., and Foiani, M. (2002). Fork reversal and ssDNA accumulation at stalled replication forks owing to checkpoint defects. *Science* **297**, 599–602.

Zhou, M. Y., Xue, D., Gomez-Sanchez, E. P., and Gomez-Sanchez, C. E. (1994). Improved downward capillary transfer for blotting of DNA and RNA. *BioTechniques* **16**, 58–59.

[29] Poly(ADP-ribose) Polymerase–1 Activation During DNA Damage and Repair

By Françoise Dantzer, Jean-Christophe Amé,
Valérie Schreiber, Jun Nakamura,
Josiane Ménissier-de Murcia, and Gilbert de Murcia

Abstract

Changes in chromatin structure emanating from DNA breaks are among the most initiating events in the damage response of the cell. In higher eukaryotes, poly(ADP-ribose) polymerase-1 (PARP-1) translates the occurrence of DNA breaks detected by its zinc-finger domain into a signal, poly ADP-ribose, synthesized and amplified by its DNA-damage dependent catalytic domain. This epigenetic mark on chromatin, induced by DNA discontinuities, is now considered as a part of a survival program aimed at protecting primarily chromatin integrity and stability. In this chapter we describe some of our methods for determining *in vivo* and *in vitro* PARP-1 activation in response to DNA strand breaks.

Poly(ADP-ribosyl)ation is a posttranslational modification of nuclear proteins induced by DNA strand-breaks that contributes to the survival of injured proliferating cells (D'Amours *et al.*, 1999). Poly(ADP-ribose) polymerases (PARPs) now constitute a large family of 18 proteins, encoded by different genes and displaying a conserved catalytic domain in which PARP-1 (113 kDa), the founding member, and PARP-2 (62 kDa) are so far the sole enzymes whose catalytic activity is immediately stimulated by DNA strand-breaks (Ame *et al.*, 2004).

PARP-1 fulfils several key functions in repairing an interruption of the sugar phosphate backbone. It efficiently detects the presence of a break by its N-terminal zinc-finger domain; the occurrence of a break is immediately

METHODS IN ENZYMOLOGY, VOL. 409   0076-6879/06 $35.00

translated into a posttranslational modification of histones H1 and H2B leading to chromatin structure relaxation and therefore to increased DNA accessibility. As an amplified DNA damage signal, auto-poly(ADP-ribosyl) ation of PARP-1 triggers the recruitment of XRCC1, which coordinates and stimulates the repair process, to the DNA damage sites in less than 15 s in living cells (Okano *et al.*, 2003). Although dispensable in a test tube DNA repair experiment, *in vivo* these three properties positively influence the overall kinetics of a DNA damage-detection/signaling pathway leading rapidly to the resolution of DNA breaks. Accordingly, poly ADP-ribose (PAR) synthesis and the accompanying NAD consumption are now considered as *bona fide* marks of DNA interruptions in the genome.

In this chapter we describe several methods for determining PARP activation in response to the occurrence of DNA breaks *in vitro* and *in vivo*.

Affinity Purification of Recombinant DNA-Damage Dependent Poly(ADP-Ribose) Polymerases (PARPs)

3-AB Affinity Resin Synthesis

ECH Sepharose 4B is formed by covalent linkage of 6-aminohexanoic acid to Sepharose 4B using an epoxy coupling method. ECH Sepharose has free carboxyl groups at the end of 6-carbon spacer arms, which are used to couple ligands containing primary amino groups with the carbodiimide coupling method. The long flexible hydrophilic spacer arm connected to the gel is particularly suitable for immobilization of small molecules such as inhibitor containing free NH_2 groups like 3-aminobenzamide (3-AB).

Materials

- ECH Sepharose 4B (Pharmacia Biotech, Piscataway, NJ) 50 ml (#17–0571–01).
- N'-(3-Dimethylaminopropyl)-N-ethylcarbodiimide   Hydrochloride, 98% (Sigma-Aldrich, St. Louis, MO) 10 g (#16,146–2);
- 3-Aminobenzamide, 99%, (Sigma-Aldrich) 1 g (#A-0788).

Methods. The method for coupling is the carbodiimide method. The N,N'-disubstituted carbodiimide promotes condensation between the free amino of 3-AB and a free carboxyl group to form a peptide link by acid catalyzed removal of water.

1. *Preparing the Gel:* ECH Sepharose 4B is supplied preswollen in 20% ethanol. Decant the ethanol and wash the required amount of gel (25 ml)

on a sintered glass filter with about 80 ml 0.5 M NaCl per ml sedimented gel added in several aliquots.

2. *Coupling the 3-AB:* The coupling reaction is performed in the cold room at 4° during 12 to 24 h in distilled water adjusted to pH 4.5–6.0 with 0.1 M sodium hydroxide to promote the acid-catalyzed condensation reaction. The pH value of the reaction mixture decreases during the first hour of the coupling. The pH must therefore be adjusted during this time by addition of 1 N sodium hydroxide solution. The reaction is performed in a 50-ml Falcon tube containing 25 ml of the gel. The volume of the solution is adjusted to 50 ml with water. 1 g of N'-(3-Dimethylaminopropyl)-N-ethylcarbodiimide Hydrochloride is added plus 0.5 g of 3-AB dissolved in 1 ml of methanol 100%. Using a rotating wheel, rotate the mixture end-over-end overnight in the cold room. During the first hour, control and adjust the pH if necessary.

3. *Stop the Coupling Reaction:* The excess of ligand is eliminated by washing the gel thoroughly in the buffer containing 100 mM acetate pH 4.0 and 0.5 M NaCl then in the buffer containing 100 mM Tris-HCl pH 8.0 and 0.5 M NaCl. Wash with distilled water. The column can now be packed and equilibrated. The column is stable for many years and can be used many times if kept in the appropriate conditions (at 4° in buffer containing 0.02% sodium azide).

Comments. This method of purification using the ECH Sepharose 4B has been found to be much more efficient than the previous method using the Affigel-10 from Biorad, which for unknown reasons had a lower stability in time. We found this column coupled to 3-AB very useful for affinity purification of PARP–1, –2, and –3; it should be potentially usable for the purification of other members of the PARP family displaying a good affinity for 3-AB.

PARP-1 Purification Protocol from Insect Cells

The following protocol is suitable for purification of the human recombinant PARP–1, –2, and –3 overexpressed in Sf9 insect cells (Ame *et al.*, 1999; Augustin *et al.*, 2003; Giner *et al.*, 1992). This system results in a high level protein expression and only one purification step is usually necessary to obtain high purity protein that can be used for activity assays or even crystallographic studies (Oliver *et al.*, 2004). This protocol can be easily adapted to other expression systems. The purified PARP-1 displays the same catalytic property compared to the classical purification scheme (Zahradka and Ebisuzaki, 1984), that is, it is free of DNA, of topoisomerase I activity, and of inhibitor contamination.

Materials

- Proteases inhibitor cocktail tablets (CØmplete Mini, Roche Diagnostic Gmbh, Mannheim, Germany).
- 3-methoxybenzamide (Sigma-Aldrich), 5 g, (#M1.005–0).

Cell Lysate Preparation. All the steps are performed at 4° ideally in the cold room or in ice.

1. Following the expression of the protein, the cells (3 × 10⁹ cells) are harvested and frozen at −80°. The cells are then quickly defrosted and resuspended in 5 ml per 10⁸ cells of 25 mM Tris-HCl pH 8.0 buffer containing 50 mM glucose, 10 mM EDTA, 1 mM Phenylmethansulfonylfluorid (PMSF), 2 proteases inhibitor cocktail tablets.

2. Add 0.2% of Tween 20, 0.2% of NP40, 1 M NaCl (weigh the necessary amount) final. For 1.5 × 10⁹ cells the final volume should be 75 ml.

3. Gently agitate the cell suspension for 20 min on a rotating wheel in the cold room. Transfer to a cold glass container.

4. Moderately sonicate 4 × 20 s on ice. Do not stop agitating during the sonication process to ensure a good lysis. The viscosity of the solution should disappear.

5. Clear lysate from cellular debris by centrifugating at 50,000g for 45 min at 4° in polycarbonate tubes (rotor Ti70 Beckman).

6. Add to the supernatant 1 mg/ml final of a 10 mg/ml protamine sulfate solution in water to precipitate the nucleic acids. Centrifuge at 50,000g for 20 min at 4° in polycarbonate tubes (rotor Ti70 Beckman).

7. Keep the supernatant and slowly add 0.226 g/ml (40%) of ammonium sulfate at 0°, mix gently for 30 min. Centrifuge at 20,000g for 20 min at 4° in polycarbonate tubes (rotor Ti70 Beckman) to eliminate the precipitated proteins.

8. To the supernatant add 0.187 g/ml (70%) of ammonium sulfate at 0°, mix gently for 30 min. At this step PARP-1 (and PARP-2) will precipitate. Centrifuge at 20,000g for 20 min at 4° in polycarbonate tubes (rotor Ti70 Beckman) and discard the supernatant. The pellet is kept on ice until resolubilization.

ECH Sepharose 4B-3AB Affinity Purification of PARP-1

1. Equilibrate the column with 4 bed volumes (4 × 50 ml) of the BB-buffer made of 100 mM Tris-HCl pH 7.5, 14 mM β-Mercaptoethanol, 0.5 mM EDTA, and 0.5 mM PMSF.

2. The 70% ammonium sulfate pellet of proteins (from 3 × 10⁹ cells) containing PARP-1 is resuspended in 100 ml per 10⁹ cells of BB-buffer supplemented with 100 mM NaCl.

3. The column is loaded overnight at low flow rates (10 ml/h) using a peristaltic pump. The flow-through is collected for further analysis.

4. The column is then washed with 4 bed volumes with successively 100 m*M*, 400 m*M*, and 800 m*M* NaCl containing BB-buffer. The flow rate is adjusted to 50 ml/h. Start collecting individual fractions.

5. PARP-1 is eluted with 4 bed volumes of 400 m*M* NaCl BB-buffer containing 1 m*M* 3-methoxybenzamide freshly prepared from a 600 m*M* stock solution in 100% methanol.

The column is regenerated by washing with 3 bed volumes of BB-buffer.

PARP-1 Activation and DNA Damage

In Vitro *Activation of Purified PARP-1*

Materials

- 5 ml Glass Assay Tubes, 25 mm diameter Whatman Glass Microfibre Filters (Whatman International Ltd., England) or equivalent, 25 mm diameter vacuum filtration unit.
- Liquid scintillation counter, high flash-point LSC-cocktail aqueous and nonaqueous samples, Ultima Gold MV–Packard (Packard BioSciences B. V., The Netherlands).
- Ice cold 5% TCA, 1% inorganic phosphate solution, ice cold 95% ethanol.

PARP Activity Incubation Mixture: 50 m*M* Tris-HCl pH 8, 4 m*M* MgCl$_2$, 100 m*M* NaCl, 1 m*M* DTT, 200 ng (6 pmol), DNase I treated DNA, 0.1 μg/μl BSA, 400 μM final NAD, 0.1 to 0.5 μCi [^{32}P]-NAD 1000 Ci/mmol, 200 to 400 ng PARP-1, H$_2$O for a final volume of 100 μl.

Procedure

1. The reaction should be done in triplicates. In a glass tube, incubate the reaction mixture for 10 min at room temperature; stop the reaction by adding 4 ml of cold 5% TCA, 1% inorganic phosphate. (This will precipitate the reaction products.)
2. Filter the reaction solution with the vacuum filtering system. (The reaction specific material [pADP-ribose] is retained on the filter); wash the glass tube 2 times with 4 ml of cold 5% TCA, 1% inorganic phosphate, then with 4 ml of cold 95% ethanol. Dry the filter and count the radioactivity in a liquid scintillation counter.

Detection of PARP-1 Activity in Activity Blots

This technique has been developed to determine on the same nitrocellulose sheet a functional PARP activity as well as immunostained active peptide(s) after renaturation of the transferred protein(s) from crude extracts (Simonin *et al.*, 1991).

Materials

- Loading buffer: Tris 50 mM pH 6.8, Urea 6 M, β-mercaptoethanol 6%, SDS 3 g/100 ml, Bromophenol Blue 0.003 g/100 ml.
- Nitrocellulose membrane: Schleicher & Schuell (Dassel, Germany) BA83 (0.2 μm).
- Reduction Buffer: Tris base 25 mM, glycine 192 mM, SDS 0.1%, β-mercaptoethanol 0.7 M.
- Transfer Buffer: Tris base 12.5 mM, Glycine 47 mM, Ethanol 20%.
- Renaturation Buffer: Tris-HCl 50 mM pH 8, NaCl 150 mM, Tween-20 0.3%, Zn-Acetate 20 μM, MgCl$_2$ 2 mM, DTT 1 mM, DNAse I treated DNA 2 μg/ml.
- NAD 0.2 mM.
- [^{32}P]-NAD 10 mCi/ml, 1000 Ci/mmol, PB 10282 (Amersham, Saclay, France).

Procedure

1. To measure PARP activity in cells, resuspend 10^6 cells in 100 μl loading buffer, sonicate 30 s, and load samples on a 10% SDS-PAGE gel. Avoid heat denaturation before loading. As a positive control, load at least 100–500 ng of purified PARP-1.
2. Incubate the gel with gentle agitation for 1 h at 37° in 25 ml of Reduction buffer.
3. Transfer the proteins on Nitrocellulose membrane in Transfer Buffer.
4. Renature the proteins by incubating the membrane 3 times for 20 min at RT in Renaturation buffer.
5. Incubate the membrane in 10 ml of Renaturation buffer supplemented with 2.5 μl NAD 0.2 mM and 10 μCi [^{32}P]-NAD.
6. Wash 3 times for 20 min at RT in Renaturation Buffer.
7. Expose to autoradiography.
8. Optional: proceed to PARP-1 immunodetection using a specific antibody.

Detection of PARP-1 Activity in Permeabilized Cells

Materials

- PBS ×1.
- H$_2$O$_2$ 30% (w/w, corresponding to 9.79 M) H1009, Sigma.
- Lysis and Activity Buffer (LA): Hepes 56 mM pH 7.5, KCl 28 mM, NaCl 28 mM, MgCl$_2$ 2 mM, Digitonin 0.01%, NAD 0.125 μM, [^{32}P]-NAD 5 μCi/ml.

Procedure

1. Spin down (1000 rpm) 10^6 cells grown in suspension or scrape cells grown subconfluently in a 60-mm Petri dish.
2. Wash twice with PBS ×1 at room temperature.
3. Spin down and resuspend the cell pellet with 100 μl of prewarmed (25°) LA buffer supplemented or not with 1 mM H_2O_2.
4. Incubate 10 min at 25°.
5. Stop the reaction by the addition of 30 μl Loading Buffer (see Section Detection of PART-1 Activity in Activity Blots).
6. Sonicate the samples 30 s before loading a 10% SDS-PAGE gel.
7. Transfer the proteins to a Nitrocellulose membrane (see Section Detection of PARP-1 Activity in Activity Blots) and expose to autoradiography.
8. Optional: Proceed to immunodetection of PARP-1 using a specific antibody.

Comments. This technique is adapted from the one described by Laslo Virag in (www.parplink.u-strasbg.fr/protocols/tools/parp_activity.html). Note that the NAD concentration used is low, far from the Km of PARP-1, but this is required since the reaction products are analyzed by SDS-PAGE.

Real-time Determination of Intracellular NAD(P)H to Monitor PARP1 Activation in Living Cells Treated with Genotoxic Agents

See Nakamura *et al.*, 2003.

Materials

- 2,3-bis(2-methoxy-4-nitro-5-sulfophenyl)-2H-tetrazolium-5-carboxa-nilide (XTT, Sigma): 251 mM XTT in DMSO.
- 1-methoxy-5-methylphenazinium methylsulfate (1-methoxy PMS, Sigma) 0.5 mM 1-methoxy PMS in DMSO (stock solution, −20°).
- XTT/1-methoxy PMS solution (−20°): 3.2 mL XTT solution + 66.6 μl 1-methoxy PMS stock solution.
- 3-aminobenzamide (3-AB, Sigma) 3.3 M in DMSO (stock solution, −20°)
- 3,4-dihydro-5-[4-(1-piperidinyl)butoxyl]-1(2H)-isoquinolinone (DPQ, Sigma) 30 mM in DMSO (stock solution, −20°).
- 1× PBS.

Protocol (for One Plate) (see Table I)

1. Count harvested cells (e.g., CHO cells).
2. Prepare cell suspension solution (e.g., 0.35 × 10^6 cells/6 ml complete medium).

4. Aliquot complete medium without cells (50 μl/well) to 8 wells/plate to prepare blank wells.
5. Load cell solution (50 μl/well) to remaining wells of the 96-well plate.
6. Incubate cells overnight in regular CO_2 incubator.
7. Add complete medium (50 μl/well) to all wells.
8. Expose cells to either test chemical or vehicle only (e.g., 1× PBS) for appropriate period.
9. Gently aspirate medium from all wells.
10. Wash all wells with 1× PBS once.
11. Mix 22 μl XTT/1-methoxy PMS solution and 11 ml of complete medium in a 15-ml tube.
12. Aliquot complete medium containing XTT/1-methoxy PMS (100 μl/well) to all wells.
13. Incubate cells in CO_2 incubator.
14. Periodically determine the amount of formazan dye in each well by a 96-well plate reader at 450 nm with 650 nm as the reference filter.
15. NAD(P)H depletion (%control) will be calculated as follows:

(Absorbance in treated group − blank)/
(Absorbance in control group − blank) × 100.

Comments

1. To test whether NAD(P)H depletion is due to PARP1 activation, add a PARP inhibitor (e.g., 10 mM 3-AB for alkylating agents; 90 μM DPQ for oxidants) 2 h prior to chemical treatment and keep in the medium during and after chemical exposure while the cells are analyzed for NAD(P)H levels.
2. Prior to formazan analysis, gently agitate the 96-well plate to mix the contents of each well taking care to avoid generating any bubbles that may interfere with absorbance readings.
3. When handling the 96-well plate avoid splashing the medium onto the lid of the plate in order to maintain the appropriate well volume.
4. The number of cells per plate should be optimized to get reasonable absorbance readings.
5. Typical placement of chemical concentrations across a 96-well plate.

PARP-1 and DNA Repair

Base Excision Repair Assay

This chapter details the BER synthesis assay using wt and PARP-deficient mouse embryonic fibroblast extracts and a plasmid containing a

TABLE I
PLATE LAYOUT FOR 96-WELL FORMAT

1	1	1	3	3	3	5	5	5	6	6	6
2	2	2	4	4	4	Control	Control	Control	Control	Blank	Blank
1	1	1	3	3	3	5	5	5	6	6	6
2	2	2	4	4	4	Control	Control	Control	Control	Blank	Blank
1	1	1	3	3	3	5	5	5	6	6	6
2	2	2	4	4	4	Control	Control	Control	Control	Blank	Blank
1	1	1	3	3	3	5	5	5	6	6	6
2	2	2	4	4	4	Control	Control	Control	Control	Blank	Blank

Numbers represent dosing solutions from highest concentration (1) to lowest concentration (6); control: vehicle only; blank: complete medium without cells.

single abasic site at a defined location, which allows fine mapping of the repair pathways. This protocol is routinely used to investigate BER using mammalian cell extracts and is adapted from Frosina *et al.* (1996).

Construction of DNA Plasmids Containing a Single Abasic Site

Materials

- PGEM-3Zf(+) single-stranded DNA prepared from the phagemid pGEM-3Zf according to the manufacturer's instructions (Promega, Madison, WI).
- *Oligonucleotides*
    5′-GATCCTCTAGAG**U**CGACCTGCA-3′ (contains a uracil, U)
    5′-GATCCTCTAGA**8oxoG**TCGACCTGCA-3′ (contains an 8-ox-oguanine, 8-oxoG)
    5′-GATCCTCTAGAGTCGACCTGCA-3′ (control).
- Enzymes
    T4 polynucleotide kinase (Biolabs, New England)
    T4 DNA polymerase, T4 gene 32 protein, and T4 DNA ligase (Boehringer Mannheim)
    E. coli uracil DNA glycosylase and *S. cerevisiae* 8-oxoguanine DNA glycosylase (kindly provided by S. Boiteux, CEA, Fontenay-aux Roses, France).
- Polynucleotide kinase buffer (5×): 500 mM Tris-HCl (pH 7.4), 50 mM MgCl$_2$, 500 mM DTT, 4 mM ATP.
- Sephadex G-50 column prepared in TE (pH 8) according to Molecular Cloning protocol (Sambrook Fritsch Maniatis, Cold Spring Harbor Laboratory Press, Cold Spring Harbor, New York).
- Annealing buffer (×10): 200 mM Tris-HCl (pH 7.4), 20 mM MgCl$_2$, 500 mM NaCl.

- Synthesis buffer (×10): 175 mM Tris-HCl (pH 7.4), 37.5 mM MgCl$_2$, 215 mM DTT, 7.5 mM ATP, 0.4 mM each dATP, dTTP, dCTP, dGTP.
- TE (pH 8): 10 mM Tris-Hcl pH 8, 1 mM EDTA.

Procedure. Phosphorylation of oligonucleotides

1. Add 2 μg of oligonucleotide to a 1.5-ml microcentrifuge tube.
2. Add 6 μl of 5× polynucleotide kinase buffer and bring the total volume to 29 μl with Milli-Q water.
3. Add 10 units of T4 polynucleotide kinase, mix, and incubate at 37° for 45 min.
4. Bring up the temperature to 65° for 10 min to denaturate the enzyme.
5. Add 20 μl TE buffer to bring the total volume to 50 μl.
6. Load the sample on G-50 sephadex column and spin for 4 min at 1500 rpm at 20°.
7. Collect the flow-through in a microcentrifuge tube.
8. Store the phosphorylated oligonucleotide frozen at −20° until further use Annealing of phosphorylated oligonucleotides to single-stranded pGEM-3Zf(+).

1. Add 10 μl of ss pGEM-3Zf(+) to a 1.5-ml microcentrifuge tube.
2. Add 3 μl of 10× annealing buffer and 2 μl of Milli-Q water.
3. Add 15 μl of phosphorylated oligonucleotide, mix, and incubate at 75° for 5 min, then allow it to cool from 75° to 30° over 30 min.

Synthesis of double-stranded pGEM-3Zf containing a single uracil (pGEM-U) or 8-oxoguanine (pGEM-8oxoG) lesions.

1. Add 30 μl of annealed DNA to a 1.5-ml microcentrifuge tube.
2. Add 3.6 μl of 10× synthesis buffer.
3. Add sequentially 2.5 μg T4 gène 32 protein, 1 unit of T4 DNA ligase, and 1 unit of T4 DNA polymerase, mix, and incubate on ice for 5 min, at room temperature for the next 5 min and finally at 37° for 90 min.
4. Add one volume of phenol-chloroforme-isoamyl alcohol (25:24:1), vortex, and spin for 5 min at 13,000 rpm.
5. Keep the supernatant and add one volume of chloroforme-isoamyl alcohol (24:1), vortex, and spin for 3 min at 13,000 rpm.
6. Keep the supernatant and precipitate DNA with 0.1 volume of sodium acetate 3 M and 2 volumes of Ethanol, overnight at −80°.
7. Spin for 30 min at 13,000 rpm at 4°, dry the DNA pellet, and resuspend in TE buffer.
8. Double-stranded plasmids are then purified by cesium chloride equilibrium centrifugation according to standard Molecular Cloning Protocol (Sambrook, Fritsch, Maniatis).

Synthesis of double-stranded pGEM-3Zf containing a single abasic site (pGEM-AP).

Abasic sites can be produced by either monofunctional DNA glycosylases (e.g.: uarcil-DNA glycosylase, UDG) that remove the altered base without cleaving the phosphodiester bond adjacent to the lesion or, bifunctional DNA glycosylases (e.g.: 8-oxoguanine DNA glycosylase, OGG1) that cleave the DNA strand 3′ of the resulting abasic site by a β-elimination reaction. Both natural and incised abasic sites are repaired in a different way by the BER machinery but in a PARP-1-dependent reaction (Dantzer et al., 2000; Frosina et al., 1996).

Incubate the plasmid molecule pGEM-U with E. coli uracil DNA glycosylase and the plasmid pGEM-8-oxoG with S. cerevisiae 8-oxoguanine DNA glycosylase in the appropriate reaction buffers according to van der Kemp et al. (1996).

Whole Cell Extracts of wt and PARP-1 Deficient 3T3 Cells

Materials

- Ice cold PBS $1\times$
- 500 mM phenylmethylsulfonylfluoride (PMSF) dissolved in acetone. Store at $20°$.
- Protease inhibitor cocktail tablets: Complete Mini EDTA-free (Roche)
- Hypotonic lysis buffer (pH 7.9): 10 mM Tris-HCl, 1 mM EDTA, 5 mM DTT, 0.5 mM spermidine, 0.1 mM spermine. Keep at $4°$.
- Sucrose-glycerol buffer (pH 7.9): 50 mM Tris-HCl, 10 mM MgCl$_2$, 2 mM DTT, 25% sucrose, 50% glycerol. Keep at $4°$.
- Dialysis buffer (pH 7.9): 25 mM Hepes-KOH, 100 mM KCl, 12 mM MgCl$_2$, 1 mM EDTA, 17% glycerol, 2 mM DTT, adjust to pH 7.9 with 5 M KOH. Keep at $4°$.
- Solid ammonium sulfate.
- Saturated ammonium sulfate solution neutralized to pH 7 with NaOH.

Procedure

1. Grow cells up to 1×10^9 cells.
2. Wash cells twice with ice-cold PBS, scrape the cells in PBS, and pellet at 1300 rpm for 5 min.
3. Carefully remove the supernatant and measure the packed cell volume (PCV).
4. Resuspend the cells in 4 PCV of ice cold hypotonic lysis buffer and add 5 μl PMSF and 1 tablet protease inhibitor cocktail per ml PCV.

5. Leave the cells to swell for 20 min on ice.

6. Pour the cell suspension in a glass homogenizer with a Teflon pestle and homogenize on ice with 20 strokes.

7. All following steps are performed in the cold room. Pour the homogenate in a glass beaker containing a magnetic stirrer and stir very slowly.

8. Add dropwise 4 PCV of sucrose/glycerol buffer under continuous stirring.

9. Add dropwise 1 PCV of saturated ammonium sulfate and stir for 30 min.

10. Pour the viscous suspension to appropriate tubes and centrifuge for 3 h at 42,000 rpm using a SW55 rotor at 4°.

11. Remove the supernatant leaving the last 1 ml in the tube and measure the volume (usually 6–7 ml for 1 ml PCV).

12. Pour the supernatant in a glass beaker containing a magnetic stirrer, stir slowly, and add 0.33 g of pure ammonium sulfate per ml of solution.

13. When dissolved add 10 μl of 1 M NaOH par g of ammonium sulfate added. Continue stirring for 30 min.

14. Pour the solution in an appropriate tube and centrifuge for 20 min at 11,000g at 4°.

15. Remove the supernatant carefully with a pipette leaving the pellet as dry as possible.

16. Resuspend the pellet carefully just by mixing with a 1 ml gilson tip in dialysis buffer (0.05 volume of the high speed supernatant).

17. Dialyze for 2 h against 500 ml of dialysis buffer. Change the buffer and dialyze again for additional 12 h.

18. Centrifuge the dialyzed extract for 10 min at 15,800g to remove the precipitate and transfer the whole cell extract to a new tube.

19. Dispense 50 μl aliquots into cryotubes, freeze on dry ice, and store at −80°.

Repair Assay. pGEM-AP or pGEMcontrol plasmids are incubated for 3 h with cell-free extracts of wt and PARP-1 deficient 3T3 cells under standard repair conditions. To measure the overall repair (short-patch repair [SPR] + long-patch repair [LPR]), repair replication is performed in the presence of α-^{32}PdTTP and plasmid DNA are digested with SmaI and HindIII to release a 33-bp fragment. To measure the long-patch repair efficiency (LPR), repair replication is performed in the presence of α-^{32}PdCTP and plasmid DNA are digested with HincII and HindIII to release a 16-bp fragment downstream to the lesion (see Fig. 1).

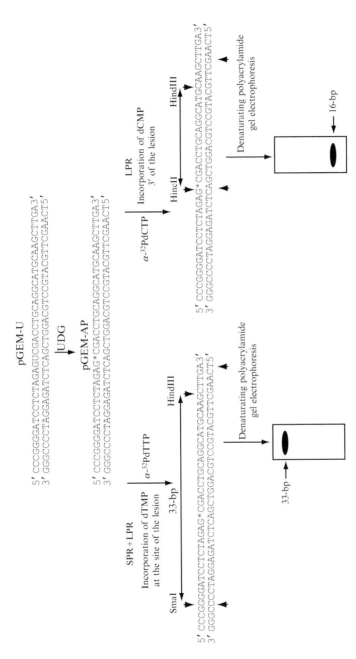

Fig. 1. Scheme of the base excision repair assay. The single natural AP-site containing plasmid, pGEM-AP, is obtained by incubation of pGEM-U with E. coli uracil-DNA glycosylase (UDG). To measure the overall repair (SPR+LPR), repair replication is performed in the presence of α-^{32}PdTTP and plasmid DNA are digested with SmaI and HindIII to release a 33-bp fragment. To measure the long-patch repair efficiency (LPR), repair replication is performed in the presence of α-^{32}PdCTP and plasmid DNA is digested with HincII and HindIII to release a 16-bp fragment downstream to the lesion. Repair products are analyzed by autoradiography after electrophoresis separation on denaturing 15% polyacrylamide gels.

Materials

- Synthesis buffer (×5): 225 mM Hepes/KOH (pH 7.8), 300 mM KCl, 37.5 mM MgCl$_2$, 4.5 mM DTT, 50 μM each dNTP, 10 mM ATP, 200 mM phosphocreatine, 90 μg BSA. Store in aliquots at $-80°$.
- Creatine phosphokinase (CPK) (Type 1, Sigma). 2.5 mg/ml dissolved in 5 mM glycine (pH 9), 50% glycerol. Store in aliquots at $-80°$.
- EDTA 0.5 M.
- RNase A: 2 mg/ml dissolved in H$_2$O. Store in aliquots at $-20°$.
- SDS 10%.
- Proteinase K 2 mg/ml dissolved in H$_2$O. Store in aliquots at $-20°$.
- Denaturing loading buffer: 80% formamide, 0.1% xylene cyanol, 0.1% bromophenol blue.
- α-^{32}P labeled dNTP solution (α-^{32}PdTTP or α-^{32}PdCTP, 3000 Ci/mmol, 10 mCi/ml, Amersham).

Procedure: Repair Assay

1. Add 400 ng of plasmid construct to a 1.5-ml microcentrifuge tube.
2. Add 50 μg of mouse whole cell extract.
3. Add 10 μl of 5× synthesis buffer and 1 μl of creatine phosphokinase.
4. Bring the total volume to 49 μl with Milli-Q water.
5. Add 1 μl of ^{32}P-labeled dNTP solution (α-^{32}PdTTP or α-^{32}PdCTP), mix, and incubate at 30° for 3 h.
6. Stop the reaction by adding 2 μl of EDTA 0.5 M and keep on ice.
7. Add 2 μl of Rnase 2 mg/ml and incubate at 37° for 10 min.
8. Add 3 μl SDS 10% and 6 μl proteinase K 2 mg/ml and incubate at 37° for 30 min.

Purification of the DNA Product

1. Add an equal volume of phenol, vortex, and centrifuge for 5 min at 14,000 rpm at room temperature.
2. Transfer the supernatant to a new tube.
3. Extract the phenol phase again with an equal volume of TE, vortex, and centrifuge as in step 1.
4. Pool supernatants from steps 1 and 2, add half volume of phenol, vortex, and centrifuge as in step 1.
5. Add an equal volume of chloroform/isoamyl alcohol (24:1), vortex, and centrifuge as in step 1.
6. To the supernatant containing the plasmid, add 1/4 volume of ammonium acetate 7.5 M and 2.5 volume of cold Ethanol, mix and keep overnight at $-80°$.

7. Centrifuge for 30 min at 14,000 rpm at 4°, carefully remove the supernatant, and air-dry the pellet.
8. Resuspend the pellet in 15 μl TE and proceed to digestion with restriction enzymes.

Digestion with Appropriate Restriction Enzymes

1. Incubate the DNA with the appropriate restriction enzymes according to the manufacturer's instructions.
2. Extract and precipitate the DNA as previously.
3. Resuspend the precipitate in 6 μl denaturing loading buffer.
4. Heat the samples at 95° for 2 min.
5. Load onto 15% polyacrylamide gel containing 7 M urea in 90 mM Tris-borate/2 mM EDTA (pH 8.8).
6. Electrophorese the samples at 60 mA for 2 h.
7. Dry the gel and expose it to autoradiography.

PARP-1 Activation and XRCC1 Recruitment in Living Cells

Materials

- 3T3 or HeLa cells (or others).
- Hydrogen peroxide 30% (H1009, Sigma).
- Methylmethane sulfonate (MMS) (Aldrich).
- *Primary antibodies*: 10H monoclonal anti-poly(ADPribose) antibody or polyclonal anti-poly(ADPribose) antibody (LP96–10, Alexis Corp.), anti-XRCC1 (Alexis Corp., Lausen, Switzerland).
- *Secondary antibodies*: goat anti-mouse or goat anti-rabbit antibodies conjugated to Alexa Fluor™ 568 or Alexa Fluor™ 488 (Molecular Probes).
- DAPI stain stock solution: 0.25 μg/μl in 50% glycerol-PBS 1×.
- Mowiol (4–88 Hoechst).
- DABCO anti-fading reagent (Sigma).

Solutions

PBS buffer
PBS + 0.1% Tween (v/v) + 0.1% nonfat dry milk
DNA staining stock solution: 0.25 mg/ml DAPI in PBS, 50% glycerol (store at −20°)
Mounting solution: Add 2.4 g of Mowiol to 4.8 ml of 100% glycerol, stir, and add 6 ml H$_2$0. Stir at room temperature for several hours; add 12 ml 0.2 M Tris-HCl pH 8.5 and incubate at 50° until dissolved.

Add 0.45 g DABCO antifading reagent at 4° [2.5% DABCO (w/v) final concentration]. Aliquote and store at −20°.

Procedure

1. *Whole Cell Damage:* Expose cultured cells (i.e., 3T3 or HeLa) to either 0.1–1 m*M* hydrogen peroxide for 10 min at 37° or 1 m*M* MMS for 30 min at 37°.

2. *Fixation:* Wash damaged cells three times with ice cold PBS, fix cells with methanol/acetone (1/1, v/v) for 10 min at 4°, and wash again three times for 5 min with ice cold PBS supplemented with 0.1% Tween (v/v).

3. *Primary Antibodies:* Cells are incubated overnight at 4° with the monoclonal anti-poly(ADPribose) antibody (10H) (1:200 dilution) or with the polyclonal anti-poly(ADPribose) antibody (1:2.000) diluted in PBS 1×, 0.1% Tween, 0.1% nonfat dry milk.

4. *Secondary Antibodies:* After washing 3 times for 5 min in PBS, 0.1% Tween (v/v), the cells are incubated for 2 h at room temperature (or for 4 h at 4°) with a 1:1,500 dilution of goat anti-mouse or goat-anti rabbit fluorophore-conjugated antibodies in PBS 1×, 0.1% Tween (v/v), 0.1% nonfat dry milk.

5. *DNA Staining:* The cells are washed once for 5 min in PBS 1×, stained for 10 min at room temperature with 1:20,000 dilution in PBS of the DNA staining stock solution, and washed twice again in PBS.

6. *Mounting:* The coverslides are dipped in water and mounted using 25 μl of Mowiol containing DABCO antifading reagent (see Fig. 2A and B).

Poly(ADP-Ribose) Synthesis and XRCC1 Recruitment at Laser Induced DNA Strand Breaks. Cells are grown onto 55 μm square size CELLocate coverslips (Eppendorf, Hamburg, Germany). The cells are incubated with 10 μg/ml Hoechst dye 33258 in DMEM medium for 10 min at 37° under 5% CO_2. Laser microirradiation is performed with a Leica DM LMD microscope (Leica Microsystems, Wetzlar Germany) fitted with a 337.1 nm laser focused through a 40× or a 63× objective. The peak performance is 7.2 mW at the maximum pulse repeat rate of 30 Hz. The laser is guided over cell nuclei using a joystick to draw the laser path. The irradiation is performed using the following features of the laser: aperture 2, power 15, speed 5, off-set 30. Cell positioning on CELLocate coverslips is recorded by phase contrast imaging of irradiated cells. After irradiation, cells are fixed and immunostained as described previously. DNA strand breaks can be detected by deoxynucleotidyl transferase (TdT)-mediated dUTP

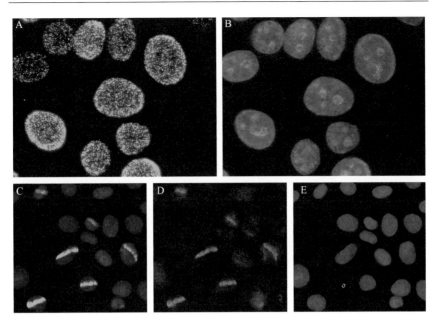

FIG. 2. Poly ADP-ribose synthesis in response to DNA breaks in HeLa cells. (A) HeLa cells treated with 0.5 mM H_2O_2 during 10 min respond by a robust synthesis of PAR visualized with the monoclonal anti-poly (ADP-ribose) antibody 10H (green). (B) Dapi staining of the same cells. (C) Laser-induced DNA damage triggers the local synthesis of PAR at the damage sites (green), that in turn induces the immediate recruitment of XRCC1 (red) dependent on poly(ADP-ribose) synthesis (D). The corresponding nuclei are stained with Dapi (E). Bars indicate 10 μm. (See color insert.)

nick-end-labeling (TUNEL) (ApoAlert DNA fragmentation assay kit, Clontech) (see Fig. 2C through 2E).

Acknowledgments

This work was supported by funds from the Centre National de la Recherche Scientifique, the Association pour la Recherche Contre le Cancer, Electricité de France, Ligue Nationale contre le Cancer, Commissariat à l'Energie Atomique, Ligue Contre le Cancer Région Alsace.

References

Ame, J. C., Rolli, V., Schreiber, V., Niedergang, C., Apiou, F., Decker, P., Muller, S., Hoger, T., Menissier-de Murcia, J., and de Murcia, G. (1999). PARP-2, A novel mammalian DNA damage-dependent poly(ADP-ribose) polymerase. *J. Biol. Chem.* **274,** 17860–17868.

Ame, J. C., Spenlehauer, C., and de Murcia, G. (2004). The PARP superfamily. *Bioessays* **26,** 882–893.

Augustin, A., Spenlehauer, C., Dumond, H., Menissier-De Murcia, J., Piel, M., Schmit, A. C., Apiou, F., Vonesch, J. L., Kock, M., Bornens, M., and De Murcia, G. (2003). PARP-3 localizes preferentially to the daughter centriole and interferes with the G1/S cell cycle progression. *J. Cell. Sci.* **116,** 1551–1562.

D'Amours, D., Desnoyers, S., D'Silva, I., and Poirier, G. G. (1999). Poly(ADP-ribosyl)ation reactions in the regulation of nuclear functions. *Biochem. J.* **342**(Pt. 2), 249–268.

Dantzer, F., de La Rubia, G., Menissier-De Murcia, J., Hostomsky, Z., de Murcia, G., and Schreiber, V. (2000). Base excision repair is impaired in mammalian cells lacking Poly (ADP-ribose) polymerase-1. *Biochemistry* **39,** 7559–7569.

Frosina, G., Fortini, P., Rossi, O., Carrozzino, F., Raspaglio, G., Cox, L. S., Lane, D. P., Abbondandolo, A., and Dogliotti, E. (1996). Two pathways for base excision repair in mammalian cells. *J. Biol. Chem.* **271,** 9573–9578.

Giner, H., Simonin, F., de Murcia, G., and Menissier-de Murcia, J. (1992). Overproduction and large-scale purification of the human poly(ADP-ribose) polymerase using a baculovirus expression system. *Gene* **114,** 279–283.

Nakamura, J., Asakura, S., Hester, S. D., de Murcia, G., Caldecott, K. W., and Swenberg, J. A. (2003). Quantitation of intracellular NAD(P)H can monitor an imbalance of DNA single strand break repair in base excision repair deficient cells in real time. *Nucleic Acids Res.* **31,** e104.

Okano, S., Lan, L., Caldecott, K. W., Mori, T., and Yasui, A. (2003). Spatial and temporal cellular responses to single-strand breaks in human cells. *Mol. Cell. Biol.* **23,** 3974–3981.

Oliver, A. W., Ame, J. C., Roe, S. M., Good, V., de Murcia, G., and Pearl, L. H. (2004). Crystal structure of the catalytic fragment of murine poly(ADP-ribose) polymerase-2. *Nucleic Acids Res.* **32,** 456–464.

Simonin, F., Briand, J. P., Muller, S., and de Murcia, G. (1991). Detection of poly(ADP-ribose) polymerase in crude extracts by activity-blot. *Anal. Biochem.* **195,** 226–231.

van der Kemp, P. A., Thomas, D., Barbey, R., de Oliveira, R., and Boiteux, S. (1996). Cloning and expression in *Escherichia coli* of the OGG1 gene of *Saccharomyces cerevisiae*, which codes for a DNA glycosylase that excises 7,8-dihydro-8-oxoguanine and 2,6-diamino-4-hydroxy-5-N-methylformamidopyrimidine. *Proc. Natl. Acad. Sci. USA* **93,** 5197–5202.

Zahradka, P., and Ebisuzaki, K. (1984). Poly(ADP-ribose) polymerase is a zinc metalloenzyme. *Eur. J. Biochem.* **142,** 503–509.

[30]   Tyrosyl-DNA Phosphodiesterase (Tdp1) (3′-Phosphotyrosyl DNA Phosphodiesterase)

By Amy C. Raymond and Alex B. Burgin, Jr.

Abstract

Tyrosyl-DNA phosphodiesterase (Tdp1) hydrolyzes 3′-phosphotyrosyl bonds *in vitro*. Because topoisomerase I, a type IB topoisomerase, is the only enzyme known to form 3′-phosphotyrosine bonds in eukaryotic cells, it was proposed that Tdp1 is involved in the repair of dead-end topoisomerase I-DNA covalent complexes that may form *in vivo*. It has also been proposed that Tdp1 may represent a novel anticancer target since known anticancer agents (e.g., camptothecin) act by stabilizing topoisomerase I-DNA covalent adducts. The importance of Tdp1 in DNA repair is also demonstrated by the observation that a recessive mutation in the human *TDP1* gene is responsible for the hereditary disorder Spinocerebellar Ataxia with Axonal Neuropathy (SCAN). Although it has been proposed that Tdp1 may be involved in the repair of multiple DNA lesions, this chapter describes the synthesis and characterization of substrates used to study the role of Tdp1 in repairing topoisomerase I-DNA adducts, and the methods used to study the catalytic mechanism and structure of this novel enzyme.

Introduction

Tyrosyl-DNA phosphodiesterase (Tdp1) is the only known enzyme to repair covalent protein-DNA adducts. Tdp1 was serendipitously discovered as a contaminating activity in a commercial preparation of 3′-terminal deoxynucleotide transferase. Preparations of Tdp1 with extremely high specific activity, lacking any transferase activity, could be purified from a variety of eukaryotic sources but not in sufficient quantities to allow direct sequencing. However, characterization of this activity showed that it specifically hydrolyzed 3′-phosphotyrosyl bonds with nanomolar affinity and would not hydrolyze 5′-phosphotyrosyl or 3′-phosphoseryl bonds (Yang *et al.*, 1996). Because a variety of DNA lesions and other agents (including the chemotherapeutic agent camptothecin) were known to stabilize 3′-phosphotyrosine linkages between topoisomerase I and DNA, the observed enzymatic activity led to the proposition that Tdp1 is involved in the repair of dead-end topoisomerase I-DNA complexes *in vivo*. Topoisomerase I, a type IB topoisomerase, is the only enzyme known to form

METHODS IN ENZYMOLOGY, VOL. 409   0076-6879/06 $35.00
        DOI: 10.1016/S0076-6879(05)09030-0

a transient 3'-phosphotyrosine linkage with DNA in eukaryotic cells (Champoux, 2001). Genetic evidence for the role of Tdp1 in DNA repair was provided when the gene encoding Tdp1 was identified from a camptothecin sensitive strain of *Saccharomyces cerevisiae* (Pouliot *et al.*, 1999). The results from this and subsequent studies have identified multiple pathways for repair of stalled Topoisomerase I-DNA complexes in yeast (Liu *et al.*, 2002, 2004; Pouliot *et al.*, 2001). Less is known about the analogous repair pathways in higher eukaryotes; however, a recessive mutation in the human *TDP1* gene was found to be responsible for the hereditary disorder spinocerebellar ataxia with axonal neuropathy (Takashima *et al.*, 2002). It remains unclear why the terminally differentiated cells of these patients are affected by the point mutant Tdp1. Biochemical analysis of the mutant Tdp1, however, leaves no doubt that DNA repair is also affected (Interthal *et al.*, 2005; Zhou *et al.*, 2005). It should be noted that it has been proposed that Tdp1 is also in involved membrane localization, although no biological substrate has been identified (Dunlop *et al.*, 2004). The *in vitro* characterization of human and yeast Tdp1 supports its role in DNA repair, including crystal structures of human Tdp1 bound to DNA and tyrosine through the phosphodiester mimic vanadate (Davies *et al.*, 2003; Raymond *et al.*, 2004). This chapter describes the biochemical tools used to study Tdp1 as a DNA repair enzyme.

Tdp1

Tdp1 is a member of the phospholipase D (PLD) superfamily of enzymes and uses a two-step catalytic mechanism to cleave 3'-phosphotyrosyl bonds. In the first transesterification step, a catalytic histidine displaces tyrosine and generates a covalent 3'-phosphohistidine Tdp1-DNA intermediate. Free Tdp1 and 3'-phosphate DNA is generated in the second hydrolysis step (Fig. 1). An alignment of Tdp1 family members shows that all members share a common C-terminal domain that is homologous to other PLD family members; however, many Tdp1 homologs contain a variable N-terminal domain that varies in both size and amino acid identity. The function of this variable N-terminal domain is not known; however, the lack of conservation suggests that this region is not critical for Tdp1 function. In support of this assumption, full length human Tdp1 and Tdp1 lacking the first 148 residues (Tdp1 Δ148) have both been expressed in *E. coli*, and the proteins have identical specific activities on DNA substrates containing a single 3'-phosphotyrosine residue (see DNA substrates later). The majority of biochemical studies of Tdp1 have been carried out using this N-terminal deletion. It is also important to note that crystal structures of both the yeast (1Q32) and human enzymes (1QZQ) do

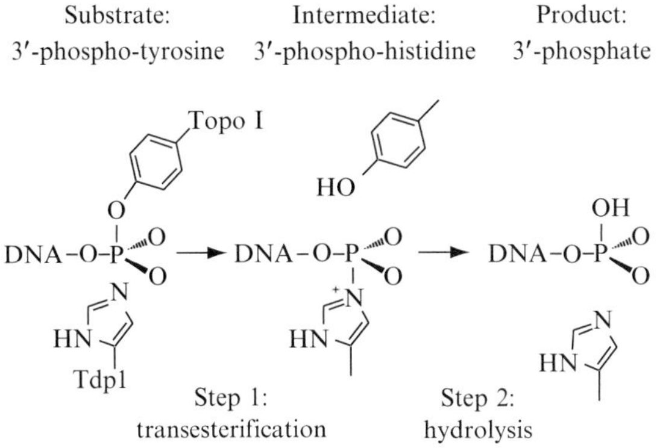

Fig. 1. Two-step reaction mechanism for Tdp1. In the first transesterification step, the Tdp1 nucleophile H263 (human Tdp1) attacks the substrate 3′-phosphotyrosyl bond. In the second step, the 3′-phosphodistidine intermediate is hydrolyzed to produce 3′-phosphate DNA and free Tdp1.

not contain this variable domain, and repeated efforts at crystalizing the full-length human enzyme have been unsuccessful. As expected, the crystal structures of Tdp1 closely resemble the structures determined for other members of the PLD family, with an α-β-α-β-α sandwich composed of two α-β-α domains that are related by a pseudo-2-fold axis of symmetry.

DNA Substrates

3′-Phosphotyrosine DNA

Tdp1 activity was first observed using oligonucleotide substrates containing a single 3′-phosphotyrosine residue. These substrates are prepared by synthesizing the oligonucleotide from tyrosine derivatized resin (Fig. 2) based on a method originally developed by Pan *et al.* (1993). The resin is first derivatized by adding 0.69 g (1.5 mmol) of F-moc-t-butyl-tyrosine in 10 ml dry dichloromethane (Aldrich, St. Louis, MO), 0.018 g (0.15 mmol) dimethylaminopyridine in 1 ml dry dichloromethane, and then 0.31 g (1.5 mmol) of dicyclohexylcarbodiimide (DCC) in 1 ml dry dichloromethane to 0.52 g (0.015 mmol) resin (TentaGel S OH; Rapp Polymere GmbH, Tübingen, Germany). Great care should be taken to ensure that all reagents are anhydrous and transferred under argon. The slurry is stirred at room temperature under argon for 2 h, and then washed sequentially with 10 ml dichloromethane, 10 ml acetonitrile, 10 ml dimethylformamide, 10 ml methanol, and then dried *in vacuo*. The t-butyl protecting group is removed

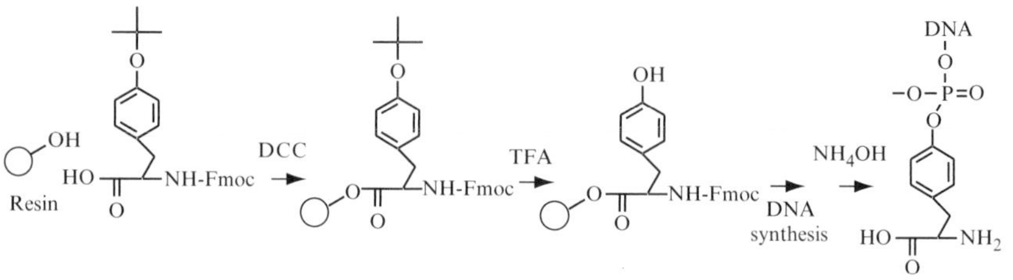

FIG. 2. 3'-phosphotyrosyl DNA substrate. Unmodified resin and Fmoc-*t*-butyl-tyrosine are shown on the left. The resin is derivatized with dicyclohexylcarbodiimide (DCC) and a free tyrosine hydroxyl is generated by treatment with a strong acid (TFA). DNA synthesis from the free tyrosine, and subsequent base deprotection (NH$_4$OH) generates 3'-phosphotyrosyl DNA.

to expose the tyrosine hydroxyl by adding 0.2 ml 50% trifluoroacetic acid (TFA) in dimethylformamide to the resin at room temperature for 10 min. Finally, the resin is washed sequentially with 5 ml dimethylformamide, 5 ml acetonitrile, 5 ml dichloromethane, and then dried *in vacuo*. The derivatized resin is divided into 1 μmol portions for synthesis, and the oligonucleotide synthesis is performed without any modification to standard protocols. The final deprotection step (using concentrated NH$_4$OH at 55° for 12 h) hydrolyzes the ester bond linking the tyrosine to the resin thereby generating the 3'-phosphotyrosyl DNA oligonucleotide (Fig. 2). There is no preference for the nucleotide at the 3'-end of the oligonucleotide, and the final 3'-tyrosine derivatized DNA is usually obtained in 75% yield relative to an unmodified oligonucleotide.

In addition to 3'-phosphotyrosine DNA substrates, synthetic 3'-phosphotyrosyl mimetic substrates have been developed for the mechanistic study of Tdp1. For example, 3'-(4-nitro)phenyl phosphate and 3'-(4-methyl)phenyl phosphate have been used in kinetic analyses of hTdp1 (Raymond *et al.*, 2004). These oligonucleotide substrates, diagrammed in Fig. 3A, share the 3'-phenyl structure provided by the 3'-phosphotyrosine substrate and presumably occupy a very similar space within the hTdp1 active site. However, the pK$_a$ of the leaving groups (4-nitro-phenol and 4-methyl-phenol/cresol) are significantly different. The lower pK$_a$ of 4-nitro-phenol would be expected to make 3'-(4-nitro)phenyl phosphate a better substrate for Tdp1. Surprisingly, these substrates have identical apparent second-order rate constants under standard conditions, suggesting that cleavage of the phosphodiester is not the rate limiting step in the reaction (see Kinetic Analyses later). Each of these substrates can be prepared in a straightforward postsynthetic condensation reaction (Fig. 3B).

The postsynthetic derivatization of 3'-phosphate containing oligonucleo-tides (3'-PO₄) also has the advantage that a large number of different functional groups can be used to replace the tyrosine leaving group without complicated chemistry to generate different modified resins. As an example, Fig. 3B schematizes the derivatization reaction that produces 3'-(4-nitro) phenyl phosphate DNA. 3'-PO₄ DNA is commercially available or can be made using standard methods and 3'-phosphate CPG (Glen Research, Sterling, VA). Typically, a 1 μmol scale synthesis of oligonucleotide is

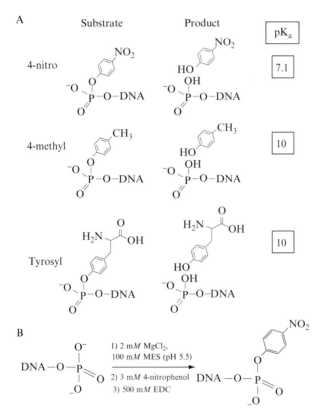

FIG. 3. (A) Derivatized oligonucleotide substrates for Tdp1 studies. 3'-derivatized oligo-nucleotide Tdp1 substrates are diagrammed on the left. 3'-(4-nitro)phenyl phosphate ester DNA is shown at the top and is abbreviated "4-nitro". 3'-(4-methyl)phenyl phosphate ester DNA and 3'-phosphotyrosine DNA, abbreviated "4-methyl" and "tyrosyl" respectively, are shown below. The second column shows the products of Tdp1 cleavage, and the third column shows the pK_a of the leaving group for each substrate. (B) Derivitization of 3'-phosphate oligonucleotide. Conversion of 3'-phosphate oligonucleotide to 3'-(4-nitro)phenyl phosphate DNA using EDC in a one-step condensation reaction is diagrammed.

prepared for each high-efficiency derivatization, while low-efficiency deri-viaizations begin with a pool of 4–6 1 μmol scale oligonucleotide syntheses. It is not necessary to purify the oligonucleotide before derivatization; following deprotection, the 3'-PO$_4$ oligonucleotide is precipitated from ethanol and dried *in vacuo* immediately before derivatization.

3'-(4-Methyl)Phenyl Phosphate DNA

Resuspend the DNA pellet (1 μmol) in 1 ml of 2 mM MgCl$_2$, 100 mM 2-morpholinoethanesulfonic acid (MES) pH 5.5. To the oligonucleotide solution add 1 ml of *para*-cresol saturated acetonitrile. Saturated *para*-cresol can be prepared by adding solid cresol to acetonitrile, shaking at room temperature for 1 h, and then removing insoluble *para*-cresol by centrifugation. Add 0.048 g of 1-ethyl-3-(3-dimethylaminopropyl)-carbo-diimide (EDC) and shake the two immiscible layers vigorously at 65° to form an emulsion. After 20–24 h, extract the aqueous layer three times with 2 ml each of ethyl acetate and recover the DNA from the aqueous phase by ethanol precipitation. Because only ~5–10% of the 3'-phosphate DNA is usually converted to desired product and the 3'-phosphate DNA remains after the reaction, the derivatization protocol can be repeated without purification from the precipitated DNA mixture. Typically, the derivatiza-tion protocol is repeated three times and 3'-(4-methyl)phenyl phosphate DNA is obtained in ~30% final yield.

3'-(4-Nitro)Phenyl Phosphate DNA

Resuspend the DNA pellet in 250 μl of 2 mM MgCl$_2$, 100 mM MES pH 5.5. Add 200 μl of 3 M 4-nitro-phenol and then 0.048 g of EDC. Initially two immiscible layers form; however, the reaction will proceed by vigor-ously shaking the mixture to form an emulsion. The 3'-(4-nitro)phenyl phosphate DNA (also known as 3'-*para*-nitrophenyl phosphate DNA) product of this condensation reaction is photolabile and should be pro-tected from direct light whenever possible. Shake the reaction vigorously for 12–16 h at room temperature. Add 250 μl water to the reaction and extract the aqueous layer three times with 500 μl ethyl acetate. A mixture of 3'-(4-nitro)phenyl phosphate DNA product and 3'-phosphate DNA starting material can be recovered from the aqueous layer by ethanol precipitation. Typically, ~60% of the 3'-phosphate DNA starting material is converted to 3'-(4-nitro)phenyl phosphate DNA product. The final prod-uct is stable under neutral and acidic conditions, and is stored at −20° in water.

The leaving group liberated by cleavage of a 3'-(4-nitro)phenyl phos-phate oligonucleotide substrate, 4-nitrophenol, absorbs 405 nm wavelength light while the intact substrate does not. This unique chromogenic property can be exploited in a spectrophotometric assay to study the enzymatic reaction in real time (Woodfield *et al.*, 2000). However, because the extinction coefficient of 4-nitro-phenol is low, this assay requires millimolar concentrations of substrate and is not suitable for many kinetic studies.

3′-(4-Methylumbelliferone) Phosphate DNA

Although 3′-phosphotyrosyl and 3′-phosphotyrosyl mimetic oligonucleotides can be used to monitor Tdp1 cleavage, fluorescent substrates can be used at a much wider concentration range and are therefore significantly more useful. 4-methylumbelliferone phosphate (7-hydroxy-4-methyl-coumarin phosphate) is a well characterized phosphotyrosine mimic used to study tyrosine phosphatases (Avrameas, 1992; Huschtscha *et al.*, 1989; Kupcho *et al.*, 2004). Hydrolysis of the electron withdrawing phosphate group generates fluorescent 4-methylumbelliferone. The 4-methylumbelliferone linked to DNA through a phosphodiester linkage is not fluorescent; however, Tdp1 efficiently hydrolyzes the phosphodiester bond linking the bicyclic tyrosine mimic to DNA to generate fluorescent 4-methylumbelliferone (Fig. 4) (Rideout *et al.*, 2004). As with the other tyrosine mimic substrates described previously, 3′-(4-methylumbelliferone) phosphate DNA (DNA-MUP) can be generated from the postsynthetic condensation of 3′-phosphate DNA.

Resuspend dry 3′-PO_4 DNA (~1 μmol) in 250 μl of 2 mM $MgCl_2$, 100 mM 2-morpholinoethanesulfonic acid (MES) pH 5.5. To the oligonucleotide solution, sequentially add 0.048 g EDC and 200 μl of 2 M

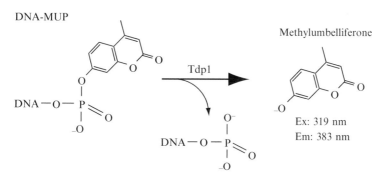

FIG. 4. Fluorescence-based study of Tdp1 activity. The 3′-phospho-(4-methylumbelliferone) (DNA-MUP) substrate shares the 3′-phospho-phenyl architecture of the Tdp1 minimal substrate. This nonfluorescent substrate is recognized and cleaved by Tdp1, generating the fluorescent reporter methylumbelliferone.

4-methylumbelliferone in DMSO. Shake the reaction vigorously at room temperature. The product of this condensation reaction is photosensitive and should be protected from direct light by covering the tube in aluminum foil. After 12 h, add 500 μl of water and extract the aqueous layer three times with 500 μl each of ethyl acetate and recover the DNA from the aqueous phase by ethanol precipitation. 3′-phospho-(4-methylumbellifer-one) derivatized DNA is usually obtained in 75% yield.

Purification of Tdp1 Substrates

The 3′-derivatized substrates described previously can be purified by PAGE; however, the size of the tyrosine or tyrosine mimetic may be very small relative to the oligonucleotide starting material. This makes the purification of longer derivatized oligonucleotides (>15 mers) very difficult. Anion exchange HPLC using a DNA PAC PA-100 4 mm × 250 mm column (Dionex, Sunnyvale, CA) is routinely used for purifying oligonucleotides as large as 50 nucleotides in length. The column is equilibrated in 20 mM sodium phosphate pH 7.0, and the DNA is loaded in water at 1 ml/min. The DNA is eluted over a linear gradient of 1 M NaCl over 15 min (1 ml/min.). For all of the substrates described previously, the Tdp1 substrate elutes after the underivatized 3′-PO$_4$ DNA. This is surprising since the underivatized DNA (3′-PO$_4$) is more negatively charged (−2) than the Tdp1 substrates containing a 3′-phosphodiester (−1). Care should be taken not to use a UV detector when purifying DNA-MUP since this can cause photobleaching of the 3′-phospho-(4-methylumbelliferone) group. Instead, small portions of the eluted fractions can be inspected for the presence of DNA-MUP.

Kinetic Analysis

Regardless of which oligonucleotide substrate is used in the reaction, the second step in the Tdp1 two-step general acid/base reaction generates the tyrosine/tyrosine mimic leaving group, free Tdp1, and 3′-phosphate DNA. The smaller size and increased negative charge of this 3′-PO$_4$ product makes it readily distinguishable from substrate in denaturing PAGE analysis (Raymond *et al.*, 2004, 2005; Rideout *et al.*, 2004) (Fig. 5). The Tdp1 substrates are 5′-end labeled with γ-^{32}P-ATP and T4 polynucleotide kinase (PNK). It is recommended to use mutant T4 polynucleotide kinase that lacks 3′-phosphatase activity (available from New England Biolabs, Ipswich, MA) for labeling Tdp1 substrates. Since Tdp1 generates 3′-phosphate DNA, residual T4 PNK in the reactions can generate 3′-OH DNA. The 3′-OH DNA has less charge and can migrate similarly to the Tdp1 substrate.

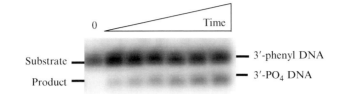

FIG. 5. Gel-based analysis of Tdp1 activity. This gel exemplifies the enzyme dependent conversion of 3′-derivitized oligonucleotide substrate to 3′-phosphate DNA product, which migrates slightly faster on a denaturing PAGE.

Tdp1 enzyme reactions (10–100 μl) contain 50 mM Tris-HCl (pH 8.0), 80 mM potassium chloride (KCl), 5 mM MgCl$_2$, 1 mM dithiothreitol (DTT), and 400 μg/ml bovine serum albumin (BSA). Individual time points are quenched by adding an equal volume of 8 M urea, 30% glycerol, 0.05% SDS, 0.25% bromophenol blue. The quenched reactions can be loaded directly onto a 20% polyacrylamide 0.2 mm slab sequencing gel (8 M urea, 0.5× TBE). As noted previously, the 3′-PO$_4$ DNA reaction product has less molecular weight and greater charge than the substrate oligonucleotide, and therefore migrates slightly faster in this denaturing gel system. An example of time-dependent cleavage by hTdp1 is shown in Fig. 5. The conversion of Tdp1 substrate to 3′-PO4 DNA product can be monitored by phosphorimager analysis of the resulting autoradiogram. The concentration of human Tdp1 in the reaction can vary between 1 pM and 1 nM, and the apparent K$_M$ for single- and double-stranded oligonucleotides is $\sim$100 nM. The enzyme is diluted immediately before use in 50 mM Tris-HCl (pH 8.0), 100 mM NaCl, 5 mM DTT, 10% glycerol, and 500 μg/ml BSA on ice and added to the reactions at 10× the desired concentration. Concentrated Tdp1 ($\sim$1 mg/ml) is stored in 50 mM Tris-HCl (pH 8.0), 100 mM NaCl, 5 mM DTT, 50% glycerol at $-20°$. The enzyme is relatively stable and can be stored for at least 1 year without a decrease in activity.

Under these standard conditions, the apparent second order rate constant (k$_{cat}$ / K$_M$) is $\sim$1 × 10^8 M^{-1} s^{-1}. This reflects the rate of cleavage at low substrate concentrations, the expected *in vivo* condition, and demonstrates that the rate limiting step is enzyme-substrate association; the enzyme is diffusion limited. This is an important feature that is often not appreciated when analyzing Tdp1 mutations. Mutations that slow the rate of catalysis may not slow the overall reaction since the rate limiting step is enzyme–substrate association. When analyzing Tdp1 mutations or screening for inhibitors, it is important to modify the reaction conditions so that the chemical steps of cleavage (transesterification and hydrolysis) become

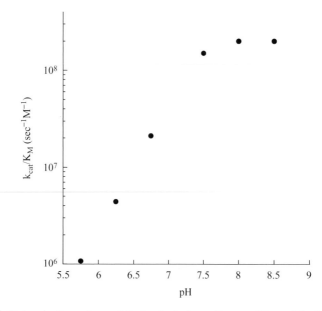

FIG. 6. Determination of non-diffusion-limited reaction conditions. The k_{cat} and K_M values were determined as a function of pH using single-stranded 3'-phosphotyrosine substrates. The observed k_{cat}/K_M values are plotted as a function of pH, showing a clear drop in second-order rate constant as the pH of the reaction is lowered.

rate limiting. This can be done easily by lowering the pH of the reaction since all PLD family members use a general acid/base catalytic mechanism. Figure 6 shows that lowering the pH of the reaction results in a corresponding drop in the second-order rate constant.

As described previously, a fluorescent Tdp1 substrate (3'-[4-methylumbelliferone] phosphate DNA; DNA-MUP) has also been developed that allows Tdp1 activity to be monitored in real time. In this assay, reactions containing 50 mM Tris-HCl (pH 8.0), 80 mM KCl, 2 mM MgCl$_2$, 1 mM DTT, 40 μg/ml BSA, and 10% DMSO are equilibrated to 37° before the reaction is started with the addition of Tdp1. The excitation maximum for the fluorescent reporter methylumbelliferone is 319 nm and the emission maximum is 383 nm. An example of an hTdp1 dependent cleavage reaction is shown in Fig. 7. One disadvantage of this assay is that the fluorescent reporter must be at pH 7.8 or greater. Under these conditions, the apparent second order rate constant for double- and single-stranded 3'-(4-methylumbelliferone) phosphate DNA is close to the rate of diffusion $\sim 1 \times 10^8$ M^{-1} sec^{-1} and care should be taken when analyzing Tdp1 mutants or inhibitors

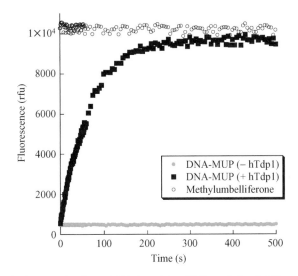

FIG. 7. Real time monitoring of Tdp1 activity. 3'-(4-methylumbelliferone) phosphate DNA (500 nM) was reacted with Tdp1 at 37° and time resolved fluorescence was measured through a 355-nm excitation filter and a 460-nm emission filter. DNA-MUP (500 nM; filled squares) or 4-methylumbelliferone (500 nM; open circles) were incubated in the presence of Tdp1 and the fluorescent signal is plotted as a function of time. Incubation of DNA-MUP (500 nM) without the addition of Tdp1 is also plotted (filled circles). In this experiment, the reactions were performed in white, flat-bottom, nontreated, 96-well polystyrene plates using Polarstar plate reader (BMG).

(Rideout *et al.*, 2004). Fortunately, this feature is not a concern during presteady kinetic analysis and this substrate is ideally suited for these applications.

DNA Binding

A conventional method for the study of protein-DNA binding interactions is an electromobility shift assay (EMSA), or gel-shift assay. In order to resolve enzyme-bound DNA from free DNA in the nondenaturing polyacrylamide gel, the protein-DNA interaction needs to be stable within the gel during electrophoresis. For reasons that are not known, several laboratories have been unable to observe a gel mobility shift of DNA by yeast or human Tdp1. Even attempts at shifts with enzyme concentrations 100× the apparent K_M of the reaction have been unsuccessful. This prompted the development of a fluorescence anisotropy method to study hTdp1 substrate binding (Raymond *et al.*, 2005). There are a number of considerations in designing a fluorescence anisotropy assay, all of which are covered in great detail elsewhere (Lakowicz, 1999). Fluorescence anisotropy indirectly measures the solution tumbling rate of the fluorescent reporter molecule. Since the tumbling rate is a function of the molecular weight, the fluorescence anisotropy and fluorescence polarization methods reflect the proportion of fluorescent population that is bound by enzyme.

The chain-terminating, 6-FAM (6-carboxyfluorescein) donating phosphoramidite (Glen Research) is incorporated at the 5′-terminal position of the oligonucleotide, which is synthesized using tyrosine derivatized resin as described previously. This results in an oligonucleotide containing a 3′-phosphotyrosine residue and a 5′-fluorescent reporter. A catalytically inactive enzyme (such as the nucleophile deficient hTdp1 mutant H263A) is used to monitor binding in the absence of catalysis. Data in Fig. 8 shows that increasing Tdp1 concentration results in an increased anisotropic signal as a result of Tdp1 binding to the fluorescent reporter. In this experiment, the equilibrated enzymatic reactions are performed in 50 mM Tris-HCl (pH 8.0),

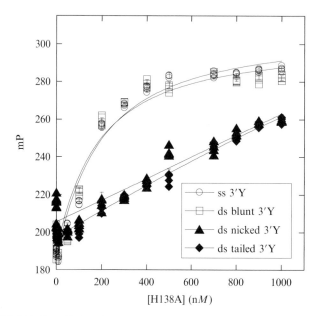

FIG. 8. Tdp1 binding affinity of fluorescein-labeled substrates. Increasing concentrations of catalytically inactive human Tdp1 (H138A) was incubated with fluorescein-labeled 3′-tyrosyl phosphate oligonucleotides (3′Y) and the resultant polarization was measured under equilibrium conditions. In this experiment, single-stranded (ss 3′Y) and double stranded (ds 3′Y) substrates were compared. Nicked and tailed substrates bind hTdp1 weakening and very high concentrations of enzyme do not saturate the polarization signal (for details see Raymond et al., 2004, 2005; Rideout et al., 2004).

5 mM MgCl$_2$, 80 mM KCl, 2 mM EDTA, 1 mM DTT, and 40 μg/ml BSA at 37°. The reactions should be monitored in black, flat-bottom, 96-well plates. The excitation and emission maxima of 6-FAM are 494 nm and 525 nm wavelengths, respectively. Polarization of the fluorescein reporter has been successfully measured by excitation with vertically polarized light at a wavelength of 485 nm and detection of fluorescence at 520 nm. It is important to note that the spectral properties of 6-FAM are pH sensitive and the experiment can only be performed between pH 7.5 and 8.5. When this method is used to determine equilibrium binding constants, it is essential that the reaction be allowed to reach equilibrium before fluorimetry measurements begin. Equilibrium conditions can be determined empirically in pilot experiments by determining the incubation time required for polarization values to plateau.

References

Avrameas, S. (1992). Amplification systems in immunoenzymatic techniques. *J. Immunol. Methods* **150,** 23–32.

Champoux, J. J. (2001). DNA topoisomerases: Structure, function, and mechanism. *Annu. Rev. Biochem.* **70,** 369–413.

Davies, D. R., Interthal, H., Champoux, J. J., and Hol, W. G. (2003). Crystal structure of a transition state mimic for Tdp1 assembled from vanadate, DNA, and a topoisomerase I-derived peptide. *Chem. Biol.* **10,** 139–147.

Dunlop, J., Morin, X., Corominas, M., Serras, F., and Tear, G. (2004). Glaikit is essential for the formation of epithelial polarity and neuronal development. *Curr. Biol.* **14,** 2039–2045.

Huschtscha, L. I., Lucibello, F. C., and Bodmer, W. F. (1989). A rapid micro method for counting cells "*in situ*" using a fluorogenic alkaline phosphatase enzyme assay. *In Vitro Cell Dev. Biol.* **25,** 105–108.

Interthal, H., Chen, H. J., Kehl-Fie, T. E., Zotzmann, J., Leppard, J. B., and Champoux, J. J. (2005). SCAN1 mutant Tdp1 accumulates the enzyme-DNA intermediate and causes camptothecin hypersensitivity. *EMBO J.* **24,** 2224–2233.

Kupcho, K., Hsiao, K., Bulleit, B., and Goueli, S. A. (2004). A homogeneous, nonradioactive high-throughput fluorogenic protein phosphatase assay. *J. Biomol. Screen* **9,** 223–231.

Lakowicz, J. (1999). "Principles of Fluorescence Spectroscopy." p. 725. Plenum Press, New York, NY.

Liu, C., Pouliot, J. J., and Nash, H. A. (2002). Repair of topoisomerase I covalent complexes in the absence of the tyrosyl-DNA phosphodiesterase Tdp1. *Proc. Natl. Acad. Sci. USA* **99,** 14970–14975.

Liu, C., Pouliot, J. J., and Nash, H. A. (2004). The role of TDP1 from budding yeast in the repair of DNA damage. *DNA Repair (Amst.)* **3,** 593–601.

Pan, G., Luetke, K., Juby, C. D., Brousseau, R., and Sadowski, P. (1993). Ligation of synthetic activated DNA substrates by site-specific recombinases and topoisomerase I. *J. Biol. Chem.* **268,** 3683–3689.

Pouliot, J. J., Robertson, C. A., and Nash, H. A. (2001). Pathways for repair of topoisomerase I covalent complexes in *Saccharomyces cerevisiae*. *Genes Cells* **6,** 677–687.

Pouliot, J. J., Yao, K. C., Robertson, C. A., and Nash, H. A. (1999). Yeast gene for a Tyr-DNA phosphodiesterase that repairs topoisomerase I complexes. *Science* **286,** 552–555.

Raymond, A. C., Rideout, M. C., Staker, B., Hjerrild, K., and Burgin, A. B., Jr. (2004). Analysis of human tyrosyl-DNA phosphodiesterase I catalytic residues. *J. Mol. Biol.* **338,** 895–906.

Raymond, A. C., Staker, B. L., and Burgin, A. B., Jr. (2005). Substrate specificity of tyrosyl-DNA phosphodiesterase I (Tdp1). *J. Biol. Chem.* **280,** 22029–22035.

Rideout, M. C., Raymond, A. C., and Burgin, A. B., Jr. (2004). Design and synthesis of fluorescent substrates for human tyrosyl-DNA phosphodiesterase I. *Nucleic Acids Res.* **32,** 4657–4664.

Takashima, H., Boerkoel, C. F., John, J., Saifi, G. M., Salih, M. A., Armstrong, D., Mao, Y., Quiocho, F. A., Roa, B. B., Nakagawa, M., Stockton, D. W., and Lupski, J. R. (2002). Mutation of TDP1, encoding a topoisomerase I-dependent DNA damage repair enzyme, in spinocerebellar ataxia with axonal neuropathy. *Nat. Genet.* **32,** 267–272.

Woodfield, G., Cheng, C., Shuman, S., and Burgin, A. B. (2000). Vaccinia topoisomerase and Cre recombinase catalyze direct ligation of activated DNA substrates containing a 3'-para-nitrophenyl phosphate ester. *Nucleic Acids Res.* **28,** 3323–3331.

Yang, S. W., Burgin, A. B., Jr., Huizenga, B. N., Robertson, C. A., Yao, K. C., and Nash, H. A. (1996). A eukaryotic enzyme that can disjoin dead-end covalent complexes between DNA and type I topoisomerases. *Proc. Natl. Acad. Sci. USA* **93,** 11534–11539.

Zhou, T., Lee, J. W., Tatavarthi, H., Lupski, J. R., Valerie, K., and Povirk, L. F. (2005). Deficiency in 3'-phosphoglycolate processing in human cells with a hereditary mutation in tyrosyl-DNA phosphodiesterase (TDP1). *Nucleic Acids Res.* **33,** 289–297.

[31] Assaying Double-Strand Break Repair Pathway Choice in Mammalian Cells Using a Targeted Endonuclease or the RAG Recombinase

By DAVID M. WEINSTOCK, KOJI NAKANISHI,
HILDUR R. HELGADOTTIR, and MARIA JASIN

Abstract

DNA damage repair is essential for the maintenance of genetic integrity in all organisms. Unrepaired or imprecisely repaired DNA can lead to mutagenesis, cell death, or malignant transformation. DNA damage in the form of double-strand breaks (DSBs) can occur as a result of both exogenous insults, such as ionizing radiation and drug therapies, and normal metabolic processes including V(D)J recombination. Mammalian cells have multiple pathways for repairing DSBs, including nonhomologous end-joining (NHEJ), homologous recombination (HR), and single-strand annealing (SSA). This chapter describes the use of reporter substrates for assaying the contributions of these pathways to DSB repair in mammalian cells, in particular murine embryonic stem cells. The individual contributions of NHEJ, HR, and SSA can be quantified using fluorescence and

METHODS IN ENZYMOLOGY, VOL. 409   0076-6879/06 $35.00
DOI: 10.1016/S0076-6879(05)09031-2

PCR-based assays after the precise introduction of DSBs either by the I-*Sce*I endonuclease or by the RAG recombinase. These reporters can be used to assess the effects of genetic background, dominant-negative constructs, or physiological conditions on DSB repair in a wide variety of mammalian cells.

Introduction

Measuring the Nature and Frequency of Repair of DNA Double-Strand Breaks Generated by the I-SceI Endonuclease

The repair of DNA double-strand breaks (DSBs) in mammalian cells occurs by nonhomologous end-joining (NHEJ) or by homologous recombination (HR), a process in which a homologous sequence acts as a repair template. Depending on the context of the DSB, another pathway involving homology can also be used for repair, which is termed single-strand annealing (SSA). In both HR and SSA, sequences adjacent to a DSB are processed to 3′ single-strand tails. Homologous recombination involves the recruitment of the RAD51 protein to the single-strand tails, which promotes a strand invasion reaction (Sung *et al.*, 2003), and eventually leads to the restoration of the initial genomic sequence if an identical template (e.g., the sister chromatid) is used in the reaction (Johnson and Jasin, 2000). By contrast, SSA involves the annealing of complementary single-strand tails formed at repeated sequences and is inhibited by RAD51 (Stark *et al.*, 2004). The SSA pathway is mutagenic, because the sequence between the repeats is deleted upon the completion of the reaction. In NHEJ, the DSB ends may be modified by the addition or deletion of nucleotides prior to ligation, making NHEJ potentially mutagenic as well. Among the NHEJ pathways, "classic" NHEJ is the most well-studied and involves the heterodimer Ku70/Ku80, the serine/threonine kinase DNA-PKcs, the XRCC4/Ligase IV (Lig4) complex, and Artemis (Lieber *et al.*, 2003).

We previously described a fluorescence-based assay for measuring the frequency of HR at a chromosomal DSB in murine embryonic stem (ES) cells (Fig. 1A), using the DR-GFP reporter (Pierce and Jasin, 2005; Pierce *et al.*, 1999, 2001). First, DR-GFP is targeted to the *hprt* locus in ES cells (Pierce and Jasin, 2005), so that it can be used to assay HR without concern for genomic position effects. DR-GFP contains an upstream green fluorescent protein (GFP) gene repeat (*SceGFP*), which is nonfunctional due to the replacement of 11 bp of *GFP* sequence to create the 18-bp recognition sequence for the I-*Sce*I endonuclease. Transfection of an I-*Sce*I expression vector *(pCBASce)* (Richardson *et al.*, 1998) will generate a DSB that can

A DR-GFP I-*Sce*I site loss assay

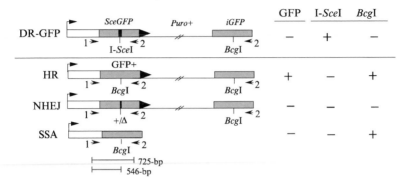

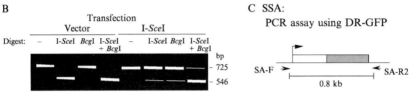

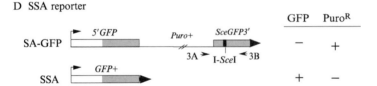

FIG. 1. Fluorescence and PCR assays to measure different pathways of repair of an I-*Sce*I endonuclease generated DSB. (A) After expression of I-*Sce*I in cells containing the DR-GFP reporter, repair of the DSB can proceed through HR, NHEJ, or SSA. Cells that repair by HR through a short-tract gene conversion without crossing-over become GFP+. The different pathways of repair can be distinguished by PCR amplification and digestion with I-*Sce*I and/or *Bcg*I. (B) Simulation of an idealized gel from the PCR site loss assay from cells containing the DR-GFP reporter transfected with empty vector or I-*Sce*I-expression vector. The 725-bp band in the I-*Sce*I-digest represents the product amplified from cells that have undergone HR, SSA, or imprecise NHEJ. The 546-bp band in the *Bcg*I digest represents the product amplified from cells that have undergone HR or SSA. The 725-bp band in the I-*Sce*I+*Bcg*I-digest represents the product amplified from cells that have undergone imprecise NHEJ. (C) PCR strategy for evaluating repair by SSA. (D) Fluorescence substrate SA-GFP for evaluating repair by SSA (Stark *et al.*, 2004). Although other pathways can give rise to a *GFP*+ gene, they appear to be rare. (See Nakanishi *et al.*, 2005 for using resistance to puromycin to remove long-tract gene conversion events.) Abbreviations: 1, DRGFP1; 2, DRGFP2; HR, homologous recombination, NHEJ, nonhomologous end-joining; SSA, single-strand annealing; 3A, SAGFP3A; 3B, SAGFP3B. Arrows, PCR primers.

be repaired by several mechanisms, including HR, NHEJ, or SSA. Homologous recombination can be further subdivided into short or long tract gene conversion, with or without crossing-over. Short-tract gene conversion without crossing-over, which appears to be the majority of HR events in mammalian cells (Nakanishi *et al.*, 2005; Richardson *et al.*, 1998), restores a functional *GFP* gene if the downstream internal *GFP* repeat (*iGFP*) is used as the repair template (Fig. 1A). If this occurs, the cells become GFP+ and acquire green fluorescence, which is quantified using a flow cytometer.

A previous chapter described the DR-GFP reporter and included protocols for ES cell culture, gene targeting, and flow cytometry (Pierce and Jasin, 2005). In this chapter, we describe polymerase chain reaction (PCR)-based assays to measure the frequencies of imprecise NHEJ and SSA at the DR-GFP reporter. Cells that have undergone repair by HR, imprecise NHEJ, or SSA will "lose" the I-*Sce*I site (Fig. 1A). Thus, the PCR product amplified from these cells will become resistant to cleavage by I-*Sce*I, allowing for the overall assessment of repair by these pathways (Fig. 1B). A *Bcg*I digest can be used to distinguish the contributions of homologous (HR, SSA) and nonhomologous (NHEJ) repair to I-*Sce*I site loss (Fig. 1B). Individual products from imprecise NHEJ can be cloned and sequenced. Furthermore, SSA products can be assayed separately using PCR (Fig. 1C) (Nakanishi *et al.*, 2005) or by a second reporter substrate (SA-GFP; Fig. 1D) (Stark *et al.*, 2004).

Measuring the Nature and Frequency of Repair of DSBs Generated by the RAG Recombinase

A substantial fraction of the diversity at antigen receptor loci results from imprecise NHEJ of DSBs created by the RAG recombinase during V(D)J recombination (Bassing *et al.*, 2002). The RAG recombinase, composed of the RAG1 and RAG2 proteins, initiates recombination by introducing nicks at recombination signal sequences (RSS elements), each composed of a conserved heptamer and nonamer sequence separated by a nonconserved spacer of either 12 bp (12-RSS) or 23 bp (23-RSS). Through a transesterification reaction, the nicks convert to DSBs, resulting in two hairpin coding ends and two blunt signal ends. The signal ends undergo precise NHEJ, whereas the hairpin coding ends undergo further processing prior to joining, resulting in a diverse set of junctions.

Genomic rearrangements in some lymphoid malignancies, including t(14;18) (q32;q21) in follicular lymphoma (Jager *et al.*, 2000) and t(11;14) (q32;q21) in mantle cell lymphoma (Welzel *et al.*, 2001), are believed to

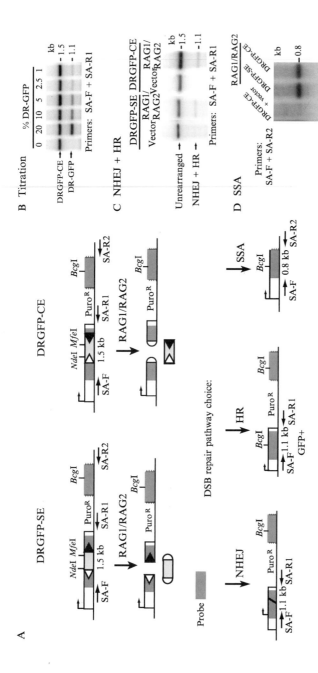

FIG. 2. Fluorescence and PCR assays to measure different pathways of repair of RAG-recombinase generated DSBs using the DRGFP-SE and DRGFP-CE reporters. (A) RAG-induced excision of the sequence between the RSS elements (light gray) in DRGFP-SE and DRGFP-CE results in DSB ends that can undergo repair by NHEJ [i.e., V(D)J recombination], HR, or SSA. For DRGFP-SE, cleavage produces two blunt, chromosomal signal ends. For DRGFP-CE, cleavage produces two hairpin, chromosomal coding ends. As for DR-GFP, HR using the downstream repair template results in a GFP+ cell. (B) A representative PCR-Southern using different mixtures of genomic DNA from untransfected cells containing the DRGFP (1.1-kb band) and DRGFP-CE (1.5-kb band) reporters. The lower band is overrepresented due to unequal PCR amplification of the two products. (C) A representative PCR-Southern demonstrating a combination of HR and V(D)J recombination of approximately 1–4% in wild-type cells. Because

result when a DSB formed during a failed attempt at V(D)J recombination joins with a concurrent DSB in a heterologous chromosome. Supporting the ability of RAG-induced DSBs to undergo pathways of repair other than canonical V(D)J recombination, we recently demonstrated that RAG-induced breaks in ES cells can undergo repair by HR and SSA (Weinstock and Jasin, 2006). Compared with NHEJ, HR of RAG-induced DSBs is rare in wild-type cells (~2% of NHEJ repair), but it is substantially increased in NHEJ mutants (Weinstock and Jasin, 2006). Another group has also reported HR repair of RAG-induced breaks in a hamster cell system (Lee *et al.*, 2004).

In this chapter, we describe reporters that can be used to quantify various pathways of repair of RAG-induced chromosomal breaks (Weinstock and Jasin, 2006). In these reporters, the upstream *GFP* repeat is nonfunctional owing to the insertion of a 12-RSS followed by 333 bp of intronic sequence from the human β-globin gene followed by a 23-RSS (Fig. 2A). In DRGFP-CE, the RSSs are oriented such that RAG-mediated cleavage will result in two chromosomal coding ends and excision of a fragment with two blunt, signal ends. In contrast, RAG-mediated cleavage of DRGFP-SE will result in two chromosomal signal ends and excision of a fragment with two hairpin, coding ends. Similar to cells containing the DR-GFP reporter, cells containing the DRGFP-CE or DRGFP-SE reporter that undergo HR by short-tract gene conversion without crossing over become GFP+ and can be quantified by flow cytometry. The percentage of cells that undergo repair by NHEJ [i.e., V(D)J recombination] is quantified using a PCR-based assay (Fig. 2). The frequency of single-strand annealing can also be estimated using the PCR-based assay (Fig. 1C).

In conclusion, our reporters can be adapted for the rapid assessment of multiple DSB repair pathways in mammalian cells. This approach is useful for assessing the effects of genetic background (Pierce *et al.*, 2001), dominant-negative constructs (Stark *et al.*, 2004), and even physiological conditions on DSB repair (Yang *et al.*, 2005). In addition, targeted ES cells can be developed into murine models for assaying DSB repair in somatic cell types (H.R.H. and M.J., in preparation).

---

the percent of GFP+ cells is typically 0.02–0.1%, the vast majority of this product arises from V(D)J recombination in wildtype cells. (D) PCR-Southern assay for SSA demonstrating approximately 100-fold higher amplification for cells containing the DR-GFP reporter after I-*Sce*I expression compared to cells containing the DRGFP-SE or DRGFP-CE reporters after RAG expression. Arrows, PCR primers.

Materials

Embryonic Stem Cell Culture

a. A well-characterized line of mouse embryonic stem (ES) cells (e.g., J1, E14).

b. The plasmids phprtDRGFP, phprtSAGFP, phprtDRGFP-CE, and phprtDRGFP-SE (available from Dr. Jasin at m-jasin@ski.mskcc.org).

c. Tissue culture incubator.

d. Laminar flow tissue culture hood.

e. 24-well and 10-cm tissue culture plates.

f. 70% ethanol.

g. Ca^{2+}/Mg^{2+} free phosphate-buffered saline (PBS).

h. ES cell medium: mix 500 ml high-glucose Dulbecco's modified Eagle's medium (DMEM), 75 ml ES cell qualified fetal bovine serum, 6 ml 100× penicillin/streptomycin (10,000 U/ml stock), 6 ml 100× nonessential amino acids (10 mM stock), 6 ml 100× L-glutamine (200 mM stock), 6 ml dilute 2-mercaptoethanol (dilution is 21.6 μl of stock 2-mercaptoethanol in 30 ml of PBS), and 60 μl leukemia inhibitory factor 10^7 U/ml (available as ESGRO from Chemicon, Temecula, CA).

i. Trypsin/EDTA solution: 0.2% trypsin, 1 mM EDTA in PBS.

j. Clinical centrifuge (e.g., Marathon model 8 K, Fisher, Pittsburgh, PA).

Measuring HR or SSA at a Chromosomal Break

a. Plasmids pCBA*Sce*, pCAG-RAG1, and pCAG-RAG2 (available from Dr. Jasin)

b. Opti-Mem I Reduced Serum Medium without phenol red (Invitrogen, Carlsbad, CA)

c. Lipofectamine 2000 (Invitrogen)

d. Flow cytometer (e.g., FACScan [488-nm argon laser], BD Biosciences, San Jose, CA).

Site Loss PCR

a. PCR primers:
   DRGFP1, 5'-AGGGCGGGGTTCGGCTTCTGG
   DRGFP2, 5'-CCTTCGGGCATGGCGGACTTGA
   SAGFP3A, 5'-GCCCCCTGCTGTCCATTCCTTATT
   SAGFP3B, 5'-ATCGCGCTTCTCGTTGGGGTCTTT.

b. Restriction enzyme: *Bcg*I (e.g., New England Biolabs, Beverly, MA), Meganuclease I-*Sce*I (Roche, Indianapolis, IN).

c. GC-RICH PCR system (Roche).

d. dNTP 10 mM each

e. DNA purification kit (e.g., QIAquick Gel Extraction Kit, Qiagen, Valencia, CA).

f. PCR cycler (e.g., Mastercycler, Eppendorf,).

g. Agarose, molecular biology grade (e.g., Invitrogen) and agarose gel apparatus, including power supply (e.g., Owl Scientific, Portsmouth, NH).

h. Gel loading buffer: mix 600 μl 50% glycerol in ethanol, 50 μl 1% bromophenol blue in ethanol, 50 μl 1% xylene cyanol in ethanol, 60 μl Tris-HCL buffer (pH 8.0), 60 μl 500 mM EDTA (pH 8.0), and 180 μl water. Store at room temperature.

i. DNA size markers (e.g., 1 Kb Plus DNA Ladder, Invitrogen).

j. Tabletop microfuge (e.g., Eppendorf 5415 D, Fisher).

k. ChemiDoc System (BioRad, Hercules, CA) or other method for measuring relative band intensities.

l. 10 mM Tris (pH 8.5).

Analyzing Individually Repaired Clones

a. TOPO TA cloning system (Invitrogen).

b. Competent *E. coli* (e.g., Top10, Invitrogen).

c. Luria-Bertani (LB) agar plates with ampicillin 50 μg/ml.

d. LB broth with ampicillin 50 μg/ml.

e. Miniprep kit (e.g., QiaPREP Spin Miniprep kit, Qiagen).

f. 40 mg/ml X-gal in dimethylformamide.

g. Restriction enzyme *Eco*RI.

PCR-Southern Assay

a. PCR primers:
   SA-F, 5′-TTTGGCAAAGAATTCAGATCC-3′
   SA-R1, 5′-CAAATGTGGTATGGCTGATTATG-3′
   SA-R2, 5′-ATGACCATGATTACGCCAAG-3′.

b. PCR SuperMix (Invitrogen).

c. PCR cycler.

d. Blotting membrane (e.g., GeneScreen Plus charged nylon membrane [NEN, Boston, MA] using the alkaline transfer instructions provided by the manufacturer).

e. Micro column (e.g., Probe Quant G-50, Amersham Biosciences, Piscataway, NJ).

f. Southern blot hybridization solution: mix equal amounts of 1 M Na$_2$HPO$_4$ and 2 mM EDTA (pH 8.0), 2% bovine serum albumin (BSA), 10% SDS. Stock solutions can be stored at room temperature indefinitely.

g. 20× SSC (3 M NaCl; 0.3 M Na citrate, pH 7.0).

h. ImageJ software (http://rsb.info.nih.gov/ij/).

Methods

Protocols for ES cell culture, gene targeting, isolation of genomic DNA, assaying the frequency of gene targeting, confirming gene targeting by Southern blotting, electroporation of the I-SceI expression vector, and flow cytometry were previously described for the DR-GFP reporter (Pierce and Jasin, 2005).

Measuring HR of a Chromosomal I-SceI DSB

Transfection of ES cells containing the DR-GFP reporter with pCBASce will result in DSB formation in SceGFP (Fig. 1A). Homologous recombination via short-tract gene conversion (without crossing over) involving the downstream iGFP repeat will generate a functional GFP gene, giving rise to cells that constitutively express GFP. The number of cells expressing functional GFP can be measured by flow cytometry. The practical limit of detection with this procedure is on the order of 0.001% fluorescent cells, if sufficient numbers of cells are analyzed by flow cytometry. Wild-type cells generally show homologous repair of a few percent. To increase the output of this assay, we now perform transfections in 24-well plates.

a. The day before transfection, seed 2×10^5 ES cells, which have DR-GFP targeted to the *hprt* locus, into a gelatinized well of a 24-well plate. Incubate at 37° in a humidified incubator with 5% CO_2.

b. Warm a bottle of ES cell medium without penicillin/streptomycin to 37°.

c. Aspirate the medium from the cells to be transfected, and add back 400 μl of prewarmed medium without penicillin/streptomycin.

d. Add 0.8 μg pCBASce to 50 μl of room temperature Opti-MEM without phenol red in a sterile tube.

e. Add 2 μl of Lipofectamine 2000 to 50 μl of room temperature Opti-MEM without phenol red in a sterile tube.

f. Incubate at room temperature for 5 min then combine the two mixtures.

g. Incubate for an additional 20 min.

h. Add the combined 100 μl to the well containing the targeted ES cells and the 400 μl of media.

i. Gently swirl the plate then return it to the incubator.

j. After 6 h, add 500 μl of ES cell medium without penicillin/streptomycin and incubate overnight.

k. The following day, aspirate all of the media from the plate (see Note 1).

l. Add 100 μl trypsin/EDTA solution and incubate at 37° for 2 min.

m. Add 200 μl ES media, gently mix the cells, and transfer to a gelatinized 10-cm plate containing 10 ml of ES media with penicillin/streptomycin.

n. Two days later (3 days after transfection), trypsinize the cells from the 10-cm plate into a cellular suspension. Analyze 1/6 vol of the cells by flow cytometry for the presence of green fluorescence (Pierce and Jasin, 2005).

o. Isolate genomic DNA from the remaining cellular suspension 7 days after transfection (Pierce and Jasin, 2005).

p. For flow cytometry, we use a FACScan (488-nm argon laser). Your settings will depend on your particular instrument.

Measuring the Frequency of Imprecise NHEJ of an I-SceI-Generated DSB

PCR amplification of pooled genomic DNA from pCBA*Sce*-transfected cells will result in a mixture of cells that have not undergone DSB formation and cells that have repaired by each of the potential DSB repair pathways. Cells that have repaired by imprecise NHEJ, SSA, or HR will "lose" the I-*Sce*I site (Fig. 1). In addition, cells that have undergone SSA or HR will replace the I-*Sce*I site with a *Bcg*I site, allowing for the discrimination of specific repair pathways.

a. Isolate genomic DNA from cells 7 days after transfection with pCBA*Sce*, as described in Methods section, step 1.

b. Follow the manufacturer's instructions for the GC-RICH PCR system (Roche), using 2 μg of genomic DNA and the PCR primers DRGFP1 and DRGFP2.

c. Perform PCR amplification as follows:

    95° for 3 min

    10 cycles of 95° for 30 s, 63.7° for 30 s, 72° for 1 min

    25 cycles of 95° for 30 s, 63.7° for 30 s, 72° for 1 min (plus 5 s for each cycle)

    72° for 7 min

d. Purify the PCR product (e.g., Qiagen gel purification kit) into 50 μl of 10 m*M* Tris (pH 8.5).

e. Digest 20 μl of the purified DNA overnight with 10 U of I-*Sce*I in a 50 μl digestion.

f. Purify the digested product into 50 μl and digest half of the volume with 2 U *Bcg*I in a 50 μl digestion (see Note 2).

g. Separate the products on a 1.2% agarose gel without ethidium bromide.

h. Stain the gel with 0.0005% ethidium bromide for 20 min with gentle agitation to ensure equal staining of the gel.

i. Destain the gel in water for 10 min with gentle agitation.

j. We quantify the bands using a BioRad ChemiDoc System with rolling disk background subtraction. This method was verified to give a quantitative linear response over a wide range of I-SceI site loss using mixtures of genomic DNA from untransfected cells and genomic DNA from a pure population of transfected cells that are GFP+ (isolated by cell sorting). Other methods of quantifying bands (e.g., ImageJ software) are likely to be accurate but have not been formally assessed by our laboratory.

k. The undigested PCR product should be a single 725-bp band. I-SceI digestion of PCR product from untransfected cells will yield 546 and 179 bp bands (Fig. 1B). Similarly, BcgI digestion of PCR product from transfected cells that are GFP+ (isolated by cell sorting) will yield 546-bp and 179-bp bands. Be sure to include control lanes to verify that the product underwent complete enzymatic digestion. Because the 546-bp band will be weaker than the 725-bp band due to its reduced length, it is necessary to include a 725/546 correction. Thus, the correct formula for site loss is:

Percent site loss = 725 bp band/(725 bp band + [725/546] × 546 bp band)

l. The percent site loss resulting from I-SceI digestion is the total percent of transfected cells that underwent repair by imprecise NHEJ, SSA or HR.

m. The percent site loss resulting from both I-SceI and BcgI digestion is the total percent of transfected cells that underwent repair by imprecise NHEJ.

n. The percent of cells that underwent SSA can be estimated as the percent I-SceI site loss minus the sum of the percent imprecise NHEJ and the percent GFP+ cells. An assay for directly comparing SSA between samples is outlined below.

Assaying the Frequency of SSA of an I-SceI-Generated DSB Using the DR-GFP Reporter

We have developed a PCR-Southern strategy for amplifying the product resulting from SSA after I-SceI expression in cells containing the DR-GFP substrate. Primers SA-F and SA-R2 will amplify a product resulting from SSA that can be compared between samples (Fig. 1C).

a. Isolate GFP coding sequence for use as a probe. Plasmid phprtDRGFP, when digested with HindIII and BamHI, will yield 7 fragments of 4825, 3344, 2298, 1009, 522, 284, and 185 bp. Gel-purify the 522 bp fragment.

b. PCR amplify 0.4 μg genomic DNA from transfected cells using 22.5 μl PCR Supermix (Invitrogen), 500 nM primer SA-F, and 500 nM primer SA-R2 in a total volume of 25 μl as follows:

    95° for 3 min
    20 cycles of 95° for 30 s, 56° for 30 s, 72° for 1 min 20 s
    72° for 7 min

c. Electrophorese the product on a 0.8% agarose gel using a suitable size marker.

d. Blot the gel onto a suitable membrane.

e. Radiolabel 25 ng of the *GFP* coding sequence probe with $\alpha[^{32}P]$ dCTP or $\alpha[^{32}P]$dATP. We find that the PRIME IT II Random Primer Labeling Kit (Stratagene, La Jolla, CA) works well.

f. Purify the radiolabeled probe from the unincorporated radionucleotides and primers using a micro column.

g. Hybridize the probe with the membrane in hybridization solution overnight at 68°.

h. Rinse the membrane using successive 30-min rinses with 2× SSC/ 0.1% SDS (twice) and 0.5× SSC/0.1% SDS (twice), all at 68°. Dry the membrane and expose to film for 5–60 min.

i. The relative intensity of each band can be measured using ImageJ software, according to the online instructions (http://rsb.info.nih.gov/ij/docs/index.html). Alternatively, a phosphorimager could be used to measure the relative band intensities.

Analyzing Individual Repair Products

We use a modified version of the PCR approach outlined above to analyze individual repair products resulting from a chromosomal DSB.

a. Amplify genomic DNA as outlined on page 533 using primers DRGFP1 and DRGFP2 and purify the PCR product into 50 μl of 10 mM Tris (pH 8.5).

b. Clone 5 μl into the pCR2.1-TOPO vector (Invitrogen), according to the manufacturer's instructions.

c. Transform 5 μl of cloned vector into competent *E. coli* and allow the transformed cells to grow on LB plates with ampicillin 50 μg/ml overnight at 37° (see Note 3).

d. The following day, pick individual colonies and grow for 8–12 h in 2 ml LB broth with ampicillin 50 μg/ml.

e. Isolate plasmid DNA by miniprep.

f. Digest approximately 1 μg of genomic DNA with 5 U I-*Sce*I and 5 U *Eco*RI in 20 μl using I-*Sce*I digestion buffer for at least 2 h.

g. Separate the bands on a 0.8% agarose gel with suitable size markers. Clones that do not have an integrated PCR fragment will have 2 bands of 3916 bp and 15 bp. Clones that contain a PCR fragment that has not undergone I-SceI site loss will have 3 bands of 3916 bp, approximately 555 bp, and approximately 185 bp. Clones that contain a PCR fragment that has undergone I-SceI site loss will have a band of 3916 bp and an additional band. If the clone contains a PCR fragment from a cell that underwent HR or SSA, the additional band will be 739 bp. If the clone contains a PCR fragment from a cell that underwent imprecise NHEJ, the band will be of variable size depending on the extent of sequence modification prior to ligation. The vast majority of recovered NHEJ junctions in wild-type cells contain deletions or insertions of less than 50 bp.

h. For the clones that have undergone I-SceI site loss, digest approximately 1 μg of genomic DNA with 5 U BcgI, 5 U EcoRI, and 20 μM S-adenosylmethionine in 20 μl for at least 2 h (see Note 2).

i. Separate the bands on a 0.8% agarose gel with suitable size markers. Clones that contain a PCR fragment that has undergone SSA or HR will have 3 bands of 3916 bp, approximately 550 bp, and approximately 180 bp. Clones that contain a PCR fragment that has undergone imprecise NHEJ will have a band of 3916 bp and a band of variable size depending on the extent of sequence modification prior to ligation.

j. Individual clones can be sequenced using the primer DRGFP1 or M13 Reverse.

Measuring SSA Using a Fluorescence Assay

To efficiently assay SSA without the need for PCR amplification, a separate reporter SA-GFP (Fig. 1D) can be introduced into cells (Stark et al., 2004). SA-GFP consists of the GFP gene fragments 5'GFP and SceGFP3', which have 266 bp of homology. I-SceI cleavage in SceGFP3' can establish a functional GFP gene if the DSB is repaired by SSA with the complementary sequence in 5'GFP. The DSB can also be repaired by HR or NHEJ, but these do not establish a functional GFP gene (Stark et al., 2004). Thus, the frequency of SSA can be compared rapidly between samples using flow cytometry. In wild-type ES cells, the fraction of GFP+ cells after I-SceI expression is typically 1–3% (Stark et al., 2004).

a. Digestion and targeting of the phprtSAGFP plasmid were previously described (Stark et al., 2004).

b. Transfection with pCBASce, flow cytometry, and genomic DNA isolation can be performed as described for the DR-GFP reporter.

c. To determine the frequencies of HR or imprecise NHEJ using the SA-GFP reporter, PCR amplify the 793-bp sequence around the I-SceI site using the same protocol on page 533 and primers SAGFP3A and SAGFP3B. Note that the SSA product is not amplified because the upstream primer is lost during SSA repair.

d. To determine the overall frequency of imprecise NHEJ and HR, perform I-*Sce*I site loss as described in step 2 and calculate the percent of the product in the 793-bp band relative to the 498-bp band using the formula:

Percent site loss = 793 bp band/(793 bp band + [793/498] × 498 bp band)

e. The percent site loss resulting from both I-*Sce*I and *Bcg*I digestion is the total percent of transfected cells that underwent repair by imprecise NHEJ.

Assaying HR of a RAG-Induced, Chromosomal Break

There is evidence that HR of extrachromosomal RAG-induced breaks may differ from HR of chromosomal breaks (Lee *et al.*, 2004), so our assays are performed with integrated substrates. Targeting and analysis of HR using the DRGFP-CE and DRGFP-SE reporters is performed using the same procedure as outlined for the DR-GFP reporter (Pierce and Jasin, 2005), except for the following differences (Weinstock and Jasin, 2006):

a. Prior to gene targeting, *Sac*I and *Kpn*I linearization of phprtDRGFP-CE or phprtDRGFP-SE should produce two bands of 10,015 and 2856 bp. If the digest was incomplete, there will be a higher band of 12,871 bp.

b. For verifying *hprt* targeting by Southern blotting, a *Pst*I digest should produce bands of 8621 and 3755 bp, corresponding to targeted integration on the 5′ and 3′ sides, respectively.

c. Transfection of cells targeted with the DRGFP-CE or DRGFP-SE reporter with RAG1 and RAG2 is performed as outlined on pages 532-533, except that 0.8 μg of pCAG-RAG1 and 0.8 μg of pCAG-RAG2 are used in place of 0.8 μg pCBA*Sce*. Wild-type cells typically show homologous repair of 0.02–0.1%. Thus, we routinely analyze at least 200,000 cells by flow cytometry for each transfection.

Determining the Frequency of V(D)J Recombination in a Chromosomal Substrate

Our assay allows for the measurement of chromosomal V(D)J recombination using a PCR-Southern strategy (Fig. 2). Nonhomologous end-joining [i.e., V(D)J recombination] or HR produces a 1.1 kb sequence that will undergo PCR amplification using primers SA-F and SA-R1. The relative intensity of this band can be compared to the unrearranged 1.5-kb band, to determine the total percent of transfected cells that underwent either V(D)J recombination or HR. Because the 1.1-kb and 1.5-kb bands may amplify unequally, it is necessary to establish a standard curve using dilutions of genomic DNA from untransfected cells containing the DR-GFP reporter and untransfected cells containing the DRGFP-CE or DRGFP-SE reporter (Fig. 2B).

a. Perform the PCR-Southern protocol using the same conditions as outlined on pages 534-535, except 200 nM each of primer SA-F and SA-RI should be used and amplification should be performed for 22 cycles.

b. Expose the membrane to film for 5–60 min.

c. The relative intensity of each band can be measured using ImageJ software or a phosphorimager. The percent of product in the 1.1-kb band is defined as:

% of product in the 1.1 kb band = 1.1 kb band/
$$(1.5 \text{ kb band} + 1.1 \text{ kb band})$$

This value should be plotted on the standard curve that was established using dilutions of genomic DNA from cells containing the DR-GFP and DRGFP-CE (or DRGFP-SE) reporters. The corresponding value is the percent of transfected cells that underwent either V(D)J recombination or HR. For wild-type cells, this number is in the range of 1–5%. The percent of cells that underwent V(D)J recombination can be estimated by subtracting the percent of GFP+ cells from this value.

d. Individual clones that repaired by V(D)J recombination can also be isolated and sequenced using the same protocol outlined on pages 535-536. To increase the percentage of colonies that contain a PCR product derived from a cell that underwent rearrangement, predigest the genomic DNA with *Nde*I and *Mfe*I overnight. These enzymes cleave within the 333 bp spacer (Fig. 2A). Thus, genomic DNA from cells that have not undergone RAG-mediated cleavage will retain these sites and will not be PCR amplified.

Assaying SSA of a RAG-Induced Chromosomal Break

The same PCR-Southern strategy outlined on pages 534-535 can be used to amplify the SSA product that results from RAG expression in cells containing the DRGFP-SE or DRGFP-CE reporter. However, the frequency of SSA in these cells is approximately 100-fold lower than the frequency recovered after I-*Sce*I expression in cells containing the DR-GFP reporter

(Weinstock and Jasin, 2005). Thus, PCR amplification should be performed for 22 cycles and the membrane should be exposed to film for 16–48 h.

Notes

1. These conditions will result in killing of 10–30% of the cells. If excessive cell killing is noted, the cells may be treated with ES media without penicillin/streptomycin for several hours prior to transfection.

2. Digestion with *Bcg*I is frequently difficult and may require prolonged incubations or multiple digests with different ratios of DNA: enzyme.

3. The pCR2.1-TOPO vector and Top10 *E. coli* allow for blue-white screening using X-gal, increasing the number of picked colonies that will contain a PCR fragment.

Acknowledgments

We thank Jeremy Stark and Andrew Pierce for assistance with developing these reporters. D. M. W. was supported by the Leukemia and Lymphoma Society (5415-05) and the Byrne Fund. This work was supported by NIH 54688 and NSF MCB-9728333 (M. J.).

References

Bassing, C. H., Chua, K. F., Sekiguchi, J., Suh, H., Whitlow, S. R., Fleming, J. C., Monroe, B. C., Ciccone, D. N., Yan, C., Vlasakova, K., Livingston, D. M., Ferguson, D. O., Scully, R., and Alt, F. W. (2002). Increased ionizing radiation sensitivity and genomic instability in the absence of histone H2AX. *Proc. Natl. Acad. Sci. USA* **99,** 8173–8178.

Jager, U., Bocskor, S., Le, T., Mitterbauer, G., Bolz, I., Chott, A., Kneba, M., Mannhalter, C., and Nadel, B. (2000). Follicular lymphomas' BCL-2/IgH junctions contain templated nucleotide insertions: Novel insights into the mechanism of t(14;18) translocation. *Blood* **95,** 3520–3529.

Johnson, R. D., and Jasin, M. (2000). Sister chromatid gene conversion is a prominent double-strand break repair pathway in mammalian cells. *EMBO J.* **19,** 3398–3407.

Lee, G. S., Neiditch, M. B., Salus, S. S., and Roth, D. B. (2004). RAG proteins shepherd double-strand breaks to a specific pathway, suppressing error-prone repair, but RAG nicking initiates homologous recombination. *Cell* **117,** 171–184.

Lieber, M. R., Ma, Y., Pannicke, U., and Schwarz, K. (2003). Mechanism and regulation of human non-homologous DNA end-joining. *Nat. Rev. Mol. Cell. Biol.* **4,** 712–720.

Nakanishi, K., Yang, Y. G., Pierce, A. J., Taniguchi, T., Digweed, M., D'Andrea, A. D., Wang, Z. Q., and Jasin, M. (2005). Human Fanconi anemia monoubiquitination pathway promotes homologous DNA repair. *Proc. Natl. Acad. Sci. USA* **102,** 1110–1115.

Pierce, A. J., Hu, P., Han, M., Ellis, N., and Jasin, M. (2001). Ku DNA end-binding protein modulates homologous repair of double-strand breaks in mammalian cells. *Genes Dev.* **15,** 3237–3242.

Pierce, A. J., and Jasin, M. (2005). Measuring recombination proficiency in mouse embryonic stem cells. *Methods Mol. Biol.* **291,** 373–384.

Pierce, A. J., Johnson, R. D., Thompson, L. H., and Jasin, M. (1999). XRCC3 promotes homology-directed repair of DNA damage in mammalian cells. *Genes Dev.* **13,** 2633–2638.

Richardson, C., Moynahan, M. E., and Jasin, M. (1998). Double-strand break repair by interchromosomal recombination: Suppression of chromosomal translocations. *Genes Dev.* **12,** 3831–3842.

Stark, J. M., Pierce, A. J., Oh, J., Pastink, A., and Jasin, M. (2004). Genetic steps of mammalian homologous repair with distinct mutagenic consequences. *Mol. Cell. Biol.* **24,** 9305–9316.

Sung, P., Krejci, L., Van Romen, S., and Sehorn, M. G. (2003). Rad51 recombinase and recombination mediators. *J. Biol. Chem.* **278,** 42729–42732.

Weinstock, D. M., and Jasin, M. (2005). Alternative pathways for the repair of RAG-induced DNA breaks. *Mol. Cell. Biol.* **26,** 131–139.

Welzel, N., Le, T., Marculescu, R., Mitterbauer, G., Chott, A., Pott, C., Kneba, M., Du, M. Q., Kusec, R., Drach, J., Raderer, M., Mannhalter, C., Lechner, K., Nadel, B., and Jaeger, U. (2001). Templated nucleotide addition and immunoglobulin JH-gene utilization in t(11;14) junctions: Implications for the mechanism of translocation and the origin of mantle cell lymphoma. *Cancer Res.* **61,** 1629–1636.

Yang, S., Chintapalli, J., Sodagum, L., Baskin, S., Malhotra, A., Reiss, K., and Meggs, L. G. (2005). Activated IGF-1R inhibits hyperglycemia-induced DNA damage and promotes DNA repair by homologous recombination. *Am. J. Physiol. Renal Physiol.* **289,** F1144–F1152.

Author Index

A

Aboussekhra, A., 30, 131
Abraham, R. T., 118
Aburatani, H., 389
Adams, A. E. M., 109, 133, 173
Adams, K. E., 119, 120, 122
Adriance, M., 184, 186
Aebersold, R., 2, 223
Agazie, Y. M., 302
Ahn, B., 56
Ait-Si-Ali, S., 330
Akerman, I., 119, 120, 122
Alberti-Segui, C., 146
Alberts, B. M., 490
Albright, L. M., 16, 27
Albright, N., 257, 272
Alers, T., 490
Alexeev, A., 375, 383
Alisch, R., 52
Allen, J. B., 114
Allers, T., 444
Almouzni, G., 358, 359, 360, 361, 362, 364,
    365, 367, 368, 369, 370, 372
Alt, F. W., 262, 269, 276, 527
Amaratunga, M., 65
Amberg, D. C., 102, 110
Amé, J.-C., 493
Amin, A. A., 13
Amon, A., 139
Anand, V., 215
Anderson, K., 213, 214, 218, 220, 222
Andre, B., 213, 214, 218, 220, 222
Andreoli, L., 85, 96
Andrews, B. J., 213, 214, 215, 217, 222, 223,
    225, 230
Aneliunas, V., 223
Anthony, K., 220
Aouida, M., 214, 215, 216
Aparicio, O., 390
Appeldoorn, E., 30
Appella, E., 56, 80
Araki, M., 426

A (continued, right column)

Arbel-Eden, A., 250, 285
Arkin, A. P., 213, 214, 219, 220, 222
Arlett, C. F., 39
Armitage, P., 207
Armour, C. D., 219
Armstrong, D., 512
Arosio, D., 80
Arrand, J. E., 45
Arrowsmith, C., 27
Arthanari, H., 91
Arthur, L. M., 56
Arunkumar, A. I., 13, 27
Ashikari, T., 390, 391, 398
Asteris, G., 200, 203, 206, 208
Astromoff, A., 213, 214, 218, 220, 222
Au, K., 40, 42
Ausubel, F. M., 16, 18, 27, 100, 310, 481
Avigan, M., 301, 309
Avrameas, S., 517

B

Baccari, C., 220
Bachant, J., 131, 136
Bachl, J., 199
Bachrati, C. Z., 86, 87
Bader, G. D., 214, 215, 222, 223, 225, 230
Bader, J. S., 223
Bai, W., 40, 45
Baker, T. A., 317
Bakkenist, C. J., 251
Baldwin, E. L., 214
Ball, H. L., 120
Baltimore, D., 256
Bambara, R. A., 30, 53, 54, 71, 72
Bando, M., 390, 391, 398
Banerjee, S., 230, 468
Bangham, R., 213, 214, 218, 220, 222
Bard, M., 219
Barlow, J. H., 119, 133, 145
Barnes, D. E., 30, 39
Barre, F. X., 330
Barrett, J. C., 236, 243

E

F

Subject Index

V

V(D)J recombination, *see* Double-strand break repair

W

Western blot
  histone γ-H2AX, 247–248
  Rad9 phosphorylation assay, 140–141
  Rad53 phosphorylation assay, 114–116
  Rad55 phosphorylation assay, 168–169, 171–173
WRN, *see also* RecQ helicases
  ATPase assay, 67–69
  electrophoretic mobility shift assay of DNA binding, 69–71
  exonuclease activity
    assays
      activity combined with helicase activity, 77–78
      activity independent of helicase activity, 76–77
      radiometric assay, 73–76
    domain, 71
    metal dependence, 73
    substrate specificity, 71–73
  helicase activity
    alternate or damaged DNA structures, 57
    assays
      activity combined with exonuclease activity, 77–78
      activity independent of exonuclease activity, 76–77
      D-loop substrate preparation, 60–62
      fluorescence resonance energy transfer, 65–67
      Holiday junction substrate preparation, 59–60
      radiolabeled duplex DNA substrate preparation, 58–59
      radiometric assay, 62–63, 65
    directionality and loading properties, 54–55
    length dependence for unwinding, 56
  phosphorylation status effects on activity, 78, 80
  protein-protein interactions
    catalytic activity modulation, 56, 80

replication protein A, 80–81
purification of human protein from baculovirus-insect cell system, 53–54

X

Xeroderma pigmentosum, phenotypes, 426
X-ray crystallography
  SWI/SNF2:DNA complexes
    complex formation, 385–386
    conformational state trapping, 386–387
    DNA
      design, 383–384
      duplex preparation, 385
    protein fragment preparation, 382–383
  tyrosyl-DNA phosphodiesterase, 512–513
XRCC1
  poly(ADP-ribose) polymerase-1 activation and recruitment assay
    DNA damage, 508
    fixation, 508
    immunostaining, 508
    laser-induced DNA strand breaks, 508–509
    materials, 507–508
  protein complex recovery with immobilized metal-chelate chromatography
    chromatography, 420
    extract preparation, 419–420
    materials, 419
    overview, 417–418
  single-strand break repair role, 410

Y

Yeast DNA damage checkpoint assays
  adaptation assays
    cdc13-1 mutant studies, 163–164
    chromosome loss assays, 161, 163
    comparison of assays, 164
    design, 151–152
    disomic strains, 156–161
    endonuclease-induced double-stranded breaks, 154–156
    ionizing radiation-induced double-stranded breaks, 153
    irreparable double-stranded breaks, 152–153
    overview, 150–151

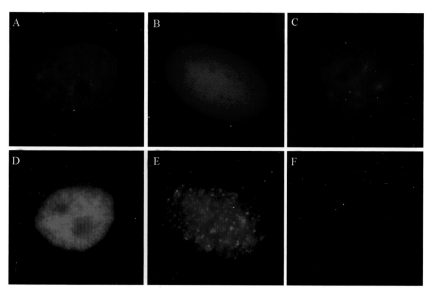

Binz *ET AL.*, Chapter 2, Fig. 2. Indirect immunofluorescence staining of RPA70. DAPI staining of an untransfected cell (A), an untransfected cell that has been treated with 0.5% Triton X solution (chromatin associated fraction) (B), and a RPA70 siRNA transfected cell (C). (D–F) Indirect immunostaining of the same cells in (A–C), respectively, with α-RPA70 antibody.

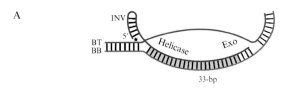

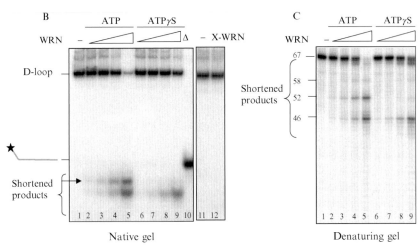

Brosh *et al.*, Chapter 4, Fig. 2. WRN helicase and exonuclease cooperate to dissociate a telomeric D-loop. (A) Telomeric D-loop schematic. A bubble was formed by annealing the BT and BB strand. The INV strand with a 5′ duplex end (hairpin) was hybridized in the melted region. The 4 tandem telomeric repeats are highlighted. See (Opresko *et al.*, 2004b) and text for details. (B) and (C) Analysis of WRN helicase and exonuclease products from a model telomeric D-loop substrate. Wild type WRN protein (lanes 2–9) or the exonuclease dead mutant (X-WRN) (lane 12) was incubated with the D loop substrate containing a labeled INV strand (0.5 n*M*) for 15 min at 37°. Reaction aliquots were run on an 8% native gel (B) and on a 14% denaturing gel (C). The reactions contained 0.75, 1.5, 3, and 6 n*M* WRN with either 2 m*M* ATP (lanes 1–5) or 2 m*M* ATPγS (S, lanes 6–9), or 6 n*M* X-WRN with 2 m*M* ATP (lane 12). Δ, heat denatured substrate. Numbers in (C) indicate product length. The arrow in (B) indicates the products generated in the presence of active helicase, but that are absent when the helicase is inactive.

A   *S. cerevisiae* cell cycle:

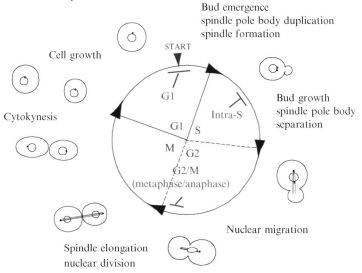

Bud emergence
spindle pole body duplication
spindle formation

START

Cell growth

G1

Intra-S

Bud growth
spindle pole body
separation

Cytokynesis

G1 / S

M / G2

G2/M
(metaphase/anaphase)

Nuclear migration

Spindle elongation
nuclear division

B   Specific cell morphologies:

Normal cell cycle forms:

Unbudded   Small budded   G2/M

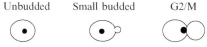

Synchronised cells:

Alpha factor   Nocodazole

G2/M checkpoint arrest:

O'SHAUGHNESSY *ET AL.*, CHAPTER 8, FIG. 1. The cell cycle of budding yeast. (A) The budding yeast cell cycle showing the points of action of the three principal checkpoint arrests. (B) Specific cell morphologies observed in the normal cell cycle, after synchronization in G1 (α factor) and G2/M (nocodazole) and at the G2/M checkpoint.

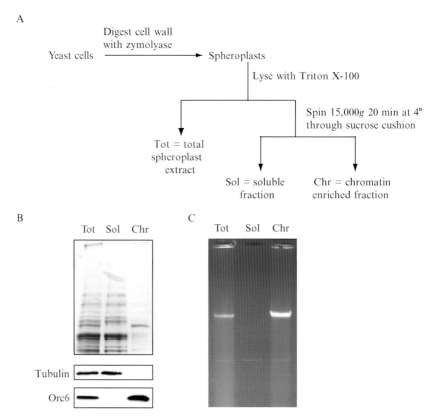

A

Yeast cells →(Digest cell wall with zymolyase)→ Spheroplasts

Lyse with Triton X-100

Tot = total spheroplast extract

Spin 15,000*g* 20 min at 4° through sucrose cushion

Sol = soluble fraction

Chr = chromatin enriched fraction

B

Tot Sol Chr

Tubulin

Orc6

C

Tot Sol Chr

O'Shaughnessy *et al.*, Chapter 8, Fig. 5. Fractionation of chromatin bound and soluble proteins. (A) Schematic of the fractionation of total yeast spheroplast extract (Tot) and soluble (Sol) and chromatin-enriched (Chr) fractions. (B) Electrophoretic and Western blot analysis of fractionated protein. (C) Electrophoretic analysis of fractionated DNA.

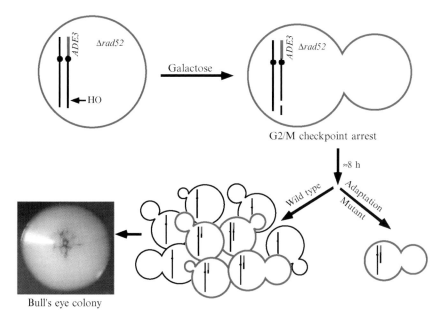

TOCZYSKI, CHAPTER 9, FIG. 3. The bull's eye assay for adaptation. A disomic strain containing an additional, and thus nonessential, copy of chromosome VII encoding the sole functional copy of the *ADE3* gene and a site for the HO endonuclease is shown. The illustrated strain (strain 19–21), is deleted for the MAT locus, such that the only copy of the HO site is on chromosome VII. The *RAD52* gene is also deleted, to prevent repair of the broken chromosome by homologous recombination off of the homolog. On induction of the HO endonuclease by galactose, the HO site on chromosome VII is cut and the cell arrests, via the checkpoint, in G2/M as a large budded cell (i.e., with 2 cell bodies). An adaptation-proficient strain will eventually continue division with the broken chromosome. The broken chromosome will continue to be degraded until it is lost, generating *ade3* progeny that do not accumulate the red pigment. This will form a "bull's eye" colony. In contrast, an adaptation mutant will remain arrested in G2/M. (Toczyski *et al.*, 1997; Sandell and Zakian, 1993; with permission from Elsevier.)

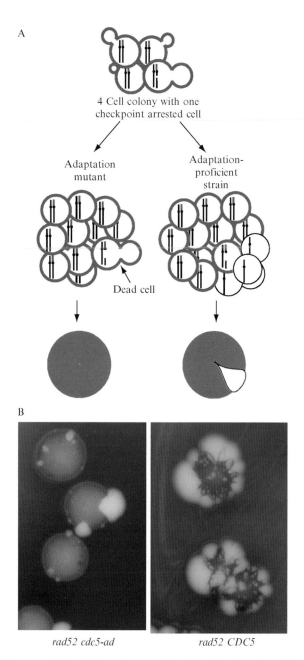

TOCZYSKI, CHAPTER 9, FIG. 5. Sectoring assay for adaptation-dependent chromosome loss. Disomic strains grown in the absence of selection for the 2 chromosomes will spontaneously lose either chromosome. If a visual marker is present on one of the disomic chromosomes, the portion of a colony that arises from a cell that has lost that chromosome will appear as a sector in the colony. The spontaneous rate of chromosome loss in *rad52* mutants is elevated by more than 1000-fold, and the observed loss events appear to be preceded by an initial checkpoint arrest, followed by adaptation to that arrest. Thus, adaptation mutants, such as *cdc5-ad*, produce fewer sectors. Cells that would have gone on to be scored as chromosome loss events will instead remain permanently checkpoint arrested. This phenomenon is cartooned in (A). (B) shows examples of such sectored colonies. (Galgoczy and Toczyski, 2001, with permission from the publisher (ASM).

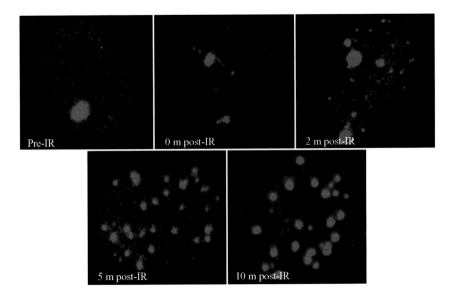

NAKAMURA *ET AL.*, CHAPTER 14, FIG. 1. Immunocytochemical detection of γ-focal growth during the first 10 min after exposure of early passage WI-38 normal human fibroblasts to ionizing radiation.

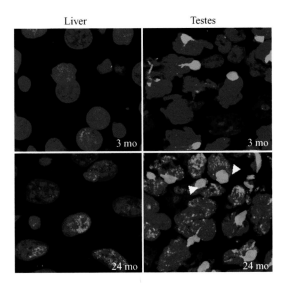

NAKAMURA *ET AL.*, CHAPTER 14, FIG. 2. Immunocytochemical detection of γ-foci in 10-μm frozen sections of liver and testes from young (3 months) and old (24 months) mice, Images shown are collapsed Z-stacks taken with 40× objective, oil immersion. The large areas of γ-H2AX in the testes sections correspond to the X-Y bodies present in late pachyteme spermatocytes.

| γ-H2AX | Merge | 53 bp1 |

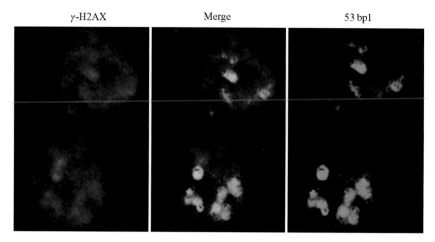

Nakamura *et al.*, Chapter 14, Fig. 3. Immunocytochemical detection of γ-foci and 53 bp1 in touchprints of mouse brain. Images shown are collapsed Z-stacks taken with 40× objective, oil immersion.

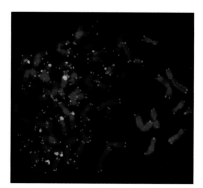

Nakamura *et al.*, Chapter 14, Fig. 4. γ-H2AX foci and telomeres in a metaphase prepared from Hela cell culture 30 min after exposure to 0.6 Gy.

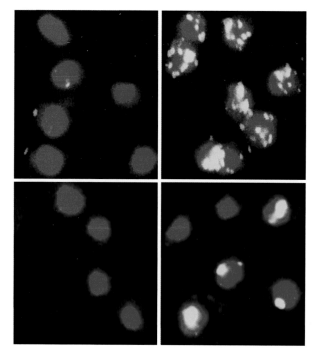

Nakamura *et al.*, Chapter 14, Fig. 5. Immunocytochemical detection of γ-foci in budding yeast formed in response to DNA DSBs. (Upper) Wild-type cells were left untreated (left) or treated with camptothecin (right) for 1 h. (Lower) Yeast cells grown in glucose (left) or galactose (right) for 2 h to induce the production of the HO endonuclease. Note that after HO induction, a single γ-focus is observed in most cells. Detection is with yeast γ-H2A peptide antibody.

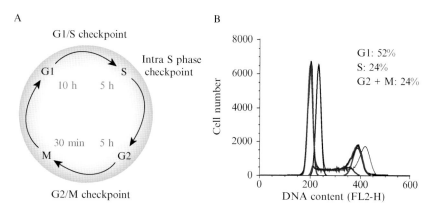

Theunissen and Petrini, Chapter 15, Fig. 1. IR-induced cell cycle checkpoints. (A) Representation of the cell cycle and the checkpoints discussed in this section. The approximate duration of each cell cycle phase (in h or min) in asynchronously proliferating early phase MEFs is depicted. (B) DNA content histogram. In this graph containing 75,000 asynchronous p3 MEFs cell number is plotted against DNA content (FL2-H). The Watson Pragmatic model in the FlowJo cell cycle platform was used to calculate the percentage of cells in G1, S, and G2 + M. Assuming that the cells in this asynchronous population are growing at about the same rate, the fraction of cells in a given phase (the ∼ percentage of mitotic cells can be derived from Fig. 4C) multiplied by the total doubling time gives you the length of that phase.

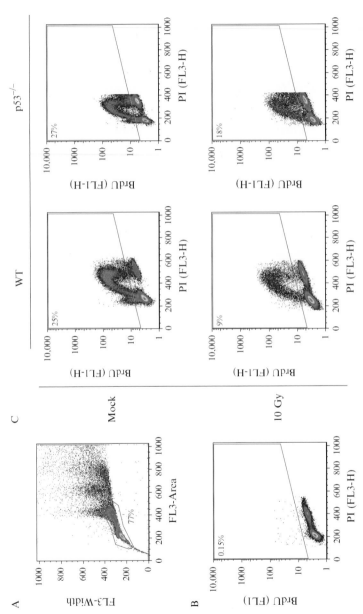

THEUNISSEN AND PETRINI, CHAPTER 15, FIG. 2. (A). FL3-Width versus FL3-Area. The red gate is devoid of aneuploid cells and nuclei doublets and the percentage of cells in this gate is depicted in the graph. (B, C) α-BrdU-FITC (FL1-H) versus propidium iodide (PI) (FL3-H). The red gate contains α-BrdU-FITC positive cells and the percentage in this gate is depicted in the upper left hand corner of the graph. (B) Isotype control sample. (C) Mock and 10 Gy irradiated wild type (WT) and p53$^{-/-}$ cells.

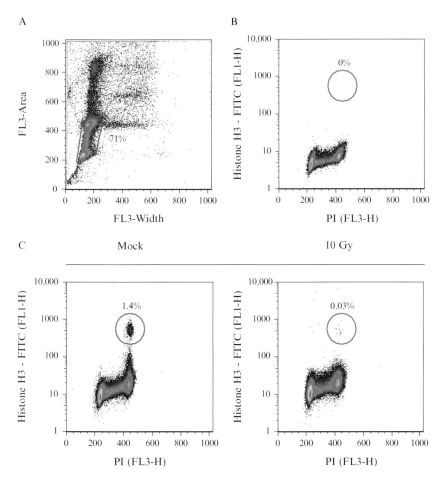

THEUNISSEN AND PETRINI, CHAPTER 15, FIG. 3. G2/M checkpoint. (A) FL3-Area versus FL3-Width. The red gate is devoid of aneuploid cells and nuclei doublets and the percentage in this gate is depicted in the graph. (B, C) Histone H3 – FITC (FL1-H) versus propidium iodide (PI) (FL3-H). The red gate encircles Histone H3 – FITC positive cells and the percentage in this gate is depicted in the graph. (B) Control sample not stained with α-Histone H3 antibody (C) Mock and 10 Gy irradiated wild type (WT) cells taken 1 h after treatment.

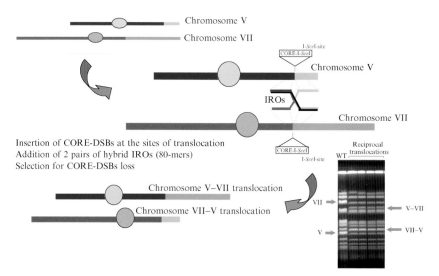

STORICI AND RESNICK, CHAPTER 19, FIG. 6. Example of genome reconstruction: reciprocal translocation between the arms of chromosome V and VII. The first step involves the insertion of one CORE-I-*Sce*I cassette at one site of translocation. (Insertion of a CORE-I-*Sce*I cassette at each site of translocation on both chromosomes [using the cassette with *kanMX4* and the one with *hyg*, respectively] is predicted to enhance translocation frequency by IROs.) The second step involves transformation with two pairs of IROs, each containing sequence from both chromosomes adjacent to the translocation point. Selection of translocation events is accomplished by isolating 5-FOA resistant clones, which also have lost the reporter marker (or both reporter markers if using two cassettes). The picture on the right shows the karyotype display using pulse-field gel electrophoresis. Bands corresponding to chromosome V-VII and VII-V translocations are indicated.

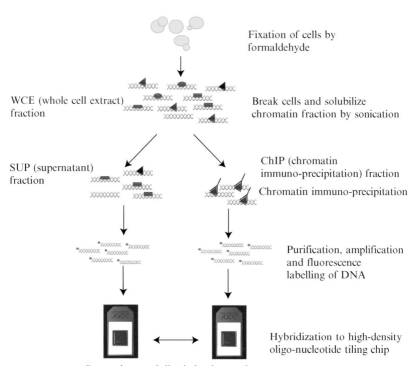

Fixation of cells by
formaldehyde

WCE (whole cell extract)
fraction

Break cells and solubilize
chromatin fraction by sonication

SUP (supernatant)
fraction

ChIP (chromatin
immuno-precipitation) fraction

Chromatin immuno-precipitation

Purification, amplification
and fluorescence
labelling of DNA

Hybridization to high-density
oligo-nucleotide tiling chip

Comparison and discrimination analyses

KATOU *ET AL.*, CHAPTER 23, FIG. 1. Schematic presentation of ChIP-chip experiment (see text for details).

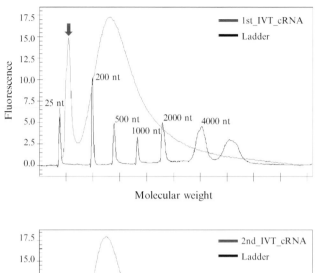

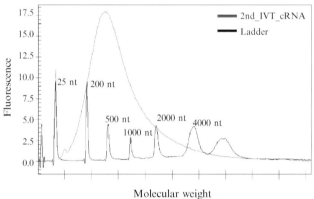

KATOU *ET AL.*, CHAPTER 23, FIG. 4. Product size analysis of cRNA by bioanalyzer (Agilent Technologies). Top; product analysis after first IVT amplification. A peak of small size fraction originated from a self amplification of T7 primer is pointed by red arrow. Bottom; product analysis after 2nd amplification. A portion of self-amplified primer disappears. Vertical axis represents fluorescence intensity (amount of DNA) and horizontal axis represents molecular weight. Red line; size distribution of cRNA amplified by IVT. Blue line; size distribution of the ladder marker. Size of each peak is shown.

A

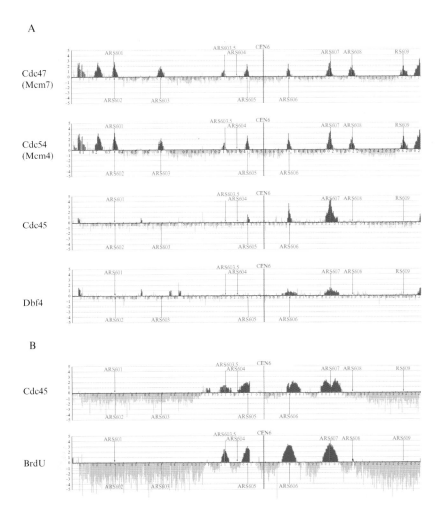

Cdc47
(Mcm7)

Cdc54
(Mcm4)

Cdc45

Dbf4

B

Cdc45

BrdU

KATOU *ET AL.*, CHAPTER 23, FIG. 5. Distribution of replication proteins during G1 and S on chromosome VI of *S. cerevisiae*. (A) Distribution of Cdc54, 47, 45 and Dbf4 in G1 phase arrested by α-factor. (B) Distribution of Cdc45 and BrdU incorporated regions in S phase challenged by HU. The blue shading represents the binding ratio of loci that show significant enrichment in the chromatin immuno-precipitated fraction. The yellow dotted line indicates the average signal ratio of loci that are not enriched in the ChIP fraction. The scale of the vertical axis is log2. The horizontal axis shows kilobase units (kb).

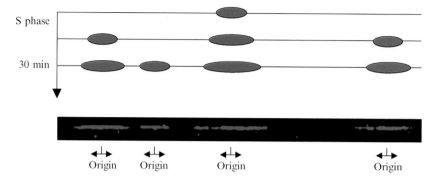

Liberi *et al.*, Chapter 26, Fig. 4. Visualization of *S. cerevisiae* replicons by molecular combing. Yeast DNA samples, treated as described in the text, were collected 30 min after the release from G1 block. The schematic diagram shows the different timing of origin firing during S phase.

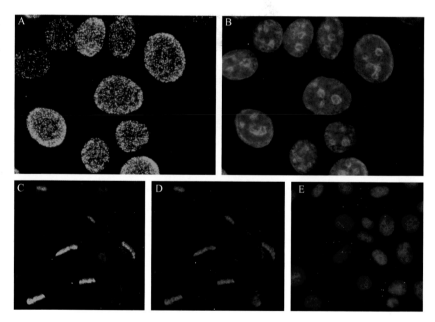

Dantzer *et al.*, Chapter 29, Fig. 2. Poly ADP-ribose synthesis in response to DNA breaks in HeLa cells. (A) HeLa cells treated with 0.5 mM H_2O_2 during 10 min respond by a robust synthesis of PAR visualized with the monoclonal anti-poly (ADP-ribose) antibody 10H (green). (B) Dapi staining of the same cells. (C) Laser-induced DNA damage triggers the local synthesis of PAR at the damage sites (green), that in turn induces the immediate recruitment of XRCC1 (red) dependent on poly(ADP-ribose) synthesis (D). The corresponding nuclei are stained with Dapi (E). Bars indicate 10 μm.